The Illustrated History of the Elements

Jan Kozák • Roger M. W. Musson

The Illustrated History of the Elements

Earth, Water, Air, Fire

Springer

Jan Kozák
Institute of Geophysics
Academy of Sciences of the Czech Republic
Prague, Czech Republic

Roger M. W. Musson
School of Geosciences
Edinburgh University
Edinburgh, UK

British Geological Survey
Edinburgh, UK

ISBN 978-3-030-21424-1 ISBN 978-3-030-21426-5 (eBook)
https://doi.org/10.1007/978-3-030-21426-5

This Springer imprint is published by the registered company Springer Nature Switzerland AG.
The registered company address is: Gewerbestrasse 11, 6330 Cham, Switzerland

Foreword

The word "elemental" carries a strong connotation of basic and primordial forces of nature. It can also evoke a vision of the ancient Greek elements of Air, Earth, Fire, and Water. And we all have strong reactions to the concept of "disaster." In the following fascinating pages you will find almost 250 of those visions illustrated and interwoven with significant events in human history. Most of the images are from Jan Kozak's Collection in Prague. While many of the events, artists, and historical authors are little known today, they have great value as evocations of disastrous events as seen by then-contemporary artists and writers. Above and beyond the sad and sobering aspects of natural disasters, Kozak and Musson have been able to illustrate and interpret a wide variety of events that resonate with human history and our understanding of how our planet operates.

An important aspect of this book, soon to be evident to the reader, is that the illustrations are far more than pretty or arresting pictures. They have the power to transport an attentive viewer back in time. We actually see portrayals of Earthly disasters in the context of their time. The text helps us understand specific points, but the images themselves virtually jump out from the past to immerse us in worldviews of the particular era.

As you view the striking images and read the informative text, you will experience a chronological and topical tour of artwork and narratives that illuminate the evolution of efforts to portray elemental Earth processes in action. There will be potent visual impacts, but readers attuned to the expansion of rational analyses of natural events will also find a rich introduction to the history of geological ideas. Many of the featured artists and naturalists are obscure today, but they made incremental contributions of significance. The power of observation, exemplified in the illustrations, served as a base for reasoned interpretations that aided the development of the earth sciences.

The Earth itself is restless. Processes of tectonism and gravity generate earthquakes and landslides that have wreaked havoc on many humans. Such power and horrific effects have caused people to seek answers, often through religious explanations. "Why us? Why there? Why then?" In due course, attempts were made to comprehend these mysteries, using the tools and reasoning of science. This book will take you on an insightful trip through art and narrative in the years prior to modern photography. You will watch humans, from the Middle Ages to 1906, attempt to artistically capture the essence of catastrophes and to then understand how a stable Earth could turn so unstable.

Water, as Lao Tsu told us, is ever soft but ever powerful. The raging power of floods, storms, and tsunamis has shaped history and impacted humans across the centuries. Visions of Noah's flood have had a huge impact on Western thought, and subsequent floods, tsunamis, and whirlpools have impacted people physically and psychologically over the years. Kozak and Musson document a number of those events and provide insight into the evolution of ideas about their nature.

Air is certainly our friend, except when it's not. The friendly, warm, still air that is the base of our metabolism can turn distinctly unpleasant when hurricanes lash our shores or tornadoes roar across the plains. Artists have depicted dangerous winds and such atmospheric visions as auroras, comets, and meteors. In addition to the mechanical power of wind, past visions of air as a carrier of disease ("malaria" and plagues) have received artistic and scholarly attention.

Fire is in the news in our own time, as climate change impacts the strength and frequency of forest fires. But its searing power was portrayed long ago, as Lot tried to escape from Sodom and as volcanoes erupted around the world throughout history. Human-generated fire also has a long record of catastrophe. Chapter 5 introduces many examples of fires through history, discusses various sparking causes, and illuminates artistic responses to elemental fire.

The appeal and power of this collection is multifaceted. Historians of art will see not only striking visions but also the evolution of pictorial practices, from woodcuts to copperplate engravings to the onset of the photographic era. Anyone interested in the sciences, particularly those devoted to understanding the Earth, will be treated to a graphic (literally and figuratively) view of the onset and evolution of rational explanations for what were long considered dark mysteries. And, happily, there are rewards for the casual reader. Not only can he or she view some amazing pictures and learn more about classic or little-known disasters, but also there is the opportunity to watch the development of interpretations of nature's harsh realities blossom over the centuries from the Middle Ages to the early twentieth century.

A key feature of this compendium is that the reader/viewer can become immersed in the pictorial representations of artists who were reflecting the level of understanding of their era. Thus, a picture not only can speak to the "elemental disaster" being portrayed but also can also conjure up an understanding of the state of science at the time. The history of science thereby becomes an active process combining visual representation and written interpretation.

Enjoy your tour. The topics concern disasters, but the pictorial narrative is an informative delight for the eye and the mind.

Department of Geosciences
Denison University
Granville, OH, USA

Kennard B. Bork

Preface

The Kozák Collection is an archive of pre-photographic depictions of natural disasters, assembled over a period of some decades. It was originally a collection of pictures of historical earthquakes, and this part of the collection is held online as part of the Earthquake Engineering Online Archive NISEE e-Library, maintained by the Pacific Earthquake Engineering Research Center (PEER) and available at https://nisee.berkeley.edu/elibrary/browse/kozak. Since it started, the collection has broadened in scope to include other natural disasters (and some non-natural ones). Previous publications based on images in the collection include *The Illustrated History of Natural Disasters* by Kozák and Cermak (2010) and *Terremoti in Italia dal 62 A.D. al 1908* by Margottini and Kozák (1999).

In this new collection, rather than presenting pictures in a purely chronological order, we take the whimsical approach of invoking the classical "four elements" of Earth, Fire, Water, and Air, and arranging the various disasters accordingly. The result, we hope, will be entertaining and readable for a wide audience without any pre-existing technical knowledge.

The emphasis throughout has been on selecting less well-known images. Many of the figures in this book are held as paper originals in the Kozák Collection in Prague, others are from other private collections, and some are public-domain images. The authors would like to thank all of the people who have helped in the production of this work, in sourcing images, and in providing advice.

Prague, Czech Republic — Jan Kozák
Edinburgh, UK — Roger M. W. Musson

Acknowledgements

For the preparation of the text of this book, as well as the processing and presentation of the large number of pertinent illustrations, we note the very welcome and much appreciated cooperation of many colleagues, friends, print owners, and collection curators: our cordial thanks go to all of them. The leadership of the Institute of Geophysics of the Czech Academy of Science and its directors, Pavel Hejda and Aleš Špičák, together with the head of the Seismology Department, Jan Šílený, supported the editing of the book and allowed the institute's infrastructure to be used in the course of the book's preparation. For treatment of the images, we cordially thank Marcela Švamberková, head of the Computer Department at the same institute. Our great thanks go also to Giuseppe De Natale, director of the Osservatorio Vesuviano, who submitted a series of images of eruptions of Vesuvius for our book, including the respective copyrights. Similarly, we were permitted to reproduce images from various other world libraries; let us name the Archivo General des Indias (Seville, Spain), National Library (Prague, Czech Republic), National Museum of Japanese History (Chiba, Japan), National Diet Library (Tokyo, Japan), Bancroft Library (Berkeley, USA), Library of Congress (Washington, DC, USA), and British Geological Survey (Keyworth, UK), to list the main donors. We give credit to all of them, and we also thank Kenji Satake of the University of Tokyo, Paola Albini and Romano Camassi of the Istituto Nazionale Geofisica e Vulcanologia (INGV; Milano and Bologna), Wolfgang Lenhardt of the Zentralanstalt für Meteorologie und Geodynamik (ZAMG; Vienna), A.A. Nikonov of the Russian Academy of Sciences (Moscow), and Michael Kühn of the Antiquariat Michael Kühn (Berlin) for providing advice. Our thanks also to Dr Kennard Bork for providing the Foreword. Lastly, we would like to remember the late Jean Vogt of Strasbourg, who contributed so much to the study of historical earthquakes, and who helped with some of the images reproduced here.

Contents

List of Figures

Introduction

1

1.1 Opening Remarks

The elements of the Earth … *Perilous natural elements! … Attention people! Save yourselves, those who can!*

Today any scientifically minded person will understand the word "element" as referring to a substance with a unique atomic number and place in the periodic table. But, for many hundreds of years, the scientific and medical orthodoxy was that Earth, Water, Air, and Fire were the four elements from which the World was made. So great was this belief that the mythical potency of this four-part arrangement continues to this day. Any survey of imaginative or magical writings, including fantasy literature, will reveal that the classical four elements still hold sway in the creative mind, so we make no apology in this text for putting aside the periodic table and adopting the classical four elements as still providing a useful perspective for thinking about the natural world.

Many readers might consider these principal elements—Earth, Water, Air, and Fire—as damaging and malicious forces, unquantifiable and highly dangerous, destroying lives and human values. Thus, the mass media present dramatic pictorial and verbal reports of a world population repeatedly threatened by the perils of natural disasters, earthquakes, tsunamis, volcanic eruptions, storms, tornadoes, floods, avalanches, landslides, fires, and other natural extremes caused by the feral planet Earth, provoked by its four furious elements.

Indeed, seen by an objective eye, natural disasters mean—and have always meant—threat and jeopardy for human populations; additionally, in most cases, they also cause considerable damage to material property. The first author of this book presented an inventory of these redoubtable natural dangers in his previous book *The Illustrated History of Natural Disasters* (Kozák and Čermák 2010), in which selected images of the effects of strong historical earthquakes and volcanic eruptions of the past were presented and discussed.

Here, we adopt a different approach to display the objective impacts of the Earth's elements. Of course, besides their unambiguously negative effects, their gentle obverse faces have provided humanity and civilizations throughout history with beneficial conditions for development and rapid advancement.

This new series of historical images of natural disasters—mostly engravings of all kinds, which the reader now holds—displays the redoubtable effects of the terrestrial forces embedded in a more complex scene, i.e., on the ground of all of the Earth's principal elements. The reader will admire new, little-known images of seismic and volcanic events and the disastrous effects caused by the extreme behavior of the other famous elements: Water, Air, and Fire. These images are discussed in the context of recalling the Earth's origin and its physical development as they relate to the extreme behavior of the elements of the Earth.

We also present some early images illustrating concepts in the natural world. These images have been selected to accompany the introductory text of the book. The first series of these images illustrates the early, foggy ideas of the universe and the Earth's creation, including the Earth's dynamic manifestations. These pictorial compositions frequently appeared as painted illustrations in manuscripts, as frescos, as oil paintings on canvas, and so on. At the end of that period, in the course of the sixteenth century, the first illustrations appeared as prints on paper by means of the newly imported printing technology of woodcuts. However, these first woodcut prints were simple and primitive in execution—generally far from what we would consider correct documentation of these subjects today. As for the standpoint of the painters, they more or less relied on a religious interpretation of the events they portrayed according to a literal view of the Bible.

The second group of these images illustrates the first efforts of European scholars to understand (and to explain) the dynamic behavior of the Earth in terms of the interaction of natural forces, abandoning church standpoints. Surprisingly enough, it was a Jesuit, Athanasius Kircher (1602–1680), who largely promoted secular explanations of extreme manifestations of natural phenomena. In the course

J. Kozák, R. M. W. Musson, *The Illustrated History of the Elements*, https://doi.org/10.1007/978-3-030-21426-5_1

of the eighteenth century, the dynamic development of the study of the natural world—what was referred to as "natural philosophy" even up to the mid-nineteenth century—opened the way for a new and important branch of science—geology—facilitating the exponential advancement of the Earth sciences in the following period.

The last dozen illustrations presented in the introductory part of this book were created approximately between the mid-eighteenth century (when the era of enlightenment started) and the end of the nineteenth century. In this period, both classical and newly established disciplines of the Earth studies began to approach the portrayal of our planet as we know it at present. (Most of these images are reproduced from originals in the collections of the National Library of the Czech Republic (Prague), except where otherwise indicated.)

In the following introductory text, subjects written with an initial capital letter relate to "our" world: our Earth revolving around our Sun in our Solar System, which is part of our Universe. The same subjects written in small letters—earth, sun, and life—are related to the general meaning of these words. Special attention is paid to the last period of the Earth's existence, since modern man appeared.

In this initial discussion, we look at the validity of the classical idea and whether the traditionally accepted strict division of all natural subjects into animate and inanimate items fits well with reality. To make sense of this query, several tools appear to be useful—namely, the all-embracing principles of cyclicity, which, in most cases, is considered to apply only to so-called animate nature.

By "cyclicity" what we have in mind is that the natural processes of so-called animate nature follow one principal time sequence pattern for each subject—namely, its "birth" or creation or initiation, followed by growth and progression, and by development and advancement, up to its culmination and, at a mature age, reproduction of itself, creating descendants for the next generation. After that, the subjects usually enter their last phase of decline and their final termination or principal transformation. The term "cyclicity" has usually been used for the "existence in cycles of generations" that is characteristic of both fauna and flora.

In the world of living subjects, individual units have complex memory: besides standard everyday memory, they also have genetic memory, by means of which their newly obtained experience may be, to a variable degree, passed on for posterity. The importance of cyclicity in the process of improvement of the given subjects, including the advancement of their instantaneous and long-term reactions or of their abilities to adapt to new conditions, is evident. However, it must be admitted that this miraculous property—to entail or to impress their new experience into their genetic code for its utilization by subsequent generations—has been granted to individual species to a widely variable extent. The largest portion of this admirable ability has been awarded to modern man. It can be concluded that the cyclicity of natural processes is the prime condition of advancement of living units, whose desired advancement is realized by a series of subsequent, young, fresh, newly born generations.

Actually, few people doubt the existence and benefit of the cyclicity of so-called living subjects. The question may be raised as to whether the principle of cyclicity can also be traced in the course of the existence of units of subjects that are usually classed as inanimate. Before discussing this possibility in more detail, let us point out one interesting feature relating the structure of recurrence time and its physical size that is valid in the world of fauna.

With regard to the time duration of any given species of individuals (that is, the lifetime of a generation), it seems that, in general, this time grows proportionally with the dimensions of the considered life-unit. For example, the cycle time of the modern human is only slightly improved on the biblical 70 years. Larger vertebrates live usually tens of years and insects usually one or two years, if that. The mayfly lives as a flying insect for one day only. The lifetime of small individuals of bacterial species is probably even shorter, although assessing this is problematic.

Another tool for our investigations may be sought in consideration of the universe's mass; we will confirm that this miraculous matter disposes of fabulous properties that are often little understood and commonly undervalued, as will be discussed below.

We hope these brief remarks on the Earth's nature will help us to properly appreciate both the positive and negative points of human life surrounded by the elements of the Earth. Many readers will probably agree that for proper description, analysis, comment, and understanding of the processes involved in the origin and further development of life, a brief survey of the history of the Earth since its origin would be reasonable. Special attention has to be paid to the last 100,000 years or so, during which one sort of primate entered the scene as modern humans.

Therefore, to better understand the complexity of the age-old struggle between mankind and angered Earth elements (which are discussed in the rest of this book), the reader is asked to undergo a short excursion through selected cosmogonic and planetary ideas, which have been obtained recently, about remote and time-distant worlds.

Care is needed, though. Inevitably, if we try to think about structures at the universal scale or about subatomic worlds, we involuntarily color our thinking with images of "our" world, apprehensible by human senses and amplified by modern observational instrumentation, or not, as the case may be. This may be a route to error, but it is hard to see how it can be avoided, since the human brain is built to analyze any situation exclusively according to human sensation and experience. Nevertheless, we must keep this limitation in mind and respect it when imagining remote worlds; it is so easy to slip into images of our well-known world, which is visible and tangible around us every day!

1.2 The Modern World: The Earth and Our Solar System

Let us focus our attention on the surface of our planet—the environment that we, the inhabitants of the Earth, have explored to a certain extent, and that is, for our reflections, most familiar. In this world we are able to conduct, evaluate, and interpret our observations from those positions that are, for the present thinking person, the most understandable.

In our later reflections about remote worlds, in which the space–time of our interest must be enlarged by many orders, we will recognize that the correctness of our terrestrial perspective—except for the situation on our own planet—can be applied also to the territory of our Solar System. However, outside this negligible corner of the Universe, as our latest cosmic observations indicate, the relations between the principles and categories that characterize the behavior of remote worlds may possibly be governed by a different set of physical laws that do not conform to those functioning on planet Earth. On the other hand, this would imply the existence of a boundary in space between a region where one type of physics applies and another region where the type of physics is different. Such a boundary is hard to comprehend.

Contemporary cosmological theories are based on the idea that the physical relations that work in small dimensions must also be true in their large-sized variants as well. This would mean that the same physical laws that rule our Earth should be true in all of the Universe as well, since the opposite situation has not been detected. It is difficult to imagine, for instance, how elements in the periodic table could be different in other galaxies.

This prospect or opinion may be open to debate: on one hand, present experience really indicates that the above opinion (that only one type of physics exists in the whole Universe) could be correct. However, we must keep in mind that we have directly explored only a negligible part of the Universe; thus, our results in this field are so minimal that we can hardly feel entitled to speak responsibly of the whole illimitable Universe. Astronomical observations, however, seem to bear out the idea that scientific principles operate in the same way throughout observable space.

Traditionally, the classical ancient concept of the four elements is linked with the name of Aristotle (384–322 BC), who distinguished them as Earth, Water, Air, and Fire. However, although we are dependent on surviving fragments, priority on this point should be ascribed rather to a predecessor of Aristotle, the Greek savant Empedocles (490–430 BC), who may have proclaimed the four elements about 100 years before Aristotle. Empedocles assumed that the four elements were the roots of all physical objects. Today we greatly appreciate these ancient Greek savants who formulated such ingenious syntheses, some of which survived into the nineteenth century. We have to keep in mind that they derived their admirable conclusions explicitly according to their incomplete sensory perceptions.

Let us examine the four elements as they may appear to us today. While we may make use of the contemporary level of astrophysical knowledge, the intention here is to discuss the elements at a symbolic level. The following text is presented not as a stringent exposition of current scientific thinking but, rather, as an imaginative exploration of the synthesis between old and new ideas.

Evidently, the classical concept of the elements—the initial formulations of which go far back to ancient times—appears obsolete today and cannot fit within modern cosmology and stellar physics, which are informed by new discoveries and by advanced technologies and knowledge. On the other hand, this ancient model of four fundamental elements appears suitable when images of strong historical natural disasters are displayed, considered, and discussed. Therefore, the pictorial section of this book, in which these extreme natural events of the past are shown, is arranged in an order corresponding to the classical order of the Earth elements commonly recognized in those historical times—or, to put it another way, historical disasters need to be understood in their historical context, according to the intellectual ideas of the period.

However, in the introductory chapter of this book, some properties of the present concepts of the Earth and our Solar System and the Universe will be outlined as well; in this modern prospect of the elements (formulated here according to the authors' opinion), some new theories will be traced. Also, some important questions, which these new theories evoke, will be formulated and presented for the reader's consideration.

1.2.1 Earth

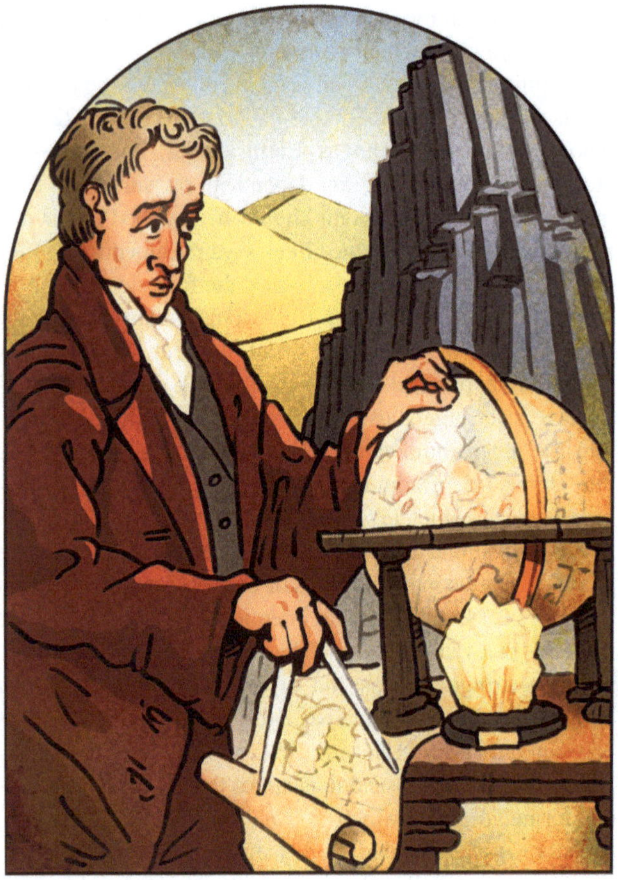

This image and the next three images are paintings by Cyril Kozák, made in 2015, as updates of traditional iconic figures of the four elements. Here, we see Earth, represented by a serious-looking geologist in nineteenth-century dress, with a globe, dividers, and a mineral specimen, standing before some columnar basalt

Earth (or mass, or solid-state matter) is the fundamental element or prime matter: nothing could exist without the element Earth. Earth/mass displays admirable properties—notably, its duality of mass and energy expressed and postulated by Albert Einstein in his famous equation $E = MC^2$, where E is energy, M is mass, and C is the velocity of light propagating in a vacuum (300,000 kilometers per second). It is evident that mass represents a concentration of an immense amount of energy, which, in nature, can be released in two ways. The first way consists of slow chemical processes in which thermal energy is one product of chemical synthesis (such as oxidation, combustion, or burning) and also of slow, subcritical thermonuclear processes. The second way is usually denoted as fast, transcritical thermonuclear fusion or explosions. In the Earth—and probably in all of our Solar System—these processes seem to operate under the control of the law of energy/mass conservation and under the thermodynamic laws defining the conditions of mass versus energy exchange and transformation.

Among the admirable properties of mass, its gravitational influence may be considered. In the late seventeenth century, Isaac Newton discovered that two masses (m^1 and m^2) attract each other by a gravitational force (G) according to the relation $G = k(m^1 + m^2)/r^2$, where r is the distance between both of the masses and k is a [gravitational] constant. During the mutual rotational movement of two or more mass objects, the gravitational attractive forces are supplemented by the additional centrifugal force, which provides the origin of quasibalanced stellar/planetary systems, such as galaxies, solar systems, and spiral nebulae.

Another principal miracle of mass (in our Solar System) can be seen in its second duality—namely, its ability to exist in an animate form, as well as in an inanimate form. The flora and fauna of the Earth demonstrate that the same chemical elements that form the basic building units of inanimate nature can additionally—if properly composited under the required circumstances and conditions—initiate the origin of simple structures. Such structures have their own metabolism and are able to realize their own reproduction: the miracle of the creation of life. The differentiation and progress of the following evolutionary process in a number of life variants and evolutionary levels may be regarded as a prolongation of the life creation prodigy.

Plants may be classified as a lower level of life: units of flora are mostly immovable or movable only passively by air, water, avalanche, or other life forms. These units (probably) have no self-knowledge and, if they act as a result of altered conditions of the medium around them, they are believed to act electrochemically, according to genetically encoded formulae.

As for the simple members of the Earth's fauna, they are mobile to varying degrees and, in parallel, they—also to a variable degree—display signs of voluntary behavior.

The evolutionary steps of fauna toward vertebrates, mammals, and primates represented a giant jump in advancement; at the end of this process, the life variant presently denoted as mankind was attained. This species is the only animal that has been able to overcome the limitations imposed by the second law of thermodynamics. This means that modern humans are able to purposefully reduce the level of entropy of the system in which they live and are the active part.

We can close our reflections on the extraordinary properties of the principal Earth element—mass—with our admiration of its nearly endless diversity, lying in countless variants of chemical compounds based on combinations of the 92 or 94 stable chemical elements, by means of which a wide spectrum of materials of diverse physical and chemical properties can be created. Additionally, there are many other alloys and mixtures that widen the spectrum of chemical compounds. Possibly, one may anticipate that it is just the polymorphism and variety of the individual variants of the element Earth (mass) that facilitated the production of life in such admirable diversity on our planet.

Earth has its negative aspects too. The rocks beneath our feet are something that, from childhood, we take as a fundamental measure of stability. When those rocks start moving like water, the effect is profoundly disturbing, which is one reason why earthquakes are such terrifying experiences. Elements should stay in their own domain! Earthquakes are great killers; no other natural disaster has killed so many so quickly in recorded history—up to 800,000 souls have perished in the space of a few minutes. In most cases, though, earthquakes do not kill directly; the idea that huge fissures open up and swallow whole towns is an old myth. Most earthquake casualties come about through the destruction of buildings, burying people in the ruins. These melancholy scenes of ruin and destruction have left a rich iconography, which we will sample in this book.

Landslides, rockfalls, and avalanches are other earth-related disasters, albeit generally localized. Often they may be triggered by earthquakes. The deadliest known avalanche occurred in Peru in 1970, when the northern wall of Mount Huascarán collapsed, destroying the towns of Yungay and Ranrahirca, with around 70,000 deaths.

1.2.2 Water

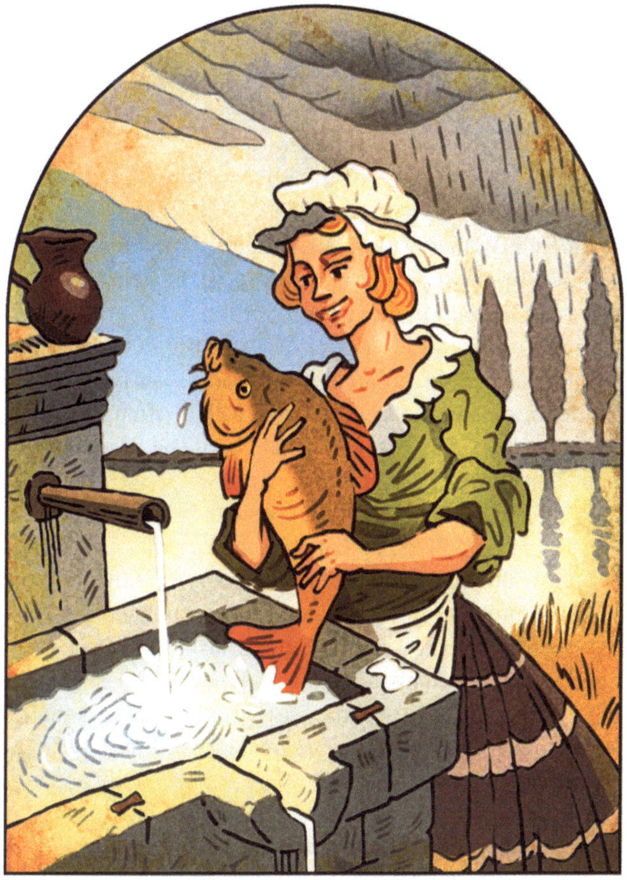

The element Water is traditionally female; this might be Aquarius the water carrier, except that she has put down her jug above the water trough and is holding a large fish instead. In the distance, rain sweeps across a lake

While the element Earth can be seen as representing mass, solidity, and—from the viewpoint of terrestrial life—the substance of living organisms, the element Water represents the principles of fluidity, movement, and transport. Water in our Solar System, besides participating in the composition and construction of living structures, has fundamental functions relating to complex communication in living structures. It regulates and transports oxygen, liquids, and nutritional material in living bodies, and replaces the remnants of somatic metabolic processes.

Water on planet Earth plays an important role also in the organisms of hominids, especially *Homo sapiens*, whose nervous system, including the brain, is composed of sensitive protein-based nerve cells. This fact, on one hand, provides possibilities for these cells to advance exponentially by means of (nearly) infinitely growth of advanced double-helix molecules. On the other hand, however, the existence of a protein basis of life is strictly conditioned by maintenance of the thermal regime of protein body cells within a narrow range of temperature—a few tens of degrees around 0 °C. In conclusion, we may state that without water, the origin and development of life on a protein basis would not be possible on planet Earth.

Other properties and functions make water essential for our planet. It enters the Earth's atmosphere as water vapor in suitable pressure/temperature conditions, and it returns to the Earth in the form of irrigating precipitation. By means of the encoded process of photosynthesis, water gives life to the Earth through terrestrial vegetation—an important link in the natural alimentary chain of the Earth's fauna. Besides this, water makes it possible for humans to pursue their everyday activities.

The crucial question—namely, how and when water appeared on Earth, enabling the creation of life—remains unanswered. Several hypotheses present well-argued explanations of how water appeared on Earth—for example, perhaps by means of accretion, or from collisions with comets and asteroids. However, none of these ideas can be confirmed by clear evidence, since this event happened during the early stages of Earth's development. Recent evidence suggests that liquid water at one time also existed on our neighboring planet Mars, no doubt as a result of similar processes, and it is presumed to have been lost by evaporation at some point, for reasons not well understood.

Water may be essential for life, but too much of it is very dangerous. Floods have plagued humanity from the earliest times, and it is notable that in the biblical story of Noah, when God chose to destroy mankind, he chose a great flood as the means of destruction. The fact that this fable is repeated with variations in many cultures demonstrates the primeval fear of the destructive power of flooding. When a river floods, its ability to move debris increases as the fifth power of the increase in the water velocity. If a flooding river bursts its banks and flows into a modern town, the force of the water can toss heavy vehicles around like toys.

As well as river floods from heavy rainfall, one must consider marine inundation—for instance, resulting from storm surges. In low-lying countries such as the Netherlands, these are a very serious hazard.

Nor should one forget the awesome power of a tsunami, a compound disaster involving both Earth and Water. (More examples of compound disasters will be seen later in this book.) When an earthquake causes a violent displacement of the seabed or a major underwater landslide (which may also occur without an earthquake), the displaced water rushes onshore with terrible force. The tsunami may be more dangerous than the earthquake—the famous recent case of Tohoku in 2011 being a case in point. The tsunami killed over 15,000 people; the earthquake killed only a few hundred at most.

1.2.3 Air

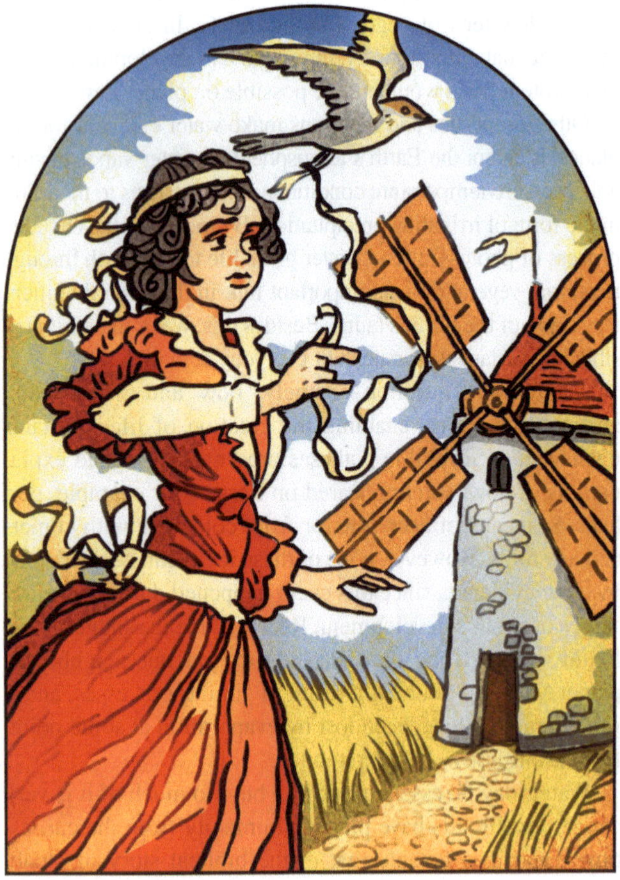

Air is also feminine; this young woman is evidently standing in a strong wind, from the way the ribbons in her dress and hair are blowing about. A bird also symbolizes Air, as does the windmill in the background

Air also plays a key role in the history of life on our Earth. Our planet is a solid-state rock with enough water on its surface, and it is situated at a viable distance (not too near and not too far) from our infernal Sun. Were it not protected by any atmosphere, the temperature on our planet's surface would differ by roughly 300 °C between day and night (as on the moon), preventing the existence of protein-based life. Fortunately, the atmosphere—Air—balances these extreme temperature variations.

Actually, the mechanism of the proper regulation of the planet's temperature variation is much more complex than that outlined above. Besides heat absorption and reflection in the atmosphere, many factors influence the "temperature play" at the Earth's surface. Let us mention at least the remnant heat propagating toward the Earth's surface from the hot center of the planet, possibly having its origin in past collisions with comets and asteroids, and also in radioactive decay of the radioactive rocks deep inside the Earth. There is also the distribution, thickness, and transport of clouds (including additional eruptive clouds due to strong volcanic eruptions) and variations of our Sun's radiation. Additionally, the requirements, activities, and impact of billions of people over the last couple of decades on numerous atmospheric and climatic factors can be considered.

Besides the protective function of the temperature shield, the Earth's atmosphere plays an essential role in the maintenance of life. Air brings the necessary portion of oxygen to terrestrial life, which, through respiration, yields the necessary portion of energy to sustain cellular function. Besides the importance of the transporting abilities of the atmosphere to distribute precipitation, air works for humanity in many other ways. For instance, it is a renewable resource of energy exploited by wind power plants, an important oxidant in the majority of industrial processes, a medium for air traffic, and so on. It is often forgotten that air plays a role in protecting the Earth's surface against smaller meteorites and asteroids, which often burn out, explode, and decompose into small, harmless pieces during their transit through the atmosphere.

What forms of destruction are visited on us by Air? Gales and storms are killers, including tornadoes, hurricanes, and typhoons, although in the latter calamities, torrential rain and flooding provide another case of compound disasters. Fortunately, casualty rates from storms today tend to be much lower than in historical times, when sailing ships with large crews were an essential part of the World's economy. Storms meant shipwrecks, with great loss of life.

One Air-related hazard that might not be readily apparent at first sight is disease: epidemics and pandemics. Today we know that these are due to bacteria or viruses, but, in historical times, when the classical four elements were regarded as mainstream science (not that those words would have been used), it was common to regard disease as a product of "bad air"—hence the derivation of the word "malaria."

1.2.4 Fire

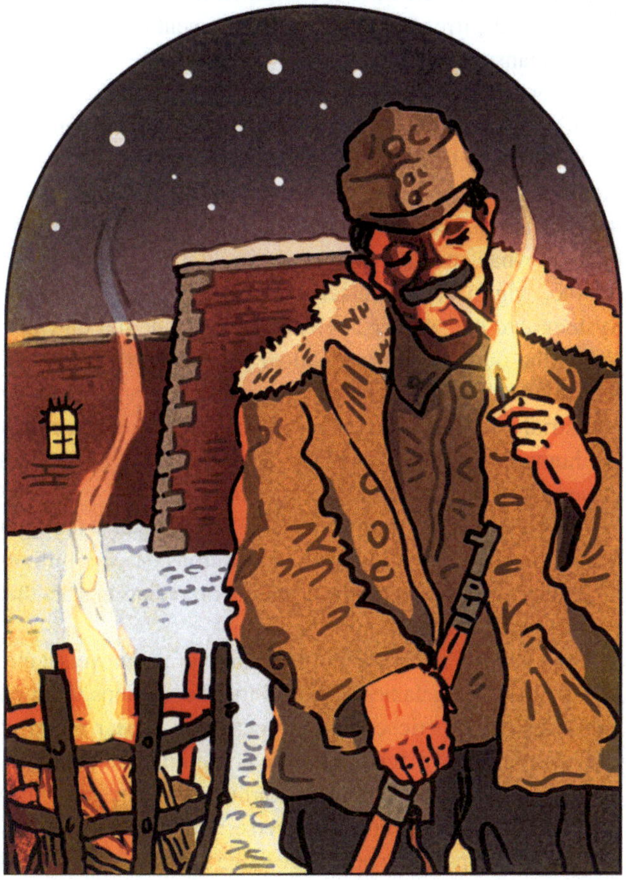

The soldier in this depiction of Fire is obviously glad of the warmth afforded by the brazier beside him on this cold night. He strikes a match to light his cigarette. Even his rifle will depend on the element Fire when he needs to use it

Before we embark on a more detailed consideration of the negative and positive impacts of the fourth element, Fire, let us point out its special character. The three previous discussed elements—Earth, Water, and Air—actually comprise one category; we can name it protomaterial or primary matter, either in a solid state or in a liquid/gaseous state. Today we know that distinguishing mass according to its instantaneous stages is a remnant from the times in which humanity did not know anything about phase transitions due to pressure and temperature alterations.

The fourth element, Fire (or burning), does not represent a mass substance; rather, it denotes a process—namely, the process of slow, fast, or explosive-like oxidation, manifested as thermal, light, electromagnetic, or nuclear radiation. This complex radiation, giving terrestrial life the necessary heat, light, and food (remember the process of photosynthesis, giving life to flora) reaches the Earth's surface from two principal sources: firstly, our planet's outer universal space (from the Sun, in particular) and, secondly, the Earth's depths, as radioactive decay is believed to play an important role in balancing the process of the Earth's warming.

Besides the welcome properties of the aforementioned system of heat transport to the Earth's surface, the irregular negative occurrence of volcanic heat attacks should not be forgotten. However, besides the occurrence of large volcanic eruptions coming from the inner parts of the Earth, also temporal irregularities and extremes of solar radiation have contributed distinctly to the extreme temperature history of the Earth (such as periods of stronger or weaker glaciation).

Volcanoes may not actually be the product of combustion, being an eruption of molten rock and ash; however, in human history, volcanoes have so often been perceived as "burning mountains" that we can fairly associate them with Fire. During large volcanic explosions, both sunlight and volcanoes interact to produce the final temperature space–time regime: because of strong volcanic explosions, large amounts of volcanic ash and small particles are launched high into the sky. Such a large accumulation of volcanic ash and volcanic dust, which is impenetrable by solar light, will (because of the Earth's rotation) spread over the temperature-regulating atmospheric shield and will immediately result in drastic cooling of the Earth's surface. This extreme situation may last as long as a few years of dense shadow and frost; it may cause an environmental calamity for vast regions. Fortunately, such disasters visit our planet rarely. During the written history of the Earth, such strong volcanic events/explosions have been recorded only a few times, examples being the eruption of Santorini Island in the Aegean Sea (in about 1600 BC), the eruption in the Sunda Strait between Sumatra and Java (in AD 535–536), and the eruption of Tambora (also in Indonesia) in 1815.

The recent model of prehistorical relations between the nutritional habits of some hominids and the process of combustion allow an interesting speculation. Let us consider an imaginary situation in which we would try to assess the present mental level of the Earth's fauna, including hominids, if future (modern) humans were absent from the Earth's rich faunal spectrum during the last hundred thousand years. Let us also presume that the geographic and climatological situation on the Earth is identical to our present condition. Most readers would agree that we might anticipate the development of an overall mental level for the Earth's fauna that would not be much different from our present situation; in other words, the rest of the Earth's fauna would be mentally unchanged, leaving aside human cultivation and domestication of some species.

This conclusion prompts a crucial question: what caused just one species of hominid—the initial mental level of which (most probably) did not differ considerably from those of other hominid species—to follow its extraordinary fast ascent on the ladder of mental abilities?

The specific explanation for the exponential rise of modern humans holds that this progress was attained with the aid of Fire or, more specifically, by early humans with the knowledge of how to kindle flammable materials and make fire, how to domesticate fire, and how to prepare thermally treated dishes. At the turn of the millennium, American anthropologists concluded that consumption of thermally treated meat enlarges the number of neurons in the center of the nervous system of an animal. Over time, this is reflected in enlargement of the inner volume of its skull and the size of its brain.

In the last two decades, a series of laboratory experiments was performed in the USA, in which animal cubs from one litter were kept in two separated groups: in the first group, the animals tested were fed exclusively with raw food, while in the second group, the cubs were supplied with thermally treated foodstuffs. The results of these tests unambiguously indicated that the number of brain neurons in the tested animals fed with cooked food was considerably greater than the number in those fed with raw meat.

Should these conclusions be found correct and convincingly confirmed, we should admit that Fire—through thermally cooked meals—initiated and supported the exponential mental rise of one species of hominids, now known as modern humans (or *Homo sapiens*) and opened the door to the building of our present world.

Fire, as the saying goes, is a good servant but a bad master. Once it gets out of control—often because of some simple incident, such as an overturned candle—whole towns and even cities can be consumed in a general conflagration, with a large loss of life. Fires are started by earthquakes so commonly that "fire following" is a standard term used in earthquake reinsurance. Consider the case of Japan: light houses of wood and paper, and numerous cooking stoves overturned by earthquake shaking. Is it any wonder that often the casualties in an earthquake are dominated by those killed by Fire?

1.3 Remote Worlds: The Milky Way, Other Galaxies, the Universe

We may conclude our considerations of the Earth's elements, which command our small subuniverse, with the following abstraction: while the three classical elements (Earth/mass, Water, and Air) provide and assure the physical conditions necessary for the existence of the Solar System, including the development and functioning of life on Earth, the fourth element (Fire) stimulates heroic development of the mental abilities of selected species of already existing simpler forms of life.

Today humans have not yet discovered all of the secrets of our small Universe, composed of our Sun and its eight planets. Not all of the questions related to energy/mass duality, mass attraction (gravity), dark energy, and dark mass have been answered or convincingly resolved. Similarly, fundamental questions related to the origin of life, the variety of its forms, and the irregularity of its advancement remain fully or partly unanswered.

Nevertheless, the present modern humans live with the Earth's elements in a careful and strongly interconnected symbiosis, and this kind of coexistence seems to be convenient and acceptable for humanity. For the wealthy part of the population, it may even be easy and pleasant. Let us now look at what situation mankind would face in the outer, remote, gigasized universe, lying far beyond our Solar System. Are the "universal elements" that rule the functioning of the outer universe identical to the Earth's elements? And do the universal elements create situations similar to our terrestrial ones? Most probably, many readers will feel that the human race has a long way to go to better understand these secrets and answer these questions.

Anyway, it is exciting to observe remote stars, the light from which has traveled to us over millions or even billions of years, carrying spectral information on radiating stars that possibly vanished millions of years ago.

For a general assessment of the nature of the universal elements in remote galaxies, it seems that our notions, built on our terrestrial experience and physics, may appear incorrect and insufficient. We must admit the possibility that the pillars of our present Earth-born and Earth-bound physics, when applied to studies of remote universe, might be replaced by new relationships fitting new knowledge and new observations and measurements with new techniques and new technologies proceeding hand in hand with new thinking. At this point, let us commemorate the glorious Newtonian physics and its apparent invincibility. We know that Newton's conclusions, and likewise Einstein's newer theoretical models, are valid—in particular, in our "small Universe" (that is, in our Solar System). When entering remote cosmic worlds, we have to reconcile these with the necessary transformation of our terrestrial physics according to (not yet generally known) new physical relations valid in superlarge worlds. This may concern the elements of the Universe, too—or the universes (plural) if, as been suggested, all possible universes exist; some of these may operate under very different rules or values.

One may recollect in passing Alexander Pope's famous epitaph:

Nature and Nature's law lay hid in night:
God said, "Let Newton be!" and all was light.

and its continuation by J.C. Squire:

It did not last: the Devil shouting 'Ho!
Let Einstein be,' restored the status quo.

What—in this confused situation—can we say about the elements of the Universe? Distant stars are capable of producing vast discharges of energy, as in supernovas, which thankfully have no equivalent in our own mundane sphere. At such vast distances, our observations are biased toward these huge events; we cannot register what small-scale processes may also be going on. Then, can we query whether the power of the elements is proportional to the size class of the Universe? Or may we ask the question of whether the strongest universal elements are—in a hyperbolic expression—more merciless and less delicate toward humans than the terrestrial elements known in our Solar System?

Or, to put it another way, the disasters we consider in this book, while frequently terrible for those who experience them, operate at a local or regional scale on our planet. When we consider what may happen in the wider Universe, we are confronted with the possibility of disasters that could consume a whole planet.

1.4 The Unresolved Question

In the course of this discussion, we have touched on various issues that might help in making better assessments of the role of the terrestrial elements in human life. The question the authors would like to leave the reader with concerns the four fundamental elements—Earth, Fire, Air, and Water—and their relationship to humanity; namely, do they play for our team, or do they act as our adversaries? The contemplations given above urge us to proceed toward the portrayals of our mysterious, simultaneously beloved and hated companions—the four terrestrial elements and their effects. Finally, in Conclusions, we will propose some answers.

1.5 The Elements in Action

Before presenting the images that comprise the rest of this volume, it is worth pausing to consider why such pre-photographic pictures of natural disasters exist at all. In doing so, we learn something about the range of these images in space and time. There are certainly some preconditions for the existence of such images. Firstly, someone must have the impulse, for whatever reason, to make an image. Secondly, they must have the technical means available, whether it be the painting of a fresco or the engraving of a woodcut. And, lastly, there must be consumers: people who wish to see such images, again, for whatever reason.

Primarily, it seems that there are three such reasons for making and viewing images of natural disasters. The first is journalistic: a desire to report on an event that has happened or to preserve its memory for posterity. The second is the making of images for educational purposes, in the study of natural history. The last is, quite simply, an appetite for horror. Death and destruction have always had a horrible fascination to them; otherwise, there would be no market for horror films in the cinema. This is often seen in images where a strong emphasis is placed on the agonies of the victims, as in a well-known dramatic image of the 1886 Charleston earthquake, where practically nothing is shown of the damage caused by the earthquake, and center stage is given to the people fleeing and screaming in terror.

Many of these images of horror also have an ulterior motive connected with the idea that these disasters of the four elements are not natural events, as we would regard them today, but are sent by God as a warning or punishment. The prime example here is the great flood of Noah, which in times past was taken as a historical event no less real than the destruction of Pompeii or any other disaster in ancient times. The awfulness of the image was thus a way of conveying the power of the Almighty to inflict suffering on his disobedient creatures.

One can therefore set some limits on the time and space domains in which these images survive, according to the intersection of those times and places where the means and motivation for making such images intersect with those parts of the world where these disasters occur. Earthquake disasters, for instance, happen mostly in defined seismic regions along the boundaries of tectonic plates. And, even then, many great earthquakes occur almost unnoticed by humans—in Kamchatka, for instance, or in the South Pacific. Great earthquake disasters happen only when the location of an earthquake coincides with areas of high population, mostly in a zone running from the Mediterranean, through Iran and northern India, to China and Japan.

But this zone where most of the high-fatality earthquake disasters have occurred is far from homogeneous culturally. In the Mediterranean region, in particular, all of the necessary components are present: a civilized society with a high regard for artistic technique, and the Christian religion, in which the divine origin of natural disasters was, for a long time, not disputed. Thus, it is not surprising that many of the images that survive today are from this region.

Contrast this with the Islamic world, stretching from North Africa through to India. Although it attained a high level of civilization and technical skills, and was certainly exposed to earthquakes, the religious prohibition of pictorial representation means that the few images we have of disasters in this region are from European travelers (or are completely imaginary).

Also, we note the complete absence of such images from the world's oldest civilization in China. Despite a very fine tradition of landscape paintings dating back many hundreds of years, there are no pictures of disasters, for cultural reasons. No one wished to see such grisly images, and there was therefore no motivation to produce them. The situation was different again in Japan, where paintings recording natural calamities from earthquakes, tsunamis, and volcanoes do exist, and some are shown here.

Elsewhere in the world, one can trace the expansion of the area from which disaster images exist with the expansion of European culture, through colonization, to the Americas, Indonesia, the Philippines, Australia, and New Zealand, where the first drawing of earthquake damage dates from 1843.

In viewing the images that make up the rest of this book, these principles will be clearly seen from the range of pictures shown, and, in many cases, they can be deduced from the nature of the pictures themselves.

Now, after a somewhat visionary introductory text full of question marks, the reader is invited to enjoy a pictorial excursion into the world of the effects and extremes of the terrestrial elements as they were portrayed all over the world during the four centuries prior to 1900. In this, we focus on the negative aspects: disasters of one sort or another, categorized by the dominant element or sometimes compounds of different elements. This is not from any morbid disposition; rather, images of elements in their beneficial aspects are extremely common and unremarkable. Any picture of a ship will show the utility of transport by water. But images of disasters are rarer and more dramatic, and it is these that we present to the reader.

Will such an excursion help us to proceed toward better understanding of the mutual position of terrestrial and universe elements on one hand and mankind on the other? Just go on and turn the page.

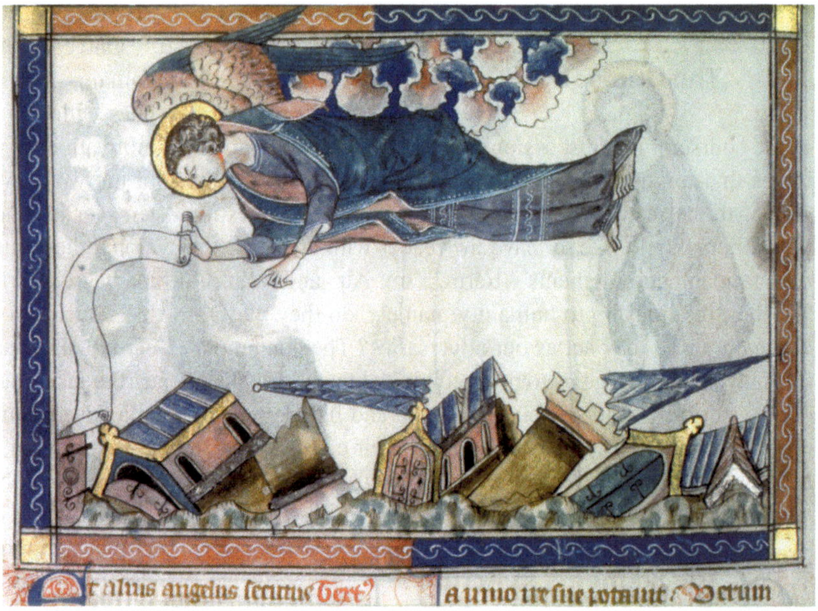

An unspecified earthquake depicted in a thirteenth-century painting on parchment, from a manuscript in the collections of the British Museum, London. In this image, the earthquake is interpreted as being sent by God as a punishment. The destructive effects of the event are shown; however, things are kept under control by an angel, who keeps watch to ensure that only the right amount of damage is caused

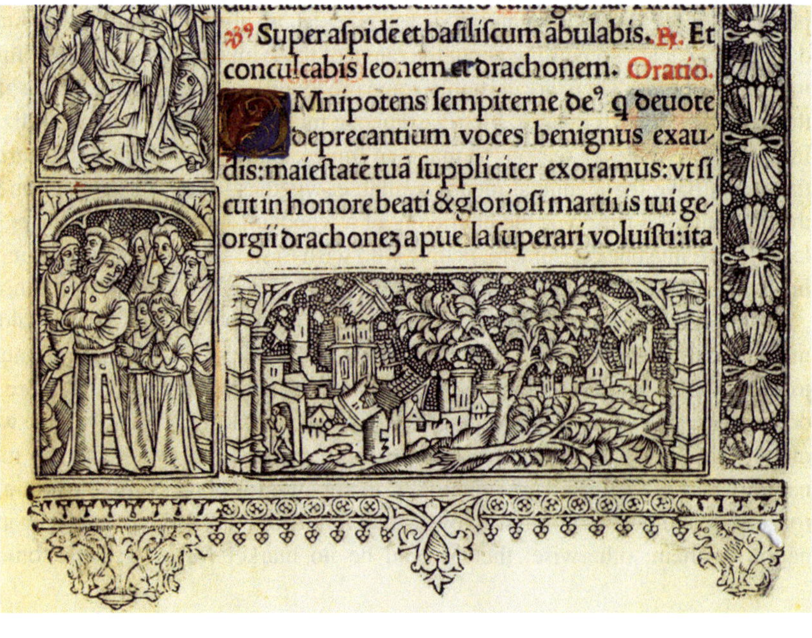

The Earthquake. This illustration is taken from an unidentified Latin incunabulum printed in the late fifteenth century. Two small images show a town devastated by an earthquake (center) and a group of people (left) praying to God for mercy. (Image courtesy of A. Boulatov, Moscow)

The Creation of the World in Six Days. This woodcut was taken from Luther's Bible of 1534 and reprinted in the Czech Melantrich Bible published in Prague in 1557. The woodcut print was originally made in Nuremberg (Bavaria) in 1532–1534. The appearance of the Earth's surface and the surrounding space appears to be strongly influenced by the cosmological concepts of the time

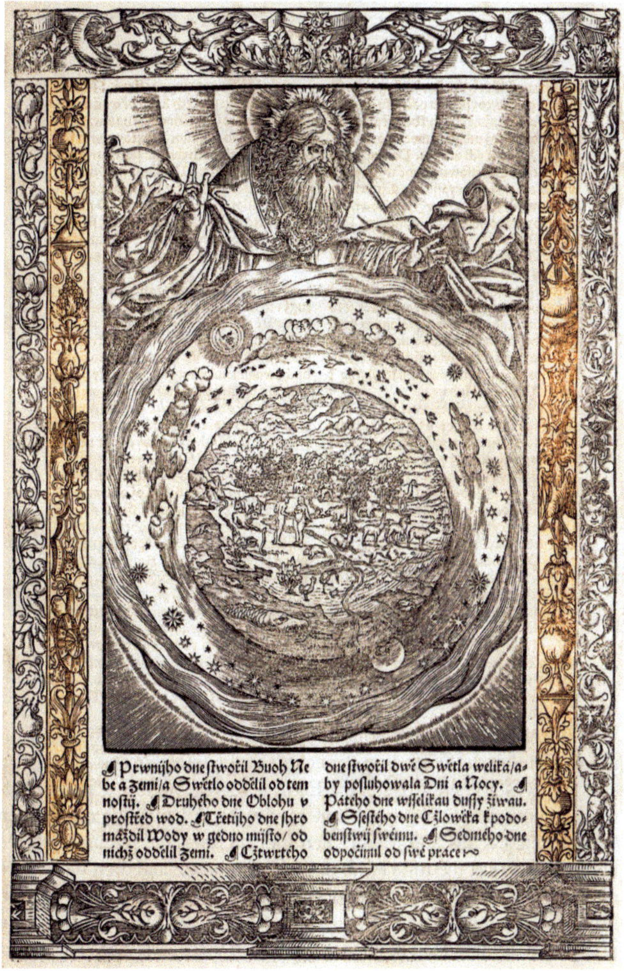

The Appearance of the Earth After the Creation. God is overlooking his handiwork from the highest Heaven; animals roam about, and the presence of a ship indicates the presence of mankind. This woodcut composition is reproduced from Sebastian Münster's *Cosmographia* (Nuremberg 1550 edition). Here, Münster's concept of the World—although presented in an easily understandable form—is not quite clear, especially the central fiery band separating the lower and upper parts of the sky. It does not seem to make sense, unless its purpose is to increase the number of elements depicted to the classical number of four. The angelic and devilish figures outside the main circle represent supernatural beings: angels above and devils below. A small curiosity in the image is that the dark side of the moon actually faces toward the sun instead of away from it, but one would have had to understand the nature of the moon's light to recognize the error

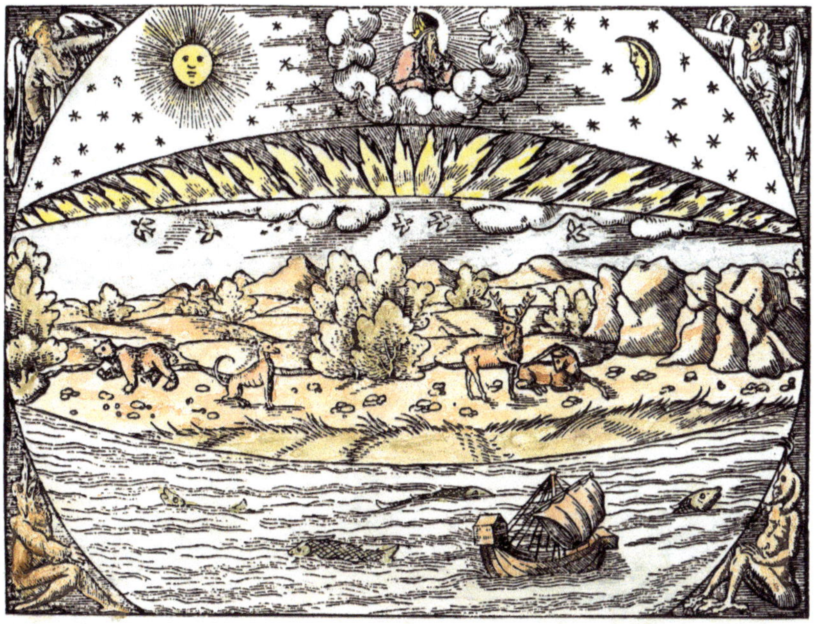

Die Erde nach der Schöpfung — Nach Sebastian Münster (1550)

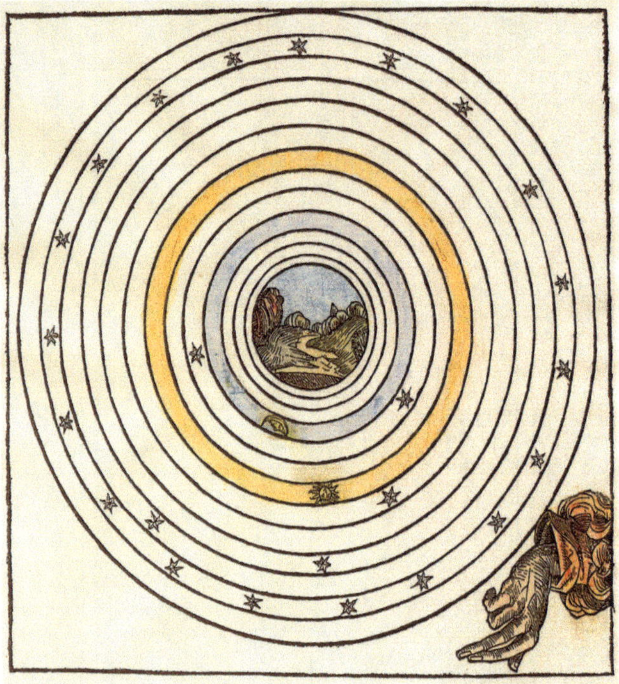

Sebastian Münster's view of the Solar System, based on the old geocentric model that the Earth is at the center of the universe and everything else revolves around it. Here, the Earth is represented by a small landscape. Above it are three blank spheres for the sky, then the moon, then two planets (Venus and Mars), then the sun, then the remaining three known planets at the time (with one blank sphere), and then the sphere of the "fixed stars." The hand of God outside keeps things in motion. From the orientation of the landscape, it is possible that this woodcut was originally intended to be viewed rotated 90° clockwise. The coloring is probably modern

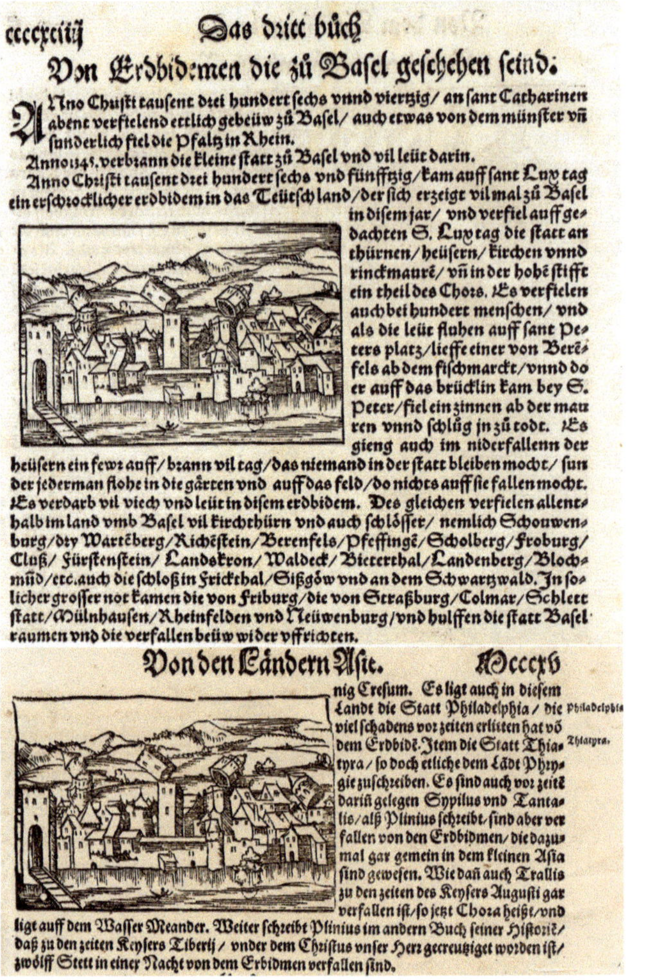

cccxciiij Das dritt bůch

Von Erdbidmen die zů Basel geschehen seind.

ANno Christi tausent drei hundert sechs vnnd viertzig/ an sant Catharinen abent verfielend ettlich gebeüw zů Basel/ auch etwas von dem münster vñ sunderlich fiel die Pfaltz in Rhein.

Anno 1345. verbrann die kleine statt zů Basel vnd vil leüt darin.

Anno Christi tausent drei hundert sechs vnd fünfftzig/ kam auff sant Lux tag ein erschrocklicher erdbidem in das Teütsch land/ der sich erzeigt vil mal zů Basel in disem jar/ vnd verfiel auff gedachten S. Lux tag die statt an thürnen/ heüsern/ kirchen vnnd rinckmaurẽ/ vñ in der hohẽ stifft ein theil des Chors. Es verfielen auch bei hundert menschen/ vnd als die leüt fluhen auff sant Peters platz/ lieffe einer von Berẽfels ab dem fischmarckt/ vnnd do er auff das brücklin kam bey S. Peter/ fiel ein zinnen ab der mauren vnnd schlůg jn zů todt. Es gieng auch im niderfallenn der heüsern ein fewr auff/ brann vil tag/ das niemand in der statt bleiben mocht/ sunder jederman flohe in die gärten vnd auff das feld/ do nichts auff sie fallen mocht. Es verdarb vil viech vnd leüt in disem erdbidem. Des gleichen verfielen allenthalb im land vmb Basel vil kirchthürn vnd auch schlösser/ nemlich Schouwenburg/ dry Wartẽberg/ Richẽstein/ Berenfels/ Pfeffingẽ/ Scholberg/ Froburg/ Cluß/ Fürstenstein/ Landskron/ Waldeck/ Bietterthal/ Landenberg/ Blochmũd/ etc. auch die schloß in Frickthal/ Sißgöw vnd an dem Schwartzwald. In solicher grosser not kamen die von Friburg/ die von Straßburg/ Colmar/ Schlettstatt/ Mülnhausen/ Rheinfelden vnd Neüwenburg/ vnd hulffen die statt Basel raumen vnd die verfallen beüw wider vffrichten.

Von den Ländern Asie. Mcccxv

nig Cresum. Es ligt auch in diesem Landt die Statt Philadelphia/ die viel schadens vor zeiten erlitten hat võ dem Erdbidẽ. Item die Statt Thiatyra/ so doch etliche dem Lãdt Phrygie zuschreiben. Es sind auch vor zeitẽ dariñ gelegen Sypilus vnd Tantalis/ alß Plinius schreibt/ sind aber verfallen von den Erdbidmen/ die dazumal gar gemein in dem kleinen Asia sind gewesen. Wie dañ auch Trallis zu den zeiten des Keysers Augusti gar verfallen ist/ so jetzt Chora heißt/ vnd ligt auff dem Wasser Meander. Weiter schreibt Plinius im andern Buch seiner Historië/ daß zu den zeiten Keysers Tiberij/ vnder dem Christus vnser Herr gecreutziget worden ist/ zwölff Stett in einer Nacht von dem Erbidmen verfallen sind.

Philadelphia

Thiatyra.

This figure superimposes two extracts from sixteenth-century volumes, depicting the famous earthquake in Basel (Switzerland) on 18 October 1356, and, below it, an earthquake in Philadelphia (Asia Minor) in AD 17. The description of the earthquake in Basel is rather brief and vague; as for the event in Philadelphia, only a few words confirming the event's occurrence are given. The most striking fact, however, is the illustrations: they show identical images of a town heavily damaged by an earthquake—a town that evidently depicts neither Basel nor Philadelphia. This "stamp method" of depicting towns using stock images was often used by sixteenth-century publishers in Europe

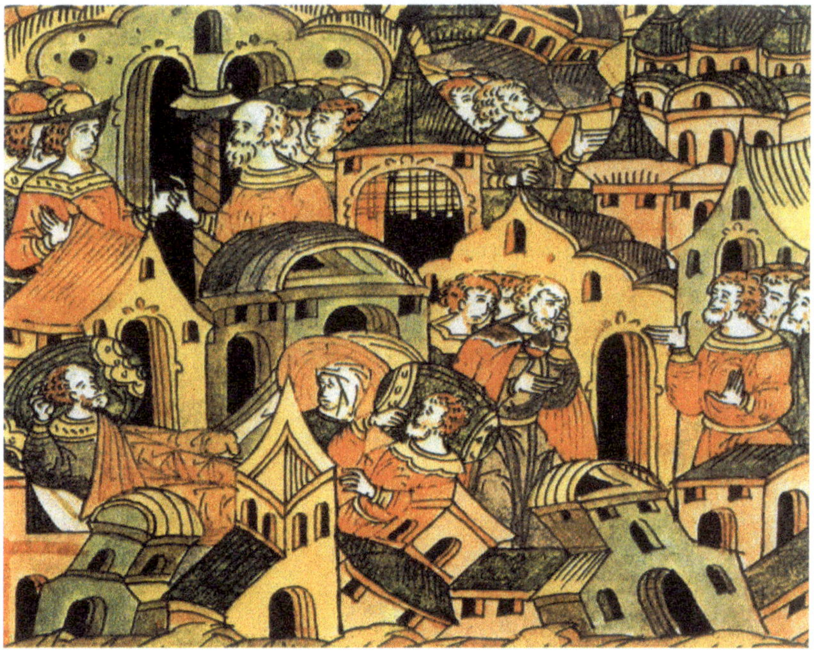

The earthquake of 10 October 1446, in Moscow. The painted illustration in this Russian chronicle was prepared for Tsar Ivan the Terrible in the second half of the sixteenth century. It is assumed that the epicenter of the earthquake was in Vrancea, Romania, and that the radiation pattern characteristic of this earthquake source directed seismic energy particularly to the northeast, toward Moscow. In fact, there are no reports of the earthquake from Romania or anywhere else other than Moscow. It is only because of the low seismicity of the Moscow region that the earthquake is assumed to be so distant. It becomes self-fulfilling: "There are no earthquakes near Moscow; therefore, this earthquake can't have occurred near Moscow." The image is greatly exaggerated; the shaking was not actually very strong (Tatevossian and Albini 2010). This must be almost unique as an earthquake depiction inspired by non-damaging shaking. (Image courtesy of A.A. Nikonov, Moscow)

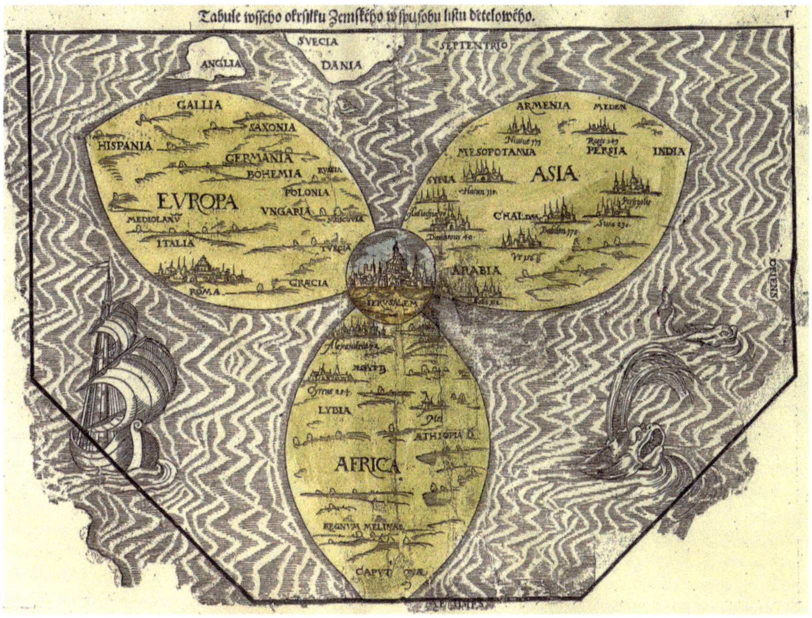

Map of the Old World shaped as a clover leaf. This is a hand-colored copper engraving from a popular book written by Heinrich Bünting: his *Itinerarium Sacrae Scripturae* (*Travel Book of Holy Scripture*), first published in 1581. It was translated into Czech by Adam of Veleslavín, Prague (circa 1593), from which this copy is taken. This is one of three figurative maps drawn by Bünting; it is possible that the three-leaved design represents the Holy Trinity. It was not imagined that this design had any geographic accuracy; Bünting's book also contains a number of conventional maps of the world as it was known in that period

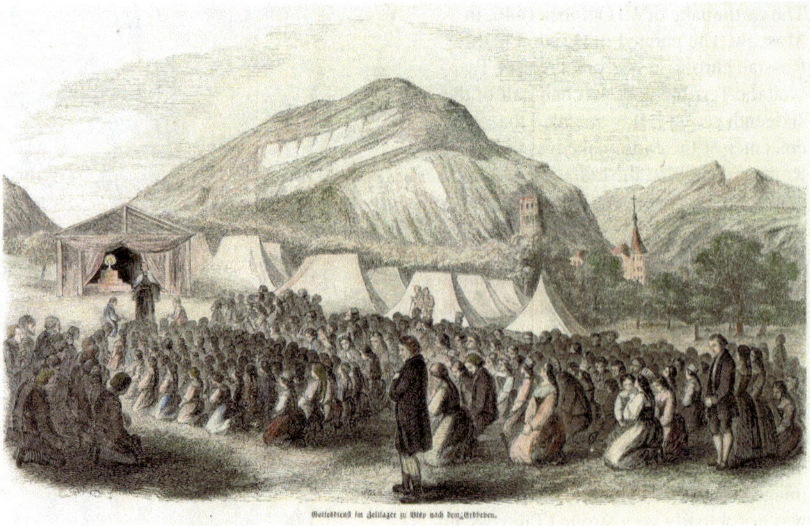

Most European naturalists had, in the course of the eighteenth century, abandoned ideas about God's responsibility for dynamic manifestations on the Earth, especially after the Lisbon earthquake of 1755. However, such beliefs still prevailed among common people in the countryside. This lithographic illustration shows a divine service being held at a tent camp near the Swiss town of Visp (seen in the background), which was damaged seriously by the earthquake of 25 July 1855. Prayer meetings after earthquake occurrences were often arranged in the eighteenth and nineteenth centuries, especially in the Mediterranean and subalpine regions

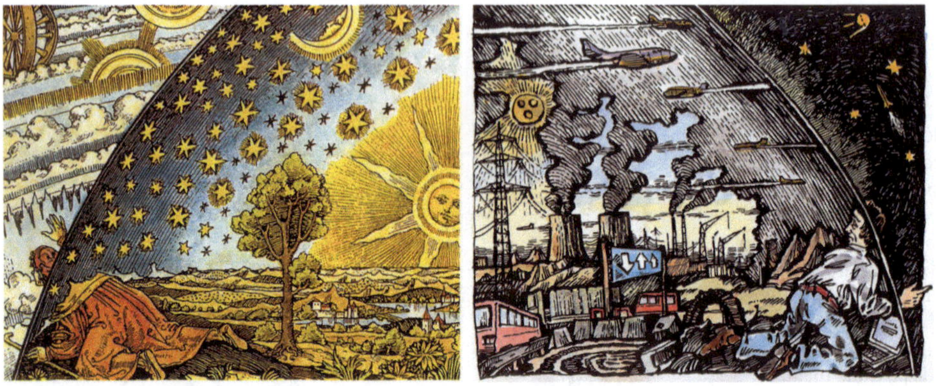

This double image shows a transformation of the relationship between modern man and nature around him, from the late Middle Ages to the present time. The first suggests an idyllic pictorial composition portraying human effort in the late Middle Ages to peep into the unknown world existing behind the limits of the sublunary realm; the inviting domesticated landscape supplies the observer with a reliable starting point in his efforts. The composition, which was believed to be created sometimes in the fifteenth or sixteenth century, was actually painted by the famous French astronomer and philosopher Camille Flammarion in 1888. The second was created as a homage to Flammarion's image in 2013 by the Czech artist Cyril Kozák and indicates the state in which the products of civilization have left the planet's surface during the twentieth and twenty-first centuries

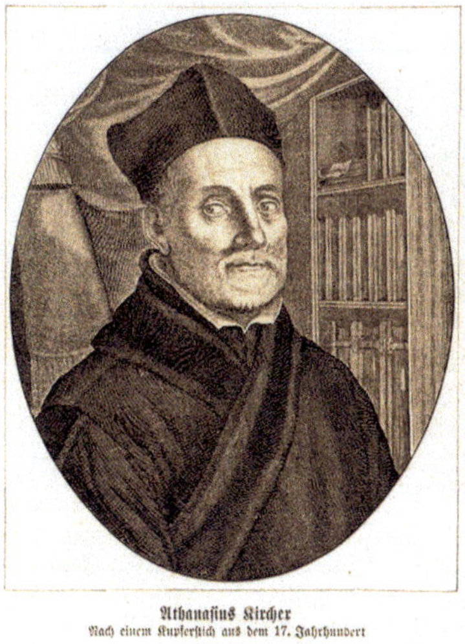

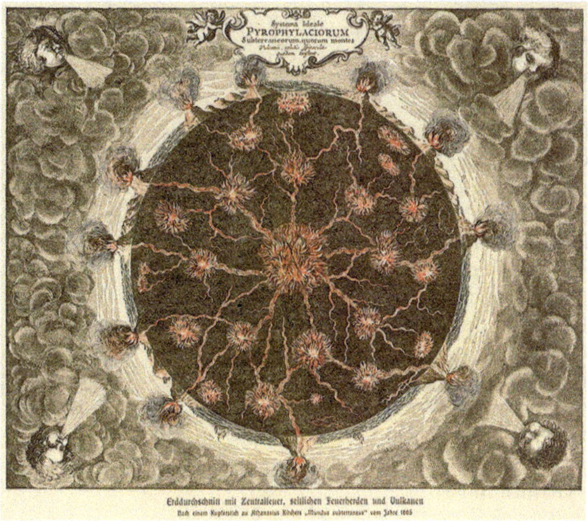

The seventeenth century and more or less the first half of the eighteenth century can be thought of as a "Pre-Enlightenment" era. Athanasius Kircher (1602–1680), the celebrated German scholar and polymath, lived in Avignon and later in Rome. The author of over 40 scientific publications, he was a dominating figure of this period and ruled a wide spectrum of scientific branches flowering in the seventeenth century. He was especially engaged in the Earth sciences all his life; in his work *Mundus Subterraneus* (1664 and 1676) he collected all of his extensive knowledge of the Earth and summarized his geophysical concept of the Earth's structure and dynamic manifestations. Despite being a Jesuit priest, he correctly argued that the dynamic extremes of the Earth do not occur according to God's whim but are due to pressure and thermal changes and variations in the Earth's interior. Because of Kircher's great reputation among his contemporaries, some of his false concepts about the composition and the state of the Earth's interior endured long after his death

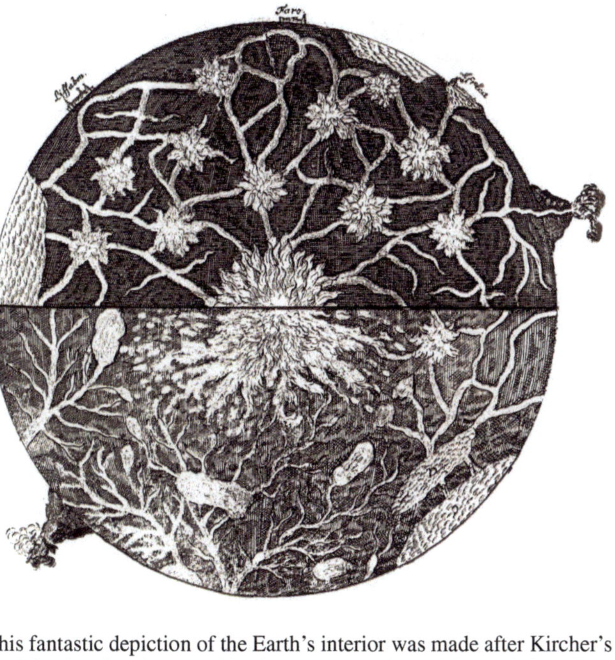

This fantastic depiction of the Earth's interior was made after Kircher's model, showing long, wide, interconnected corridors of fire. This unidentified print was created after 1755, at least 75 years after Kircher's death; it suggests that the earthquake in Lisbon was caused by a volcano lying in an antipodal location

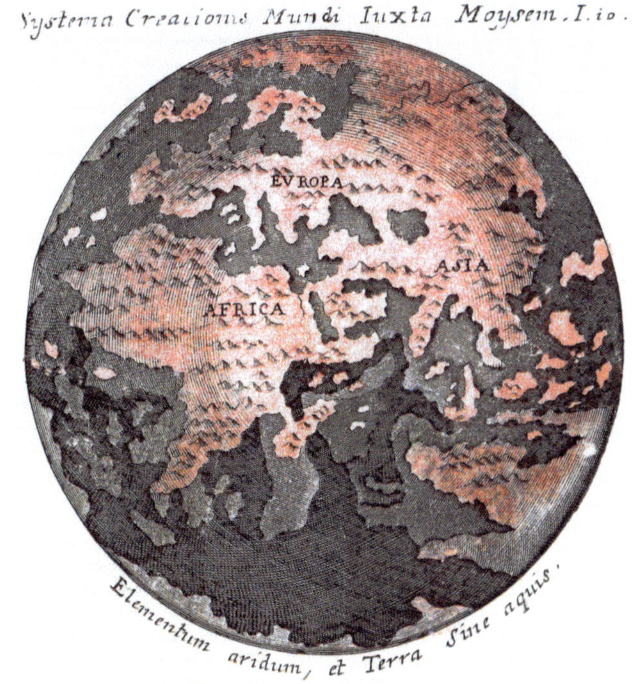

In the seventeenth century, religious literature in Europe slowly adapted to scientific viewpoints regarding the description of the natural environment; nevertheless, Holy Scripture remained its principal source. These oscillations between church and natural standpoints are responsible for this image explaining the creation of the Earth's dry land ("elementum aridum" in the caption) without water ("sine aquis"), according to Genesis 1:10 (although the implication of Genesis is that the water existed first). The three continents of the Old World—Europe, Asia, and Africa—are more or less correctly expressed in relation to the globe. This is particularly true for the Mediterranean area; in depicting more distant areas, some vagueness of the map maker is apparent. Some relief of the ocean floor is depicted, making this perhaps the first ever bathymetric map of the world, even if the undersea topography is only imagined. This solitaire, hand-colored, copper engraving print is taken from an unidentified late-seventeenth-century Bible

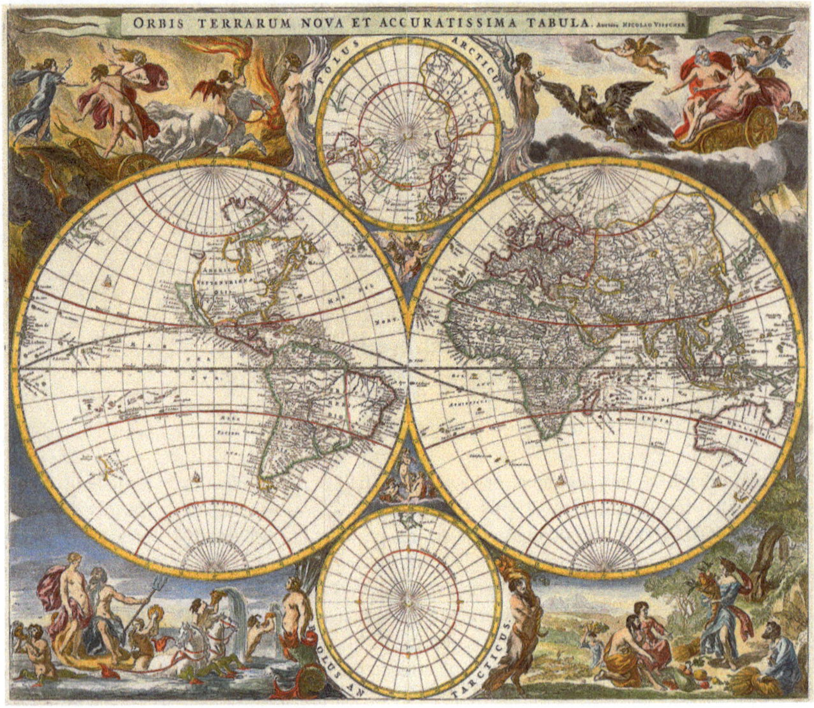

In comparison with religious illustrations, world cartography in secular atlases improved considerably in this period, as this reproduction of Nicolas Visscher's map *Orbis Terrarum Nova et Accuratissima Tabula* (circa 1658) confirms. The world's topography is already well expressed except for some not-yet-explored distant regions. The two additional small projections of the polar zones are an interesting detail. Note also the four corner scenes, portraying the four classical Earth elements—clockwise from top left: Fire, Air, Earth, and Water. Biblical illustrations soon caught up, and versions of Visscher's map soon found their way into Dutch and German Bibles

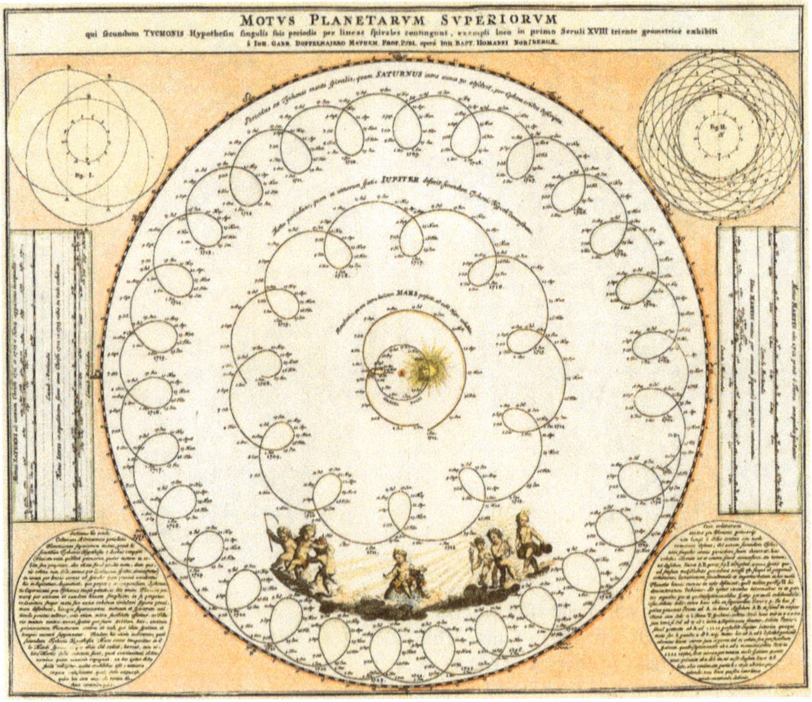

Astronomy, the observation of the motions of stars and planets, belongs among the oldest efforts of humanity, since ancient times, to make accurate measurements of natural phenomena. Let us examine the development of astronomy in the Pre-Enlightenment period for which we have appropriate illustrations at our disposal. We see that regardless of the new heliocentric concepts of Kepler, Galileo, and others, geocentric models were still kept alive up to the beginning of the eighteenth century. The scheme reproduced here for the solar/planetary system is that presented by the Bavarian cartographer Johann Baptist Homann (1664–1724), titled *Motus planetarum superiorum*. It shows apparently looping motions of Mars, Jupiter, and Saturn, and the geocentric motion of the inner planets Venus and Mercury. These looping motions were known as epicycles and seemed to be the only way to reconcile the motion of the planets, as observed from Earth, with a geocentric model. Homann worked in collaboration with the mathematician Johann Gabriel Doppelmayr on several works; this chart appeared in two atlases: the *Atlas von hundred Charten* (1712) (as plate no. 10) and the *Grossen Atlas* (1716). The chart shows essentially the model of Tycho Brahe, some 140 years old

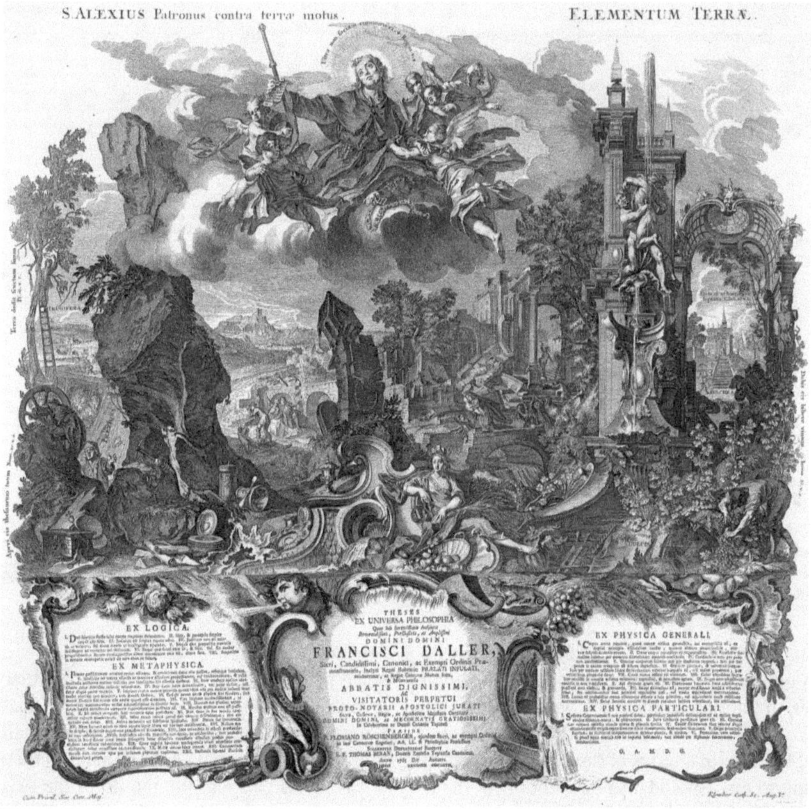

The Lisbon earthquake and tsunami of 1 November 1755, which killed some 100,000 people, shocked and frightened the European population; nobody appeared to have been prepared for such a terrifying seismic impact in what was seen as a relatively safe part of the European continent. What was particularly difficult was the fact that the earthquake occurred on the solemn feast of All Saints at a time when all good Christians were at worship, and many were killed by the collapse of churches and cathedrals; on the other hand, sinners enjoying themselves in parks escaped. This more or less quashed any idea that the earthquake could have been sent by God. Thus, philosophers and naturalists across the continent involved themselves in numerous studies of the event, and soon the first serious tracts about this enigmatic phenomenon appeared. Simultaneously, in numerous European universities, lectures on the Earth's dynamic manifestations were read and disputed; also, theses on earthquakes and volcanic eruptions were read and defended for higher university degrees. The illustration here is the degree diploma of a rare university thesis on geological phenomena and earthquakes. It was defended by the doctoral candidate Francisci Daller in front of the university commission at the Teplá Monastery in west Bohemia in August 1765, ten years after the Lisbon earthquake. Details about the thesis and its defense are given at the bottom of the diploma. In the allegorical composition decorating the upper part, damaging earthquake effects are apparent, as are other symbols of the "elementum terrae" (top right). The composition is dominated by St. Alexius, shown here as the patron saint of protection against earthquakes, a charge given elsewhere to the equally mythical St. Emidius

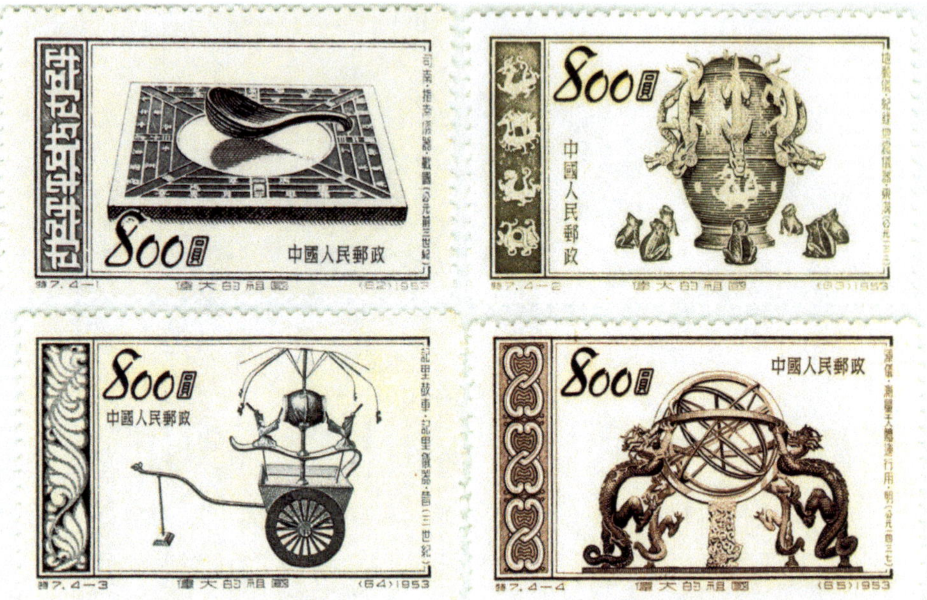

In the Modern Age it was undoubtedly European naturalists, later with American colleagues, who pioneered the advancement of the natural sciences. This fact, however, did not mean that they were alone in the world in this effort. Chinese scientists who advanced their scientific, technical, and technological levels with a series of discoveries—such as the compass, gunpowder, and so on—should also be mentioned. In the eighteenth century, the Empire of Heaven (or Middle Kingdom) gradually opened up toward Western civilization. This picture shows a series of four modern (1953) postage stamps showing historical instruments from the oldest period of Chinese research in the natural sciences. These are a magnetic spoon (actually a compass), a seismoscope for detecting an arriving seismic wave from a distant earthquake, a distance-measuring device, and an astrolabe. The seismoscope was invented in AD 132 by the Chinese mathematical and astronomer Zhang Heng; eight dragons hold bronze balls in their mouths, the intention being that even a slight tremor would open the mouth of a dragon, releasing its ball into the mouth of the frog below. Which dragon was activated would show the direction of the shock and thus inform the Emperor as to which part of his dominions had been struck. Many attempts have been made to make modern reconstructions of the device, but it is not known for sure exactly how the internal workings were arranged

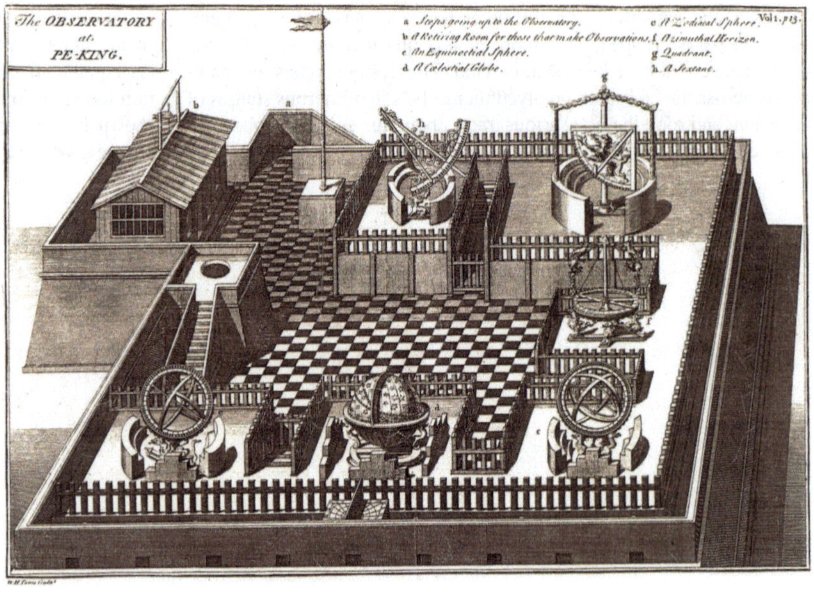

Astronomy was taken seriously in China from ancient times. The first observatory in Beijing was established in the earlier part of the thirteenth century and reconstructed in 1442. From the seventeenth century onward, contacts between China and the West become more frequent, and in the 1670s the responsibility for running the observatory was put in the hands of the Jesuit Ferdinand Verbiest. This depiction of Beijing Observatory was made by Jean Baptiste Bourguignon d'Anville and published in 1737. This ancient observatory still exists today

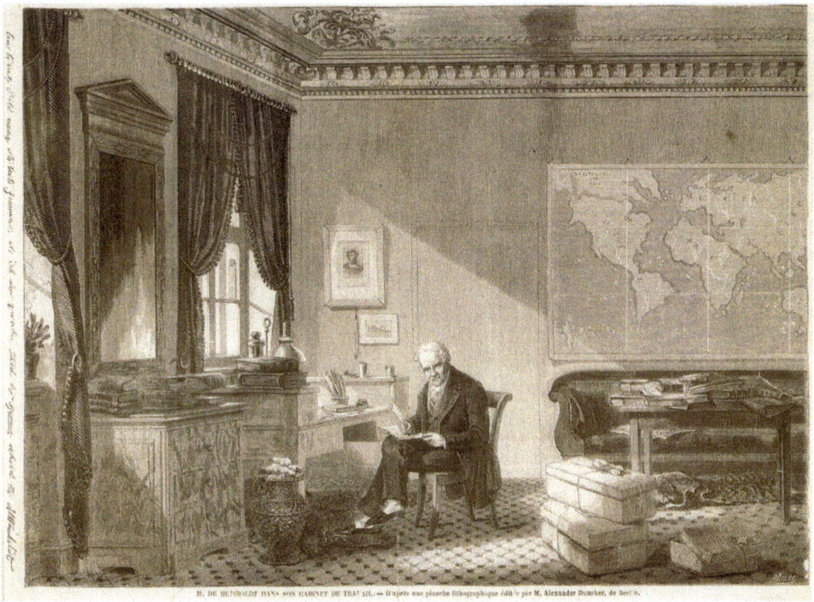

One can place the development of the natural sciences in Europe as starting sometime around the mid-eighteenth century, at the beginning of the Age of Enlightenment, when the famous French *Encyclopédie* by Diderot and d'Alembert was composed and published. The outstanding figure of the period was undoubtedly the German naturalist Alexander von Humboldt. After early travels in the Americas, he returned in 1804 with 60,000 plant specimens and hundreds of pages of maps, charts, and research notes. During his long life (1769–1859) he published 41 voluminous works on both animate and inanimate nature, together with narratives of his travels. His fame rests particularly on his encyclopedia *Kosmos*; Humboldt is reckoned to have been the last person ever to be capable of writing a complete encyclopedia on his own; after him, the sum total of human knowledge was too great for one person to know. In our illustration, Humboldt is shown at a mature age in his study in Berlin. This print is signed "A. Mare" and captioned as edited by Alexander Duncker, but in fact it is a copy of a painting by Eduard Hildebrandt, made in 1845. Note particularly the boxes stuffed with correspondence on the floor by his chair. It is estimated that Humboldt wrote around 50,000 letters in his career and received twice as many (Wulf 2015)

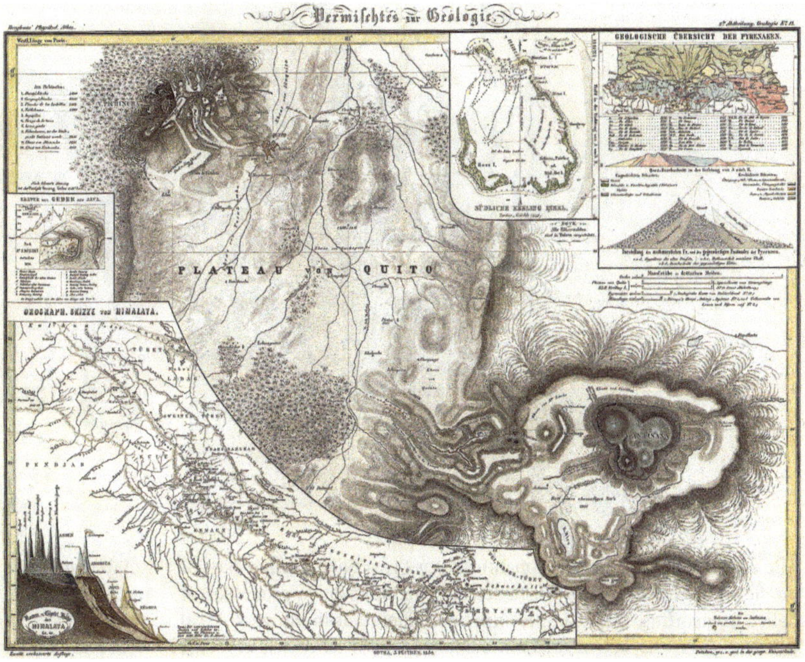

Humboldt collaborated for many decades with the famous German geographer, map maker, and naturalist Heinrich Berghaus (1797–1884), who prepared and gradually published his *Physikalischer Atlas* between 1838 and 1848, a work consisting of some 90 plates. In his atlas, Berghaus used many sketches, diagrams, and drawings made by Humboldt, always correctly cited, as the basis of local, regional, and world maps. A typical specialty or curiosity of cartographic plates by Berghaus is his practice of supplementing the main map with small, informative maps, diagrams, and side views, which increase both the cartographic information and also simultaneously the attractiveness of the final cartographic compositions. The composition reproduced here, titled *Miscellanea of Geology*, presents a map of the Antisana volcano, together with the Pichincha volcano and the city of Quito (Ecuador). The main map is complemented by insets consisting of seven small maps, tables, and figures of mountain ridges and volcanoes of the world

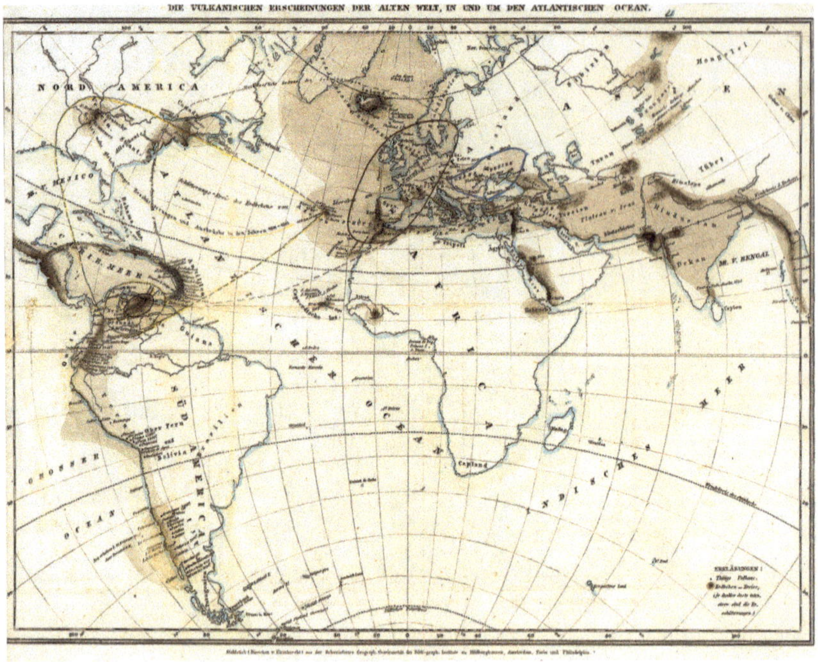

This map, titled *The Volcanic Regions of the Old World, and of the Atlantic Ocean*, shows volcanically and seismically active areas across most of the globe, together with the felt areas of four recent strong earthquakes. This was the first attempt to depict seismicity over such a large area, predating the famous map made by Robert Mallet ten years later. However, the first map of seismicity over the entire world was made by a Scottish cartographer, Alexander Keith Johnston, in 1849. Both Berghaus and Johnston used the earthquake catalog of Karl Ernst Adolf von Hoff as their data source (Musson 2013)

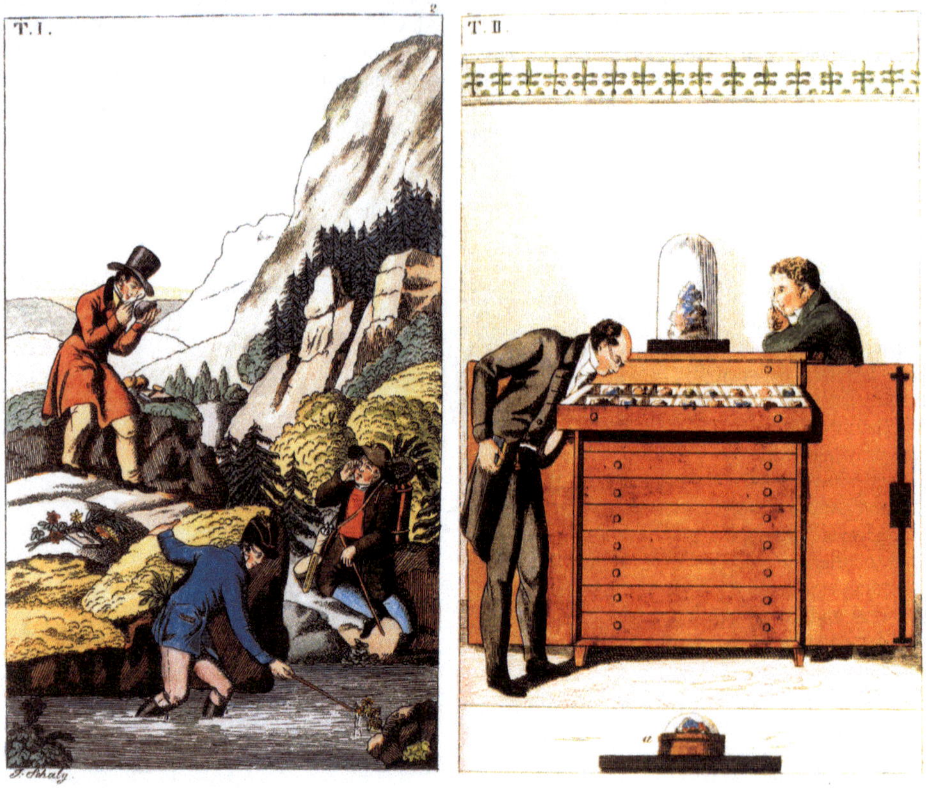

The study of Earth's dynamic manifestations was only a minor part of the development of geology as a science; the descriptive processes of collecting, sorting, and classification were important in the geological sciences from the late eighteenth century onward as rocks, crystals, minerals, metals, and sand were tested and examined. Our two depictions give the reader an idea of what this process looked like: a group of enthusiasts armed with a magnifying glass, a mortar and pestle, and a small round net for catching insects (and not forgetting the stovepipe hat!) are visiting a locality for the purpose of collecting the products of nature. Then, later, the fruits of their endeavors can be studied, classified, and stored in the quiet atmosphere of the cabinet rooms. The original hand-colored copper engravings are taken from Wilhelm (1828) and are published here with kind permission of M. Široký (Dobřichovice, Czech Republic)

Volcanism and volcanic rocks were a subject of common appeal in Europe in the late eighteenth and early nineteenth centuries; one thinks of Mendelssohn's overture "Fingal's Cave" and the many paintings of this spectacular example of columnar basalt. The other type example, almost equally popular as a tourist destination, is the Giant's Causeway in County Antrim on the north coast of Ireland, which is shown here. The illustration comes from the *Bilderbuch für Kinder*, a children's encyclopedia published by Friedrich Justin Bertuch of Weimar as a monthly part-work between 1790 and 1830. The image is taken from a print made by the Stark Brothers of Lower Sackville Street, Dublin, a firm that published many popular views of Irish scenes (a copy of the Stark Brothers' original print is held by the National Museum of Natural History, Paris). The lower part of the image illustrates in detail the hexagonal structures formed by the slow cooling and separation of lava, separated also along horizontal planes

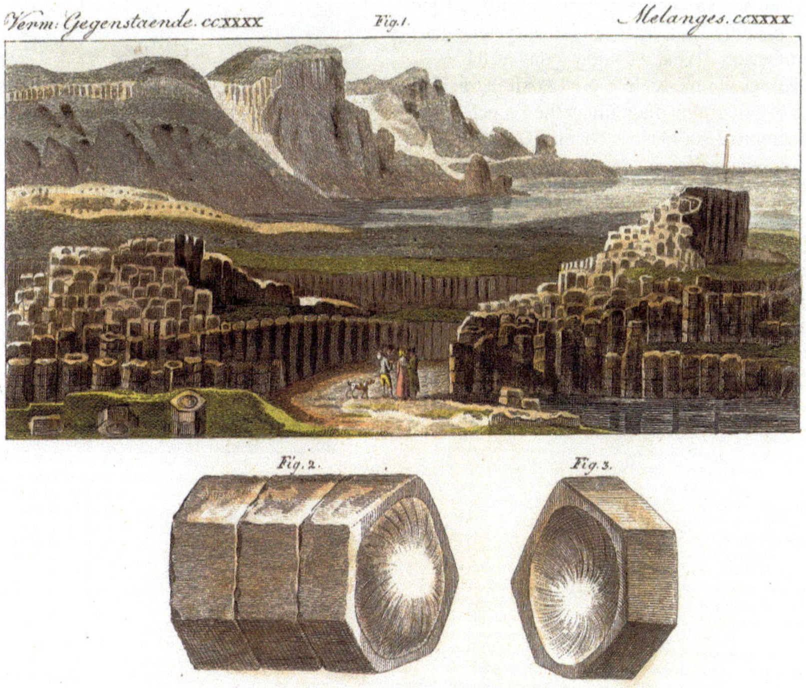

The study of basaltic cleavage seems to have been of considerable interest in the early nineteenth century. This somewhat naïve illustration of different geometries of basalt columns is taken from a work by the French naturalist Jean Baptiste George-Marie Bory de Saint-Vincent (1778–1846), derived from his studies on the island of Réunion (published in 1804)

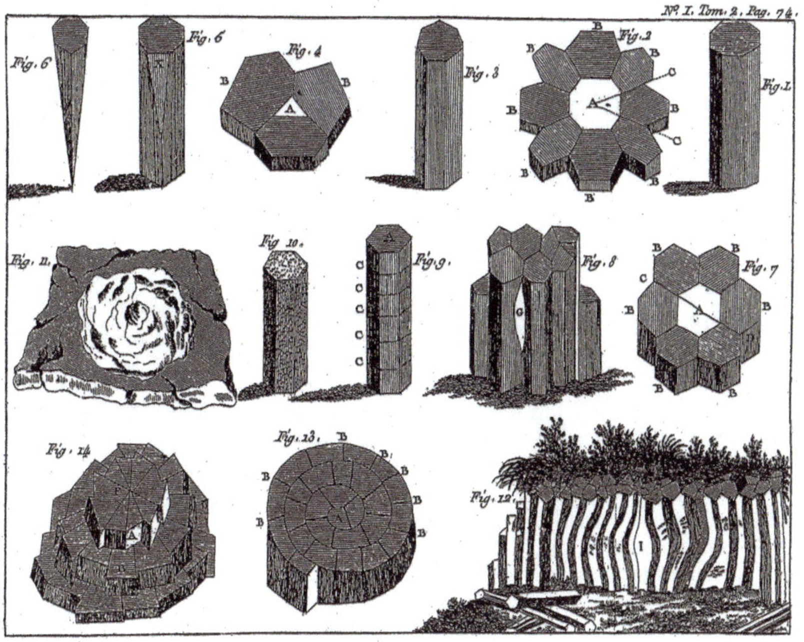

THÉORIE DES BASALTES

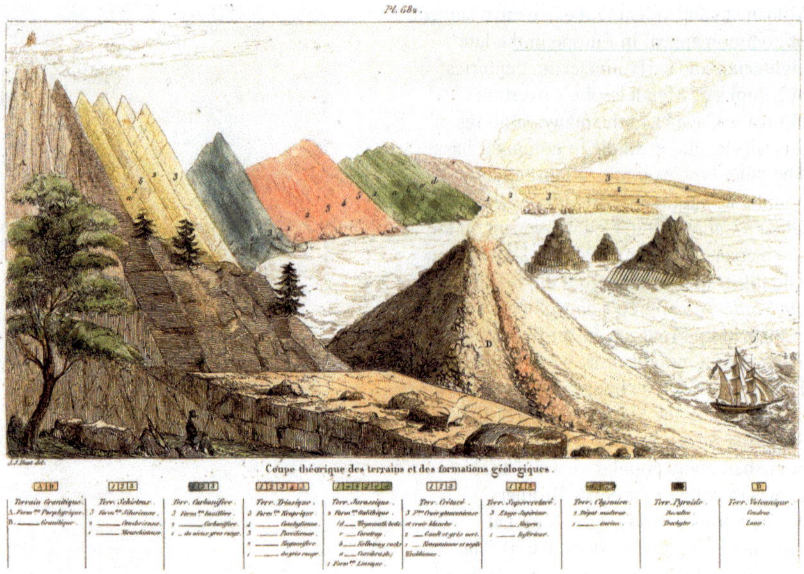

As geological data were gathered, they had to be classified and arranged into a systematic framework. What we know today as the geological time scale is prefigured in this rather charming diagram by the French geologist Jean Jacques Nicolas Huot. The different rocks are depicted in an idealized landscape, but the key demonstrates that granitic rocks lie at the bottom, overlain by Cambrian and Silurian schists, then Carboniferous, Triassic, and Jurassic rocks, and so on. Intruding into this arrangement is a smoking volcano, providing the most recent rocks

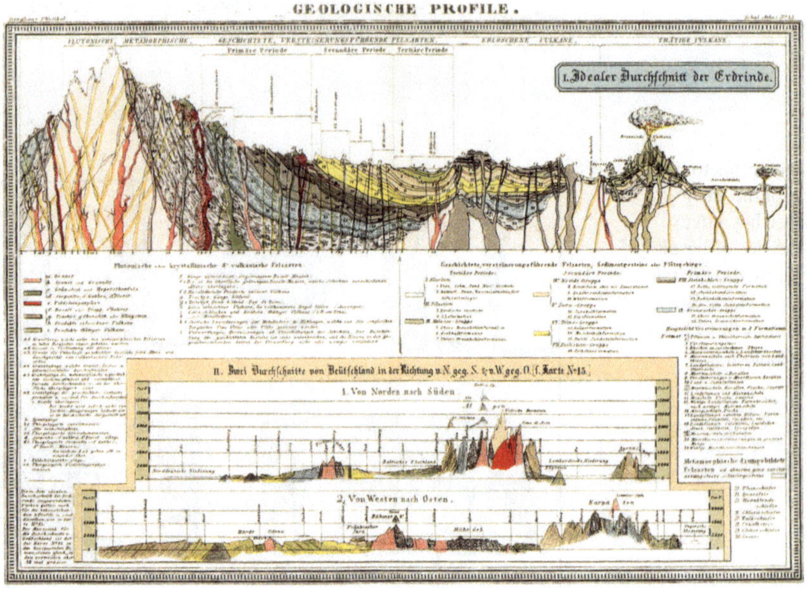

The systematic study and classification of the rocks found on the Earth's surface resulted in the first detailed geological maps that appeared in the first decades of the nineteenth century, which, in some cases, were of very high quality. This achievement, however, sparked off a new question: how much, and in which way, did this surface geological pattern change with depth? It was not easy to solve this puzzle, since direct observation, from drilling or mining, was limited to a depth not exceeding one kilometer. However, regardless of this limitation, some simple sketches of shallow geological profiles had already been plotted before 1850 as supplements to regular geological maps. Of course, as the depth increased, so did the map maker's reliance on fantasy. This figure shows an early attempt to look deeper into the Earth's interior, again from Berghaus's atlas. In the upper part of the plate, an ideal cross-section through the Earth is proposed. Here, the artist's fantasy and invention have no limits, since the cross-section is an "ideal" one. However, the real geological profile shown below, based on actual data, is much more modest and limited in the depth depicted

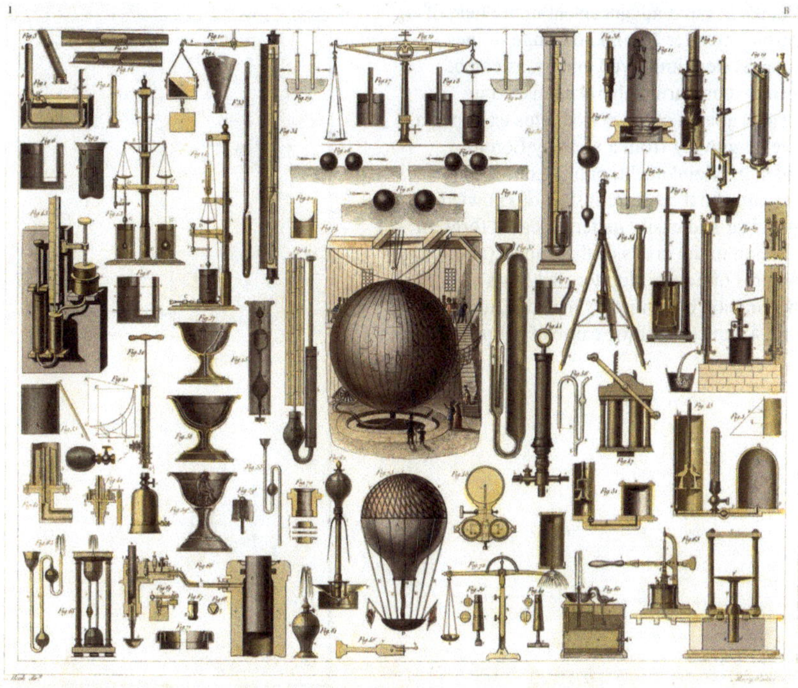

This image shows a selection of different measuring devices that the natural philosopher of the mid-nineteenth century might have had at his disposal. This fascinating illustration comes from the *Iconographic Encyclopaedia of Science, Literature, and Art* by Johann Georg Heck, which appeared in ten parts between 1849 and 1851. The work includes 500 prints like this, covering many different subjects, with over 12,000 separate illustrations. The engraving was the work of Henry Winkles, an Englishman who was a co-founder of the first steel-engraving studio in Germany. This popular work also appeared in an English four-volume edition translated by Spencer Fullerton Baird, published in New York in 1851

From early times, and not just in Europe, one of the chief tasks of astronomers was to observe solar eclipses and to be able to forecast them. Since these were widely viewed as awesome phenomena, foreknowledge was important. Both the Babylonians and the ancient Chinese were able to predict solar eclipses as early as 2500 BC; the Chinese believed that failure to predict an eclipse would put the Emperor in danger. This painting of a total solar eclipse in northern Prussia (now part of Poland) was one of a number made by the scientific artist Wilhelm Kranz in 1887. In fact, because of cloud cover, Kranz failed to see the eclipse for himself, and he painted the details according to descriptions from others. The image here is a copy by Kraemer (1910)

Thanks to rapid advancement in the field of astronomy after 1800, knowledge of the Solar System improved greatly, as can be seen by comparing charts published in various atlases of the period. This example, a steel engraving from a work by Lorenz Valentin Kleinknecht, published in 1844, presents the then current state of knowledge in a clear and easily understandable form. In addition to the orbits of the planets (and pictures of Saturn, Jupiter, and Mars), information on comets is included, together with a depiction of the cause of eclipses

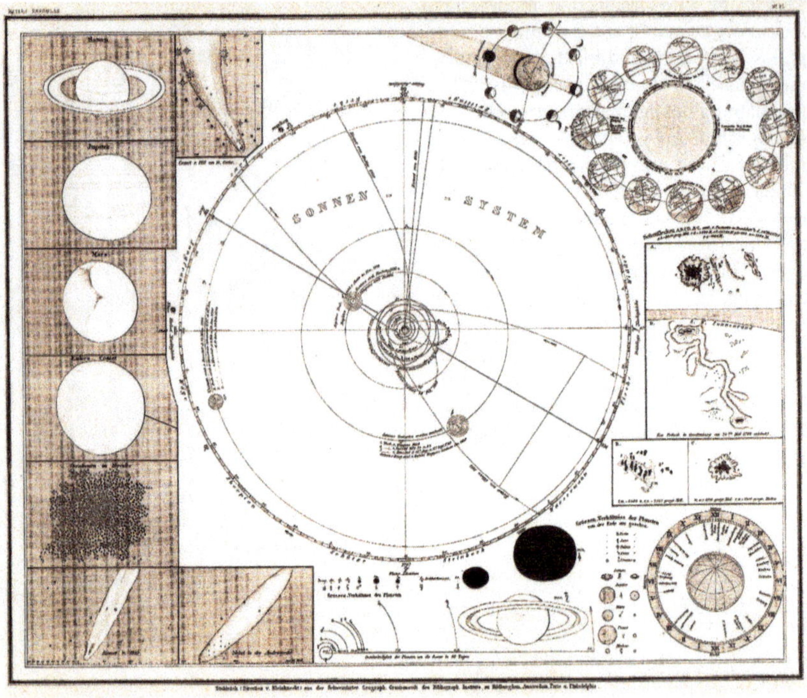

Studies of volcanic rocks had a practical aspect, as well as a scientific one. Basalt is an excellent homogeneous stone suitable for building, especially as a decorative stone—for instance, for monuments. Its exploitation was made particularly easy thanks to the prismatic columnar form of basalt, which made further processing substantially easier. Basalt was valued and quarried as long ago as ancient Egypt. Besides the Mediterranean region, in Europe, basalt quarries are found in Scotland, the Massif Central in France, the mid-German volcanic zone, the Bohemian Massif, Iceland, and elsewhere. Today the majority are found in China. Shown here is the basalt quarry at Dubičná village (Česká Lípa, north Bohemia) as it was in the late nineteenth century (in about 1880); numerous quarries in Bohemia have continued to operate up to the present day

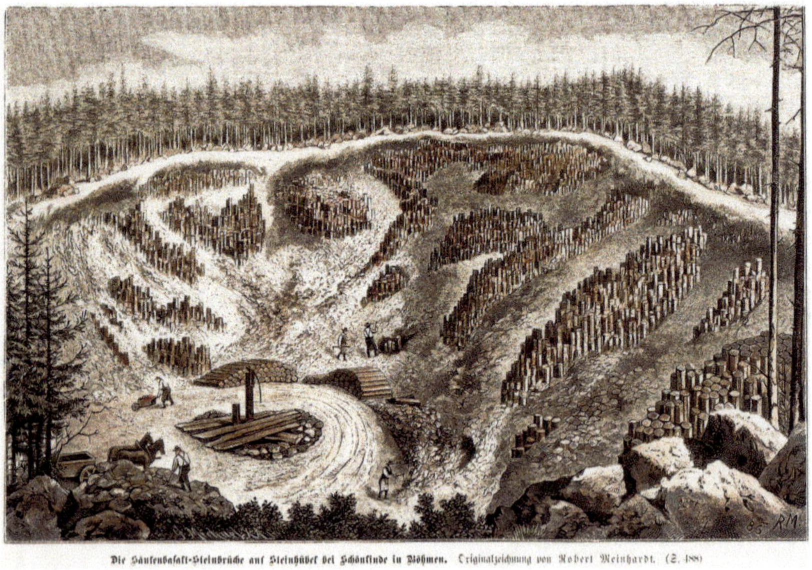

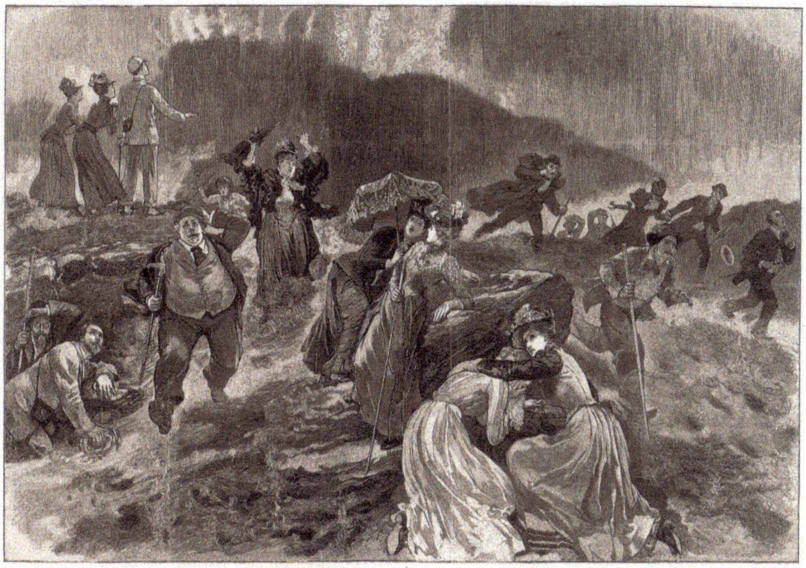

VISITORS TO MOUNT VESUVIUS—AN UNEXPECTED ERUPTION
DRAWN BY WILLIAM SMALL

One more aspect of the common interest in volcanic phenomena should be noted: it became an attraction for tourists. This was particularly true of Vesuvius: rising above the city of Naples, it was an important location on the "Grand Tour" made by wealthy Northern Europeans from the later seventeenth century onwards. The gentleman artist could be expected to bring back a painting of Vesuvius, preferably in eruption and viewed from a safe distance across the bay. That Vesuvius was still a tourist attraction in the late nineteenth century is shown by this engraving by William Small, published in the *Illustrated London News* of 1892. The excitable locals are running in panic from the falling ash and pumice, while the unflappable British visitors (a man and two women, upper left) are unperturbed and calmly watch the magnificent natural spectacle

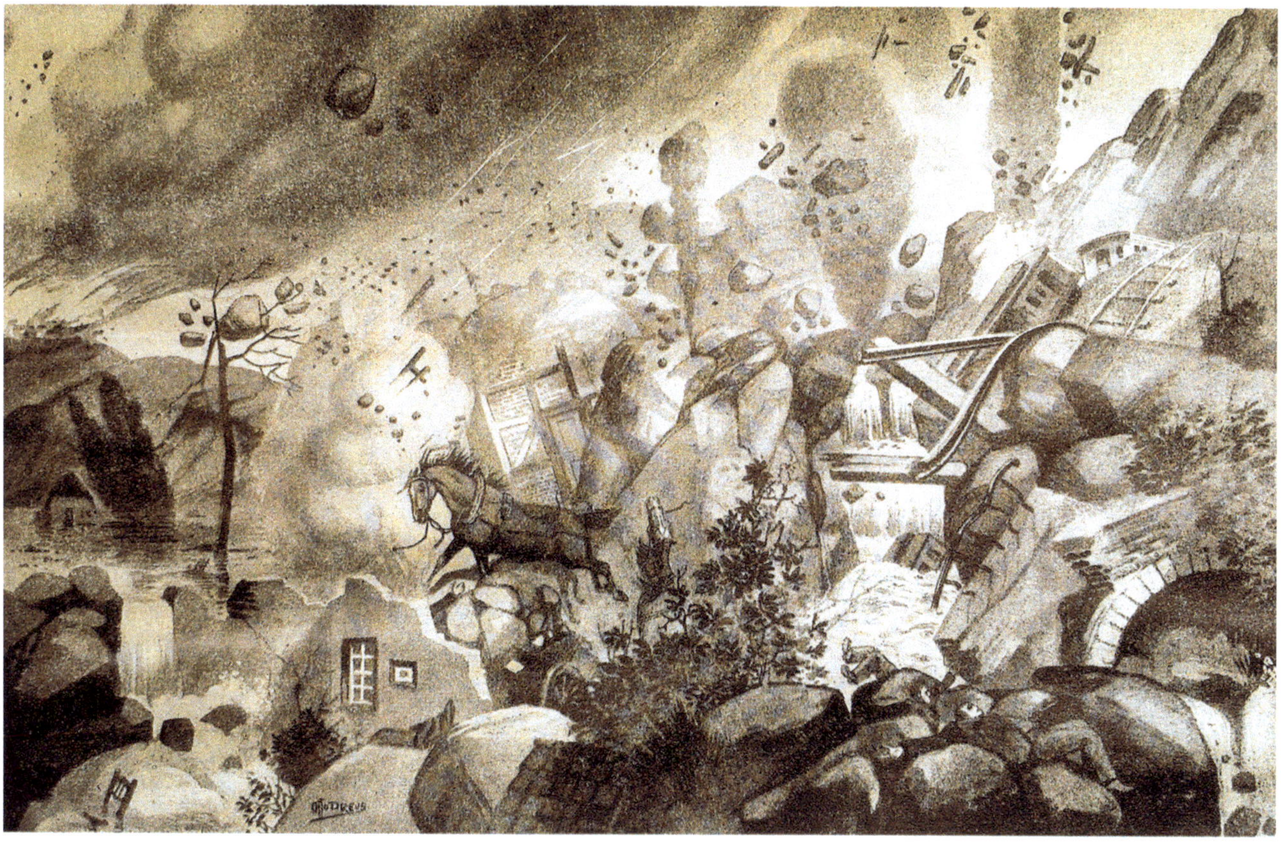

The ultimate natural disaster is surely the end of the world. The destruction of the Earth at the time of the Second Coming, as foretold in the Book of Revelation, has proved a popular subject for artists with a flair for the dramatic. This little-known example is taken from a book titled *His Glorious Appearing: An Exposition of Matthew Twenty-Four*, by the American evangelical writer James White (1821–1881). This popular book went through many editions after its first printing in 1895, after the author's death. Many of the illustrations used were poor-quality reproductions of well-known images. The depiction of the "day of wrath" varies between editions; this one from the 1916 edition appears to be an original composition

One more aspect of the cognitive interest in volcanic phenomena should be noted: it became an attraction for tourists. This was particularly true of Vesuvius, rising above the city of Naples. It was an important location on the Grand Tour made by wealthy Northern Europeans from the later seventeenth century onwards. The [illegible] area could be expected to bring back [illegible] souvenirs, preferably [illegible] eruption observed from a safe distance across the bay [illegible] was still [illegible]

[illegible]

2 Earth

2.1 Introductory Remarks

For the first of the four elements to come under inspection for its violent—and indeed disastrous—extremes, we choose Earth. There are two chief manifestations by which this element produces fatalities among human populations: the earthquake, and the landslide or landslip, in which gravitational forces cause mass movements, with ruinous results for any settlement below.

The earthquake is of particular interest. Firstly, its killing power is unequalled. Thousands of people can be killed in the space of a couple of minutes. In fact, the grisly record for the most casualties stands at 800,000 for a mid-sixteenth-century earthquake in China, and because China was a civilized empire with a well-functioning bureaucracy, there is no reason to doubt this figure. Seismologists have long speculated as to whether there will ever be an earthquake with a death toll topping one million (Musson 2012). Secondly, earthquakes cause particular terror because they mostly happen without the least warning. One can see a hurricane coming as the weather worsens; a volcano usually rumbles extensively before it blows its top. But, with an earthquake, one moment you are enjoying a normal day, and then suddenly there is chaos and ruin all around you. Thirdly, earthquakes lend themselves well to pictorial illustration, at least with regard to dramatic ruins.

Landslides and similar gravitational collapses often go hand in hand with earthquakes, as the shaking is enough to bring down a cliff slope weakened by weathering, and if this happens on a large enough scale, whole villages can be buried. In addition, extensive earthquake-induced landslides in mountainous areas can block lines of communication, interfering with relief efforts. Often the full death toll from an earthquake is not known for a considerable period of time because of the difficulty in reaching remote areas. After another Chinese earthquake in 2009, it was necessary to send troops in by parachute to reach the valleys that had been cut off by landslides blocking the roads.

Oddly enough, there is another link between landslides and earthquakes that may seem strange to the modern reader. In the English language (and also in Latin), the term "earthquake" (*terra motus* in Latin) used to be quite flexible in meaning, and early writers would enumerate up to ten types of earthquake, most of which we would now think of as landslips of one sort or another. What today we think of as an earthquake was just one sort: the vibratory kind. German and French are more precise: *erdbeben* and *tremblement de terre* both mean shaking of the earth, whereas *terra motus* simply means the moving of earth, and that movement may be downward under gravity. Even as late as the 1830s, one could find a major cliff fall being referred to as an earthquake.

Earth, as an element, has always been associated with solidity and stability. We expect it to bear the weight of our dwelling places; we talk about something being "as solid as a rock." It is the antithesis of the other three classical elements—flickering Fire, flowing Water, and gusting Air. Earth is something dependable, which is another reason why earthquakes are so upsetting. Our most fundamental belief about the world, instilled from childhood, is that we can rely on the Earth beneath our feet to let us stand safely on it. When the Earth starts shaking, it strikes at a very core belief about the nature of things.

So perhaps it is not surprising that when early savants attempted to explain earthquakes, they looked to the other three elements. Thales of Miletus compared the bucking Earth to the movement of a ship and concluded that water must be responsible, Aristotle thought that underground gusts of wind venting from caves produced earthquakes, and the eighteenth-century philosopher Immanuel Kant concocted a scheme whereby fire travelling through subterranean channels was the root cause.

It was in the 1760s that an English clergyman, John Michel, published a paper that recognized for the first time that earthquake shaking was produced by elastic waves travelling through rocks. Scientists in the nineteenth century

J. Kozák, R. M. W. Musson, *The Illustrated History of the Elements*, https://doi.org/10.1007/978-3-030-21426-5_2

looked to try and explain the origin of these waves but skirted round the idea that earthquakes and geological faults were related. After Charles Lyell's study of the 1855 Wairarapa earthquake in New Zealand, it was clear that movement of rocks along a fault line happened at the same moment as an earthquake—but did the fault movement cause the earthquake, or did the earthquake cause the faulting? It was not until the San Francisco earthquake of 1906 that it was definitely proved that faults became strained to breaking point by geological forces until they snapped, producing an earthquake. And it was not until the 1960s that the nature of those forces was finally understood—that the Earth's crust is made up of plates of lighter rocks that float on the semi-molten and denser mantle, and that earthquake zones are regions where these plates scrape past each other or are pushed one under the other.

So the element Earth is not as stable as it first appears. In presenting our images of this element, we will deal with earthquakes first, arranged on a regional basis, and gravitational movements second.

The image here—a sixteenth-century painting—presents a situation in the aftermath of a weak earthquake visiting some unknown Swiss settlement. No seismic damage is visible on either the church or the three houses shown. In this period, earthquakes were considered God's punishment if they were severe, or just a warning if they were not. This picture was kindly supplied to us by the late Jean Vogt of Strasbourg

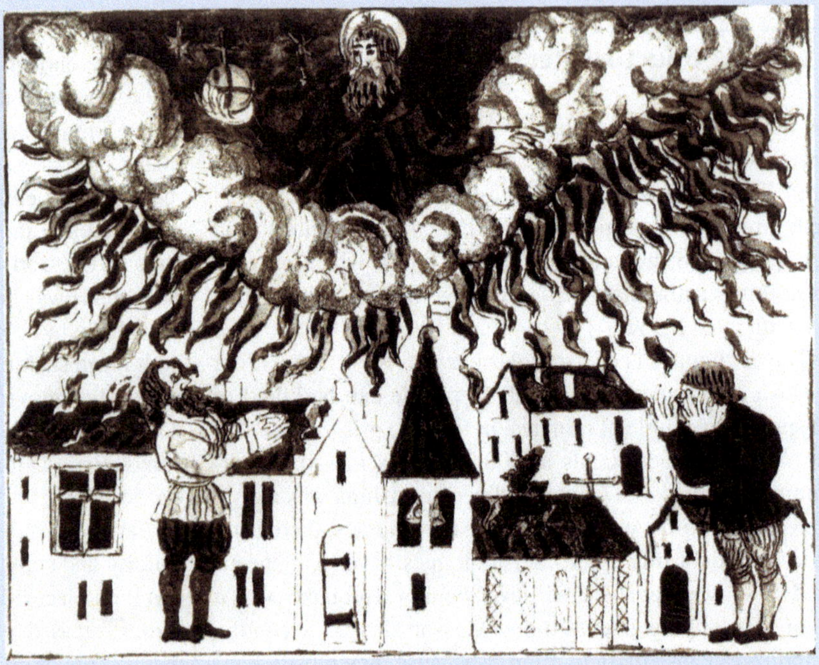

2.2 Europe—The Mediterranean

The first section of images deals with the earthquakes of Europe and the Mediterranean countries. This particular zone—regardless of the fact that the strongest earthquakes have generally occurred elsewhere—plays a unique role in Earth studies; it was largely Europeans who developed scientific studies of seismic phenomena before 1900, and Europe's rich cultural heritage provides a wealth of depictions going back over 500 years.

Not only that, but—as will be seen throughout this book—virtually all of the images are made by Europeans or those of European ancestry. Only one country in the world has a tradition of disaster images that is entirely separate from the European canon, and that is Japan.

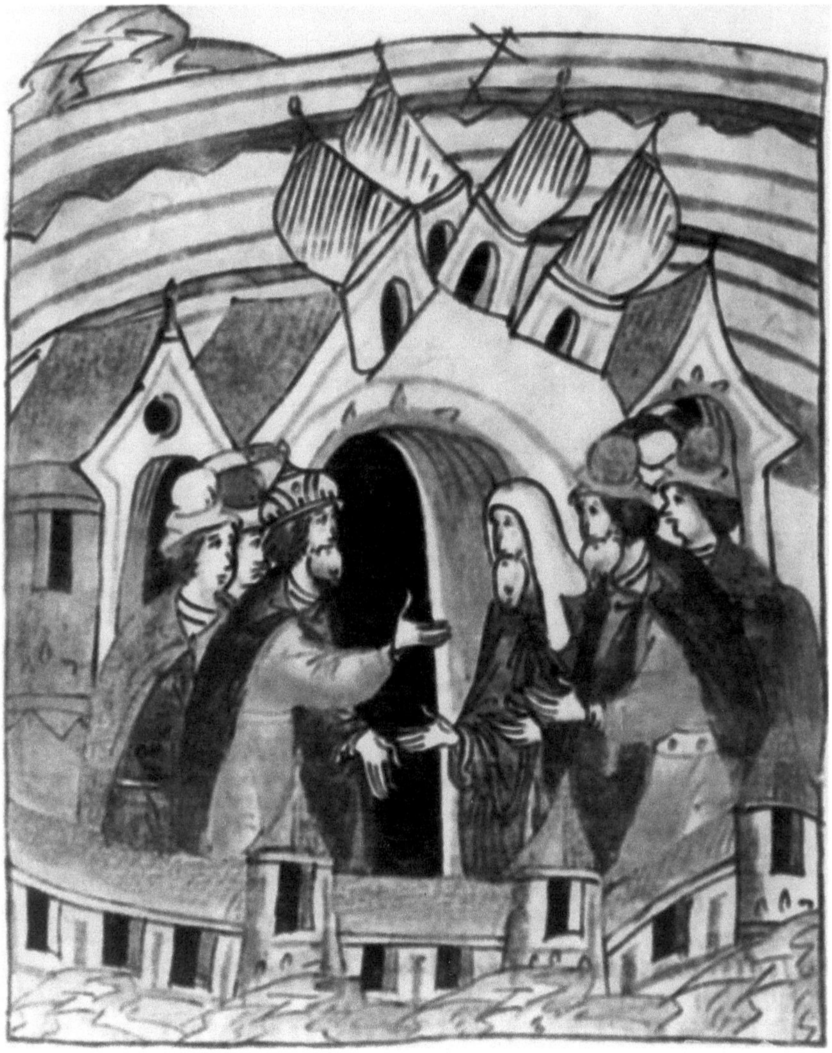

Fig. 2.1 An earthquake felt in Kiev in 1230

One of the earliest pictorial compositions of seismic effects, and therefore the first chronologically in our survey, comes from the sixteenth-century Russian illustrated manuscript chronicle titled *Povest' vremennykh let,* often translated into English as *The Tale of Bygone Years*. The work (Fig. 2.1) is a compound of works by various authors of different periods, but the last version was composed and illustrated (by allegorical compositions) by historians and artists employed at the court of the Russian Tsar Ivan the Terrible in the 1570s. In the illustration dealing with an earthquake in 1230, shown here, we see the excited population of Kiev (or Kyiv) discussing their remarkable experiences. Seismic damage to the city is expressed by the splitting and collapsing of the top cupolas of the church towers.

The Ukraine and European Russia are seismically very stable, and no earthquakes worth mentioning are known in these regions today. However, in northeast Romania there is the region of Vrancea, which repeatedly produces earthquakes of a rather peculiar nature: they are large, they are very deep (hundreds of kilometers deep), and they radiate energy much more to the northeast than in other directions. They can therefore be strongly felt in Kiev and even Moscow. Therefore, whenever the historian finds accounts of an earthquake felt in Kiev or Moscow, the immediate conclusion is that it must have been a Vrancea earthquake, even in the absence of any information from Romania. It may be true; but there is a danger of a self-fulfilling argument: there are no earthquakes in Kiev; this is an earthquake in Kiev; therefore, it must be Romanian; therefore, we see no quakes in Kiev.

The illustration from the chronicle reproduced here is kept in the Russian State Library in Moscow. The authors are grateful to the historical seismologist A.A. Nikonov from Moscow, who kindly arranged for us to use this image.

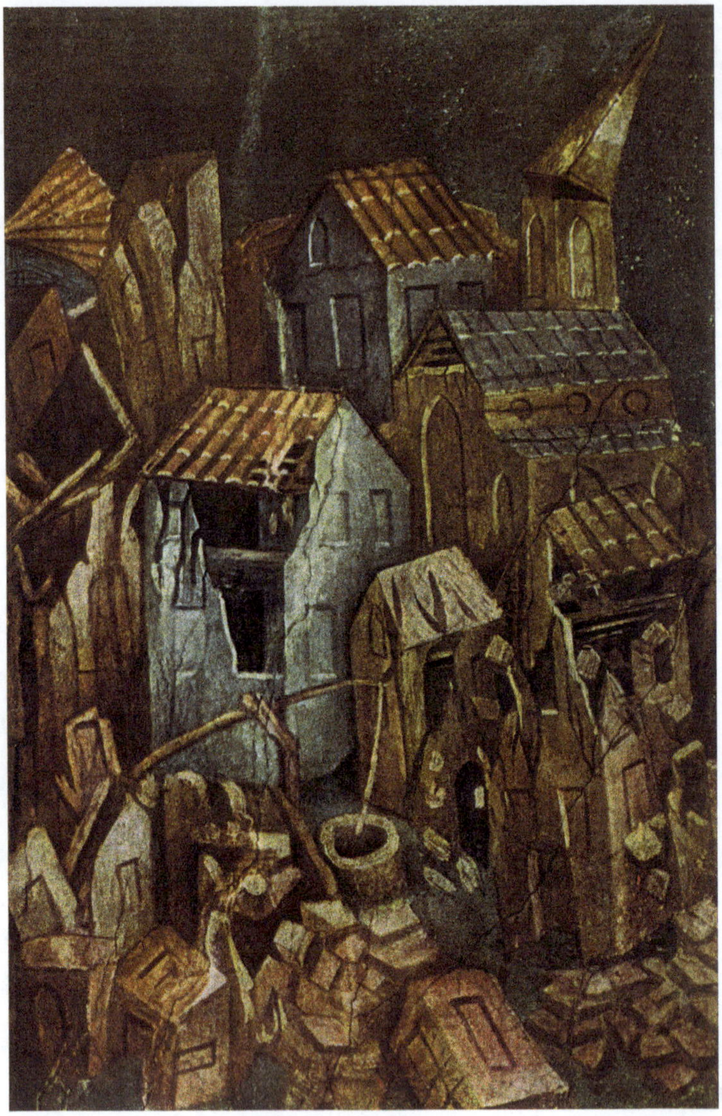

Fig. 2.2 The 1348 Villach earthquake

It might be surprising to find, from the fourteenth century, one of the few images of a strong European earthquake outside the main seismic zones of the Mediterranean. It shows the seismic disaster that ruined the Carinthian town of Villach in Austria on 25 January 1348. The collapsed structures shown are part of Arnoldstein Castle, built on two adjacent hilltops. The illustration (Fig. 2.2) is the central segment of the wall painting in St. Mary's chapel in Karlstein Castle, some 20 kilometers southwest of Prague, Czech Republic. The tempera wall painting is ascribed to the Strasbourg painter Nikolaus Wurmser, who personally visited Arnoldstein three years before the earthquake and thus knew the setting. Wurmser's composition—part of a series of apocalyptic images—was intended to show God's punishment of sinners. The seismic damage, however, is shown quite realistically in comparison with other Renaissance images of seismic effects. The earthquake was clearly felt over a large part of transalpine Europe and in north Italy and Slavonia. Seventeen villages were ruined, and some 5000–10,000 people were killed.

Some technical terms need to be used in this chapter. *Earthquake magnitude* is a measure of the size of an earthquake (what is known to journalists as the "Richter scale" although this is not actually a scientific term) and reflects the amount of fault movement and energy released. For a modern earthquake, this is calculated from seismograms. Obviously, this is impossible for a historical event, but the extent of the area over which the quake was felt, and the maximum strength of the shaking, allow a reasonable estimate, in this case a little less than 7. To put things in context, it is also worth bearing in mind that the scale is logarithmic and increases in steps of 30; so a magnitude 7 earthquake is 30 times more energetic than a magnitude 6 one—not one sixth bigger, as might be thought.

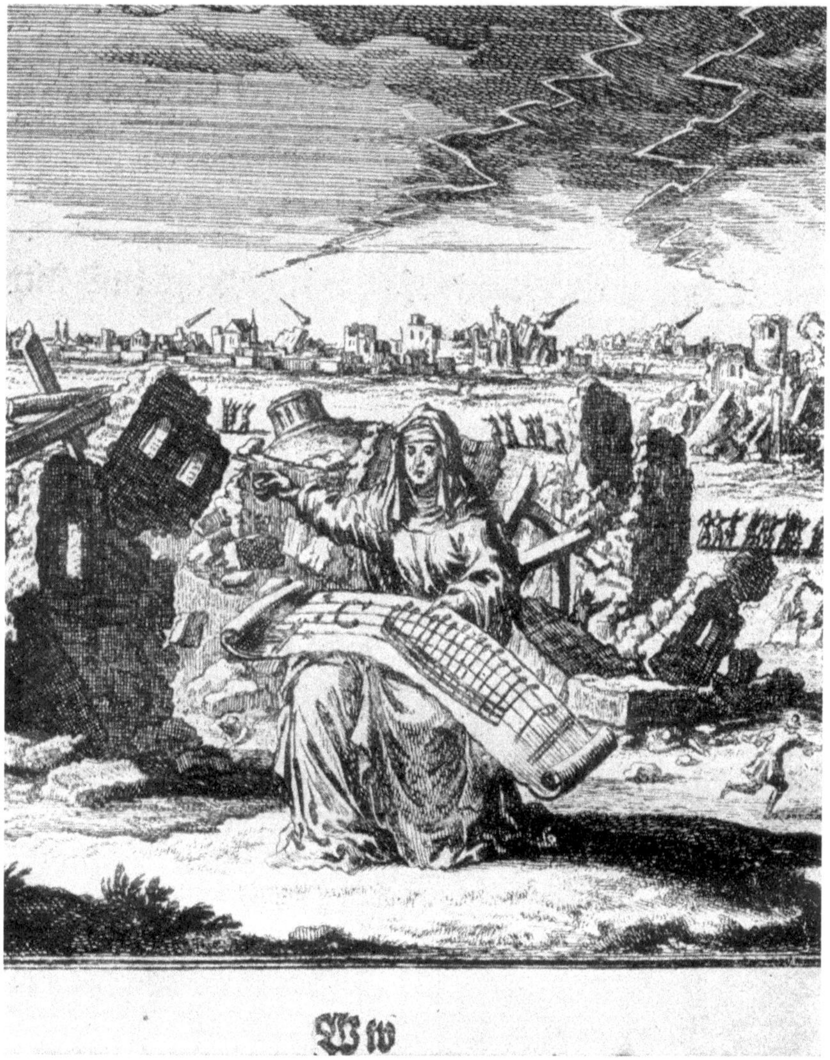

Fig. 2.3 The 1356 Basel earthquake

The 1356 earthquake in Basel, Switzerland, is believed to be the strongest ever seismic event in central Europe—in fact, the strongest event in Europe north of the Alps in historical times. The strength of the event has been the subject of much dispute, with magnitude estimates ranging from 6 to 7. The problem rests, to some extent, on the vagueness of some contemporary descriptions; various castles are said to have been "ruined" but it is unclear what condition they were in before the earthquake. The epicenter is also a subject of disagreement. It is widely considered that the Reinach fault, southwest of Basel, caused the earthquake, but some studies argue for a source further to the west. What is clear, though, is that the city of Basel was extensively damaged (partly by fire) and that a recurrence of the earthquake, now that Basel is a large industrial center, would be extremely expensive. Historical records confirm 300 casualties in the town itself, while the total number of victims approached 1000.

Concerning the illustrations of ruined Basel, we have to face certain difficulties. On one hand, we have several impressive depictions of the town in ruins at our disposal. In them, the steeples, towers, and houses are often at an oblique angle to the horizon in order to heighten the dramatic effect.

However, the "oldest" depictions of the event were created by the Alsatian humanist and encyclopedist Conrad Lycosthenes (1518–1561) and Sebastian Münster (1488–1552) much later, some 200 years after the Basel earthquake. These depictions *could* be free reconstructions of the 1356 earthquake. However, it is really rather unlikely that these illustrations are actually portrayals of damaged Basel, since the same prints appeared in works by Münster and Lycosthenes illustrating other earthquakes. Additionally, numerous and often anonymous portrayals of Basel damaged by the 1356 disaster appeared. These often more or less copied the popular woodcuts that illustrated Münster's *Cosmography* (which ran to 44 editions—a popular work). Sometimes artists created their depictions according to their imagination, such as the nineteenth-century Swiss artist Karl Jauslin, who specialized in historical subjects (for more details, see Mayer-Rosa and Wechsler, 1991). The image shown here (Fig. 2.3) is one of the less well-known ones, being a German copper engraving, probably seventeenth century, and perhaps based on an earlier woodcut. The allegorical figure in the foreground (a giant compared with the man fleeing on the right) possibly symbolizes the restorative power of the arts.

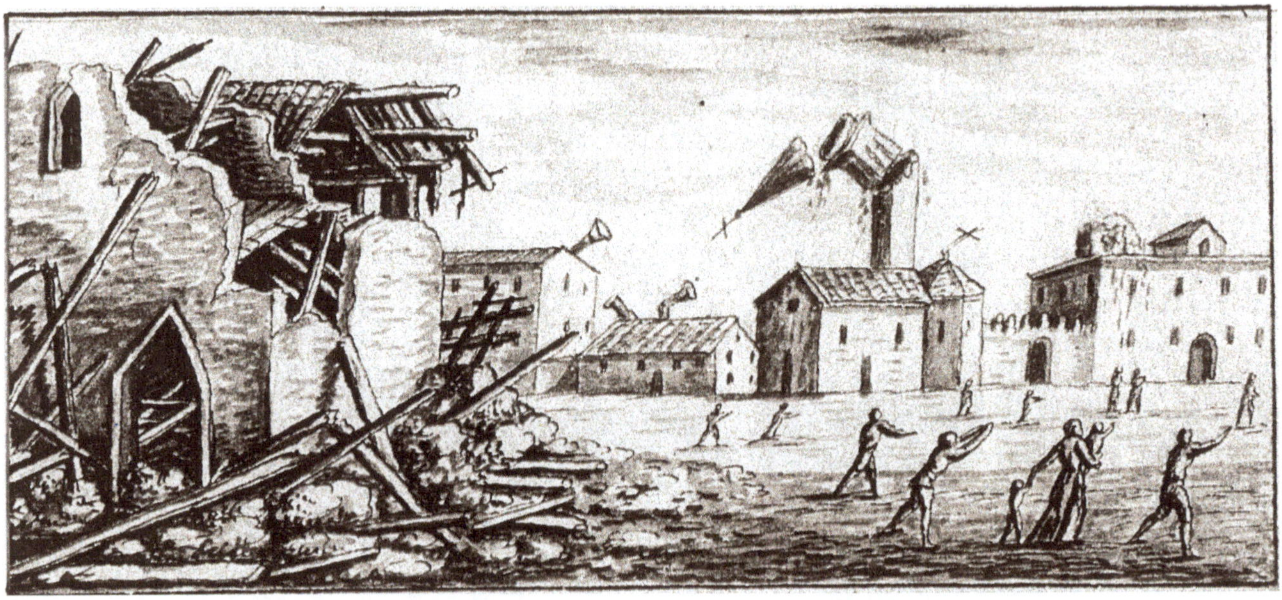

Fig. 2.4 The 1511 Idrija earthquake—damage in Udine

The start of the sixteenth century was an unlucky time for the region of Friuli in northeast Italy (part of the Republic of Venice at that time), since several disastrous blows hit the region.

Firstly, the Republic of Venice was at war with the House of Hapsburg between 1508 and 1516 (the War of the League of Cambrai). Fighting raged across most of northern Italy as Venetian, Imperial, Papal, and French armies all competed for territory.

Secondly, plague broke out in 1510 and 1511. Victims notably included the painter Giorgione. Venice seems to have been particularly susceptible to plague because of its status as a port, and the outbreak in 1510 was just the first of a number of outbreaks in the sixteenth century.

Additionally, on 27 February 1511, the so-called Fat Thursday Revolt erupted in Friuli, as the locals protested against the injustices visited on the region by the Venetian administration; Venice saw Friuli mostly as a buffer against the Ottomans and treated the region less favorably than other Venetian territories. For weeks, warring factions of mutineers pillaged and destroyed castles. The rebellion was violently suppressed.

Finally, on 26 March 1511, a strong earthquake struck. This was probably the largest to affect the region, and damage occurred over a wide area stretching from Vicenza in Italy to Novo Mesto in what is now Slovenia. There was also damage to St. Stephen's in Vienna, and in Zagreb the effects were so strong that the citizens were absolved of taxes for a period of three years. The town of Bovec in western Slovenia was reported to have been totally destroyed. The complexities and obscurities of the source materials have plagued studies of the earthquake, leading to many errors of date and time. The definitive account is that written by Camassi et al. (2011). The location of the epicenter has not been conclusively determined, but somewhere near Idrija in modern Slovenia is likely; the magnitude may have been around 6.8.

The illustration here (Fig. 2.4) portrays the effects in Udine; the damaged building on the left is prominent, and the roof and upper floor have suffered much from the seismic impact. In the middle part of the depiction are a broken bell tower and falling house chimneys. The source is an eighteenth-century manuscript chronicle preserved at the Museo Correr in Venice and thus is not in the least contemporary; the text below the image gives completely incorrect dates.

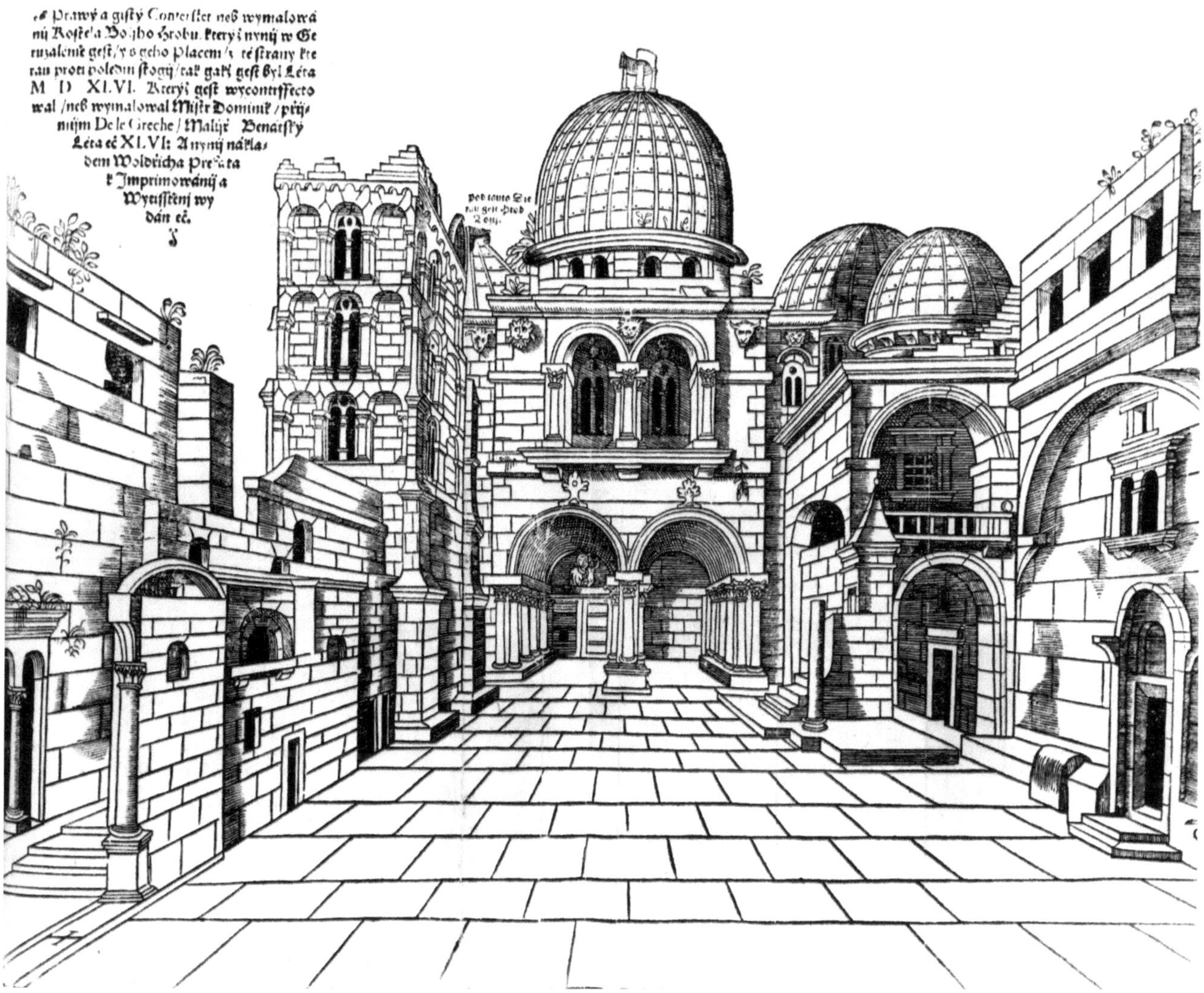

Fig. 2.5 Damage to the Church of the Holy Sepulcher in 1546

On 14 January 1546, inhabitants (and visitors) in Jerusalem were startled by strong earthquake shaking. Contemporary reports of the event circulated far and wide, stimulated by interest in what was happening in the Holy Land. It was said that there was not a tall house in Jerusalem that was not left destroyed or fissured, that the dome of the al-Aqsa (the principle mosque and the third most holy site in Islam) collapsed, and that many public buildings were destroyed, with 500 lives lost under the debris. More recent investigations (Ambraseys and Karcz 1992) have shown that these reports are wildly exaggerated to the point of being totally false. Undoubtedly, this was a moderately strong earthquake on the Dead Sea rift (the major fault running along the Jordan Valley), but it was not a great earthquake and has been repeatedly overestimated.

Our interest here is in the fine perspective drawing (Fig. 2.5) of the Church of the Holy Sepulcher, made by the Venetian painter Dominik de la Greche a few months after the earthquake and subsequently published by his travelling companion, the Czech mathematician Oldřich Prefát z Vlkanova (1523–1565). The only visible damage is the loss of the bell tower cupola to the left of the main dome, confirmed by another drawing by de la Greche of the church with the bell tower intact before the earthquake. The weeds seen sprouting from the walls are not an indication of ruination but simply the run-down condition of the building. There is also a panoramic image of the whole city, made by de la Greche on the same visit, and the damage to the bell tower is the only visible sign of the earthquake. Even the decrepit city walls seem to have escaped any damage.

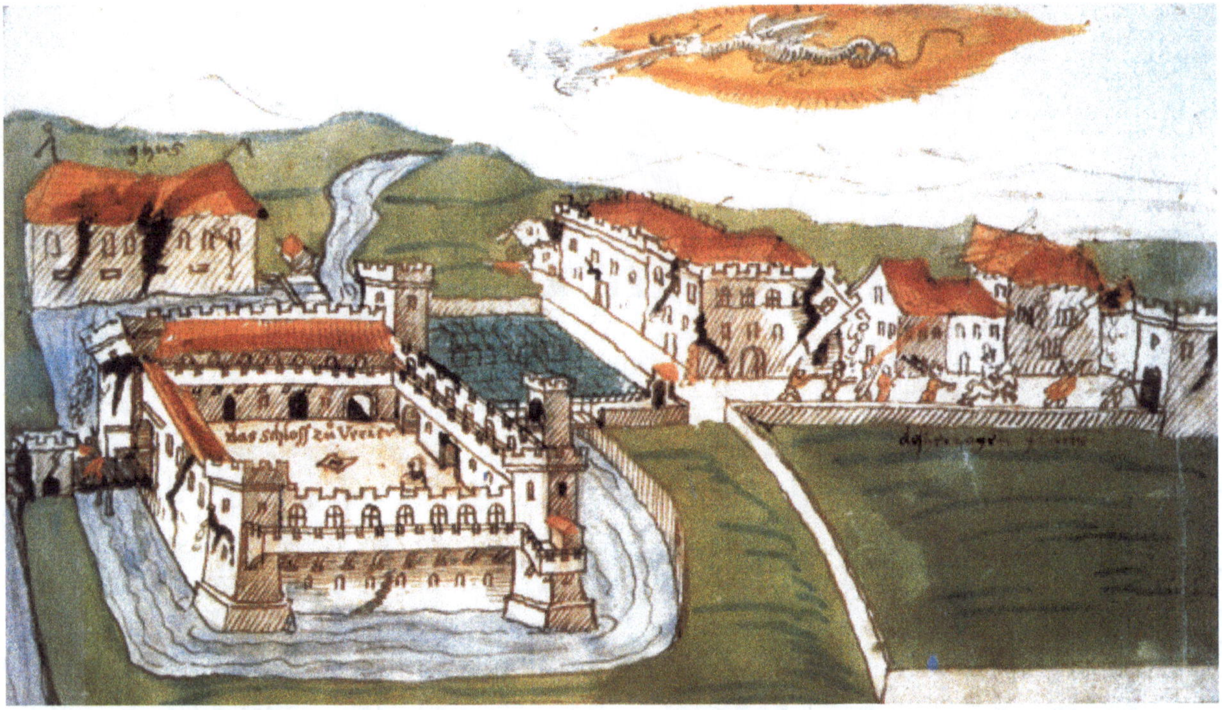

Fig. 2.6 Earthquake damage to Ferrara Castle in 1570

The earthquake that struck central Italy on 17 November 1570, caused significant damage in both Ferrara and Florence. A double image of damage to the two cities will be presented in Chap. 4 for reasons that will become apparent. The image here (Fig. 2.6) concentrates on damage to Ferrara Castle and the neighboring buildings; the damage is shown as deep cracks and partial collapse, and looks to be quite accurately drawn. Doubt about the cause of earthquakes is reflected in the drawing of a dragon in the sky, who looks to be responsible for it all. This drawing was kindly supplied by the Municipal Museum, Zurich.

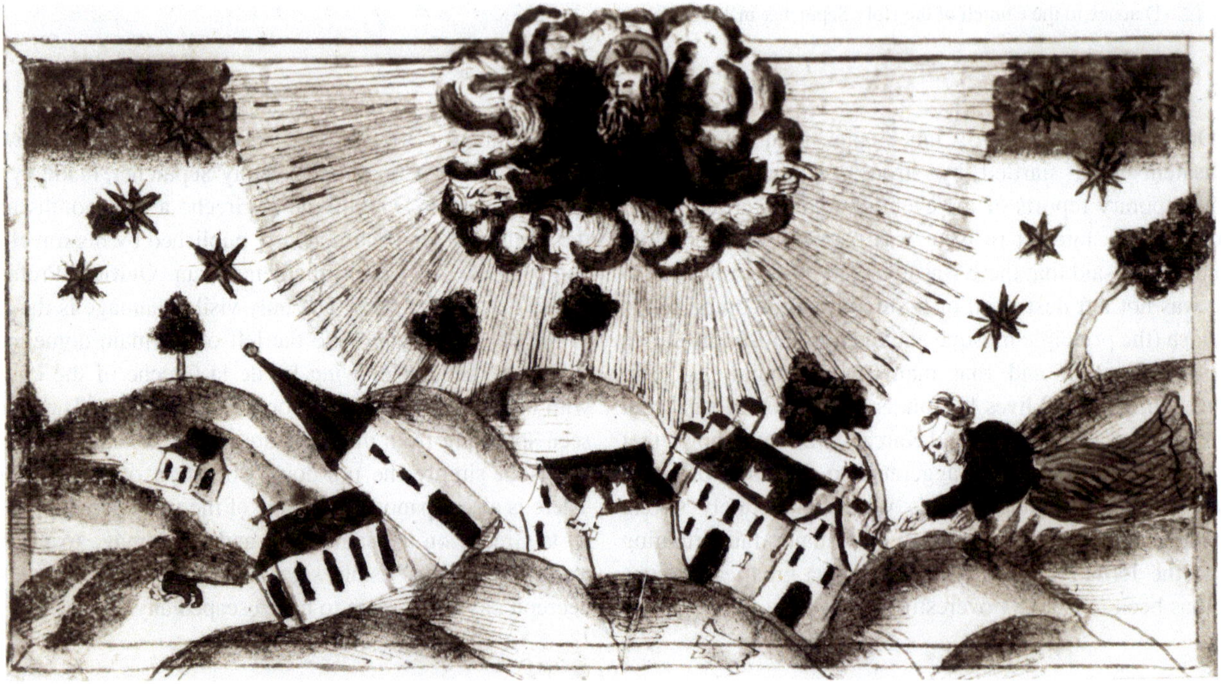

Fig. 2.7 The 1584 Aigle earthquake

Talking about Swiss tsunamis might seem like a history of Hungarian lighthouses, but they do occur when an earthquake dislodges a landslide that falls violently into a lake. Such a thing occurred on 1 March 1584, when an earthquake of around 6 in magnitude near Aigle damaged towns and villages, and caused a major landslide, which entered Lake Geneva and caused a damaging tsunami affecting Lausanne and Geneva itself. The Scottish author John Pinkerton (1758–1826), visiting the area in the first decade of the nineteenth century, wrote of the village of Yvorne (just north of Aigle) that it was "ruined, in 1584, by the lapse of a mountain occasioned by an earthquake." Evidently, it was still unreconstructed over 200 years later.

The earthquake and its effects were painted and described in an anonymous report titled *Wytere bestrybung von diesem Erdbidem*, a copy of which was found in the Zentral Biblothek Zurich by Jean Vogt. The picture itself (Fig. 2.7) carries typical marks of such sixteenth-century images: the town that is depicted cannot be identified from the picture, the damage shown is unrealistically expressed by obliquely inclined houses and towers that are not actually damaged by shaking or mechanical disintegration (but possibly this is an attempt to convey them being struck by the landslide?). Furthermore, individual parts of the image are out of scale; compare the size of the figure of the woman on the right and the size of the house just nearby. Finally, as in most sixteenth-century images of seismic effects, the story is interpreted from a religious standpoint. Notice the angry figure of God, portrayed in the black cloud in the upper center of the picture, sending punishment on human sinners.

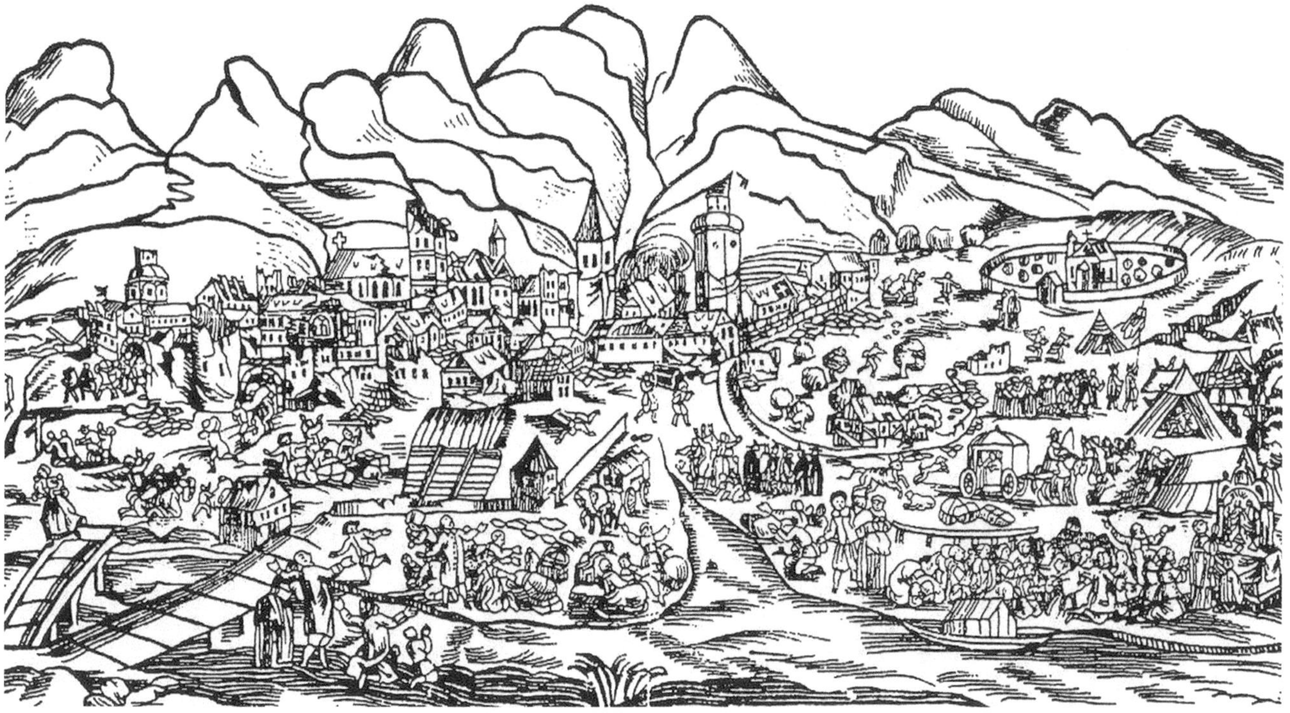

Fig. 2.8 The 1670 earthquake in Hall in Tirol

Austria is not among the most seismically active countries in Europe, the two most earthquake-prone countries being Italy and Greece. Yet, it has still suffered some notable seismic events in different parts of the country, one of which introduced this chapter.

The illustration here (Fig. 2.8) shows the town of Hall in Tirol, situated some 25 kilometers east of Innsbruck, in the valley of the River Inn. The town was struck by a strong earthquake on 17 July 1670, causing serious damage, and collapsing houses resulted in a number of casualties. The main shock—which was felt in all of Austria, in northern Italy, in Switzerland, and in Bavaria as far north as Nuremberg and even Frankfurt—was followed by numerous aftershocks for the next 50 days.

In this woodcut illustration, kept in the Stadtmuseum in Hall in Tirol, the inhabitants of the town can be seen milling around in the fields, where many of them camped in tents. There are numerous detailed descriptions of the effects of the earthquake—for instance, that by the Tyrolean nobleman Johannes Sigmund von Rost zu Kehlburg und Aufhofen

(1653–1729), who visited the town three days after the earthquake. He noted that many houses were devastated and the tower of the parish church was partly collapsed. Nine days after the event, a survey of the damage was made, so that today one can make an exact reconstruction of the effects in the town (Hammerl 2017).

This depiction was kindly made available to illustrate our book by Wolfgang Lenhardt.

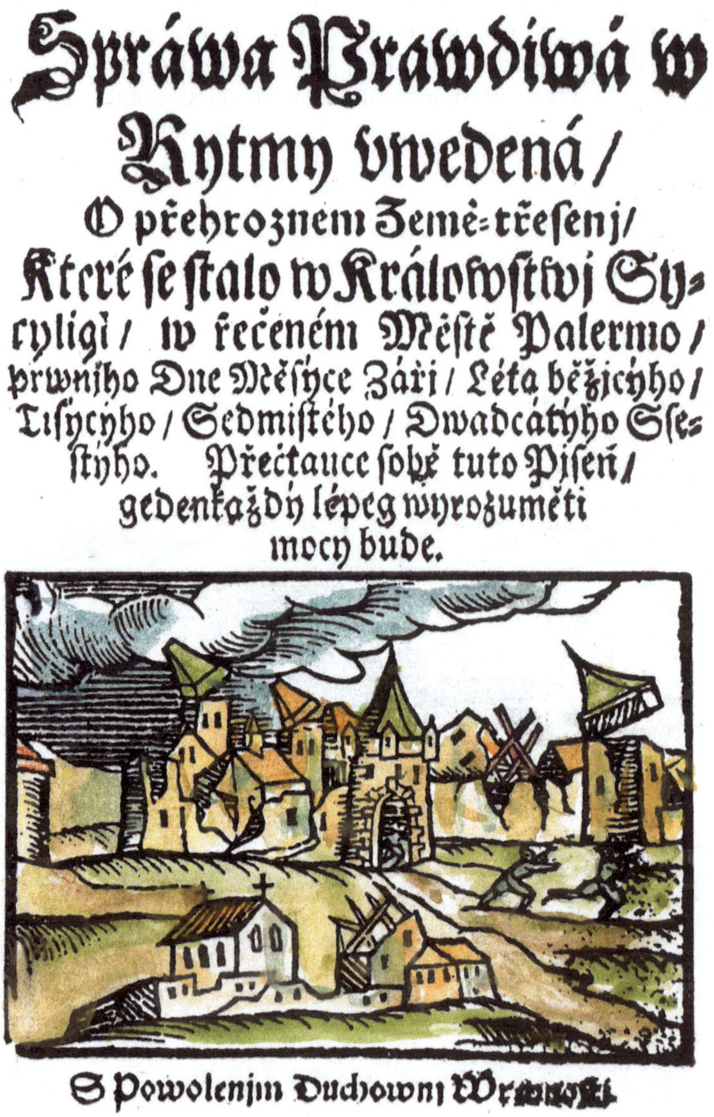
Spráwa Prawdiwá w
Rytmy vwedená/
O přehroznem Země=třesenj/
Které se stalo w Králowstwj Sy=
cyligi/ w řečeném Městě Palermo/
prwnjho Dne Měsýce Zářj/ Léta běžjcýho/
Tisycýho/ Sedmistého/ Dwadcátýho Sse=
stýho. Přečtauce sobě tuto Pjseň/
gedenkaždý lépeg wyrozuměti
mocy bude.
S Powolenjm Duchownj

Fig. 2.9 The 1726 Palermo earthquake

The seismicity of Italy is ultimately due to it occupying the collision zone between the Eurasian and African tectonic plates. Sicily has been affected on numerous occasions, and what was probably the strongest ever Italian earthquake in historical times occurred in eastern Sicily in 1693. On the north coast, Palermo has been visited by strong earthquakes on three occasions: 1 September 1726; 5 March 1823; and 15 January 1940.

The image here (Fig. 2.9) purports to show damage from the first of these, in 1726. It comes from a contemporary leaflet published in Brno, Moravia, by the bookseller Jakub Maxmilian Swoboda; the title translates as *True Message in Verse About the Horrible Earthquake that Happened in the Kingdom of Sicily in the Said Town Palermo on the First of September of the Current Year One Thousand Seven-Hundred Twenty Six*. In fact, the image shows nothing more than the fantasy of an anonymous artist, who knew roughly what an earthquake should look like but had certainly never seen Palermo. However, it is interesting to see that news of a remote Sicilian earthquake in this period was considered worthy of publication by someone far away in Moravia.

The earthquake itself—which damaged the king's palace, together with a series of city churches and monasteries, killing about 250—is interesting on account of the rational approach to reconstruction taken by the city's senate. Although Sicily at this date was ruled by the Austrian Hapsburg dynasty, the senate was capable of acting independently when the need arose, as it did here. They called an emergency meeting and appointed a group of engineers and architects who would decide how badly damaged buildings were and if they were in danger of collapse. In repairing or rebuilding damaged structures, a real attempt was made to make buildings safer by the use of lighter materials and appropriate reinforcement. The senate also outlawed the old baroque heavy stone balconies, which were considered a risk, thus changing the look of the city and also improving earthquake safety. This was despite the widespread belief, in this Catholic country, that earthquakes were instruments of God's wrath (Sciblia 2015). It was not a great earthquake (the magnitude was perhaps about 5.6), but it was a reminder that even a moderate event can cause heavy damage if it makes a direct hit on a city.

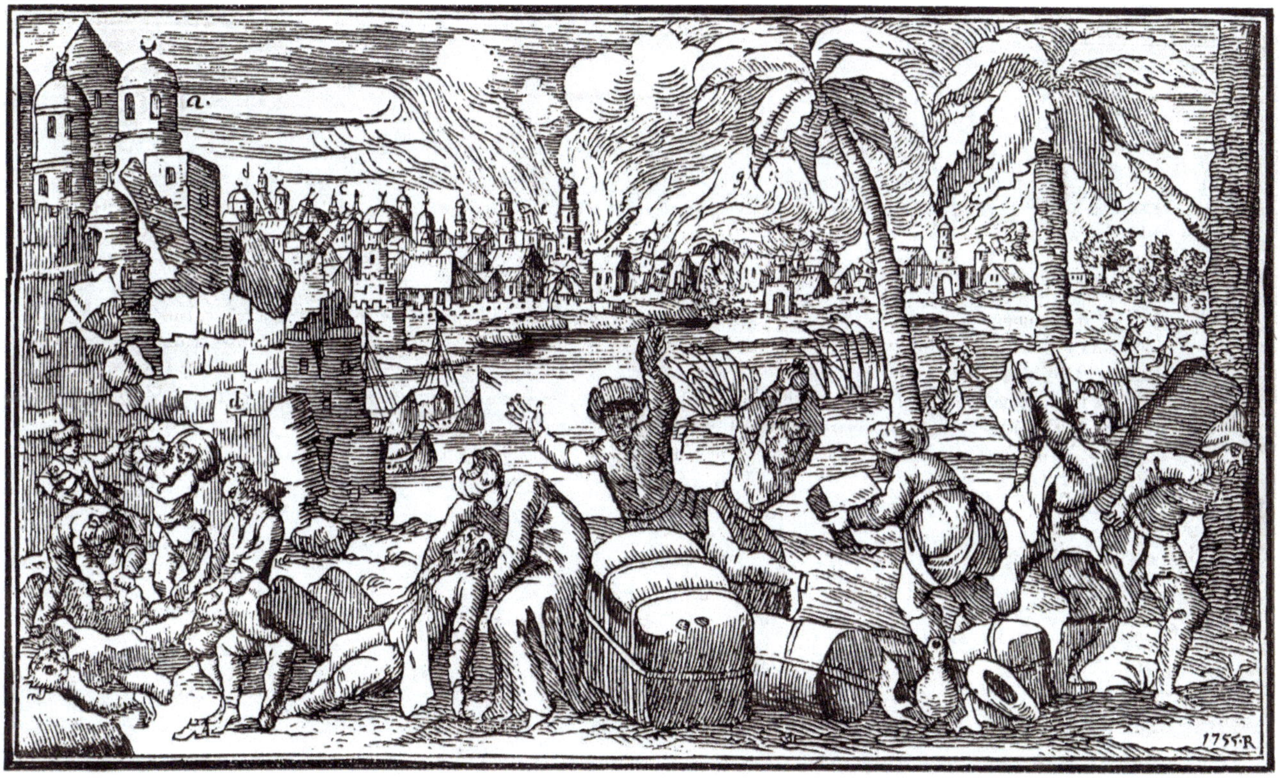

Fig. 2.10 The Izmit earthquake of 1755

The image presented here (Fig. 2.10) is an instructive woodcut, signed "1755 R," which depicts the havoc caused by a strong earthquake centered on the Gulf of Izmit on 2 September 1754, just over a year before the great Lisbon earthquake. The picture is titled *Grand ravage causé à Constantinople par des violents tremblements de Terre, suivis d'un incendie fort considerable* and appeared in a Swiss almanac called *Le Messager Boiteaux de Berne* (1756).

The foreground of the illustration, made with some poetic license, is crowded by despairing inhabitants of Constantinople/Istanbul, assembled in front of the damaged city walls. Some of them are treating the wounded; others are trying to save their property. In the background we can see the town itself, damaged not only by the seismic impact but also by the subsequent conflagration, which started at the palace of the Imrahor and destroyed around 2000 houses. The damage in Istanbul was widespread, but mostly involved old stone or brick buildings. The walls and one or two towers of the fortifications suffered, and some mosques and minarets were also damaged. The image here has a key to some of the features: (a) the fortress of the Seven Towers; (b, c) the palace and mosque of Hagia Sophia (both damaged); (d) the collapse of one of the towers of the outer walls; and (e–g) the fire.

The earthquake was preceded by a series of foreshocks, and strong aftershocks compounded the damage by finishing off some buildings that had been weakened by the main shock. The numbers of fatalities were around 50 in Istanbul itself and perhaps 800 in the region overall; the effects in Izmit were more severe, but Istanbul was of more interest to European readers. The magnitude is estimated at around 7.

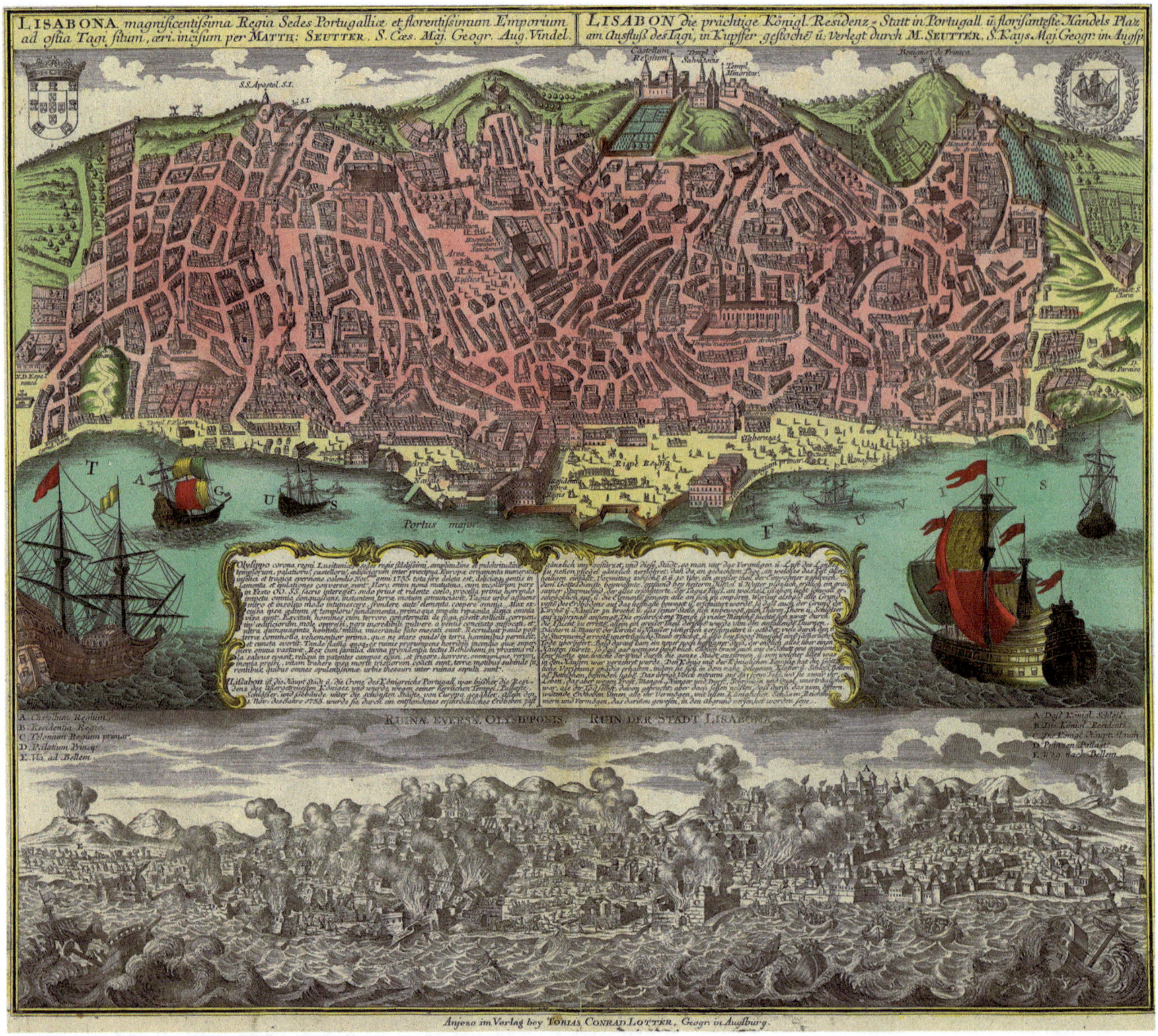

Fig. 2.11 Lisbon in 1755 before and after the earthquake

The great Lisbon earthquake of 1 November 1755 (All Saints Day), is one of the most significant earthquakes on record. Besides the severe damage it inflicted, it caused a revolution in thought about man's place in the natural world and sparked a wave of scientific investigation as learned people strove to come to grips with the forces that might cause such devastation. Yet, strangely, this most important event remains a seismological mystery; experts are unable to agree on its size or location, or even whether it was one earthquake, or two together, or even three. The fact that a major tsunami occurred is evidence of an offshore epicenter, and several locations have been proposed in the scientific literature. But the large number of aftershocks felt in Lisbon itself suggests that a separate quake must have occurred nearby, in the Tagus Valley. The earthquake not only was felt all over the Iberian Peninsula and slightly into France but also caused damage along the coast of Morocco. The tsunami, besides wreaking havoc in Lisbon itself, traveled across the Atlantic and was very perceptible in the West Indies and even Brazil.

Lisbon was devastated; the damage to the city came in three forms. The town itself suffered much damage from the earthquake shaking, which ruined or heavily damaged a large part of the thriving town: people ran out from the collapsing houses, and, unfortunately, many went down to the shore of the River Tagus and were caught when the tsunami swept up the estuary. Lastly, fires raged across the city as the ruins were set alight by upset stoves, ovens, and open fires. To better appreciate the shock occasioned by the destruction of Lisbon all over Europe, one must remember that at this time, Lisbon was still the capital of a world empire; the international impact was akin to what might be felt today if Washington, DC, or New York City were wiped out.

A large series of depictions of the earthquake appeared almost immediately. At present, the largest series of depictions of its effects are in the Municipal Museum in Lisbon and in the Prague Archives (NKC), where there are some 110 pictures. Mostly, they are contemporary depictions or their copies; there are also a number of later pictures and illustrations from the eighteenth and nineteenth centuries, often based mostly on imagination. Among all of the pictures, there are about 25 that seem to have been made in situ and truly reflect the reality. They can be roughly divided into two groups. In the first group, we can find depictions portraying the town from south to north, from a standpoint on the Tagus River or in Almeda on the left bank of the Tagus. Sometimes, these pictorial views take a before-and-after approach, showing both the ruined city and what it looked like beforehand. The image shown here is one example (Fig. 2.11). The Bavarian printer and cartographer Johann Baptist Homann (1664–1724) had come up with the idea of combining a map and a panorama of a city into a single print. Tobyas Lotter, Homann's successor in Augsburg, continued this tradition, making such double depictions of most European capitals, including Lisbon. After the 1755 earthquake, Lotter played with the idea of updating his double portrayal of Lisbon, showing the seismic damage and city fires in the second, bottom part of the print. He made several copies of the updated print but, for reasons unknown, did not include the result in his atlases, and the modified print exists exclusively in these few sample pieces; luckily, one of them is kept in the first author's private collection of historical earthquake prints in Prague.

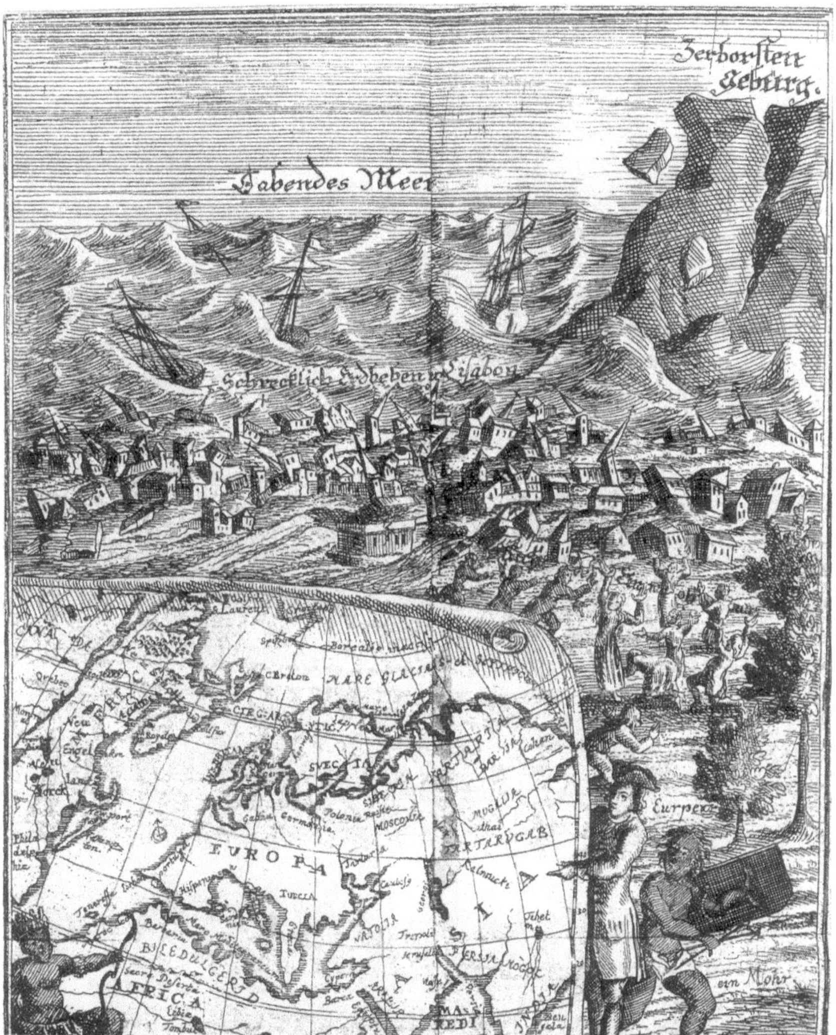

Fig. 2.12 A fantasy engraving of the Lisbon earthquake

Besides the other effects, the earthquake produced what is called a seiche—a long-period wave that travels long distances and causes sloshing in bodies of water such as lakes and ponds. This was very perceptible as far away as Scotland and Scandinavia; reports that the earthquake was actually felt at such distances are false and really refer to these effects on water. Something of the huge reach of this earthquake is suggested by the wildly inaccurate German copper engraving shown here (Fig. 2.12), in which a European and two "Moors" are consulting a map of Europe, while the city behind them is falling this way and that, the sea is heaving all the way to the horizon, and a nearby mountain is split in two. The image is an illustration from a book titled *Die Denkwürdigsten Geschichte der Erdschutterrungen,* or *Memorable History of Earthquakes*. The artist obviously had no idea of what Lisbon and its surroundings looked like. The extension of the map to North America probably reflects the mistaken idea that the earthquake was felt there—a confusion with the Cape Ann earthquake, which will feature later in this chapter.

The cultural impact of the earthquake was certainly felt across Europe. The collapse of churches killed people attending holy service on one of the most sacred days of the Catholic Church's calendar. On the other hand, those who were enjoying themselves sinfully in parks or fields were unharmed. If this earthquake was sent by God, who was He trying to punish? The idea of a supernatural cause of earthquakes was fatally undermined. Instead, there was a surge of scientific attempts to understand the nature and cause of earthquakes, of which the most farsighted was that of the Englishman John Michell (1724–1793), who realized that shaking took the form of elastic waves radiating out from some causative feature.

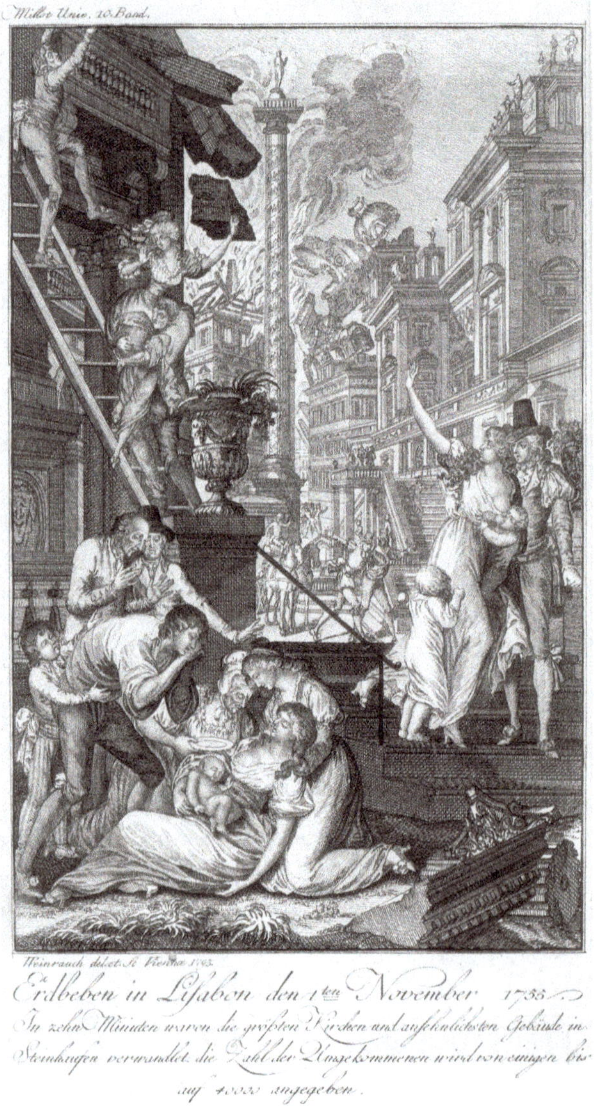

Fig. 2.13 Rescuing victims of the Lisbon earthquake

This picture (Fig. 2.13) shows the fearless, unflinching efforts of rescuers entering damaged houses to try and save the inhabitants, at the risk of their own safety. Women and children can be seen being rescued with ladders. This is one of very few images showing rescue efforts in any detail. In the background, one house is on fire, one is in the midst of collapse, and the house on the right appears undamaged. However, this copper engraving dates from 1793 and originated in Vienna, so it may not be reliable in terms of detail.

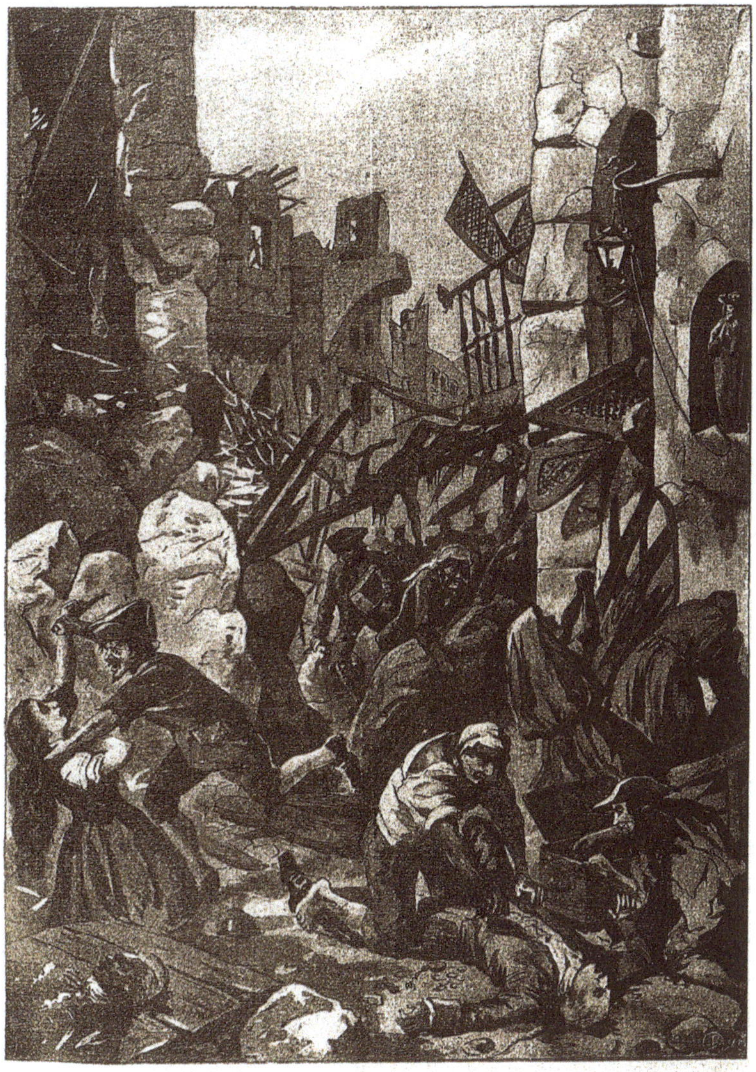

LISBOA SALTEADA POR MALFEITORES DEPOIS DO TERREMOTO

Fig. 2.14 Looters in Lisbon

One great disaster—and two human approaches to it. In contrast to the previous image of heroic rescuers (Fig. 2.13), one notes here a further misfortune visited on the inhabitants of Lisbon: an invasion by a large number of thieves and robbers from Portugal and Spain, who arrived in the ruined city with the aim of stealing and robbing (Fig. 2.14); sometimes these looters set fire to the plundered houses in order to mask their robbery. Many of these bandits, when caught in the act, were sentenced and hanged immediately. The rather sensational image here is an illustration taken from a late nineteenth-century Portuguese historical romance.

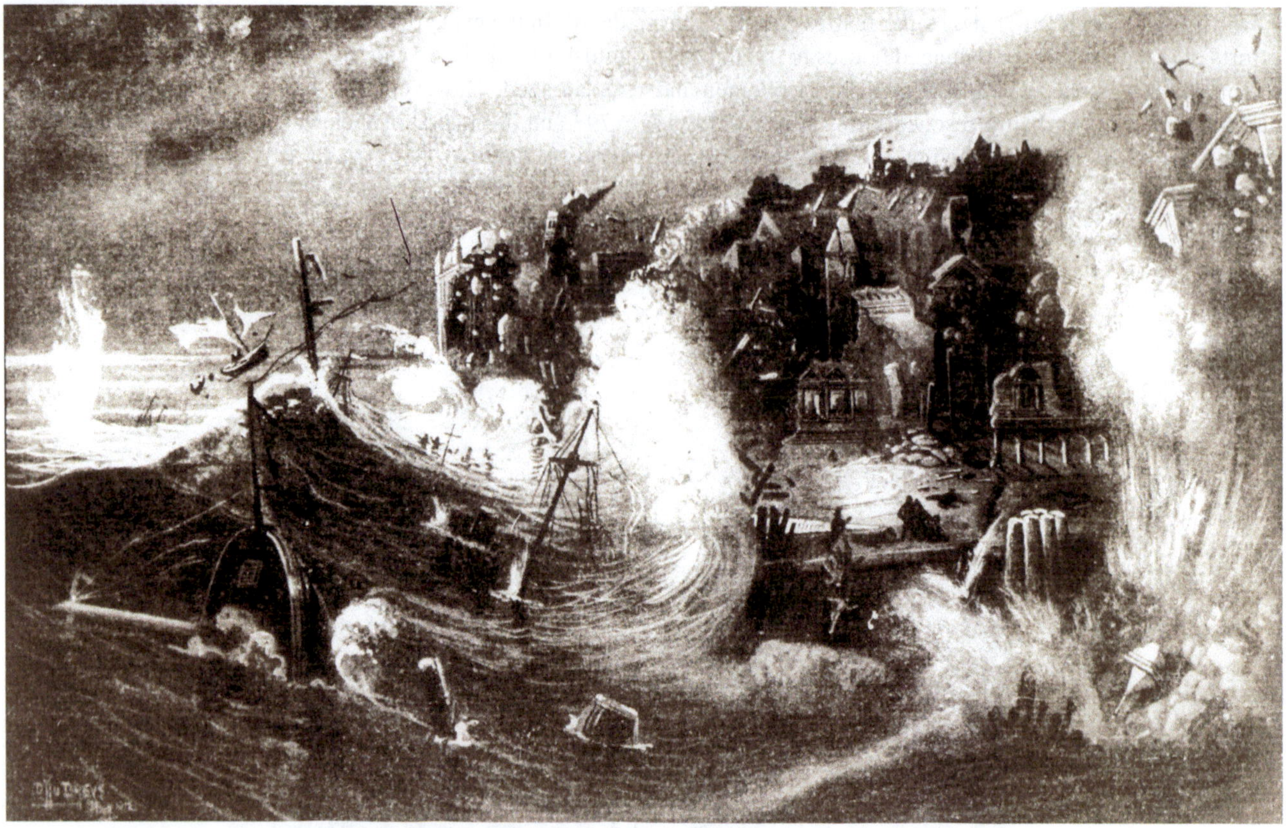

Fig. 2.15 A late imagining of the Lisbon earthquake and tsunami

Many of the images of the Lisbon earthquake are extremely well known and are usually found decorating any book or paper about the event. Therefore, we show here some little-known depictions of the event—omitting, for instance, the six large depictions (copper engravings) made by Jacques Phillipe le Bas, the court artist and engraver to the French King Louis XV. The king sent the artist to Liston in 1757 to record the damage to the town. The pictures of le Bas's series, dated 1757, are as exact and true as present-day photographs; they represent top-quality pictorial material for the seismologist's appraisal. All have been republished previously by Kozák and Čermák (2010). Naturally, the artist's attention was focused especially on the individual principal buildings that suffered most from the earthquake.

In contrast, the view presented here (Fig. 2.15) is distinctly noncontemporary. It comes from an anonymous, Czech-language religious book published in 1916, long after the earthquake, by the Seventh Day Adventist International Tract Company of Budapest. The title, *Slavný příchod Kristův*, can be translated as *Glorious Advent of Christ*. The book still suggests that earthquakes are God's will; some ideas die hard. In fact, in the immediate aftermath of the earthquake, a law was passed by the Marquis of Pombal (who took control of managing the disaster in Lisbon) that anyone preaching in the city that the earthquake was sent by God should be arrested immediately!

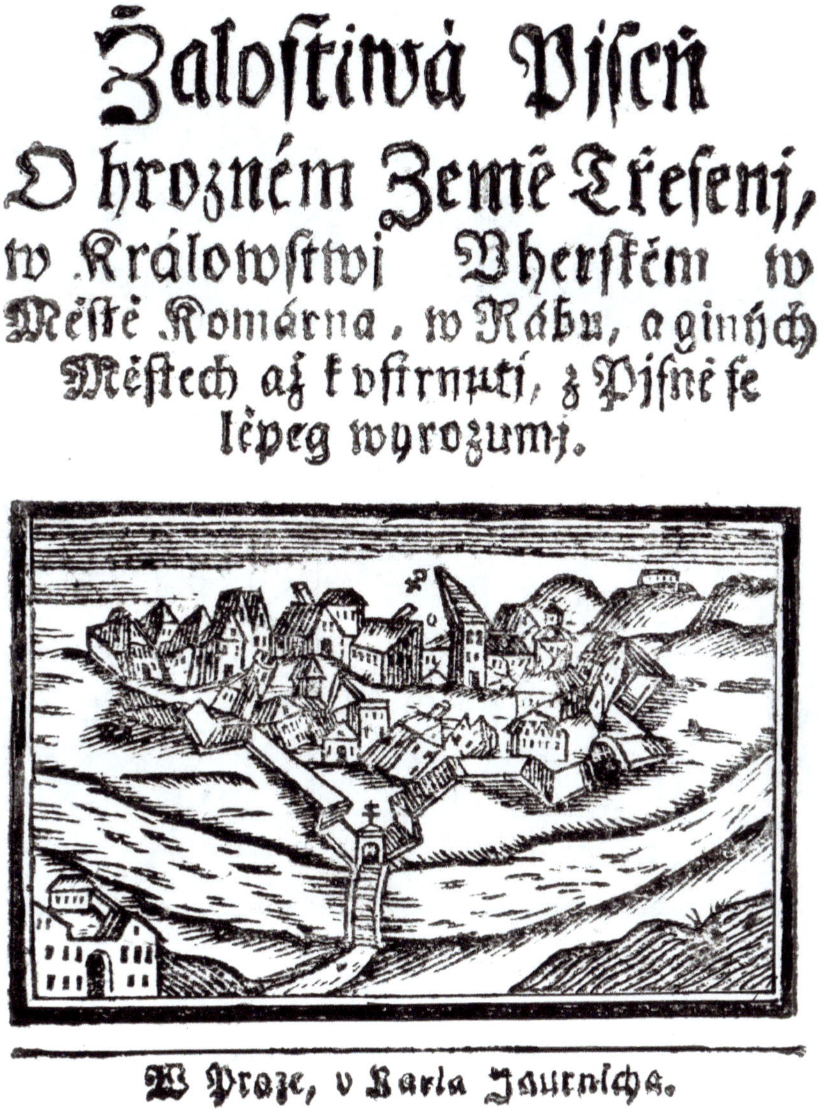
Žalostiwá Pjseň
O hrozném Země Třesenj,
w Králowstwj Uherském w
Městě Komárna, w Rábu, a ginjch
Městech až k vstrnutj, z Pjsně se
lépeg wyrozumj.

W Praze, v Karla Jaurnicha.

Fig. 2.16 A woodcut of the 1763 Komárno earthquake

The earthquake of 28 June 1763, that struck the town of Komárno—a location now more or less on the border between Slovakia and Hungary—is another example of an earthquake that, despite its modest magnitude (about 5.7), manages to find a place in the iconography of historical earthquakes, thanks to its location with respect to cultural centers. It was felt over most of what now constitutes Slovakia and Hungary, and was felt slightly as far west as Vienna. The famous Prague editor of illustrated leaflets Karel Jaurnich prepared, edited, and published a leaflet titled *Lament About a Terrible Earthquake in the Town of Komárno in the Hungarian Kingdom, in Rab and Other Towns*. The title page of his leaflet (Fig. 2.16), carrying a report on the disaster in verses (apparently intended as a song), was decorated by a raw woodcut view of the damaged town. The image carries no correct information—the creek crossed by a footbridge should be the mighty and wide River Danube. Also, the town itself is not recognizable.

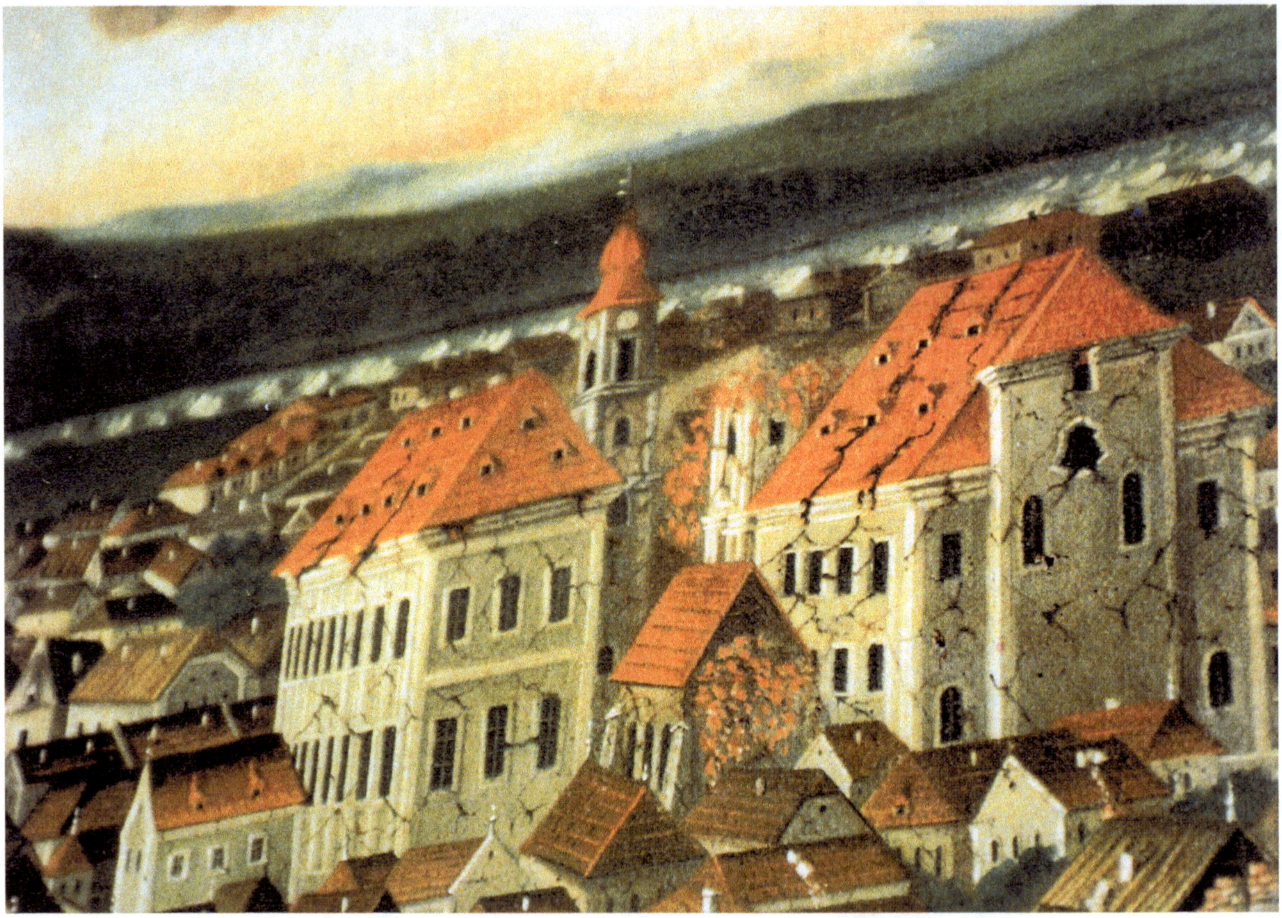

Fig. 2.17 A rare painting—the 1763 Komárno earthquake

The Komárno earthquake on 28 June 1763, also features in this picture (Fig. 2.17), a painting in the Franciscan church at Frauenkirchen-Burgenland (Austria). It is far more realistic than the woodcut in the previous image despite being somewhat stylized. It was composed in 1766 as an oil painting on canvas, apparently for votive purposes, and shows damage to ecclesiastical buildings in Komárno.

The damage may well be exaggerated, although the pattern of cracks reflects what is actually a typical earthquake effect. Records show that seven churches and some 279 houses were more or less damaged in Komárno, with 63 fatalities (Eisinger et al. 1992).

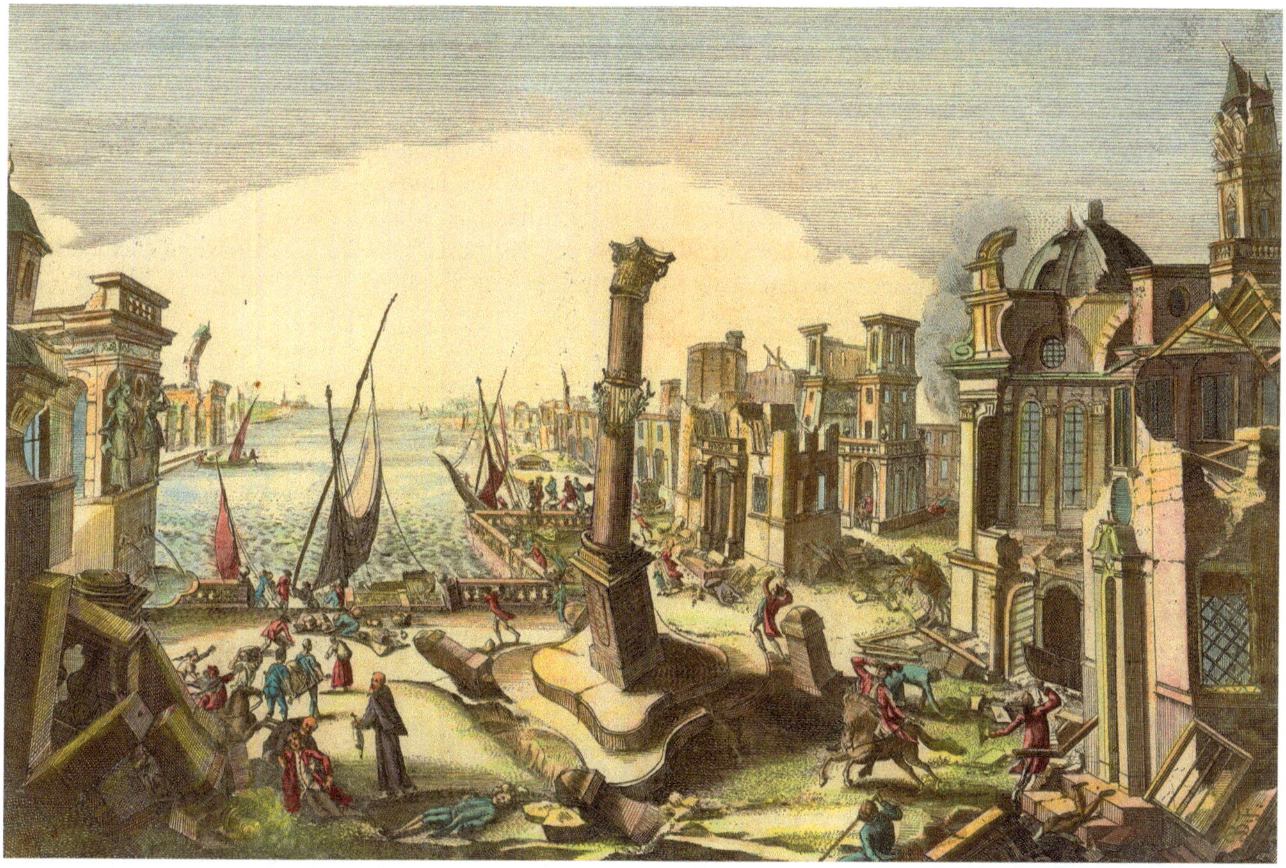

Fig. 2.18 The destruction of Oppido in 1783

The Calabrian earthquake of 5 February 1783, is—together with its predecessor, the 1755 Lisbon quake—one of the two outstanding seismic events of the eighteenth century in Europe. In general, we may state that it was especially these two strong European seismic events that accelerated the process of the Enlightenment as it applied to seismology. From this period on, it was generally understood that natural disasters, including earthquakes, should be studied as natural phenomena created through the interplay of natural physical processes.

The disaster in Calabria was unusual in that it constituted a veritable seismic crisis, being not just one strong earthquake and its aftershocks but a sequence of five main shocks between 5 February and 28 March 1783, with epicenters strung out along the "toe" of the Italian "boot," with magnitudes ranging between 6 and almost 7. The first event occurred at Bagnara on the Calabrian coast just north of the Straits of Messina, and the last occurred on the shore of the Gulf of Squillace, some 80 kilometers to the northeast. The first event damaged more than 380 villages, of which 180 were almost totally ruined. Over 25,000 people were killed. A number of tsunamis occurred; that on 6 February was due to a massive rockfall causing a large portion of Mount Campallà to collapse into the sea (Graziani et al. 2006).

In fact, the 1783 sequence was particularly distinguished by its effects on the landscape, such as a large earth block collapse, massive landslides, wide cracks opening, and so on. All of these features attracted the scholars and savants of the day to study this dramatic phenomenon more closely. As a partial result of this situation, a rich pictorial documentation of the event survives. At the end of the eighteenth century, reproduction techniques made significant progress in Europe: apart from traditional ways of representation, such as woodcuts, new techniques for engraving and etching were invented and widely utilized. This rapid technical development resulted in a large series of depictions of the 1783 Calabria quake. At present, there exist about 130 such images; no other historical earthquake is portrayed in so many pre-photographic depictions as the 1783 Calabria event. We select here some of the less well-known ones.

Here, the destruction of the Calabrian town of Oppido is depicted (Fig. 2.18). The image is a so-called *vue d'optique*, designed to be viewed through an optical toy called a zograscope, which gave an illusion of depth (see also Fig. 3.12).

Such images are generally characterized by being mirror images and having bright hand-coloring, as shown here. The undated, unsigned image of Oppido was produced by the Academic Editorial House in Augsburg and was created shortly after the main seismic disaster of 1783, when the news was still current. Above the image, the title of the event became readable after conversion on being reflected in the zograscope's mirror; under the image, there is a more detailed text in German and French. The damage to the principal edifices in Oppido is shown, together with the confusion of the inhabitants searching for safety outside in the open air. The unrealistic image of the damaged column in the center of the image indicates that the artist did not have much experience of earthquake damage. Although one might not think it from the image, it was reported that the damage to Oppido was so severe that the survivors decided not to rebuild it on the same spot and instead constructed a new town in another, hopefully more safe, locality.

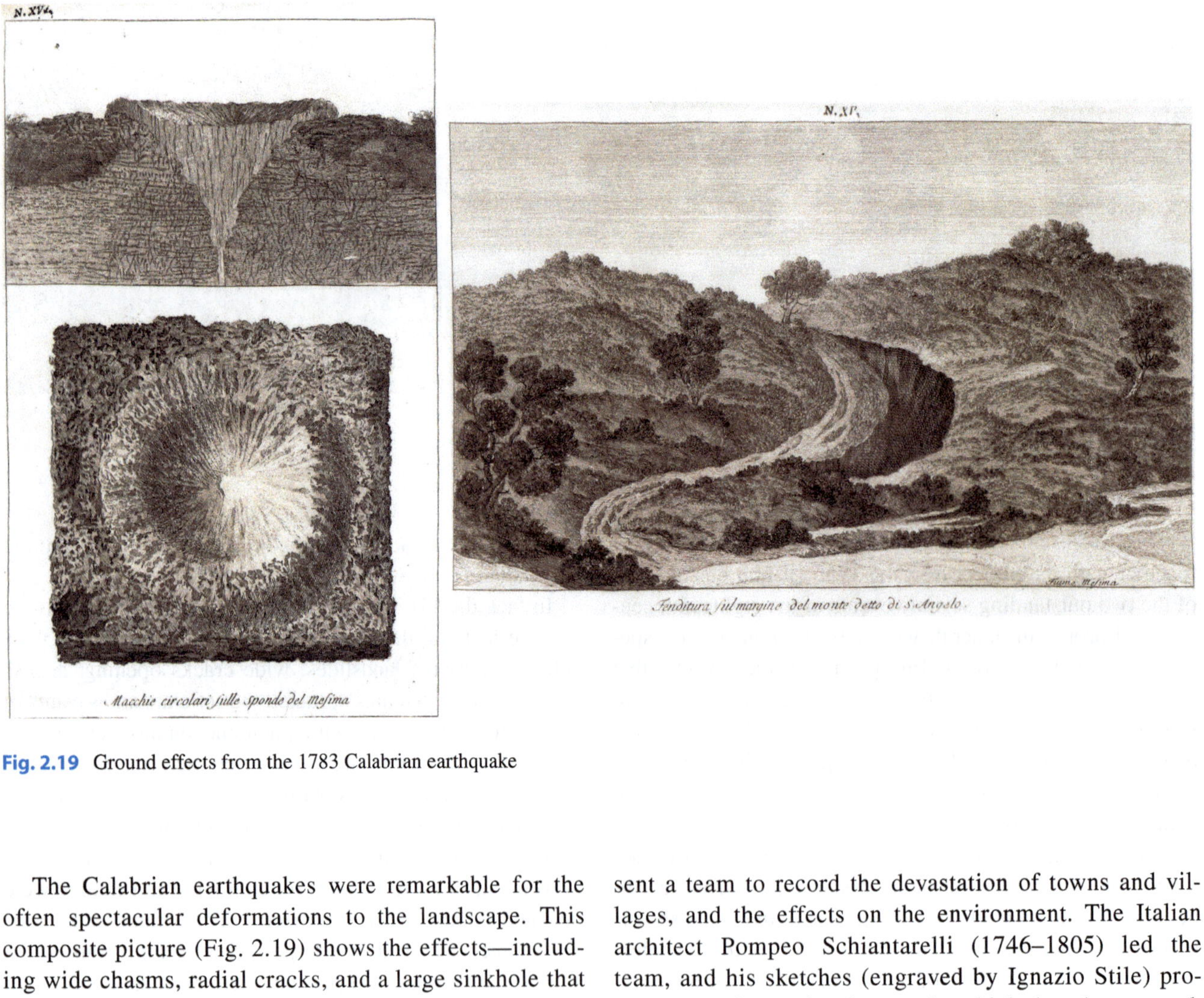

Fig. 2.19 Ground effects from the 1783 Calabrian earthquake

The Calabrian earthquakes were remarkable for the often spectacular deformations to the landscape. This composite picture (Fig. 2.19) shows the effects—including wide chasms, radial cracks, and a large sinkhole that has filled up with water—in Polistena, Gerocarne, and Rosarno. The Neapolitan Academy of Science and Letters sent a team to record the devastation of towns and villages, and the effects on the environment. The Italian architect Pompeo Schiantarelli (1746–1805) led the team, and his sketches (engraved by Ignazio Stile) provide a copious visual record, which has been much reproduced.

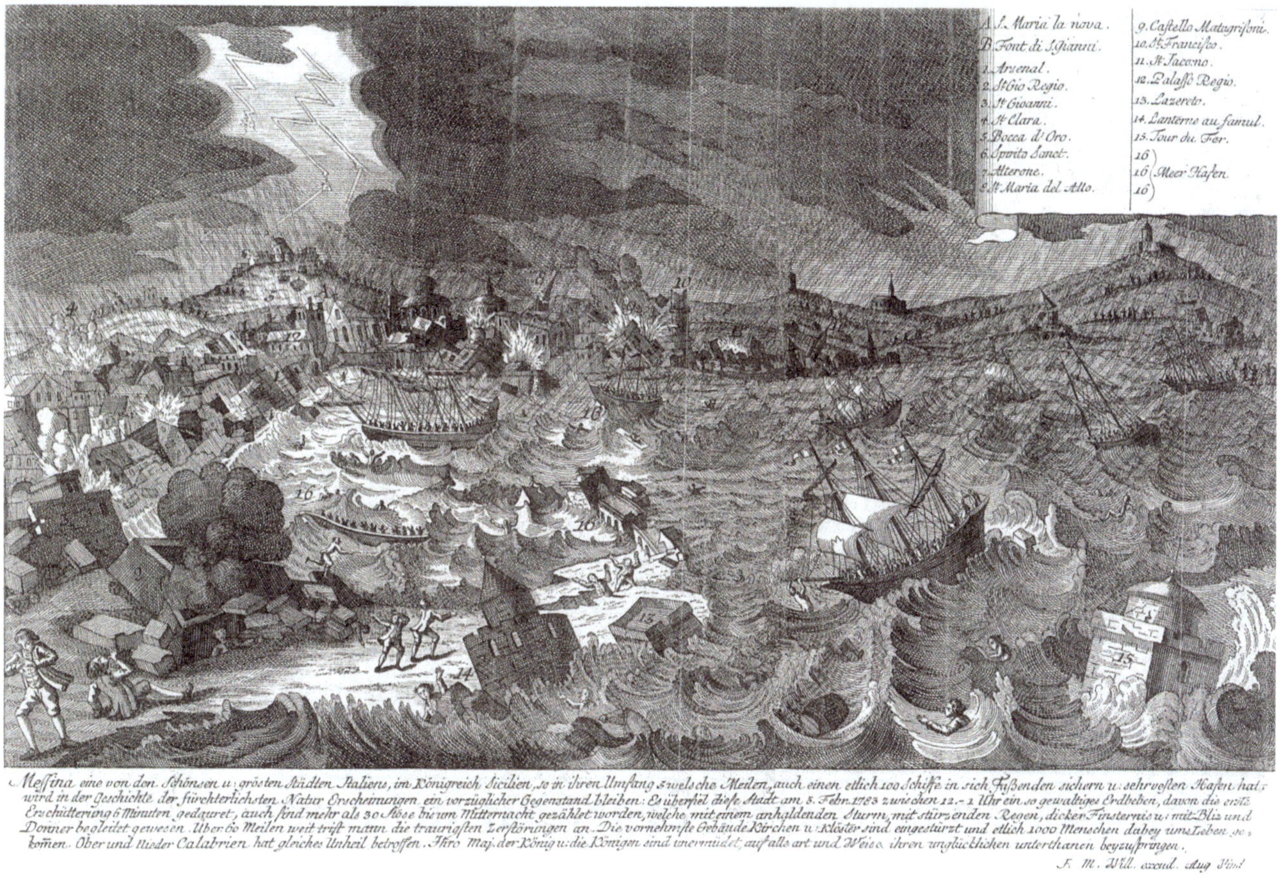

Fig. 2.20 The destruction of Messina in 1783

This dramatic image (Fig. 2.20) of the destruction of Messina has practically everything in it: buildings toppling, the sea raging with the tsunami, ships being tossed about, fires breaking out in all quarters, and thunder and lightning, to top it all off. No wonder the poor people in the foreground are panicking. Some are even caught up in the rough sea. This contemporary German engraving provides a useful key to the details shown; it is signed F.M. Will in the bottom right corner.

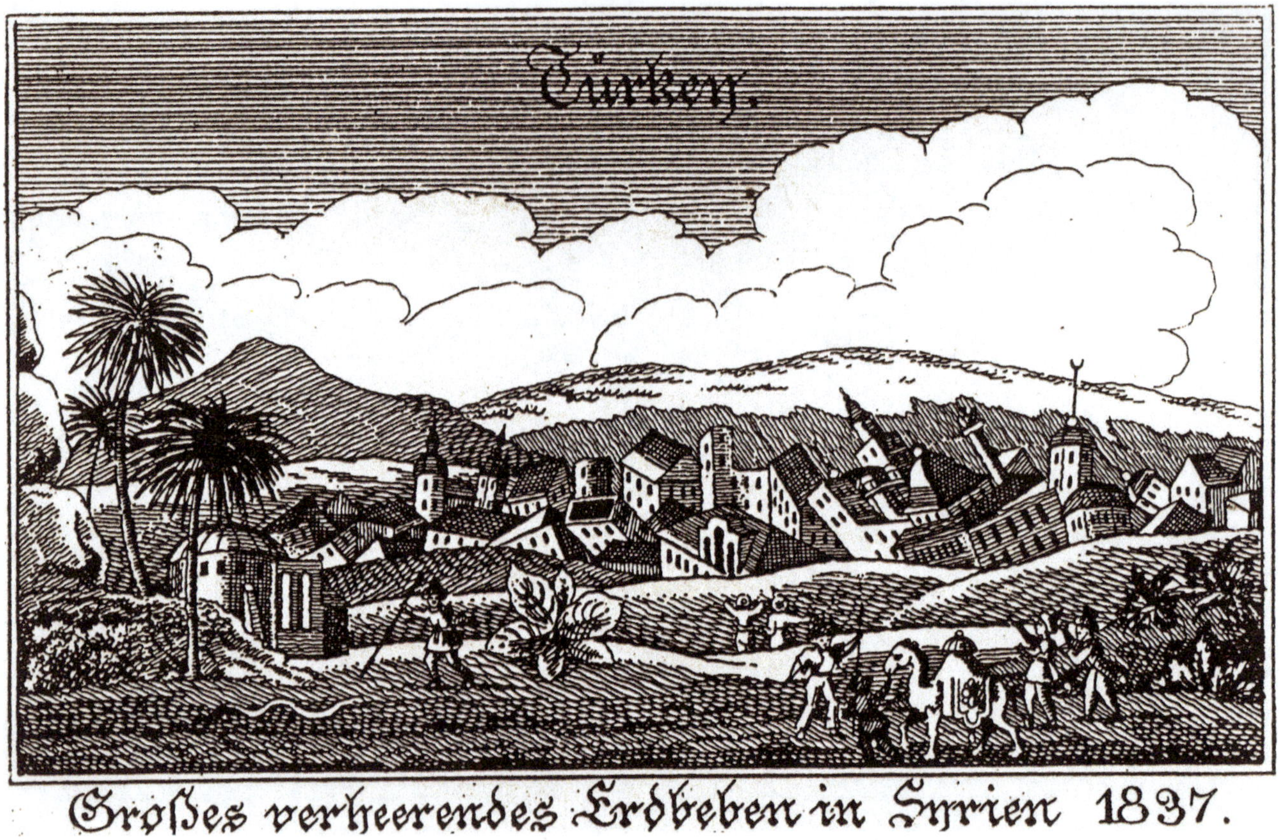

Fig. 2.21 An earthquake in Syria in 1837

On the eastern littoral of the Mediterranean and eastward through the Middle East toward Persia, even where the seismicity is high, we have few or no historical pictures of earthquakes.

This state of affairs is due in some part to the attitude of Islamic cultures toward images; for instance, for Syria—a prosperous country but nonetheless subject to strong earthquakes—we have only scarce illustrated reports on local earthquakes. Presented here is one leaflet created by J. Engelmann, with a text describing "a strong, devastating earthquake that visited Syria in 1837" (Fig. 2.21). In this report, the seismic event was localized in the region of Aklin el Tiffa, Shana (Shara?) and Mevigaoun, for which the seismic manifestations were briefly characterized. The number of casualties was not given, as was a habit in the reports on natural disasters composed in the Orient. In fact, some 6000–7000 people were killed. In the report by Engelmann, strong damage to Jaffet and Tiberias is mentioned; both localities were said to be quite ruined. In fact, Syria was in a state of turmoil at the time, being occupied by the Egyptians, which hindered the transmission of information about the event. The earthquake occurred on 1 January 1837, and the fact that it happened in the evening, during a wet period in winter, meant that most people were indoors having dinner, which undoubtedly contributed to the large number of casualties.

This leaflet was kindly donated to the authors by the late K. Stopp of Mainz, Germany. The image is a typical fantasy of the period, with European-looking buildings leaning at crazy angles. The fact that the name "Turkey" appears in large letters at the top is not an error, since Syria was a province of the Ottoman Empire at the time.

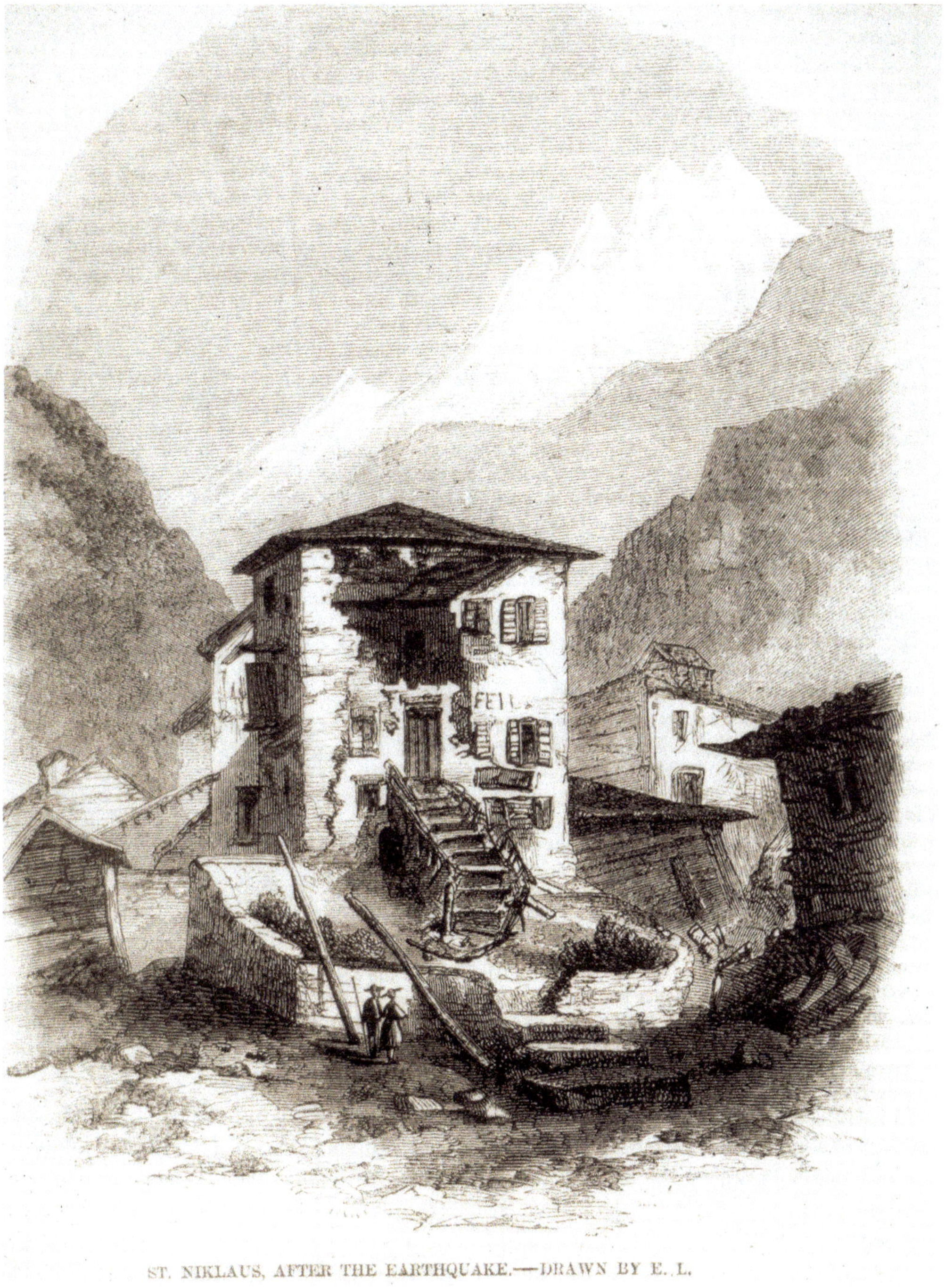

Fig. 2.22 The Visp earthquake—damage in St. Niklaus

Even in early studies of earthquakes as natural phenomena, it was noticed that quakes seemed to be more common in mountainous areas—and not without reason, as we now understand. Massive mountain chains such as the Alps occur when two continents collide, pushing up strata into a crumpled series of ridges. So Switzerland is a country not without its earthquakes. One of the most notable such events was the Visp earthquake of 25 July 1855. It had its focus in the lower part of the Mattertal, and, as there are no larger towns between Visp and Zermatt (which are about 40 kilometers apart), the damage was mostly confined to small settlements located in the Mattertal. Among the most heavily damaged places, the village of St. Niklaus was often mentioned, located 14 kilometers south of Visp. The picture here (from the *Illustrated London News*-Fig. 2.22) shows typical seismic damage to a stone-built house in St. Niklaus. Luckily, the Mattertal Valley was not densely inhabited in the mid-nineteenth century, so the various rockfalls, landslides, and snow avalanches triggered by the 1855 earthquake frightened mostly a few shepherds living in remote and isolated alpine herders' huts.

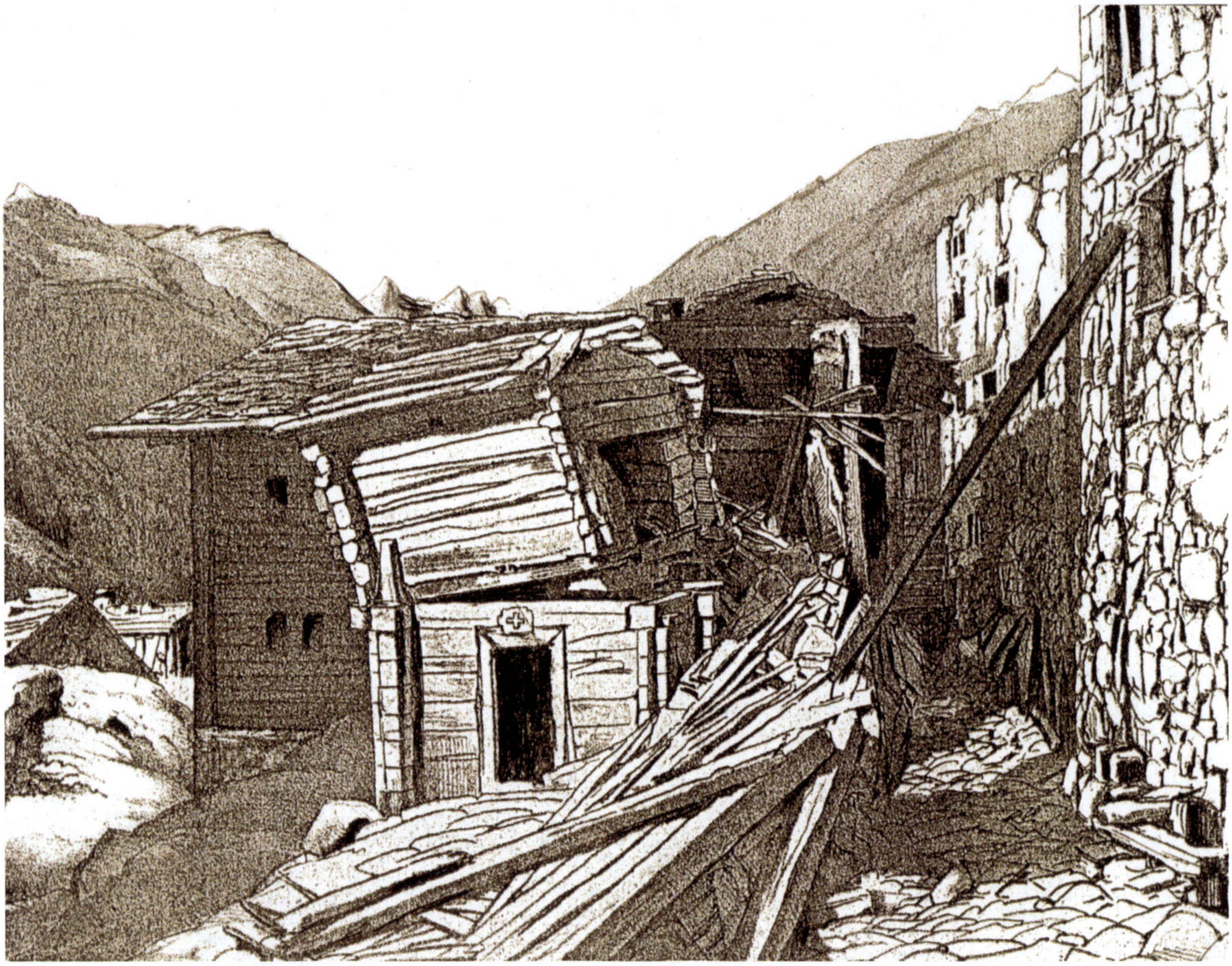

Fig. 2.23 A drawing of damage in St. Niklaus in 1855

The Visp earthquake is also interesting in the development of seismology. In the absence of seismic recording instruments, one can make a numerate assessment of an earthquake by assigning numbers to each settlement where effects were reported, with values increasing according to the strength. The classification of the shaking strength, in terms of its effects (including damage) and the corresponding table of numbers for each degree, is called an intensity scale. The intensity values can then be plotted on a map. The first such scale comparable to those in modern use was devised by the German mathematician Peter Egen in his study of an earthquake in Belgium in 1828, but this innovation did not catch on. The concept of the intensity scale was reinvented in 1846 by Johann Nöggerath and again after the Visp earthquake by the German geologist Georg Volger (1822–1897). In collaboration with his friend the cartographer August Petermann (1822–1878), Volger was able to plot the Visp earthquake data on a geographic map. Seismic intensity was still a work in progress, though; the idea was not widely adopted until the publication of the Rossi–Forel scale, the work of the Italian Michele de Rossi and the Swiss François Forel, in 1883.

The image here (Fig. 2.23), which also exists in a colored version, again shows the village of St. Niklaus. It appeared in a book by Christian Heusser, titled *Das Erdbeben im Visperthal vom Jahr 1855*, published in Zurich in 1856.

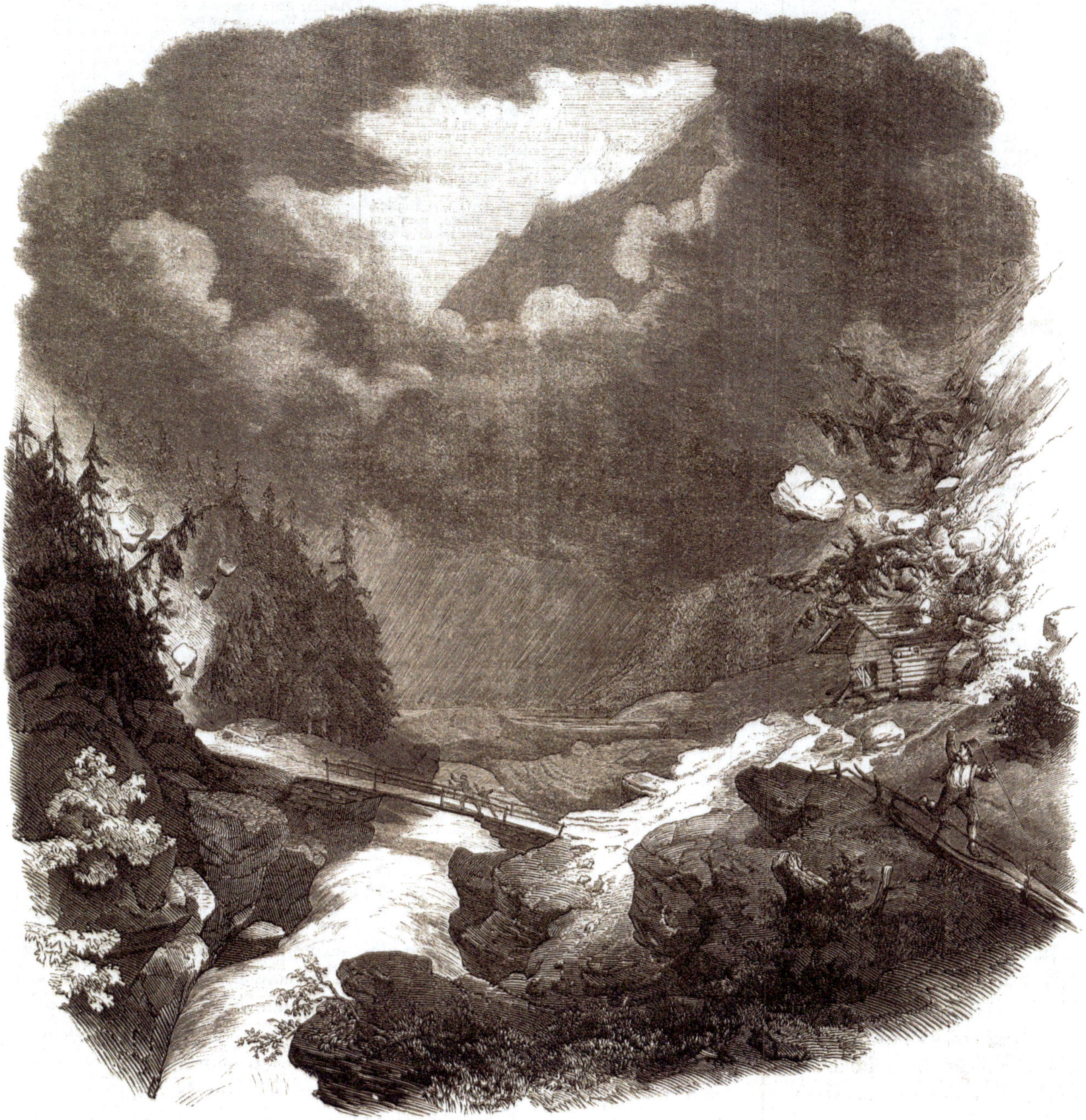

Fig. 2.24 An imaginative engraving of the Visp earthquake in a rural setting

This last image from the 1855 Visp earthquake is an engraving attributed to R. Kummer, depicting the effects in a remote part of the Mattertal near St. Niklaus (Fig. 2.24). The earthquake is shown as triggering avalanches and rockfalls, much to the distress of the lone figure in the lower right area of the image.

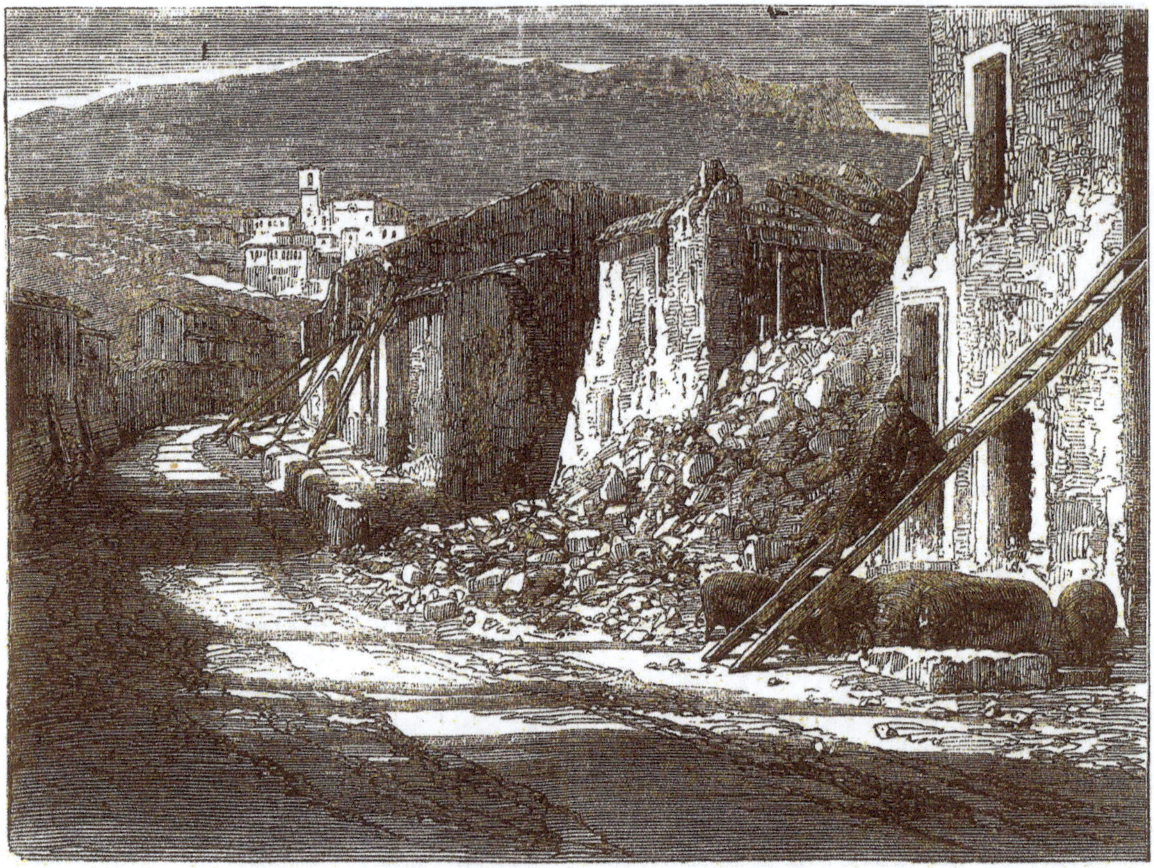

Fig. 2.25 The ruins of Polla in 1857

The earthquake of 16 December 1857, in the Basilicata-Campania region of Italy is important in the development of seismology. The Irish engineer Robert Mallet, who had recently published a long paper on the nature of earthquakes (in which he rediscovered the ideas of John Michell), had been looking for an opportunity to put his ideas into practice. This came when reports reached Britain of a severe earthquake near Naples. He obtained funding for an expedition and spent a week in the most affected area, recording details of the damage. His idea was that by plotting the direction from which the shock had arrived, over a large number of locations, he could obtain the position of the earthquake source (Mallet 1862; Ferrari and McConnell 2005; Musson 2013). This is essentially the same process of triangulation that has been used in seismology ever since; the difference being that Mallet had no instrumental records of the shock but had to rely on such directional information as was provided by damaged buildings.

Mallet also engaged the services of two French photographers, Alphonse Bernoud and Grellier, who took over 150 photographs of the damage, some of the earliest photographic records of earthquake damage. Many of them were even stereoscopic—a technique that Bernoud had pioneered, which provides an amazing view into the damaged villages of over 150 years ago. Most of the extant drawings of earthquake damage from this quake are based on these photographs.

One of the most severely damaged villages was Polla, of which Mallet wrote: "Months of bombardment would not have produced the destruction, that the awful shudder of five seconds involved, when thirteen hundred houses fell together with deafening crash, and overwhelmed the two thousand of their sleeping inmates, and with clouds of suffocating dust, choked the cries of horror and anguish, that rose from the startled and often wounded survivors. In three different directions, conflagration soon added its terrors to the scene, and beamed up, a flickering and ominous light, into that dreadful night of cold and wailing, throughout the lingering hours of which, in helpless agony, they listened to the passionate entreaties for relief, the dying sobs, of relatives and friends entombed around them, and dreaded for them, more than for themselves, the recurrence of other shocks. The cold grey light of winter's dawn, obscure with smoke and dust, revealed hundreds bruised or with broken limbs without a roof to shelter them, many without a garment to cover them."

The engraving here of the ruins of Polla is from the *Illustrated London News* of 16 January 1858 (Fig. 2.25), and is based on a photographic original.

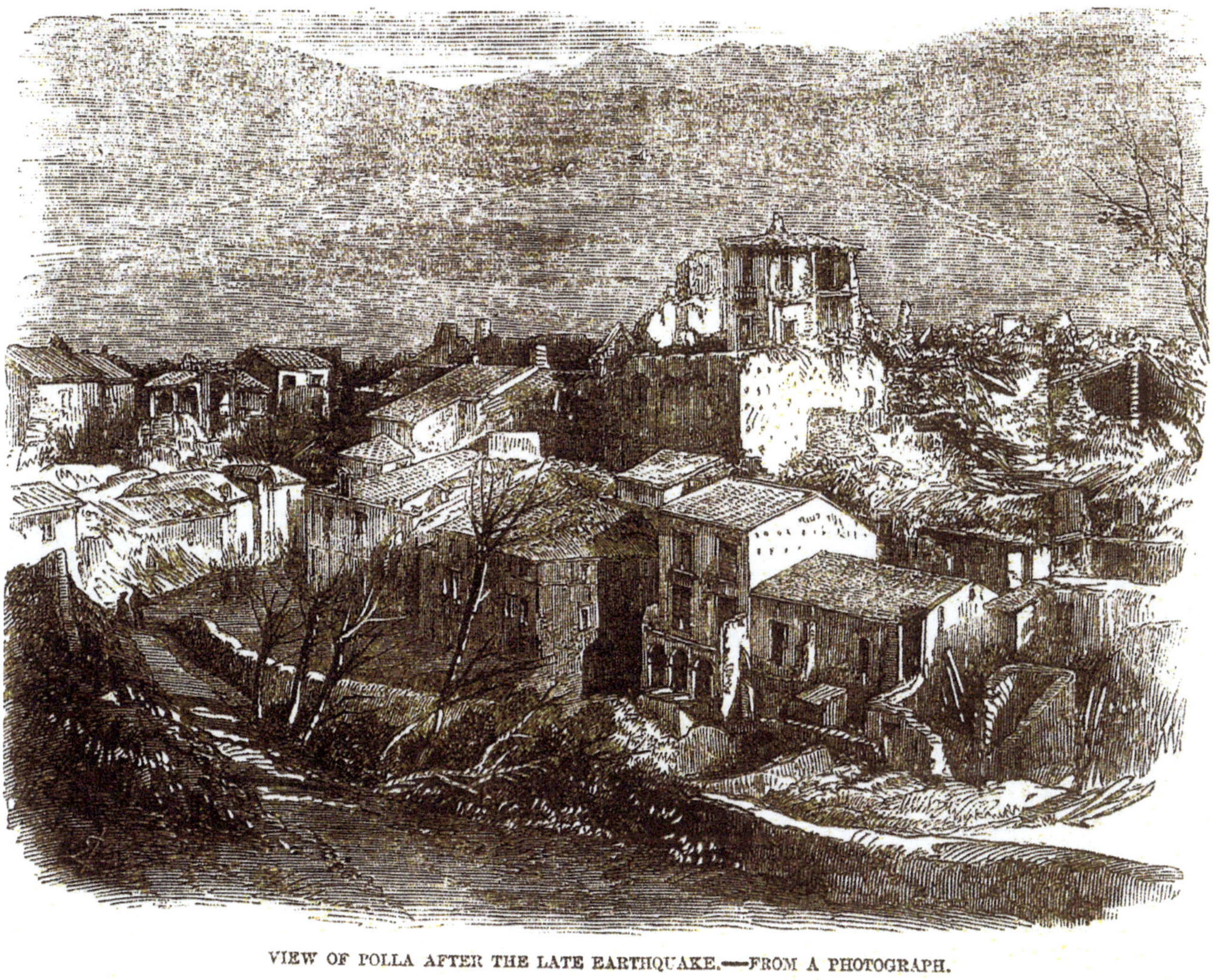

Fig. 2.26 A second view of Polla

This second picture (Fig. 2.26) of Polla is from the same source as the previous one (Fig. 2.25). The correspondent of the *Illustrated London News* wrote, "I send you in haste two sketches taken on the site of the late horrible disaster in the province of Basilicata. They represent the state of Polla … beautifully situated; in the extreme angle of the valley is the town of La Polla with a population of 7000 souls. So says 'Murray's guide for Southern Italy'. That town is now a ruin, and much of it has entirely disappeared. 'When I approached the melancholy scene', said a gentleman who had just arrived from the spot, 'I found the Earth seamed with fissures; at first they had been much wider, but they were gradually closing'. In some districts they had opened sufficiently to swallow up entire buildings; and in this way Tito, Briarza, Marsica-Nuovo and Lamarzana are said to have nearly disappeared. During the whole period of my visit the Earth was quaking like flour under the action of leaven. Few people were to be met with in the outskirts of the town—it was a scene of desolation and abandonment; but amidst the ruins people sat in a state of grief and consternation. 'We have lost everything', they said to us, 'our property and our friends'. The great cause of grief, however, appeared to be that there were none to dig out the dead."

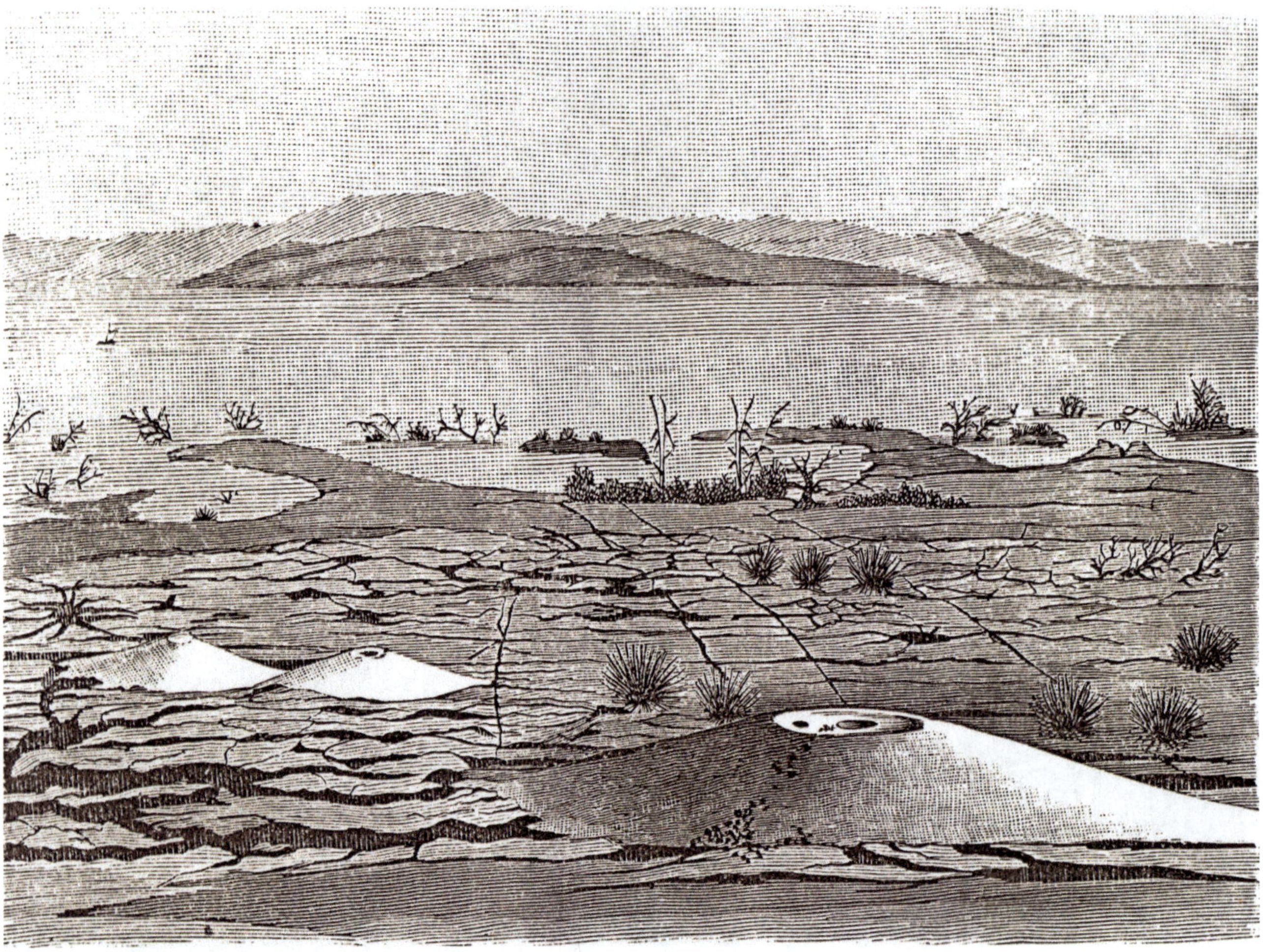

Fig. 2.27 The Aigio earthquake of 1861

The most seismically active country in Europe is not Italy, nor Portugal, but Greece. However, Greece does not feature so heavily in the historical iconography of earthquakes, because of the Ottoman occupation for so many centuries. It is only after Greek independence in the nineteenth century that one finds illustrations of Greek earthquakes; even written accounts are relatively sparse before then. After the previous depictions of ruined buildings, the reader may find the image here, from the Gulf of Corinth earthquake on 26 December 1861, underwhelming (Fig. 2.27). But it illustrates an interesting seismic effect that is perhaps not so well known.

The town of Aigio, at the western end of the Gulf of Corinth, has been struck by earthquakes repeatedly, in 1748, 1817, 1861, 1888, and 1889. The 1861 earthquake was the largest of these, with a magnitude of slightly less than 7. It was studied at the time by the German astronomer Johann Friedrich Julius Schmidt (1825–1884), who, at the time of the earthquake, was the director of the Observatory of Athens. The most recent study is the very full account of the earthquake by Albini et al. (2017). While the damage was substantial, fatalities fortunately seem to have been few. The earthquake was notable for its effects on the ground surface, although these were not as spectacular as the large ground translocations, such as deep trenches and large open cracks, that accompanied the 1783 Calabria event. Schmidt mapped a prominent ground crack for some 12 kilometers, which some authors have interpreted as a fault break, but it is now considered to be the result of slumping, and it was no doubt slumping in the waters of the gulf that produced a small tsunami (a reminder that tsunamis are not solely the consequence of great earthquakes).

Strong earthquake shaking of sandy soils can produce a phenomenon known as liquefaction, when the soil effectively turns to quicksand. This is often accompanied by the ejection of sand and water in fountains, leaving behind small cones called sand boils, as seen in this image taken from the book *Erdbebnkunde*, by the Austrian geologist Rudolf Hoernes (1893), but based on a sketch by Schmidt. Traces of old sand boils can often be found in excavations on the sites of ancient earthquakes, allowing the dating of large prehistoric earthquakes.

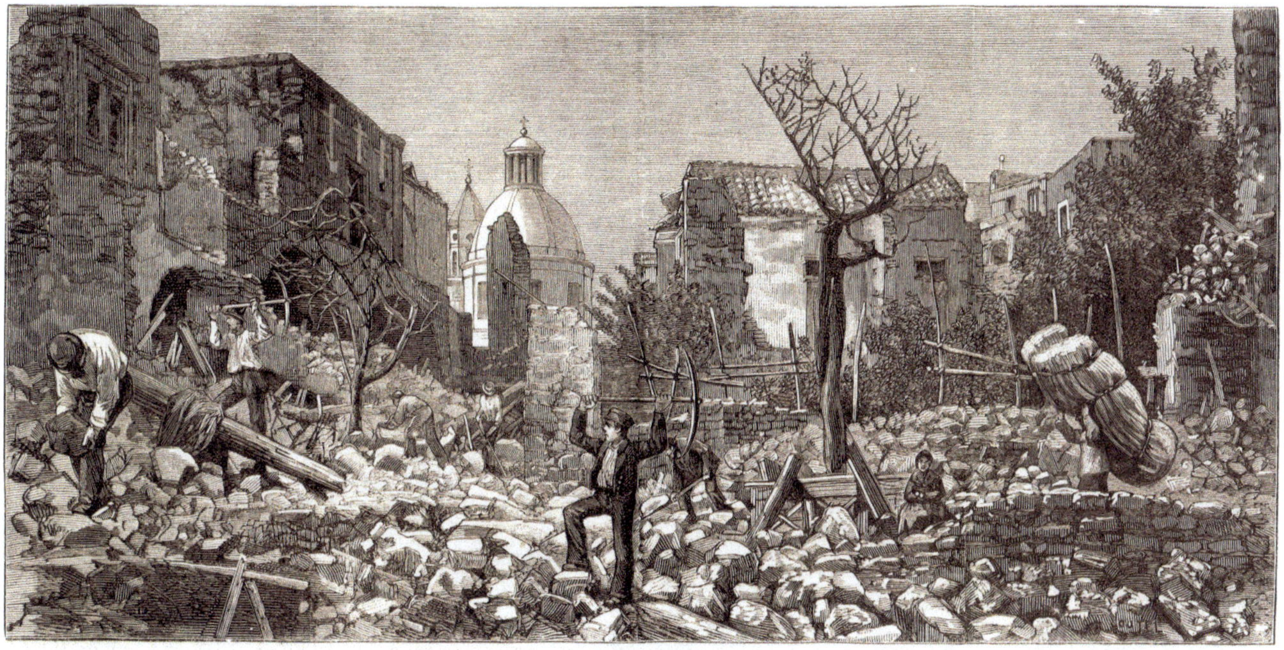

CLEARING AWAY THE RUINS.

Fig. 2.28 The 1881 Ischia earthquake—clearing the ruins

The Ischia earthquake on 4 March 1881, is another example of how an earthquake of a modest magnitude (a little less than 5.5) can cause extensive damage when it is shallow and hits poorly built structures directly. Ischia is a volcanic island at the northern end of the Gulf of Naples and is known today as a tourist destination. When the earthquake struck in 1881, the dominant building style for both houses and churches used undressed, round stones linked together with lime mortar; walls of this kind have low resistance to tensile stress, regardless of the wall thickness. In addition, the chimneys and roofs were massive and heavy, and the rafters were so poorly inserted into the walls that they were drawn out by the rocking of the houses, leading to complete destruction of the houses. The damage extended over most of the island but was worst in the north, in the villages of Fango and Casamicciola. At the latter place, 249 houses collapsed, with 121 killed and 140 injured. In Fango, to the west, nearly all of the houses collapsed or were rendered uninhabitable.

The earthquake was documented in great detail by the volcanologist Henry James Johnston-Lavis (1856–1914), who was inspired by Robert Mallet's study of the 1857 Basilicata earthquake to try and apply Mallet's methods to the earthquake of Ischia. The image here is titled *Clearing Away the Ruins* and was published in *Harper's Weekly* on 30 April 1881 (Fig. 2.28). The few figures armed with pick-axes seem quite unequal to the task ahead of them.

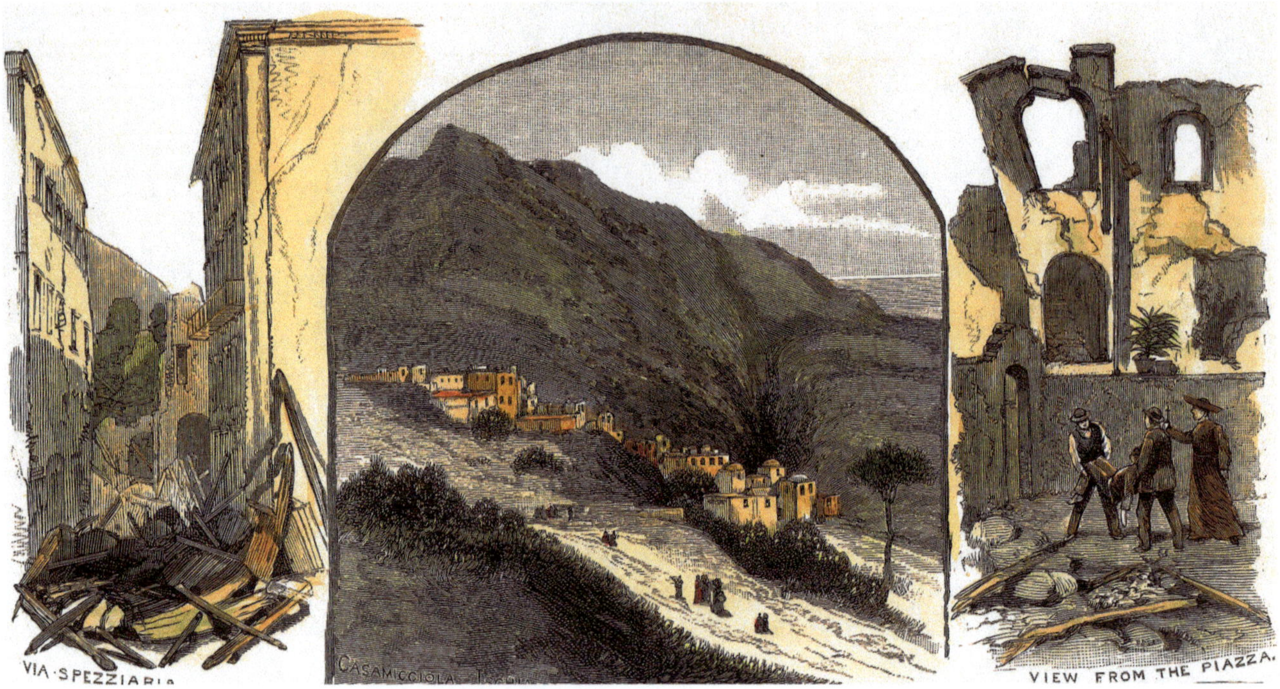

Fig. 2.29 A compound image of damage at Casamicciola in 1883

The Ischians' woes were not over after the 1881 earthquake; on 28 July 1883, an even stronger earthquake (magnitude 5.8) hit in exactly the same place, and villages that had been badly damaged before were now totally destroyed, with around 2300 people being killed. Numerous houses whose owners had failed to reconstruct, repair, or reinforce them adequately did not survive, especially in the main town of the island, Casamicciola. The compound image here of five wood engravings was published in *Frank Leslie's Illustrated Newspaper* on 11 August 1883, two weeks after the disaster (Fig. 2.29). That pictures could be sent across the Atlantic (presumably from local artists) and published so quickly prompts reflection about the speed of pictorial journalism during this period.

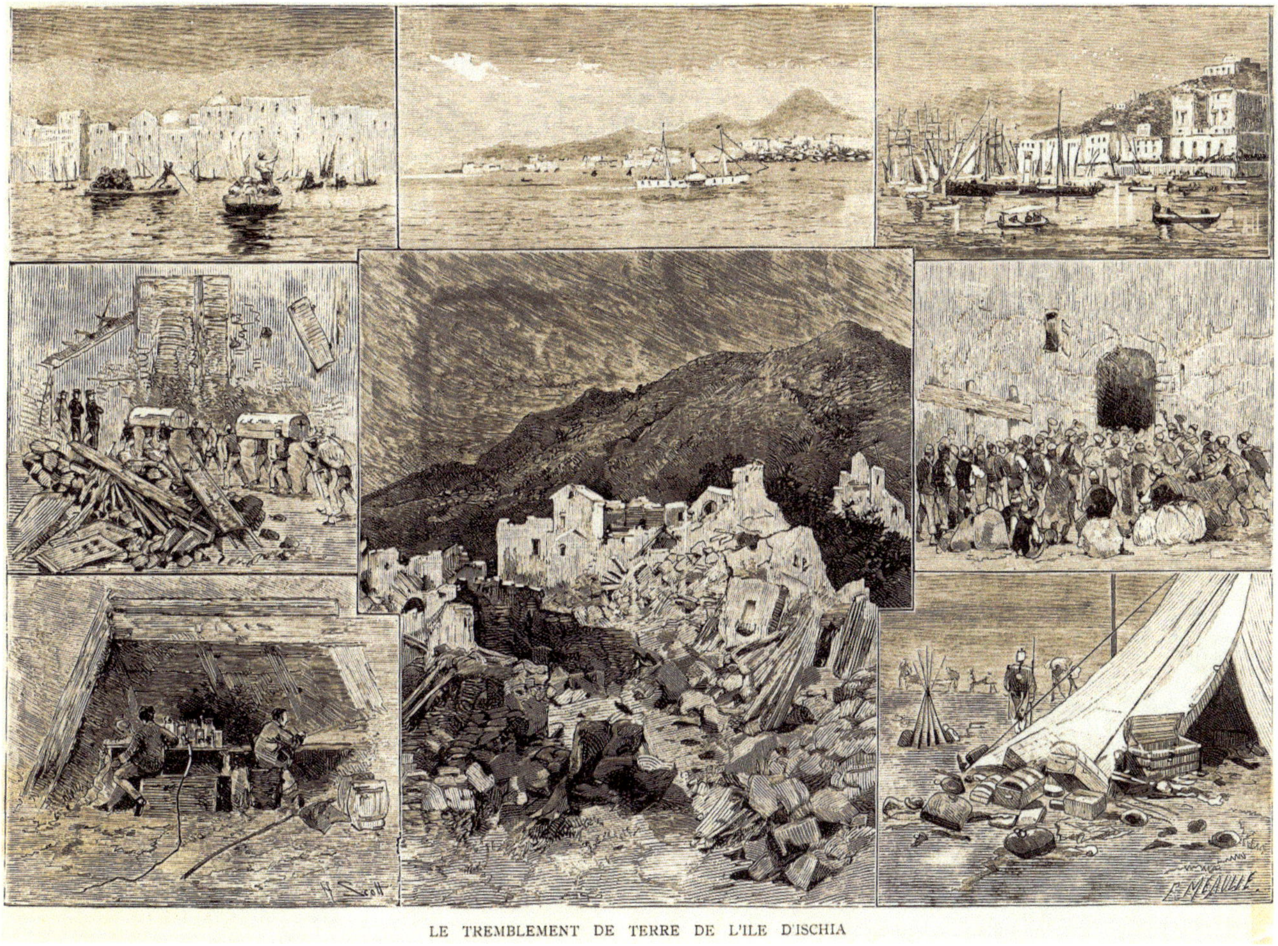

Fig. 2.30 The ruins of Casamicciola

The *Illustrated London News*, like *Frank Leslie's Illustrated Newspaper*, had a fondness for combining several scenes into a single large print. Such a survey of damage from the 1883 Ischia earthquake was published on 18 August 1883, with seven scenes of ruins and two of prominent buildings as they were before the earthquake. The segment reproduced here shows how extensive the damage in Casamicciola was; almost all of the town is ruined (Fig. 2.30).

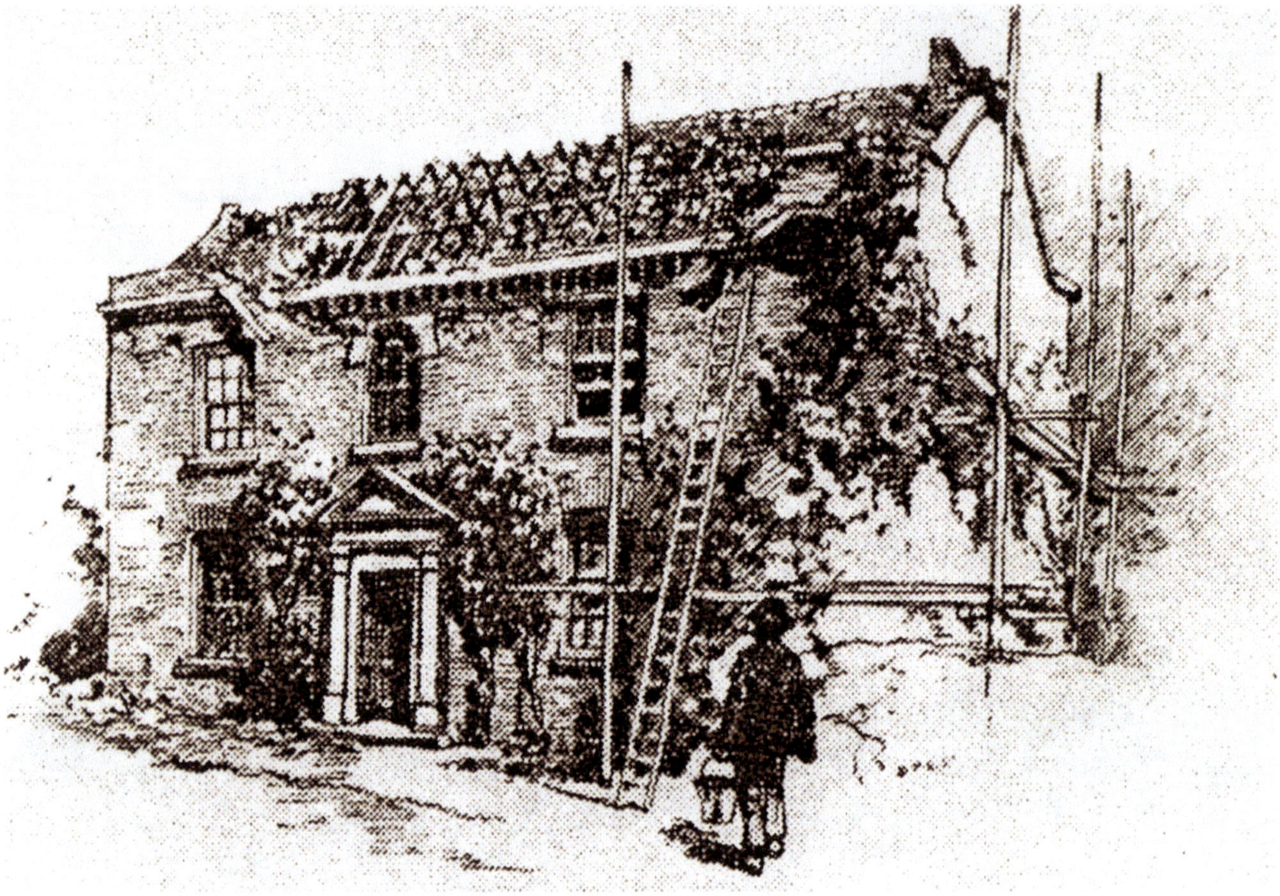

Fig. 2.31 Damage from the 1884 Colchester earthquake

"Surely there are no earthquakes in England" is a common refrain heard whenever an earthquake does occur in England. It seems to be the case that in countries that are mostly subject only to small or moderate shocks, earthquakes quickly fade from the memory. But nowhere in Europe is completely free from potential earthquakes, even though in the least seismic countries—such as Ireland, Poland, and Finland—any sort of seismic activity is very rare. The most damaging earthquake in Great Britain in modern times occurred near the town of Colchester in southeast England on 22 April 1884. It was probably not very big in terms of magnitude (about 4.5), but, because of its shallowness, it delivered a concentrated shock over a limited area, causing much damage to houses and churches. The picture here, of Mr. Charles Harvey's house in the village of Great Wigborough, is a typical example of the cases of more severe damage (Fig. 2.31). The chimneys have been broken off at the roof line, most of the tiles have gone from the roof, and there is a large crack in the gable wall. This drawing is taken from an excellent book-length report on the earthquake, which was prepared by Meldola and White for the Essex Field Club, although it initially appeared in the *Illustrated London News*. Many photographs of the damage also exist, although in some cases the photos were doctored to make the damage look worse than it was, by removing the chimneys from the images by painting them out with black ink in the photographic negatives. Drawings may be exaggerated, as has been shown in many of the images in this book, but one should not imagine that the camera never lies.

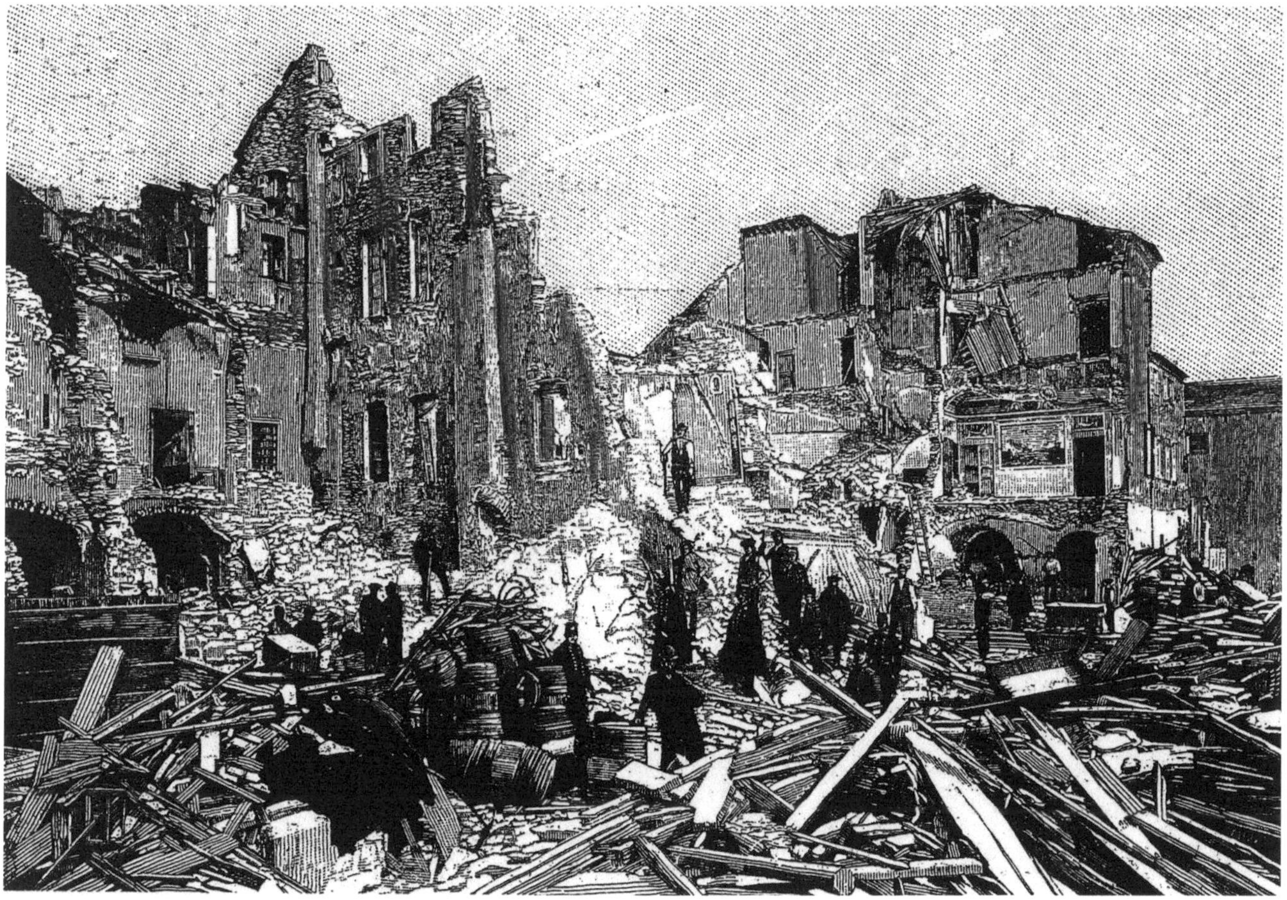

Fig. 2.32 The ruins of Diano Marina in 1887

The labels that get attached to earthquakes are really for ease of reference and are seldom exact indicators of the location of the epicenter. They can also vary, especially in the case of events occurring near an international border, where a different reference name may be used on each side. The earthquake of 23 February 1887, seems to have attracted more names than usual, and one can find it referred to as the Nice earthquake or Menton earthquake (on the French side), or as the Imperia or Western Liguria earthquake (on the Italian side). Jean Vogt, the author of the *Catalogue of French Earthquakes*, decorated the cover of his work with an engraving of the heavily damaged (and much photographed) Ecole Municipale des Jeunes Filles du Quartier Saint-Etienne, Nice—although he was well aware that anyone standing on the spot would not actually see any other damaged buildings nearby.

However, the same earthquake actually should preferably be thought of as an Italian event; in the coastal Italian town of Diano Marina located on the Italian coast of the Ligurian Sea, some 75 kilometers east or northeast of Nice, the damage was catastrophic, and it was even more severe inland at Diano Castello. But Nice, Monte Carlo, and Menton on the French side of the border were all places much better known to international travelers than the more badly damaged localities on the Italian side; thus, the French locations were reported on much more widely in newspapers around the world.

The image here shows the ruined town of Diano Marina in a woodcut probably made from a photograph (Fig. 2.32). The people shown searching for their family members and remnants of their property convey well the hopeless situation of those who have lost their homes. It comes from *L'Illustrazione Italiana* of 1887 (volume 1, page 193).

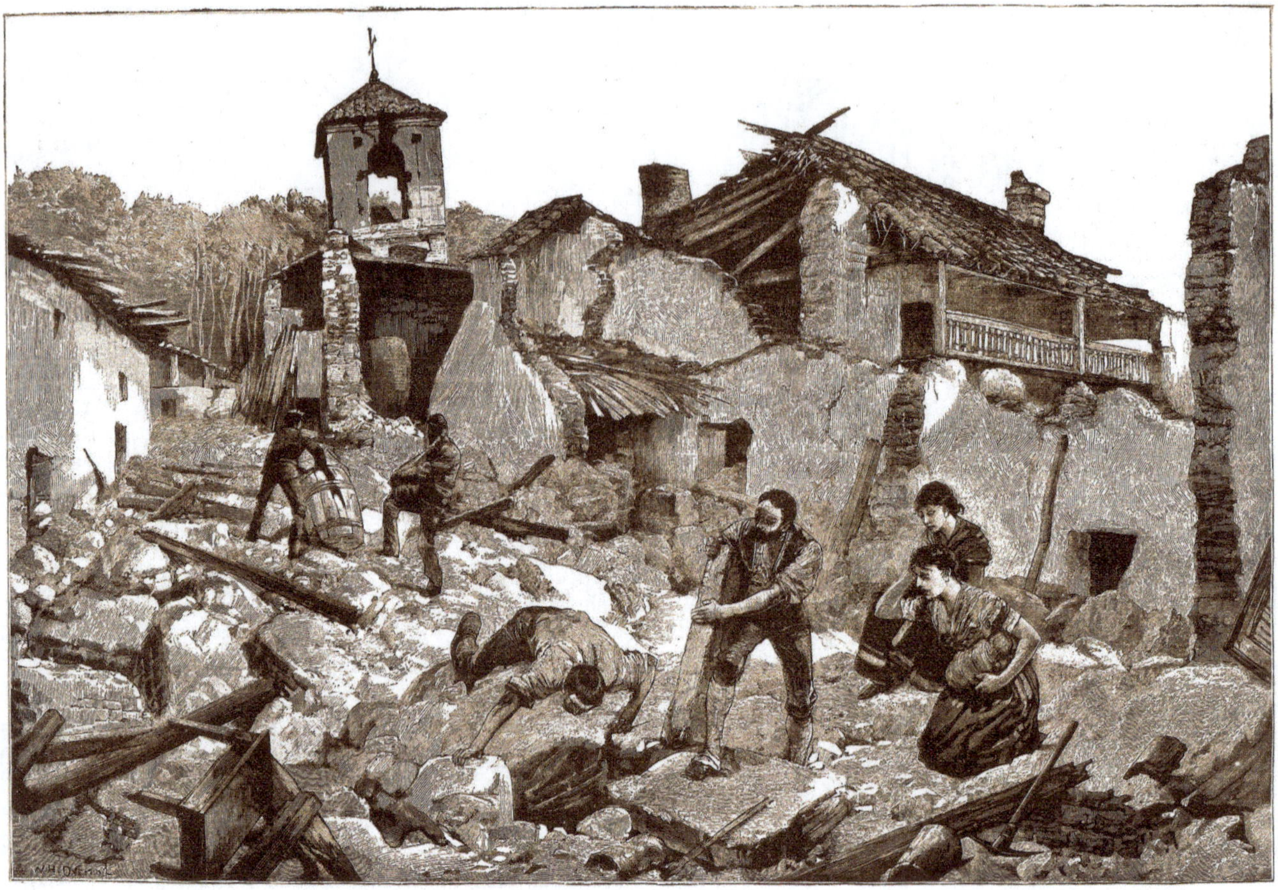

SEARCHING THE RUINS AT DIANO-MARINA ON THE COAST OF ITALY.
THE EARTHQUAKE ALONG THE RIVIERA.—[See Page 230.]

Fig. 2.33 Diano Marina in 1887

This second view of the ruins of Diano Marina comes from an as yet unidentified English-language journal. Again, the engraving seems to be based on a photograph (Fig. 2.33). Two men seem to be attempting to free someone from the rubble with their bare hands, while two women look on anxiously.

The epicenter of the earthquake was actually offshore, around 50 kilometers south by southeast of the town of Imperia. The earthquake was felt over a very large area: across Switzerland to the north, east as far as Udine, throughout Corsica, even at Sassari on the north coast of Sardinia, and almost as far west as the French–Spanish border. It was the strongest earthquake in this area over at least the last 1000 years, and a notable tsunami was caused by submarine slumping, with a maximum run-up height of two meters on the Italian coast. Two strong aftershocks occurred on the same morning, compounding the damage at the worst-affected places, and around 200 aftershocks in all were felt over the following 12 months. The magnitude of the earthquake has been a matter of dispute for years, but a value a little over 6.5 seems likely (Larroque et al. 2012).

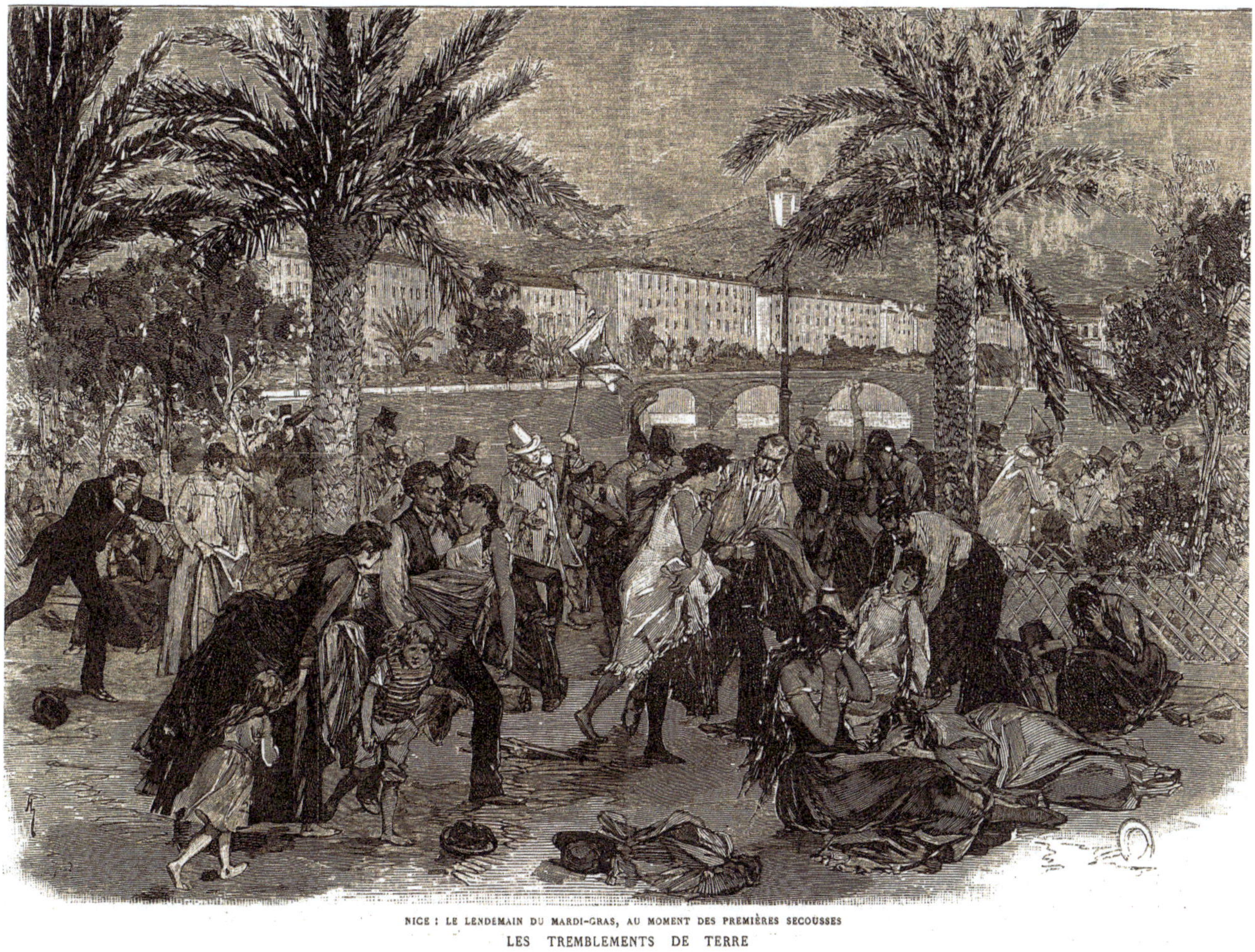

Fig. 2.34 Nice during the 1887 earthquake

The engraving here (Fig. 2.34), from an unidentified French journal, shows a great contrast with the previous two images (Figs. 2.32 and 2.33). The Mardi Gras revelers on the promenade at Nice are startled and alarmed, but there is a complete absence of any indications of damage. There were a number of isolated cases of damage in Nice, but none so severe as the collapse of the girls' school at St. Etienne, burying the schoolmistress, Mme. Cheylon, who was found to be dead when she was recovered by the local fire brigade. For most people in the French part of the affected area, the experience was more alarming than injurious. One woman in Menton was moved to verse:

Crash went the ceilings, down came walls,
Whilst all were still in bed;
Each on the "Bon Dieu" loudly calls,
Then frantically fled.

Seizing their jewels, silver, gold,
Rushing they knew not where;
Helpless alike, both young and old—
Presence of mind was rare.

Before the dawn the streets and shore
Were peopled far and near,
Quaintly were mixed both rich and poor
Distinction, none was there?

With this event, our survey of European and Mediterranean earthquakes is brought up almost to the end of the nineteenth century and into the period in which photography dominates pictorial recording. It is time to look at other parts of the world.

2.3 The Americas

At the risk of stating the obvious, to have pictures of earthquakes from a time and place, one needs both earthquakes and the means of making pictures. It is well known that Native Americans in pre-Columbian times did not have any tradition of pictorial representation; they created only a few symbolic paintings of earthquakes, not intended to be representational. Thus, the first images of earthquake effects are closely linked with the advancing waves of European colonization, drawn or painted by European incomers, colonists, surveyors, and other visitors. For making images they used the reproduction methods that were used in Europe at that time: woodcuts in the sixteenth century and subsequently engravings and other kinds of printing.

When considering the first pioneer period of the gradual European conquest and settlement of the continent, we become aware that in the early stages of conflict between the Native Americans and bands of (mostly Spanish) conquerors, survival was of pre-eminent importance, and so there was no great interest in making portrayals of natural phenomena. The pattern of Spanish conquest was conditioned largely by the search for treasure, including the fabled El Dorado. The fabulous treasures of the Incas were a major lure; these treasures, preferably gold and silver, were discovered by Francesco Pissarro in 1522 and shortly afterward plundered in the early 1530s. It is understandable that gold and silver mines (such as Potosi, one of the world's largest at that time) attracted many prospectors and adventurers to make their arrangements in this golden paradise.

Thus, the first images showing the effects of American earthquakes—simple woodcut prints—have their origin on the Peruvian coast. However, in most cases, they do not carry reliable pictorial data.

For the first American depiction of an earthquake, we therefore turn to the British colonies of New England.

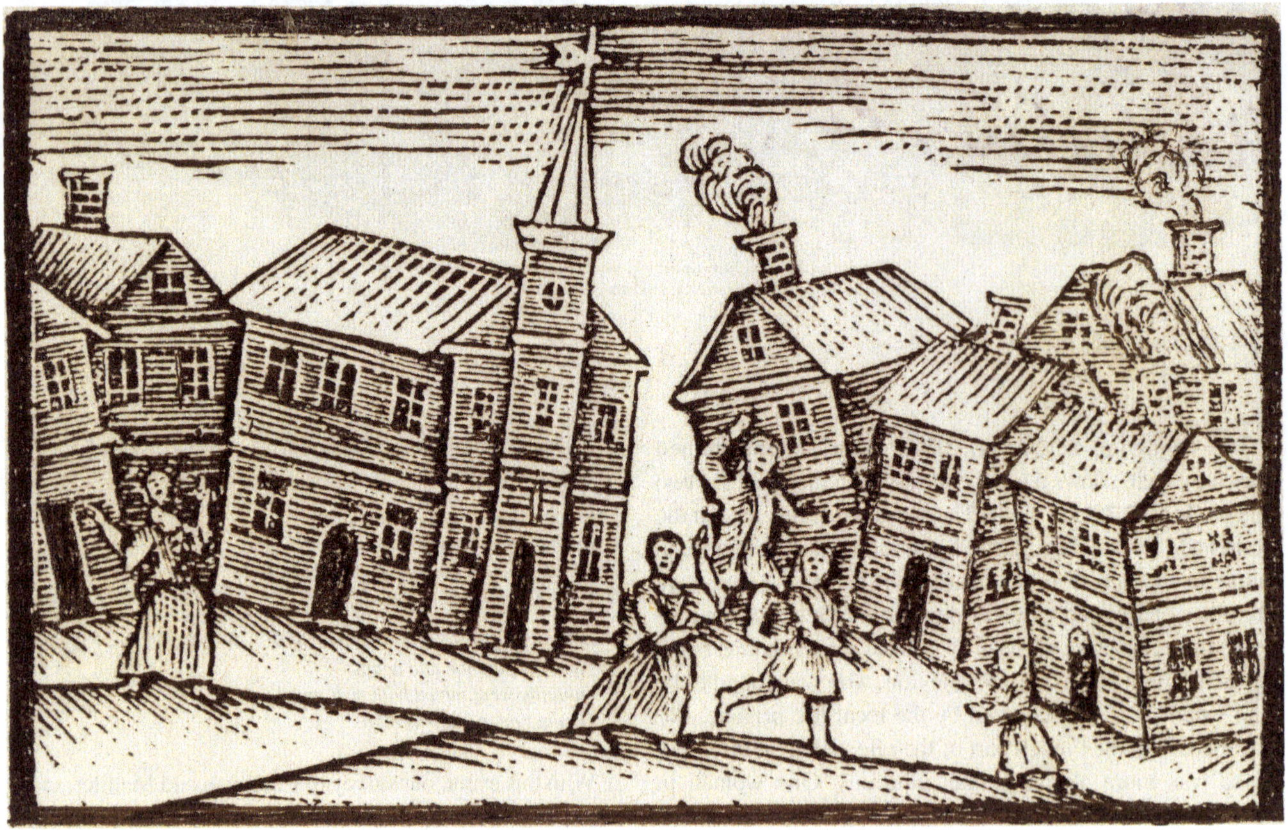

Fig. 2.35 The Cape Ann earthquake of 1744

Around the mid-eighteenth century, the northern part of New England was seismically active; within 28 years (1727–1755), three medium-sized earthquakes visited the region of Cape Ann and its vicinity. The first of the larger events was the Cape Ann earthquake of 29 October 1727 (the date often appears as 10 November, as in the Gregorian calendar, but the Julian calendar was still in use in New England at the time). The epicenter of the 1727 earthquake was located off or on the New Hampshire and Massachusetts coast, and the shock was felt from Maine to Connecticut, even into Canada, and south as far as New York City. No one was hurt, but there were cases of damage to brick chimneys and cracks in walls; springs were rendered turbid, and at Newbury, Massachusetts, cracks in the ground were observed.

The second strong Cape Ann event occurred at 10:15 a.m. on Sunday 3 June 1744, just after church services had begun. Many people in the churches in Boston and neighboring towns ran out into the streets, fearing that the buildings would fall upon them. The damage was mostly limited to the fall of bricks from the tops of chimneys and the overthrow of field walls. The shock was felt strongly as far away as Falmouth, Maine. The contemporary woodcut reproduced here (Fig. 2.35) is from a religious tract issued in 1744, with the title *Earthquakes, Tokens of God's Power and Wrath.* Inevitably, there were sermons proclaiming the event to be a warning from God. The way the buildings are leaning at different angles could conceivably be an attempt to convey the rocking motion of the earthquake, but, equally, it is likely that the draftsman made no real attempt to do anything more than convey the idea of an earthquake.

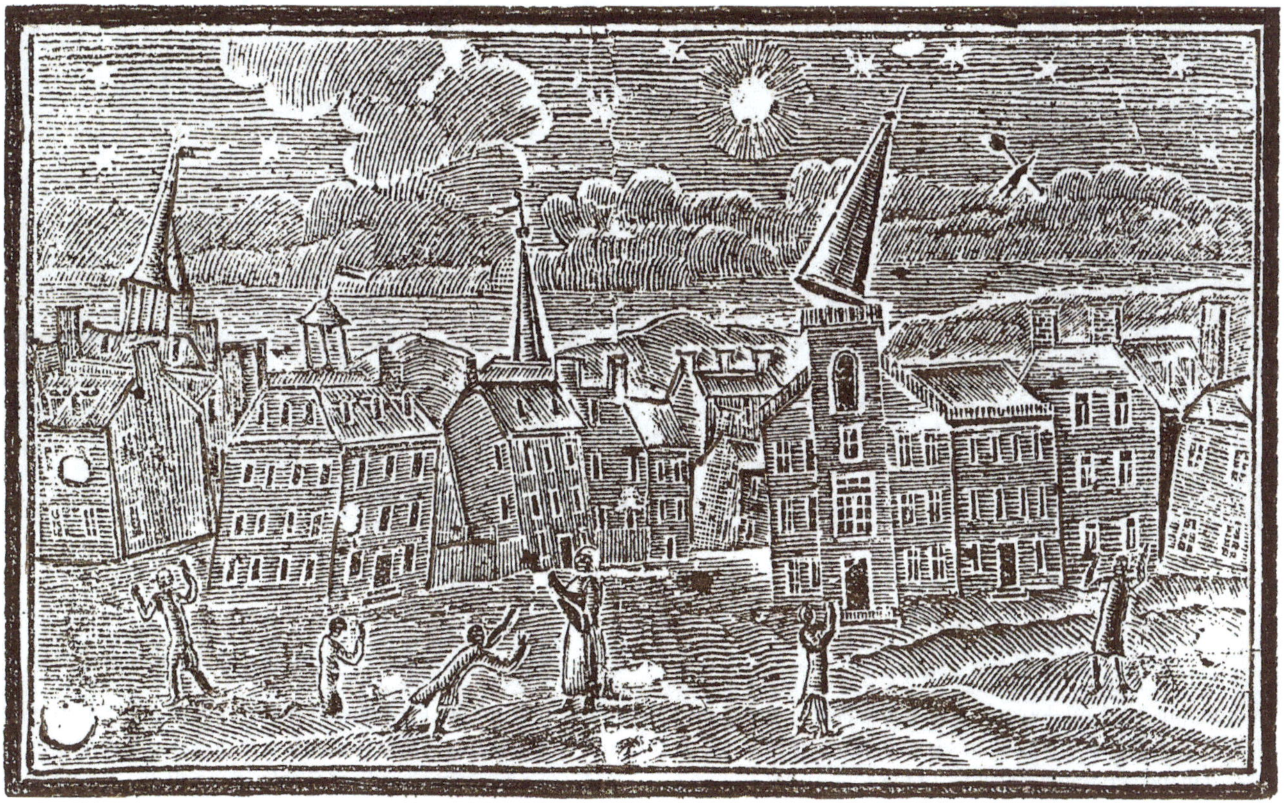

Fig. 2.36 The Cape Ann earthquake of 1755

After the 1727 and 1744 earthquakes came the last and strongest of the bunch, on 18 November 1755, which caused damage to chimneys (as many as 1500, according to reports at the time) and church steeples around Cape Ann and Boston. New water springs formed and old springs dried out, and "there were several chasms or openings made in the earth, from some of which water has issued" according to John Winthrop (1755) in the first scientific investigation of an American earthquake. Chimneys were also damaged as far away as Springfield and New Haven; in Boston, streets were littered with debris from fallen chimneys, many of which had damaged roofs on the way down. The magnitude was probably just under 6, making it the strongest earthquake in the history of Massachusetts; a recurrence today would be immensely costly for modern-day Boston.

Coming so close in time to the devastating Lisbon earthquake on 1 November 1755, and with communications in those days being at the speed of a sailing ship, a certain amount of confusion appears to have occurred. One can find studies of the Lisbon earthquake that draw a map of the felt area extended so as to take in the New England seaboard—which is false (see Fig. 2.12, for instance)—and accounts that the Cape Ann earthquake caused a tsunami—also false.

The woodcut illustration here (Fig. 2.36) is similar in style to the previous image (Fig. 2.35), with buildings leaning this way and that. The source is a pamphlet titled

Earthquakes Improved: or Solemn Warning to the World: By the Tremendous Earthquake Which Happen'd on Tuesday Morning the 18th of November 1755, Between Four and Five O'Clock, issued by John Green (1727–1787), the publisher of the *Boston Weekly Advertiser*. The view is of Boston, and the building on the right with its spire launching into the air is probably the Brattle Square meetinghouse.

This earthquake also has one very obscure and improbable claim to fame, being the only pre-twentieth-century earthquake (at least to the authors' knowledge) to feature in a computer game, *Fallout 4* by Bethesda Softworks.

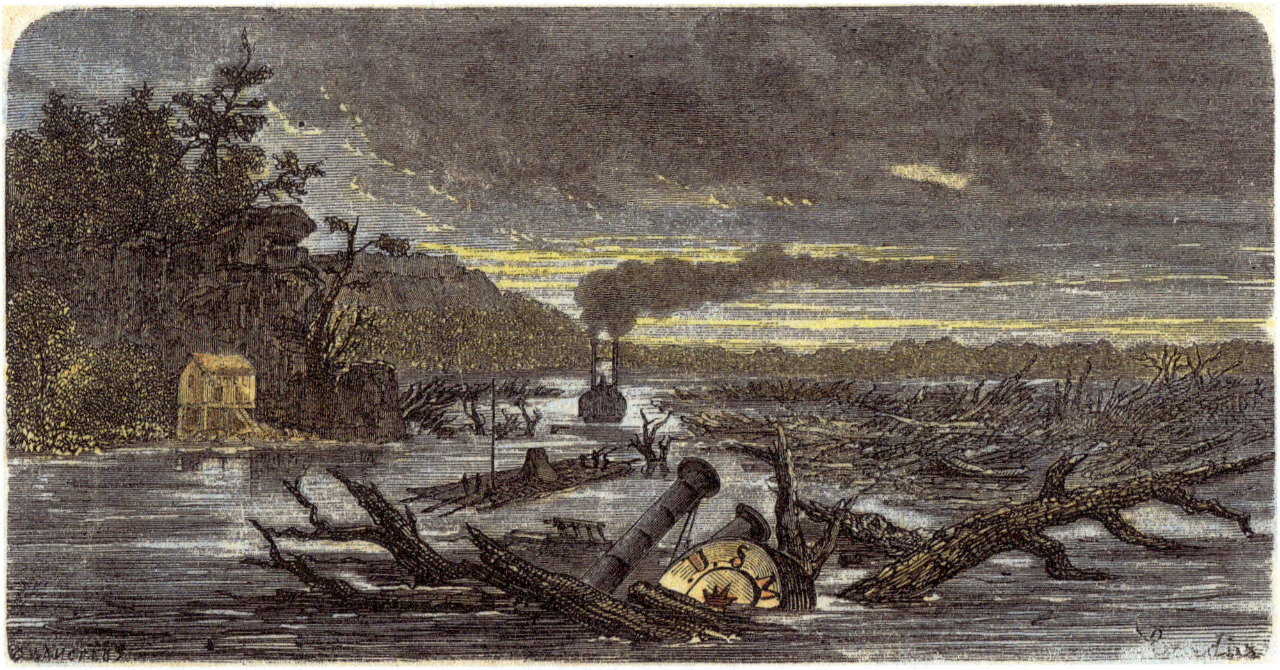

Fig. 2.37 A supposed image of river clearing after the New Madrid earthquakes of 1811–1812

Looking at a modern map of earthquake hazard in the USA, it is perhaps surprising to see an area shaded red for high hazard in the geographical center of the country, far from plate boundaries. This bullseye is on account of three remarkable earthquakes that occurred over a short period in the years 1811–1812, centered on the area around the town of New Madrid, Missouri (pronounced "New MAD-rid"). The shocks occurred on 16 December 1811; 23 January 1812; and the strongest on 7 February 1812, although aftershocks continued for a long while afterward. In this period, the Mississippi River was about the western limit of the territory claimed by the USA, with few settlements, and these only along the riverbanks. Thus, while the earthquakes were felt over most of the continental territory east of the Mississippi, no one knows what happened to the west. The huge felt area was partly due to the way that the rocks of the continental interior easily transmit seismic waves, and was partly influenced by the tendency of the soft alluvium along the riverbanks to amplify seismic vibration. For this reason, estimates of the magnitudes of the three largest earthquakes have been the subject of much disagreement, and values as high as 8, proposed in some quarters, are now considered untenable; none of the earthquakes exceeded a magnitude of 7 (Hough and Page 2011).

There exists a dramatic eyewitness account of the effects of the first shock on the steamboat *New Orleans*, traveling down the Mississippi at the time, by William Pierce (published in the *New-York Evening Post*). The earthquake toppled many trees, and sunken branches in the river made navigation more hazardous than usual. There are very few images of the earthquake effects, and this one, published in Prague many years later, supposedly represents clearing a passage through the river after the earthquake (Fig. 2.37). It is very doubtful that this is genuine; although some boats on the river were sunk by the earthquake (as mentioned by Pierce), the wreckage looks like a steamboat, and no steamboat was sunk as a result of the earthquake; in fact, the only such boat on the river at that time was the *New Orleans*, on which Pierce was traveling.

The letters “USM” (which stand for “US Mail”) on the wreckage suggest a much later date for the picture. If the picture was drawn in Prague, it is wholly an invention; alternatively, it could be a reproduction of an earlier American print, perhaps from the 1860s, but we have been unable to trace an original.

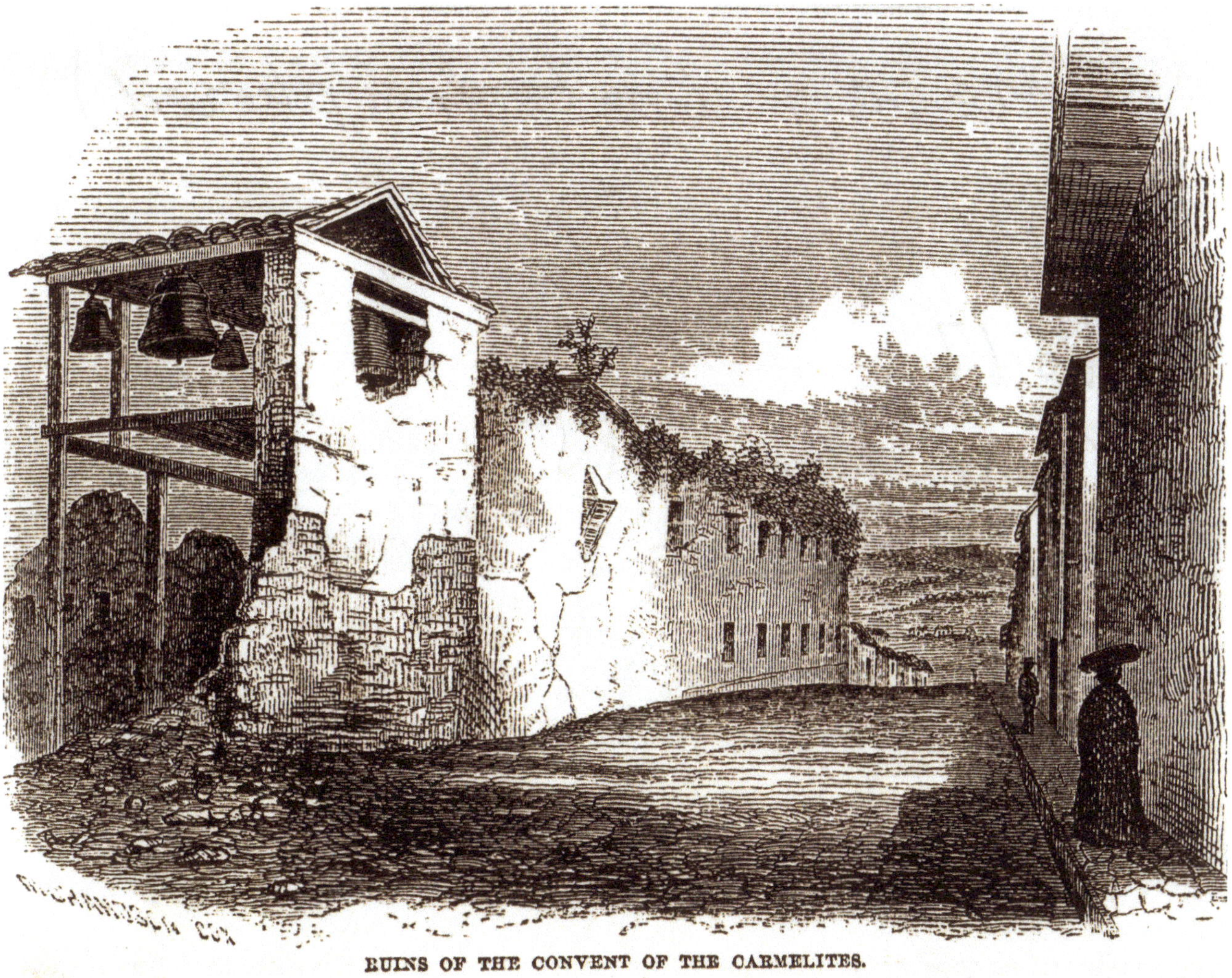

Fig. 2.38 Relics of the 1812 Caracas earthquake

Earthquakes can occur at any time of the day or night, and they take no regard of human affairs. The Battle of Lake Trasimene in 217 BC, a major conflict in the Second Punic War between the Romans and Carthaginians, was famously interrupted by an earthquake, although, according to the historian Livy, both armies were so heavily involved in the fighting that neither noticed it. Another earthquake that has found a place in military history is the Caracas earthquake on 26 March 1812, which occurred during the Venezuelan War of Independence. This major earthquake (magnitude about 7.7) struck the northeastern part of Venezuela and caused serious damage to several Venezuelan towns—namely, Caracas, La Guaira, Barquisimeto, and San Felipe. At Barquisimeto, a Venezuelan army under Diego Jalón was gathering and was caught under the ruins of the town; Jalón himself was injured. Simón Bolívar was in San Jacinto and, seeing the ruins of the church there, exclaimed in horror, “If nature oppose us, we will fight against her and force her to obey us.” Predictably, since the worst-affected areas were under rebel control, the Spanish claimed that it was divine punishment. The archbishop of Caracas, Narciso Coll y Prat, referred to the event as “the terrifying but well-deserved earthquake.”

The seismic event itself seems to have been a multiple or twin earthquake: the first and largest shock devastated Caracas, and the second, which followed the first one after 30 minutes, destroyed the town of Merida located some 550 kilometers southwest of Caracas. Altogether, 15,000–20,000 lives were lost. Changes to the landscape occurred; in the region of Vallencillo a new lake was created.

Caracas was so strongly damaged that it was suggested that a new Caracas should be rebuilt in a safer location; this plan was not realized.

The engraving here (Fig. 2.38) is much later, published in *Harper's New Monthly Magazine* (June–November, 1857; page 190), and shows the ruins of a Carmelite convent, which was never rebuilt.

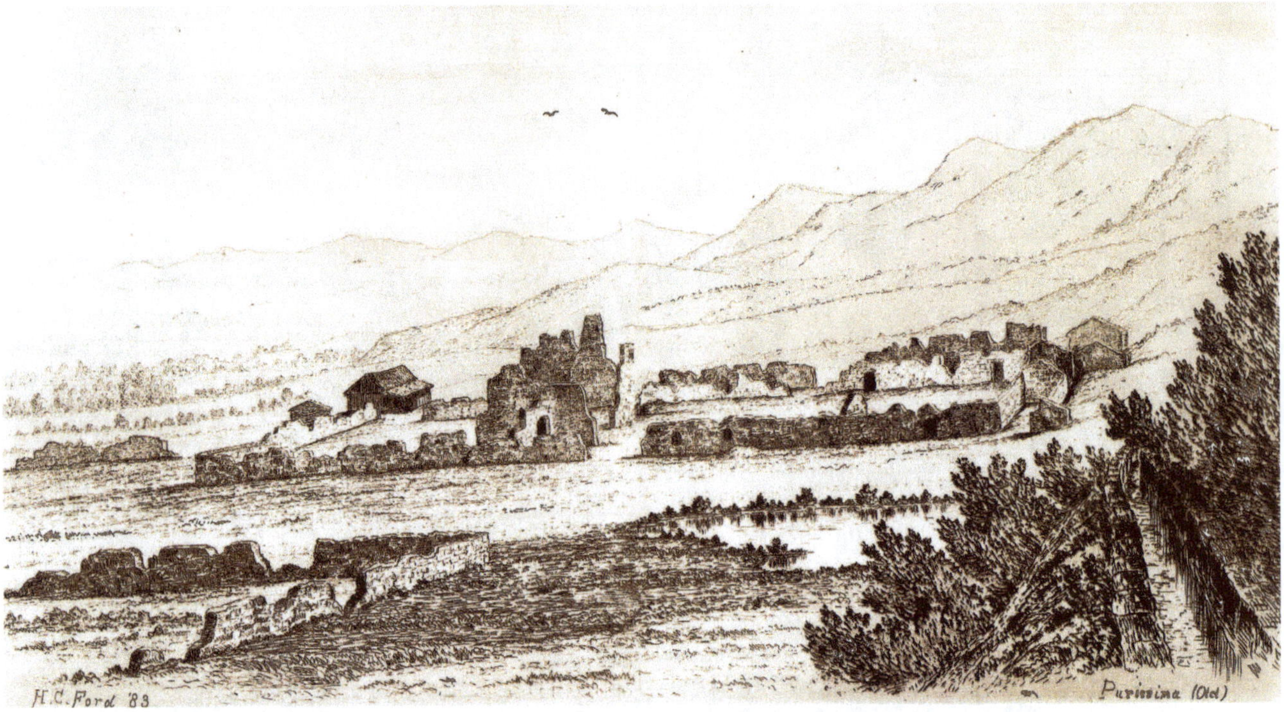

Fig. 2.39 The San Juan Capistrano earthquake of 1812

Although in 1812 the still-young United States of America stretched only as far west as the Mississippi, California was being settled by the Spanish, moving up from Mexico. Thus, the earthquake in southern California of 8 December 1812, is quite well documented. The area of damage was centered on San Juan Capistrano, and the event is known today as the San Juan Capistrano earthquake or sometimes the Wrightwood earthquake. The affected area covered a wide swath, down to San Diego in the south, and the magnitude has been estimated as between 7 and 7.5. The source of the quake is uncertain; there are several major faults there, of which the San Andreas is but one.

The Spanish colonization was led by missionaries, who established mission buildings as bases for preaching to the local native tribes; actual colonial settlement was sparse, and, for all the size of this earthquake, only 40 people were killed. There were about 21 Spanish missions in California at the time; one of them, La Purísima Mission, was situated in what is now the city of Lompoc in Santa Barbara County, northwest of Los Angeles. The mission was established in 1787–1788 to convert the Chumash Indians to Christianity. Although it was successful in attracting converts, the impact of European diseases soon began to take its toll, and deaths from smallpox and measles were common. Thus, the community of La Purísima was already strained when, after a short series of foreshocks starting on 8 December 1812, the main shock struck, followed by a second half an hour later, which brought down any buildings that had survived the first event. To make matters worse, the earthquake was followed by heavy rains, which reduced the now unroofed adobe bricks of the mission to mud.

Eventually, the mission was rebuilt, but not in the same spot, as the local Indians were too afraid of a repeat disaster. Thus, the ruins of the original mission were more or less as they had been after the earthquake when the artist Henry Chapman Ford (1828–1894) made this drawing in 1883 (Fig. 2.39).

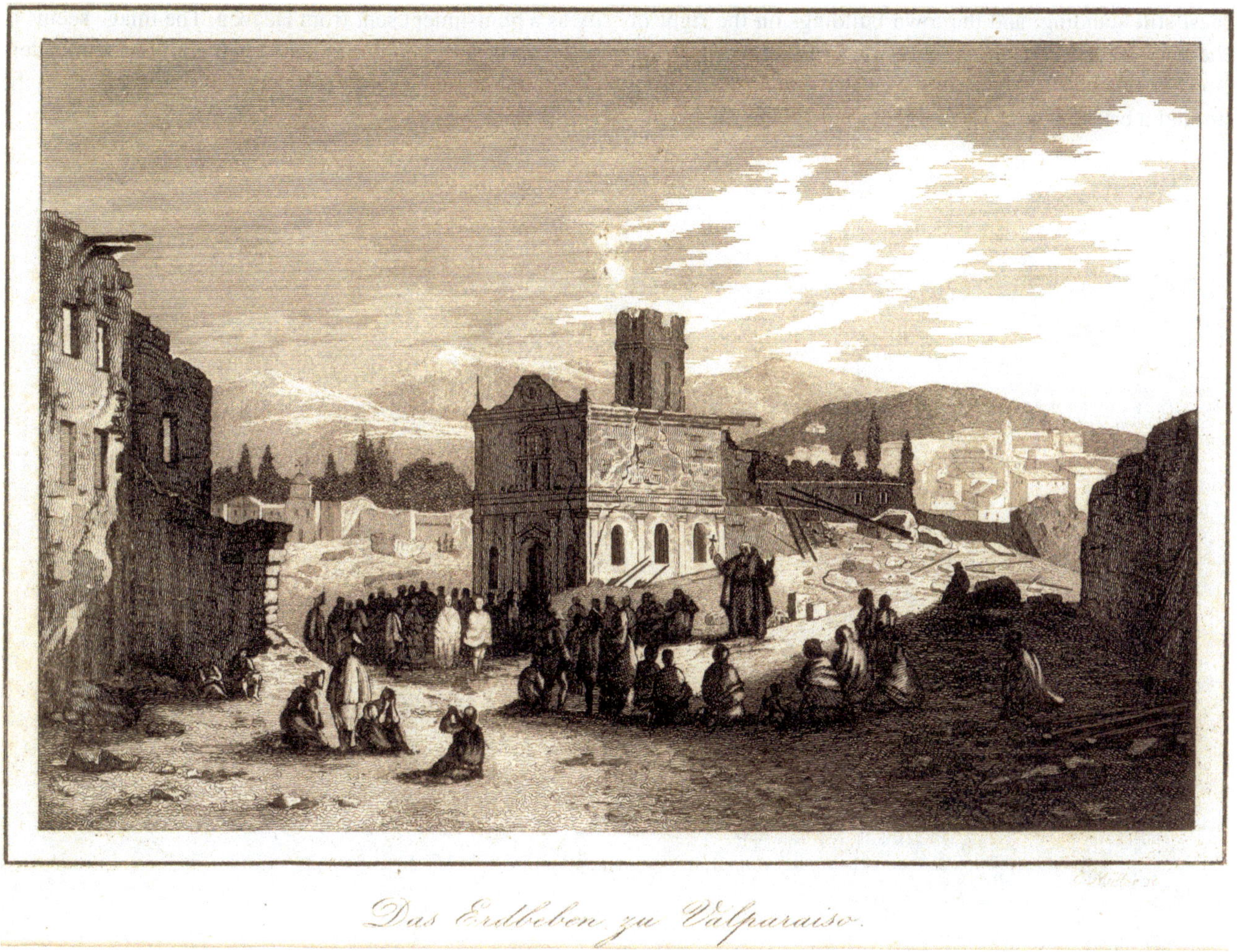

Fig. 2.40 The Valparaiso earthquake of 1822

A survey of earthquake disasters would not be complete without mention of Chile, one of the most seismically susceptible countries in the world. Indeed, it was Chile where the largest earthquake ever recorded occurred on 21 May 1960, with a magnitude of 9.5—quite possibly the limit to the largest earthquake possible. Some historical earthquakes may be equally large, but estimating the size of "megaquakes" from the historical record is difficult. The earthquake of 20 November 1822 (19 November, local time), was certainly strong, with a magnitude of around 8.5 and an epicenter in the La Ligua area. Damage was severe in Valparaiso, where all of the churches collapsed, being built of heavy masonry. Houses built of adobe also suffered, but wooden houses escaped such damage. A tsunami also occurred, and ships at anchor were damaged; on warships, cannons rolled about. A detailed account of the earthquake was written by the English traveler Maria Graham (1785–1842), who was living near Valparaiso at the time. "We hear that the large and populous town of Quillota, is a heap of ruins, and Valparaiso is little better," she wrote. In her journal, she kept a detailed register of the many aftershocks that followed the earthquake (Graham 1824).

Immediately after the shaking, looters got to work in the ruins of Valparaiso—too soon, as it turned out, as, when the rubble was eventually excavated, the bodies of looters were found with stolen items still clutched in their hands. Local priests wasted no time in using the earthquake as the basis for propaganda against the government of Bernardo O'Higgins, who at that time was carrying out anticlerical reforms. O'Higgins himself was nearly a casualty of the earthquake, as he spent the night of the earthquake in Valparaiso in the government palace, which was one of the monumental buildings that collapsed. O'Higgins had to be virtually dragged out. On the other hand, "God should not be imagined so irrational as to ruin His own temples," as the Jesuit Camilo Henriquez remarked.

The image here (Fig. 2.40) is an unidentified steel German engraving (signed by C. Müller) published in the 1840s, depicting the ruinous condition of Valparaiso, although the church tower in the center of the picture is at least still standing, and the town buildings on the right in the background carry no visible signs of seismic destruction. Locals can be seen sitting on the ground in a circle around a priest, who is undoubtedly portraying the calamity as a punishment sent from Heaven. The image seems to be taken from a popular almanac, which informed readers about interesting events that had happened all over the world.

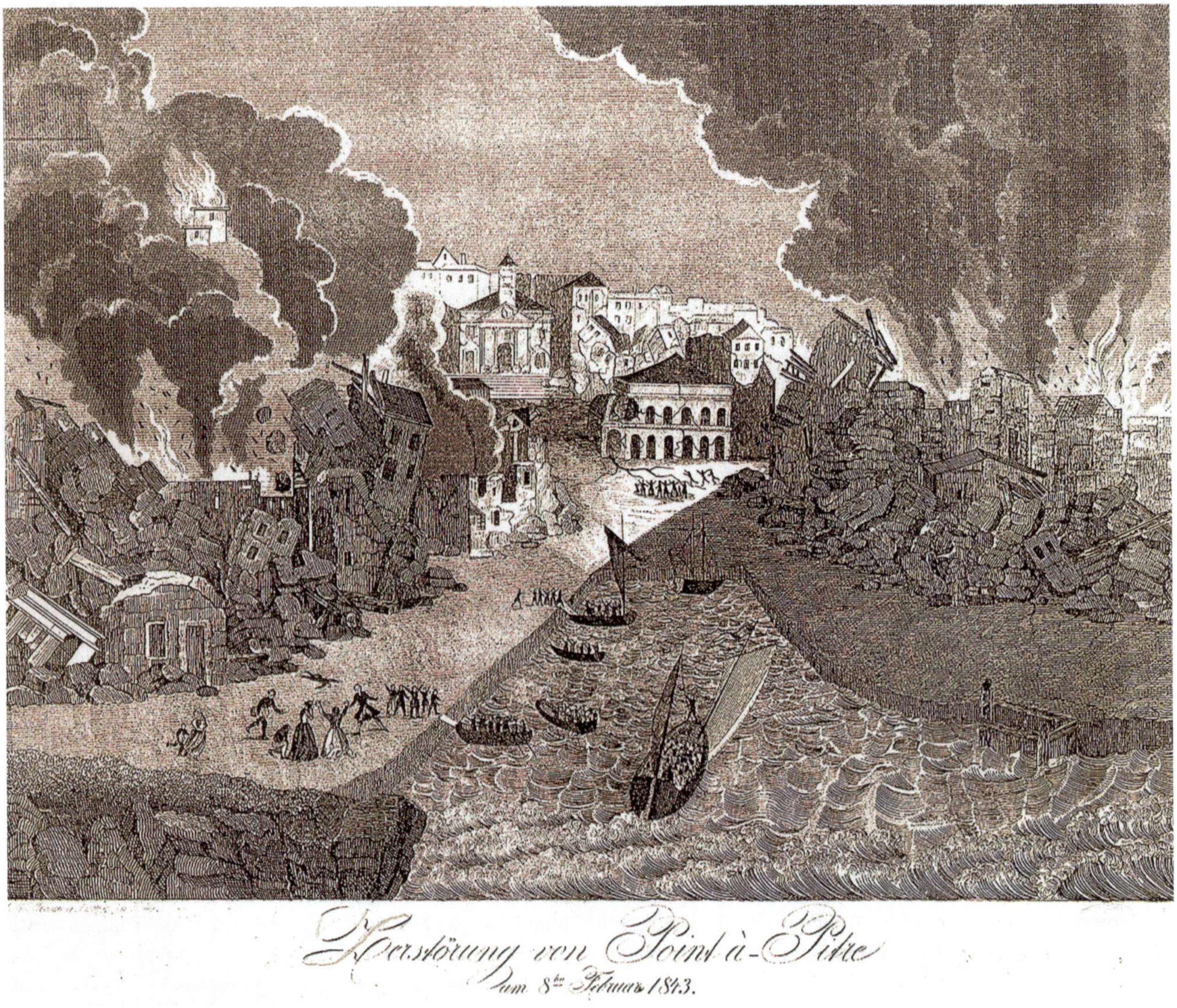

Fig. 2.41 The Guadeloupe earthquake of 1843

The Lesser Antilles Arc—the easternmost group of islands in the Caribbean—runs along the line of a subduction zone where the Atlantic Plate pushes beneath the Caribbean Plate. The largest recorded earthquake in this zone occurred on 8 February 1843; it was felt over a very large area (extending even to the east coast of the USA), but the maximum impact was on the French island of Guadeloupe. The city of Pointe-à-Pitre was completely destroyed, with the loss of 1500 lives. The wharfs subsided by a third of a meter, and there were extensive fires, as can be seen in the picture. One can see people attempting to flee in boats. The magnitude of the earthquake has been estimated recently as 8.5, making it a great megathrust event, although there was no significant tsunami. The copper engraving reproduced here has had a convoluted history: slightly different variants were published first in France (in *L'Illustration*) and then in Britain (in the *Illustrated London News*), and this version (Fig. 2.41) is German in origin.

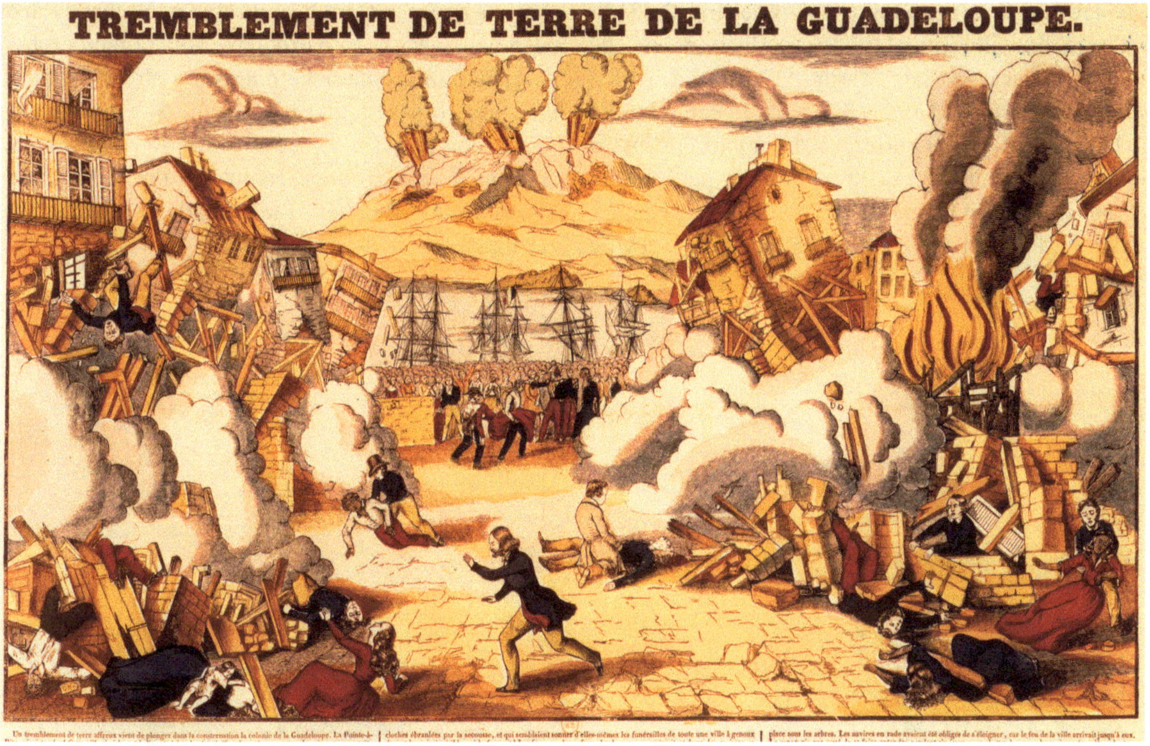

Fig. 2.42 A colored broadsheet of the Guadeloupe earthquake

This second image of the 1843 Guadeloupe earthquake is very different in style, being rather cruder in execution (Fig. 2.42). It was published as a broadsheet by Pellerin, in Epinal, with descriptive text under the gaudily colored lithograph. As in the previous image, one can see the town on fire along with tottering buildings and people fleeing toward the harbor. What appears to be a volcanic eruption in the distance is actually clouds of dust from landslides on the volcano's flanks, which might have been mistaken for eruptive columns by witnesses.

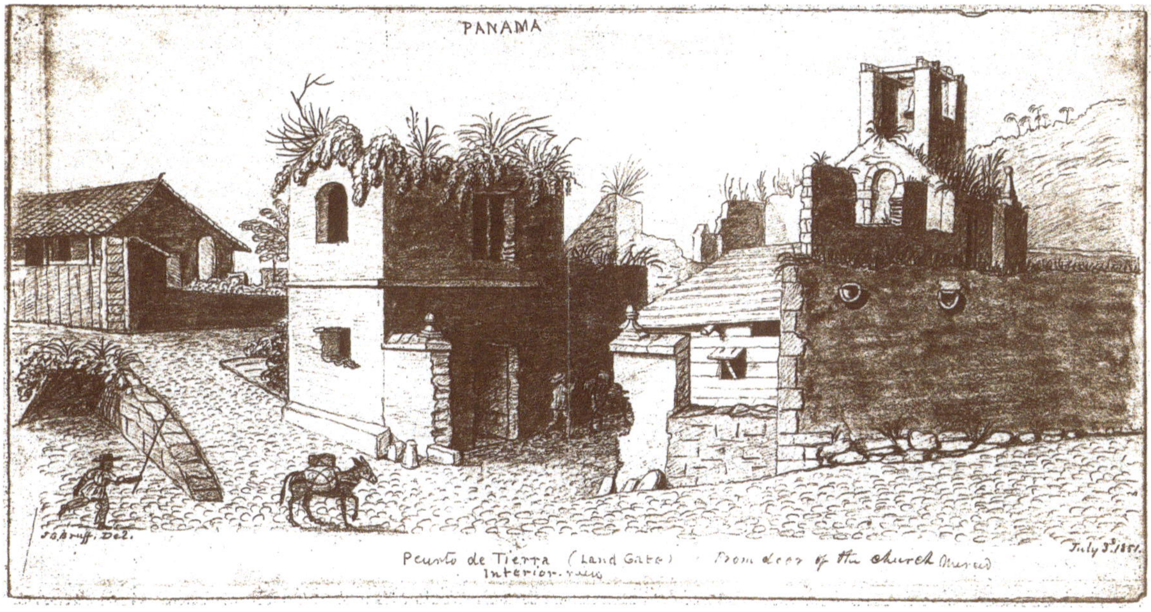

Fig. 2.43 An unknown earthquake in Panama, circa 1850 (description on page following)

The image here is rather a mysterious one (Fig. 2.43). It shows a rare drawing found in the collections of the Huntingdon Library in San Marino, California—titled *Earthquakes in Panama, About 1850*—and is dated by the artist as 1851–1852. The historical seismicity of Panama is very poorly documented, and, apart from a particularly strong event in 1621, it has not been possible to trace any earthquake in Panama before the 1870s. The drawing shows a house, a church, and some walls that evidently had been damaged by an earthquake some time before the artist visited the spot, since plants are already growing over the ruins. It seems that no attempt was made to repair or restore the properties, and the site was just abandoned. It remains a subject for research into the seismic history of Panama to discover the details of this enigmatic earthquake.

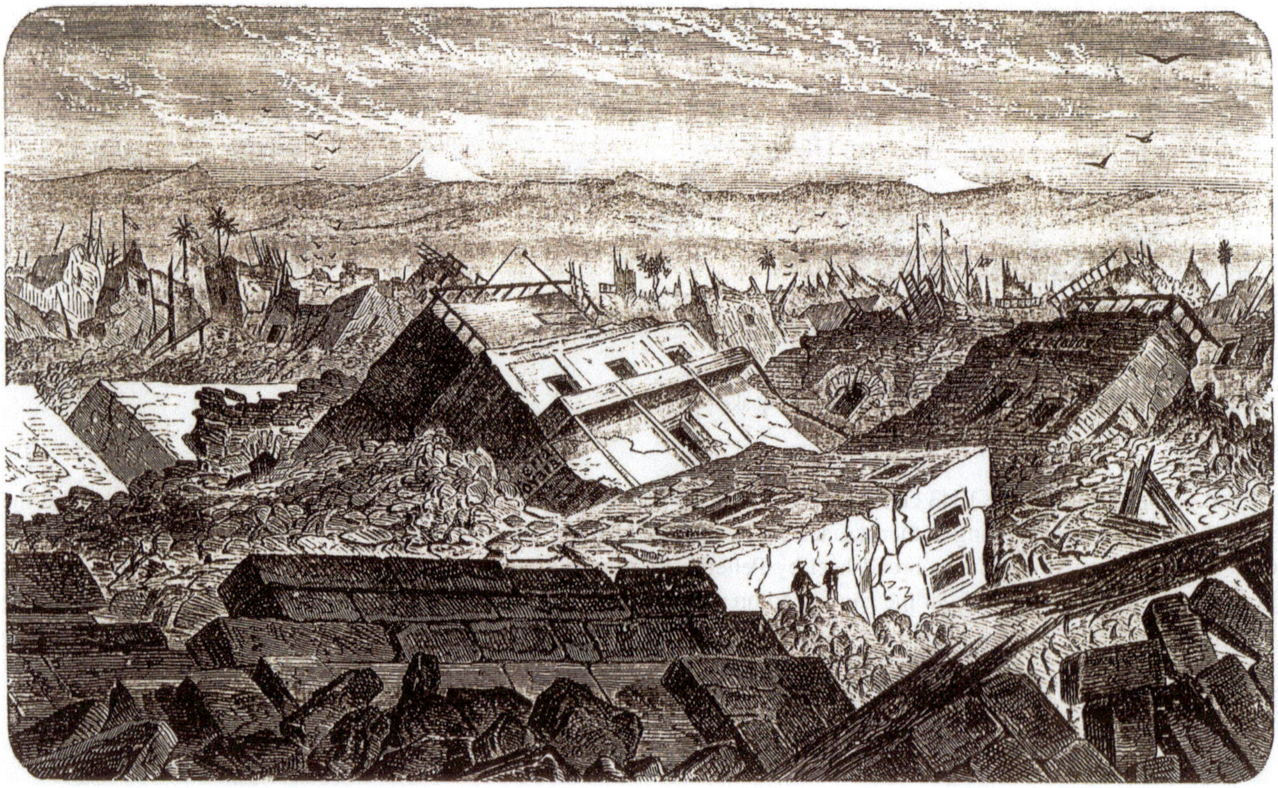

Fig. 2.44 The 1868 Arica earthquake

The greatest earthquakes in the world occur in specific parts of the globe where one tectonic plate is pushing below another. The magnitude of an earthquake is related to the area of the fault that breaks, so if a fault is vertical—and taking the Earth's crust to be 30 kilometers thick, more or less—600 kilometers of fault break gives $600 \times 30 = 18{,}000$ square kilometers of fault rupture. If the fault is gently dipping, as is the case in these crustal collision zones, one can have $600 \times 400 = 240{,}000$ square kilometers of fault rupture—and an earthquake that is literally an order of magnitude larger. This type of crustal collision, where an oceanic plate is forced under a continental one, is a process known as subduction. This is the situation in Sumatra and Japan, and also especially all along the west coast of South America. This very long subduction zone is capable of truly massive earthquakes, and one of the largest in the historical record was the earthquake of 13 August 1868, that occurred near the Peruvian–Chilean border. The magnitudes of very large historical earthquakes are hard to estimate accurately, but that of the 1868 event must have approached 9, given the extent of the area of damage along the coasts of southern Peru and northern Chile. This earthquake is particularly famous for the Pacific-wide tsunami it created, which will be discussed later under the heading of watery disasters (Chap. 3). Here, we concentrate on the effects of the earthquake itself, which are not always easy to disentangle from the compound calamity of an earthquake and a tsunami.

Estimates of the death toll vary greatly; at Arica itself, figures as disparate as 300 and 25,000 deaths can be found from different sources. The German wood engraving here purports to show Arica completed ruined (Fig. 2.44); however, the artist visited the town later after its reconstruction, and this picture showing all of the buildings lying at strange angles is not at all what seismic damage looks like; clearly, it is a work of imagination rather than reportage.

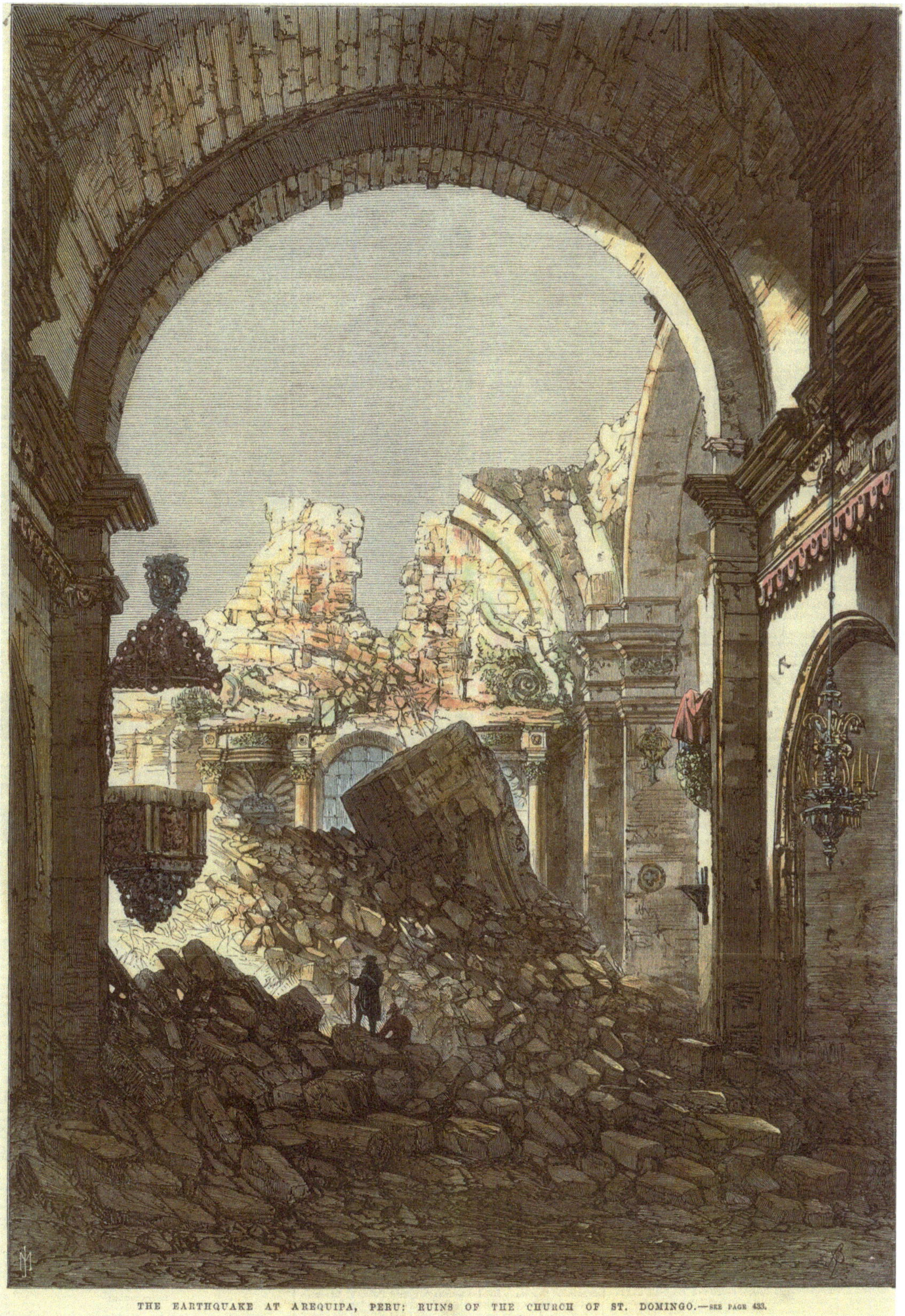

Fig. 2.45 Damage at Arequipa in 1868 (description on page following)

The Arica earthquake is sometimes referred to as the Arequipa earthquake; Arequipa is away from the coast, so the damage observed there was undoubtedly due to the earthquake and not the tsunami. A contemporary photograph in the US Library of Congress shows heavily damaged buildings all around the principal square of Arequipa after the earthquake, with the open area being used for makeshift tents to house the homeless inhabitants. The engraving here is taken from the *Illustrated London News* of 31 October 1868 (page 420), and shows the ruins of the church of St. Domingo (Fig. 2.45). It is actually based on a photograph (also kept in the US Library of Congress), with just a slight change in the position of the figures standing on the rubble.

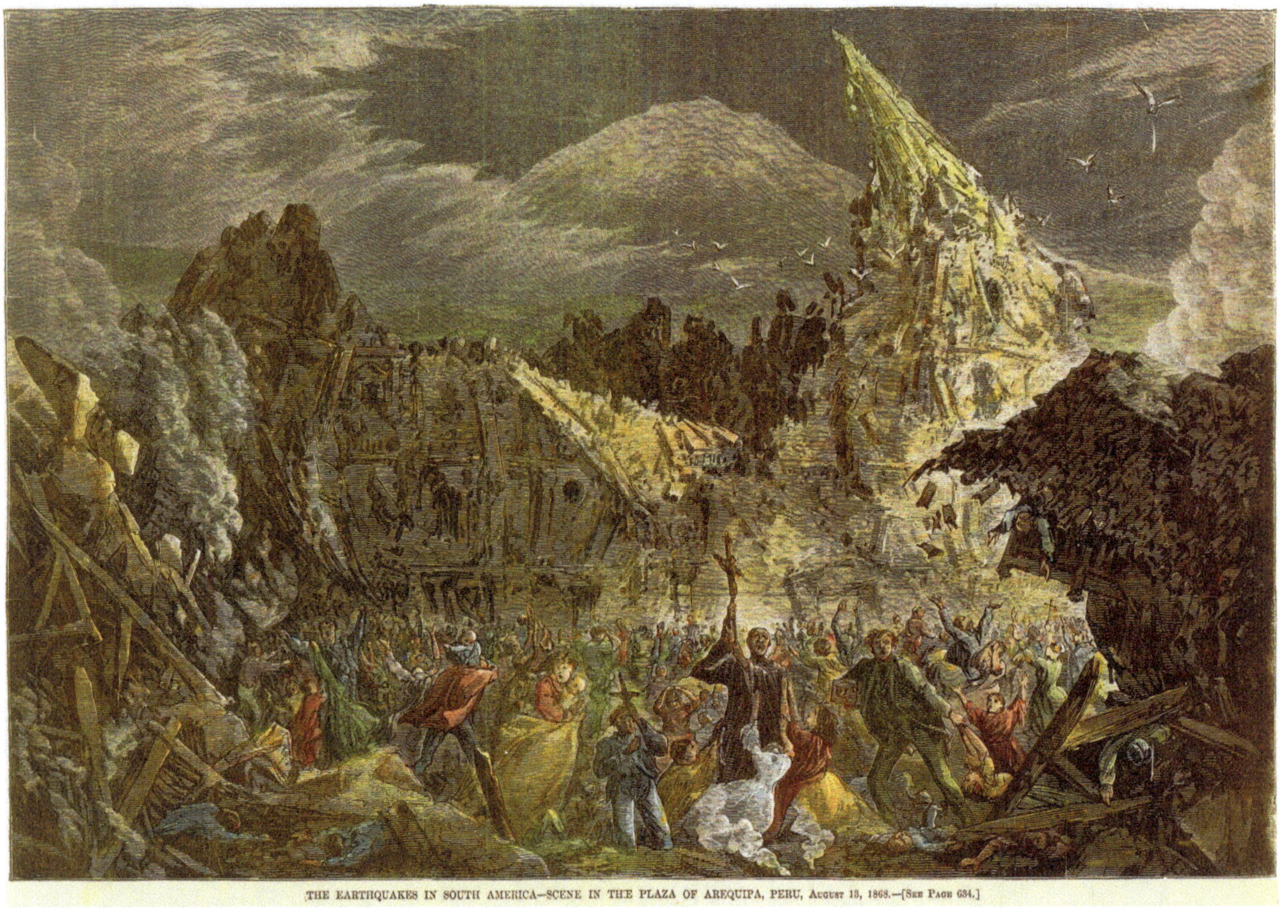

Fig. 2.46 A fantasy picture of Arequipa in 1868

This xylographic hand-colored illustration of the 1868 earthquake in Arequipa is taken from *Harper's Weekly* of 3 October 1868 (Fig. 2.46). The scene purports to capture the moment of the earthquake, showing the inhabitants in a state of panic, as the buildings collapse around them. The town is not at all recognizable; the collapsing church does not look very Spanish. The draftsman undoubtedly worked according to his imagination.

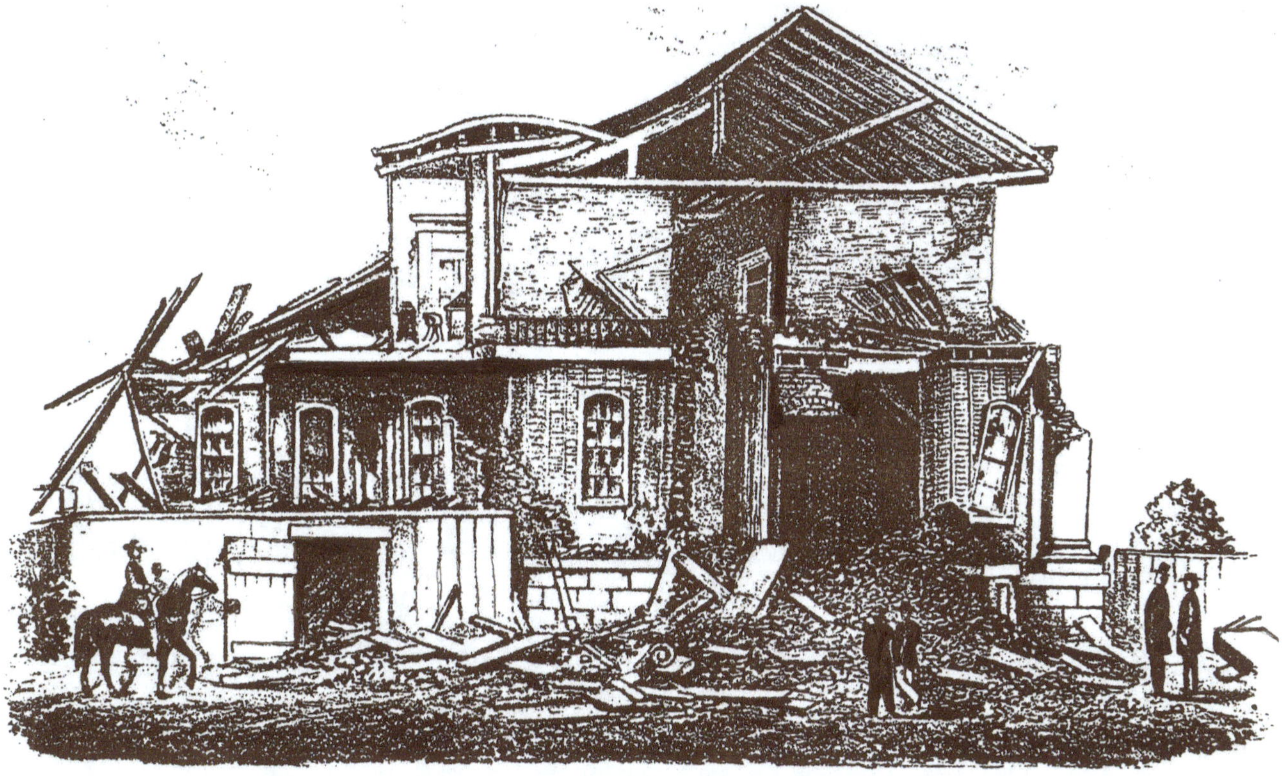

Fig. 2.47 The San Leandro Courthouse after the 1868 Hayward earthquake

Mention earthquakes in California, and people think of the San Andreas fault—but in fact there are quite a number of dangerous seismogenic faults in the state. One that causes particular worry is the Hayward fault, a structure running up the Hayward Valley northeast of San Francisco. The valley is heavily built up, and a large earthquake there would be likely to cause much damage. The last strong earthquake there was back in 1868—on 21 October 1868, to be precise. An earthquake of around magnitude 7 occurred, causing extensive damage in San Leandro, Fremont, and Hayward itself. Back in 1868, these were small communities of a few thousand souls, and the total population of the Bay Area was only 260,000—it is now bigger by a factor of 27. A total of 30 people were killed, and the amount of damage would come to over US$9 million in today's money. Seismic damage was not limited to the settlements located along the Hayward fault: all of the Bay Area was affected.

One of the most picturesquely damaged buildings was the courthouse at San Leandro, which features in a number of illustrations. The drawing here is attributed to a local artist, George Holbrook Baker (1827–1906); it was published in *Harper's Weekly* on 28 November 1868, and seems to have been issued also as a pictorial letter sheet (Fig. 2.47).

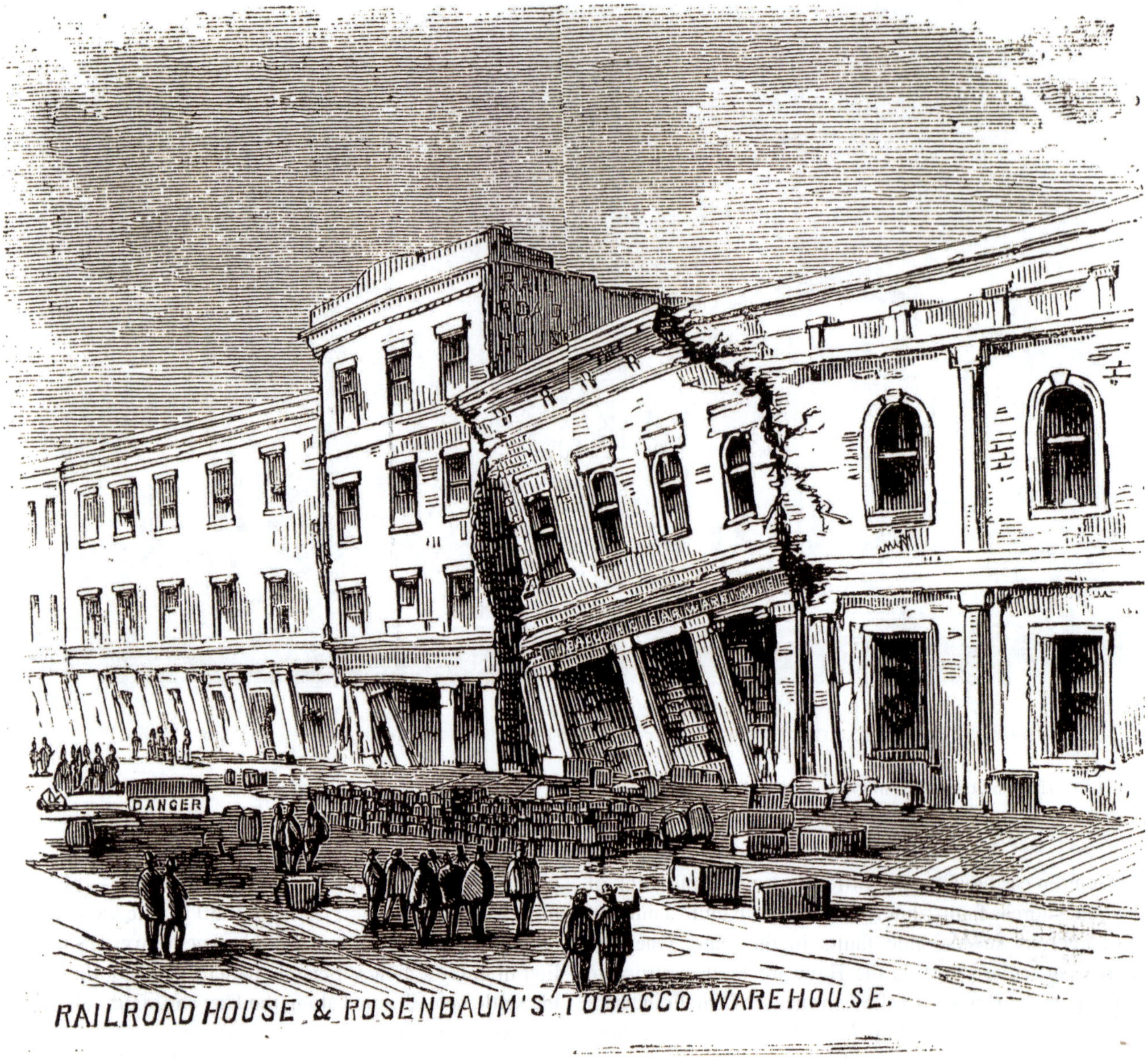

Fig. 2.48 Damage in San Francisco in 1868

Before 1906, if someone in San Francisco referred to the "great earthquake" it was the 1868 event they were referring to. As an example of damage in San Francisco itself, this engraving shows the Railroad House and Rosenbaum's Tobacco Warehouse in San Francisco, heavily damaged by the 1868 Hayward earthquake (Fig. 2.48). It is difficult to judge entirely from the picture, but it looks rather as if the damage here is an example of what is called "soft-storey collapse." In building a warehouse, the architect has maximized the open space on the ground floor for public access and storage, and so the rather heavy upper part of the building, containing the offices, rests on just a few supporting columns. The horizontal motion of the earthquake pushes the columns out of the vertical, and the result is that the building is unbalanced and leans to one side, as shown here. If it had not been constrained by its neighbors, it would have collapsed completely. *Harper's Weekly* of 28 November 1868, is again the source of the image.

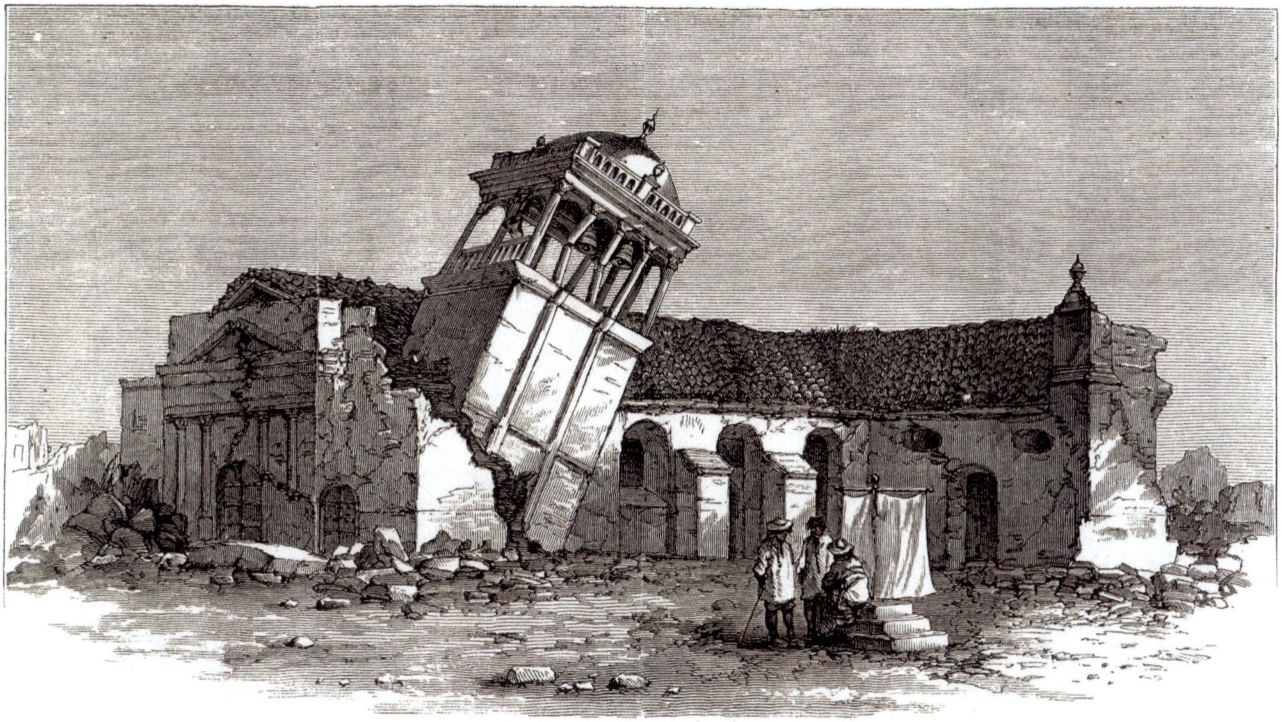

SAN SALVADOR—RUINS OF THE CATHEDRAL AFTER THE EARTHQUAKE.—[FROM A SKETCH BY CAPTAIN W. R. KENNEDY.]

Fig. 2.49 Damage to the cathedral in San Salvador in 1873

When Spanish colonists first started settling in Latin America, they brought with them the Spanish style of building with which they were familiar, regardless of the fact that it was highly unsuited to countries subject to strong earthquakes. This was remarked upon even as early as the eighteenth century by no less a person than the historian Edward Gibbon in his *Decline and Fall of the Roman Empire*. This was particularly the case with the churches built in Latin America in the Spanish style, with their ornate towers constructed in heavy, unreinforced masonry. This engraving (Fig. 2.49) shows a typical result after the San Salvador earthquake of 19 March 1873 (magnitude 7.1). Among other damage, the tower of the cathedral has sheared off diagonally above the base. This striking case of damage was the subject of several pictures; this one was published in *Harper's Weekly* on 10 March 1873.

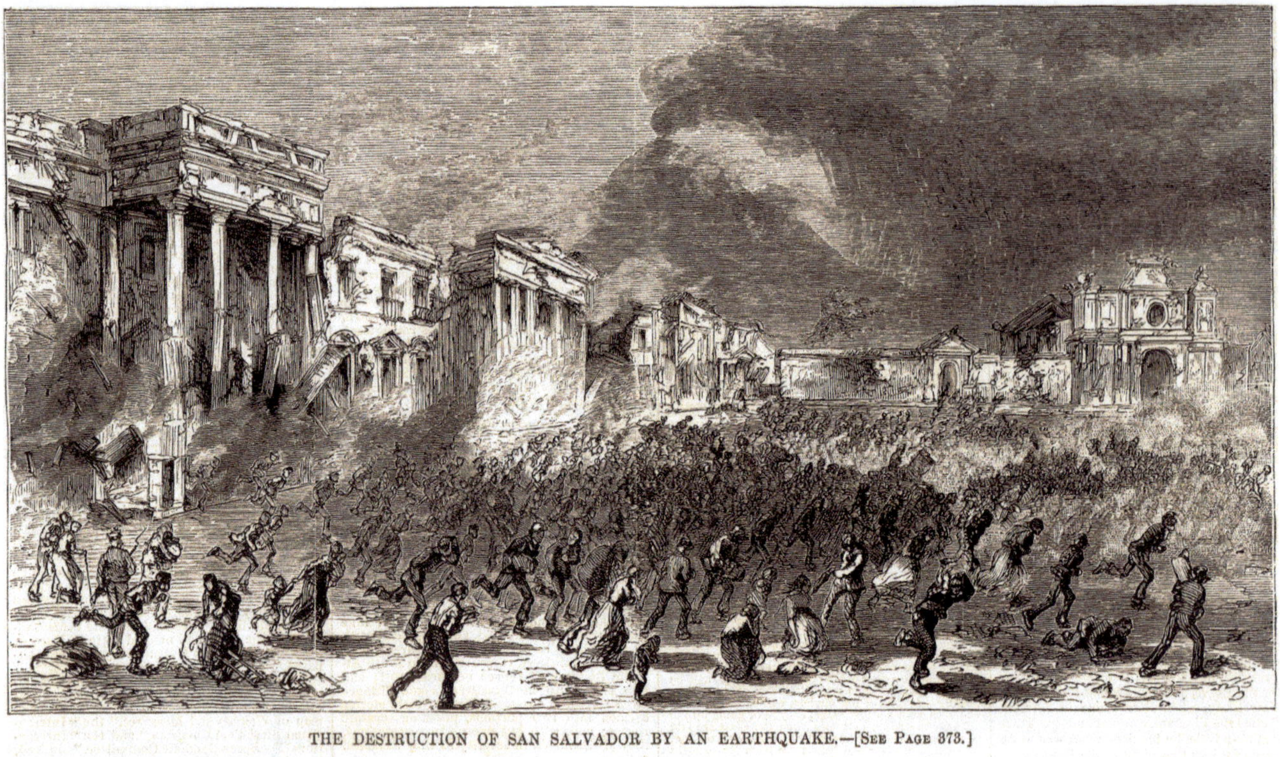

Fig. 2.50 The 1873 San Salvador earthquake

This dramatic image from *Harper's Weekly* shows crowds fleeing in panic during the 1873 San Salvador earthquake (Fig. 2.50). The artist has used his imagination to add in an erupting volcano in the background, to heighten the drama. Needless to say, no such eruption occurred to coincide with the earthquake. Although monumental buildings such as the cathedral were badly damaged, lightweight local houses generally survived the earthquake quite well.

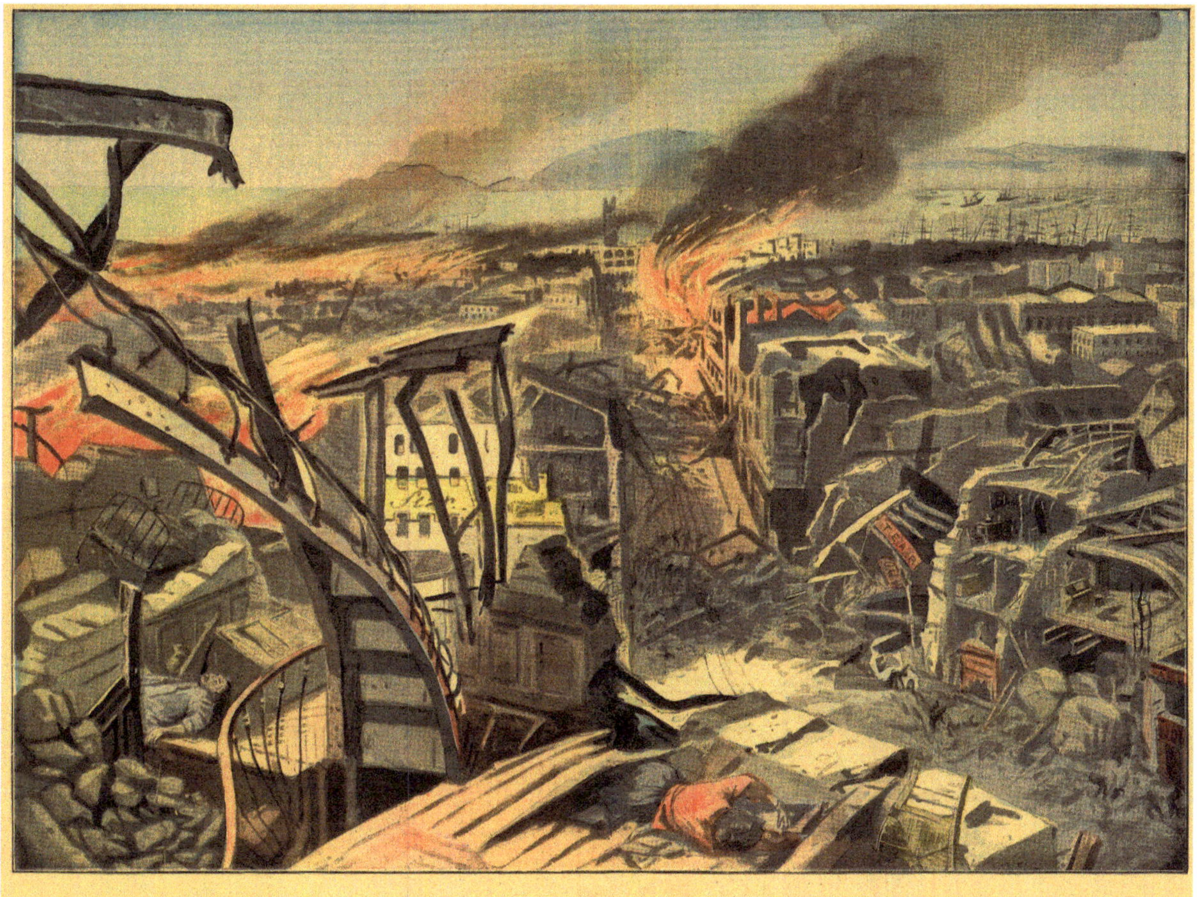

Fig. 2.51 A French image of San Francisco in 1906

The great San Francisco earthquake on 18 April 1906, was remarkable from several points of view. This event is the most devastating in the history of California to date. The number of casualties of the earthquake itself and of the resulting fire has recently been assessed as around 3000; previous estimates of about 700 are considered to be gross underestimates, partly because many of the bodies were carbonized by the subsequent fire, which burned for days. To many people, the 1906 earthquake is *the* earthquake—the one that everyone has heard of and, perhaps, the one that people most fear a recurrence of. It was also this earthquake that established the San Andreas fault as the one geological structure everyone can name; it broke for a lengthy rupture of 477 kilometers, mostly to the north of San Francisco. The magnitude has been assessed as 7.9, making it just short of the seismologist's definition of a "great" earthquake as being one of magnitude 8 or larger.

The earthquake mobilized all leading American geophysicists, especially seismologists, who used the seismic data obtained for a detailed examination of the event, with the aim of explaining the occurrence of the event. The results of this effort appeared in a three-volume report, which was published two years after the earthquake and was edited by Professor Andrew C. Lawson, chairman of the Geology Department at the University of California, Berkeley. In section II of this voluminous work, the landmark "theory of elastic rebound" was introduced by Professor Harry F. Reid of Johns Hopkins University, who served on Lawson's commission. This made it abundantly clear that earthquakes were caused by faults and not vice versa. His theoretical conclusions were approved as quite correct and commonly accepted, and Reid's work is a foundation stone of modern seismology.

The rather fanciful image here depicts the burning town seen along Market Street (in the center), over the Bay Area, and toward Berkeley (Fig. 2.51). It is reproduced from the French periodical *Le Petit Parisien* of 6 May 1906 (page 144).

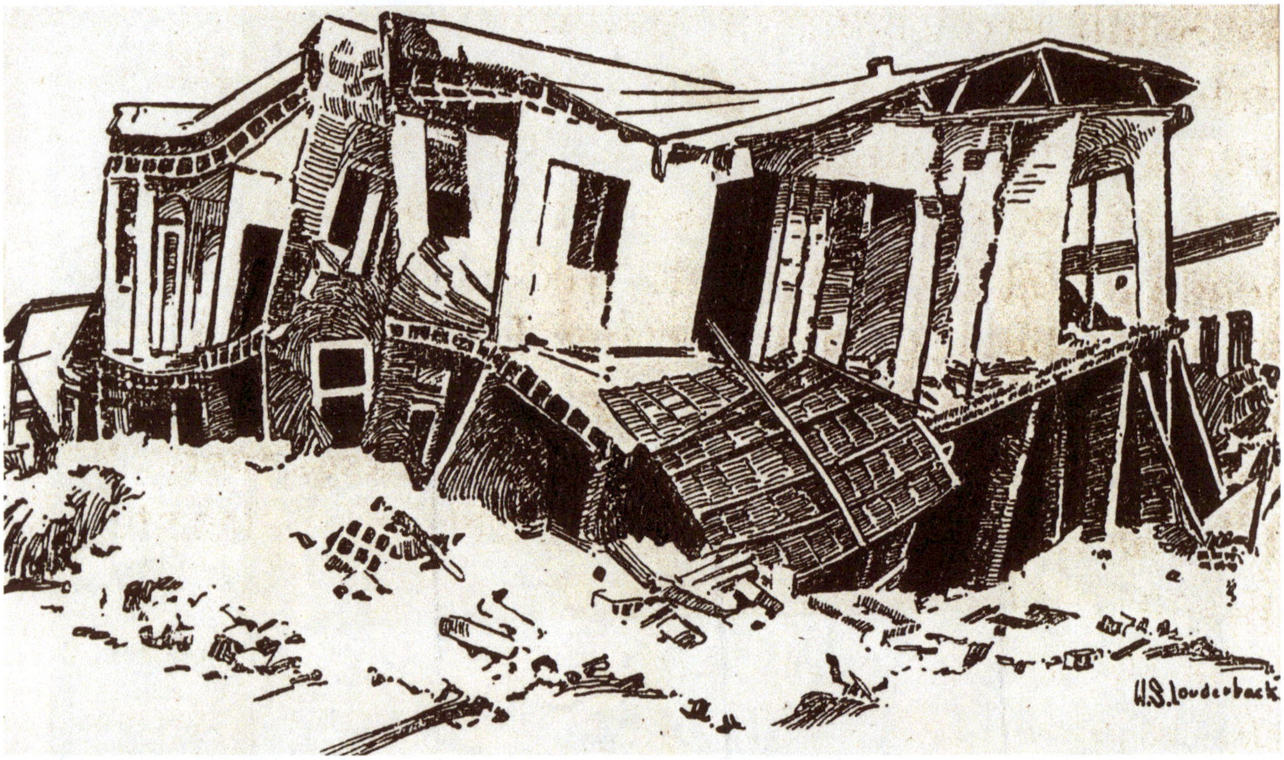

Fig. 2.52 Damage at San Mateo in 1906

The damage was not confined to San Francisco. This drawing shows damage to a house in San Mateo, 25 kilometers to the south (Fig. 2.52). It comes from the work of the American geologist William Herbert Hobbs (1864–1953), who published his *Earthquakes: An Introduction to Seismic Geology* a year after the earthquake. The subject was topical!

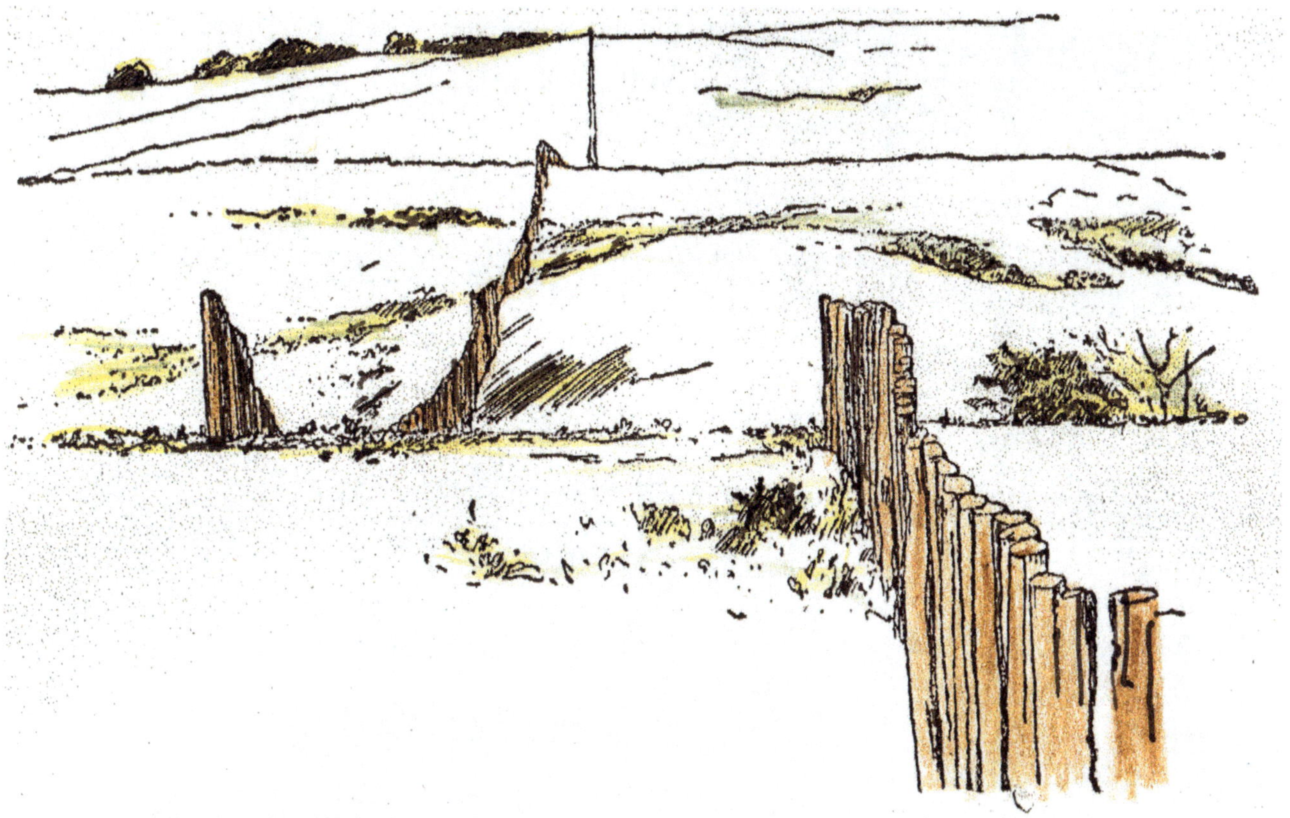

Fig. 2.53 A fence break in the 1906 earthquake—the clue to earthquake processes

This drawing (Fig. 2.53) may look very unspectacular after all of the images of ruined buildings and devastated cities, but, in fact, to the seismologist, it is one of the most interesting. It shows what looks like two incomplete fences, five meters apart. In fact, it was originally one continuous straight fence when it was built (for who would build a fence differently?). But the farmer was unaware that he was building it right across the San Andreas fault. Under the strain of the relative movement between the Pacific Plate on one side and the North American Plate on the other, the fence would have progressively warped, and then when the strain on the fault exceeded the force of friction holding it together, the ground moved with a jerk in the 1906 great San Francisco earthquake, and the two parts of the fence separated. It was observations such as this that underlay Harry Reid's idea of elastic rebound and our understanding ever since of how earthquakes occur.

The drawing is taken from *An Introduction to Geology* by William B. Scott, published in 1924.

2.4 Asia and the Pacific

In comparison with Europe and the Americas, relatively few images are available for the rest of the world, which chiefly means Asia (Turkey and the Levant, while technically on the continent of Asia, have been considered here as Mediterranean). There is seismicity in the north of Africa, especially in Algeria and Morocco, but few images of Algerian earthquakes exist. South of the Sahara, the seismicity of the continent before 1900 is virtually unknown, except for a few relatively minor earthquakes around Cape Town and some localized seismicity in Ghana. Australia has largely escaped picture-worthy earthquake damage; New Zealand, however, is much more earthquake prone.

In Asia, earthquakes follow a belt of activity from Turkey through Persia/Iran, through Afghanistan and northern India, up into China, and into Japan and the Philippines. But, as mentioned before, only in Japan do we find indigenous paintings or drawings of earthquake effects. We can start our selection in Japan—but not with a Japanese image.

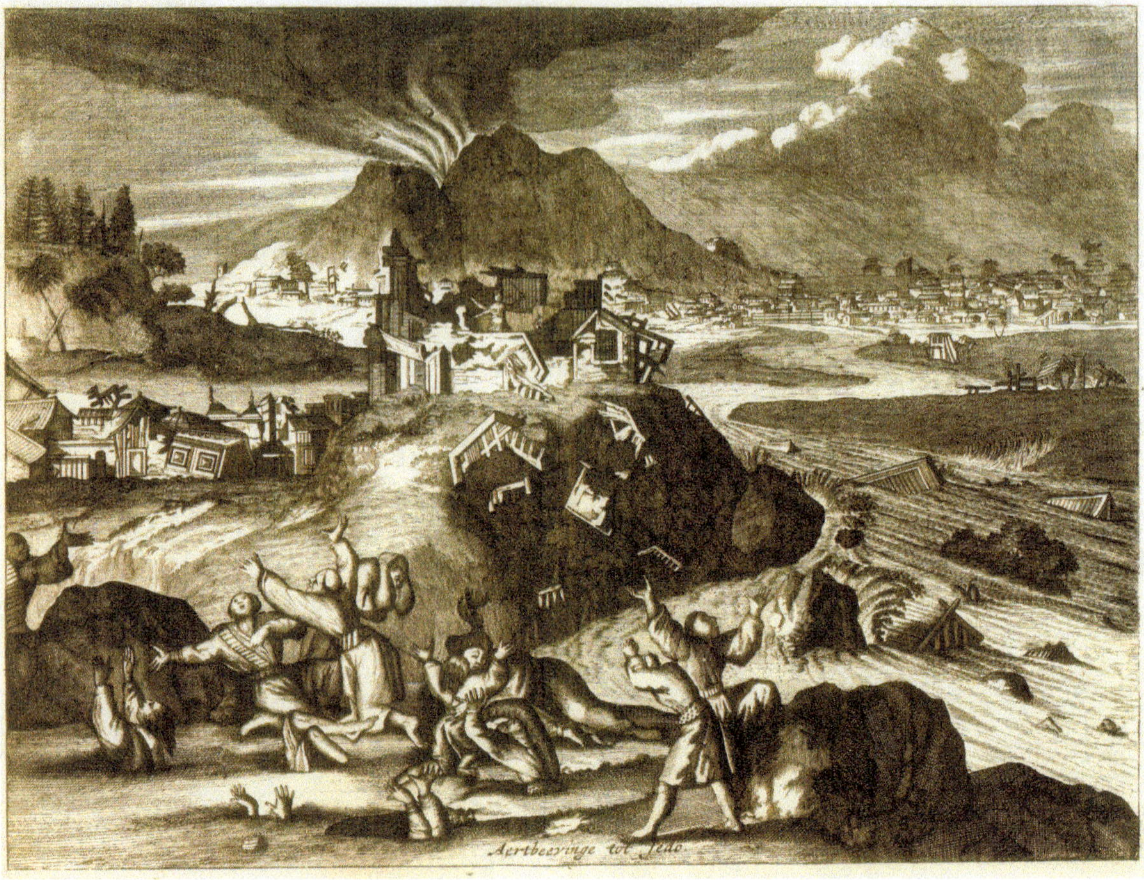

Fig. 2.54 The Saitama Prefecture earthquake in 1649

It is correct to state that there were two basic ways of representing earthquakes occurring in the Japanese archipelago. The first group of these seismic images can be denoted as the European style of execution (such as copper engravings, etchings, and so on), the second being images created by Japanese artists in a Japanese style. European-style images were mostly created by Portuguese, English, Spanish, Dutch, and German visitors to the country, coming as merchants, colonists, and especially Christian missionaries. Japan was, however, more or less a closed country until the Meiji restoration in the second half of the nineteenth century. Missionaries were particularly unwelcome. Visitors often made sketches during their brief stay in the country, which could be used for making engravings after their return home to Europe.

One of these examples shows an earthquake sometime around 1650, possibly near Edo (now Tokyo). This copper engraving (Fig. 2.54) is reproduced from a report by the Dutch minister Arnoldus Montanus (circa 1625–1683), whose name is a Latinization of his birth name of Arnold van den Berg. In the mid-seventeenth century he acted as an agent for the Dutch East India Company; his 1669 report to them was published under the title *Gedenkwaerdige Gesantschappen der Oost-Indische Maetschappy in't Vereenigde Nederland, aen de Kaisaren van Japan*, republished the following year in English as *Atlas Japannensis: Being Remarkable Addresses by Way of Embassy from the East India Company of the United Provinces, to the Emperor of Japan*.

The scene does not seem to be portrayed authentically; one might wonder how ready his Japanese hosts would have been to let him approach a disaster scene. The seismic effects are shown chiefly in the foreground of the illustration; people have left their collapsing houses, seeking safety in the open air. However, they are not safe even there, see the figures sinking into deep holes and newly opened cracks (something that generally does not often happen in reality). Also, the light Japanese houses do not seem to be very resistant against the seismic loading, given the appearance of the broken constructions in the central half of the composition. To add to the drama, a volcano erupts in the background.

Which earthquake was it? Montanus drew on an embassy to Japan in 1649–1650, so these are the likely dates. Almost certainly, it shows the very damaging earthquake of 30 July 1649, in Saitama Prefecture and a little to the northwest of Edo/Tokyo. This had a magnitude of about 7 and was followed by strong aftershocks into the next year. The erupting volcano seems to be Nikko-Shiranesan, which is known to have erupted in February 1649 and may have still been in eruption in July—or else the artist has merged the two disasters.

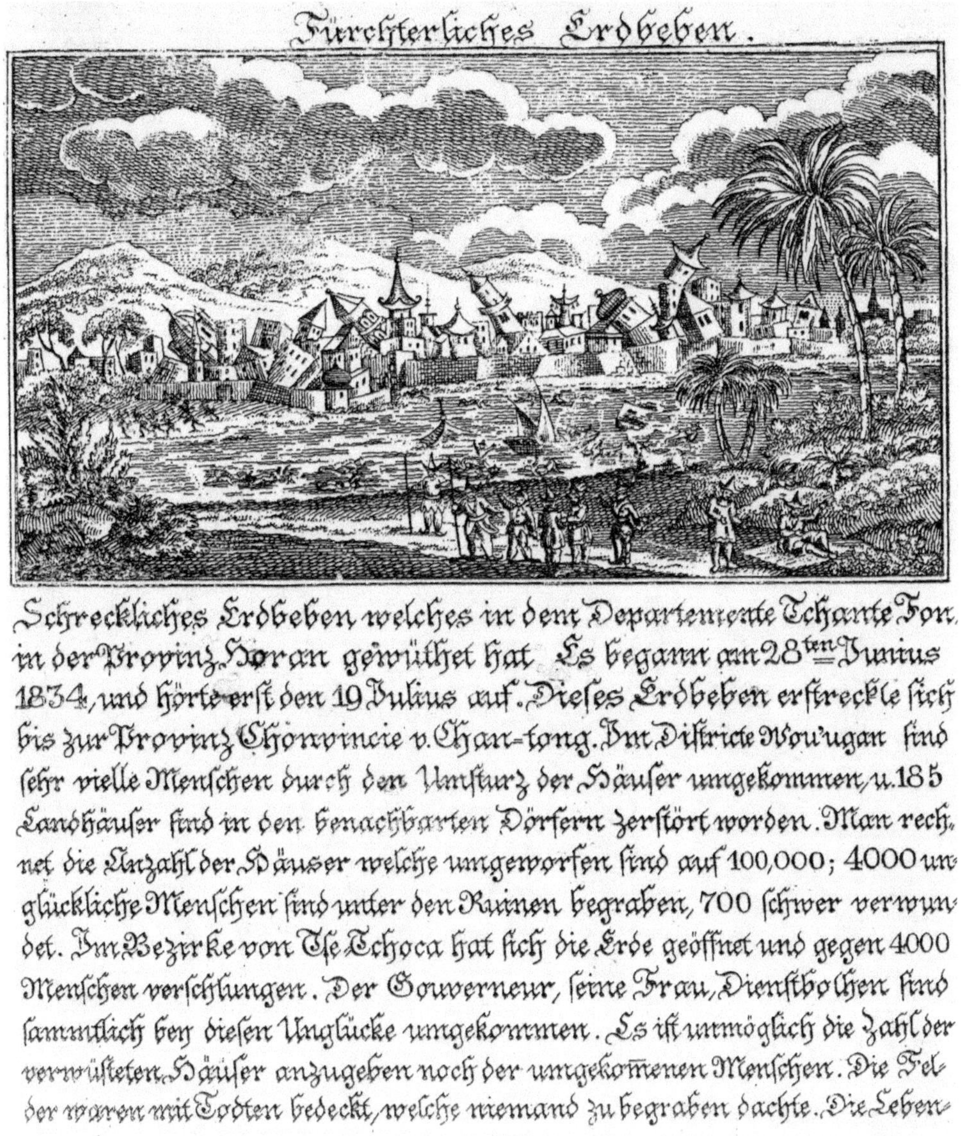

Fürchterliches Erdbeben.

Schreckliches Erdbeben welches in dem Departemente Tchante Fon. in der Provinz Horan gewüthet hat Es begann am 28ten Junius 1834, und hörte erst den 19 Julius auf. Dieses Erdbeben erstreckle sich bis zur Provinz Chonvincie v. Chan-tong. Im Districte Wou'ugan sind sehr vielle Menschen durch den Umsturz der Häuser umgekommen, u. 185 Landhäuser sind in den benachbarten Dörfern zerstört worden. Man rechnet die Anzahl der Häuser welche umgeworfen sind auf 100,000; 4000 unglückliche Menschen sind unter den Ruinen begraben, 700 schwer verwundet. Im Bezirke von Tse Tchoca hat sich die Erde geöffnet und gegen 4000 Menschen verschlungen. Der Gouverneur, seine Frau, Diensbothen sind sammtlich bey diesen Unglücke umgekommen. Es ist unmöglich die Zahl der verwüsteten Häuser anzugeben noch der umgekomenen Menschen. Die Felder waren mit Todten bedeckt, welche niemand zu begraben dachte. Die Leben-

Fig. 2.55 The supposed Shantung earthquake in 1834

The picture here represents a leaflet illustration attached to text carrying information on Chinese earthquakes on 28 June 1834, continuing till 10 July 1834, in "Tchante-Tong" (Shandong), China (Fig. 2.55). According to the text in the leaflet, during the earthquake, 100,000(!) houses were demolished, 4000 people died, and 700 others were seriously injured. There is a problem, though. No earthquake meeting this description—even slightly—ever occurred, according to Chinese sources. It raises an issue with early journalism before the invention of the telegraph: sometimes there was no news. A newspaper editor had to fill his space somehow, even if it meant inventing a story. And if he did invent a story, it had to concern some remote place about which no reader would have a source of recent information. China fits the bill. So one can perhaps imagine a German printer; his press is lying idle, he has a family to feed, but there is nothing to print. So he says to his assistant, "Let's have an earthquake in China!" The readers who buy the leaflet go away happy to read that exciting and terrible things are happening in a far-off land, and the printer is happy that he has made some money. Everyone benefits except for the poor seismologist in the twenty-first century who tries to make sense of this amazing earthquake that somehow has been missed out by the Chinese earthquake catalogs.

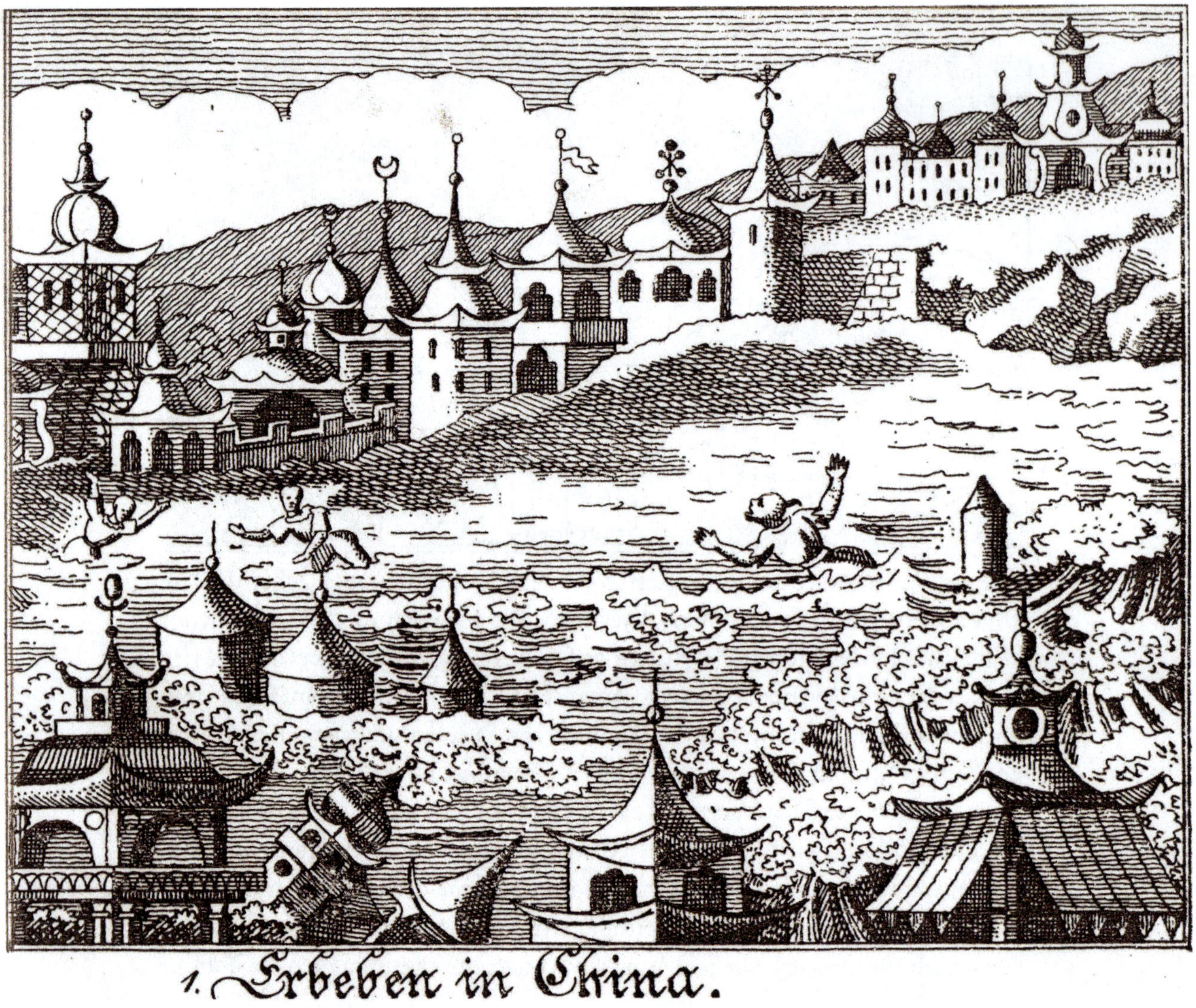

Fig. 2.56 The supposed Beijing earthquake in 1836

This is another woodcut image from a German-language broadsheet issued in 1837 (Fig. 2.56). It is, of course, completely fanciful. The artist had some vague idea about oriental architecture and then gave full rein to his imagination. The center ground of the image is given over to a depiction of flooding in which four out-of-scale figures are floundering about, and there is actually very little sign of any earthquake damage beyond two small leaning towers at the very front of the scene.

According to the text, it shows an earthquake in Beijing in 1836; during the earthquake approximately 100,000 (again!) houses were ruined, and the ruins of the houses covered the black waters of a newly created lake. In the following text of the leaflet, the sad stories of the refugees, individuals, and families are recounted.

Like the previous leaflet, this seems to refer an earthquake unknown to Chinese sources. It is just faintly conceivable that underlying this story is the Cixian earthquake in Hebei province, about 400 kilometers south of Beijing, which occurred on 12 June 1830, with a magnitude of 7.5. It was generally felt in Beijing, though without damage. Of course, the European readers of the broadsheet had no idea what a Chinese city looked like; nor would they know much (or care much) about contemporary Chinese affairs, beyond a certain schadenfreude that terrible things happened far away from home. But, most probably, this is another case of a printer with time on his hands and nothing to print.

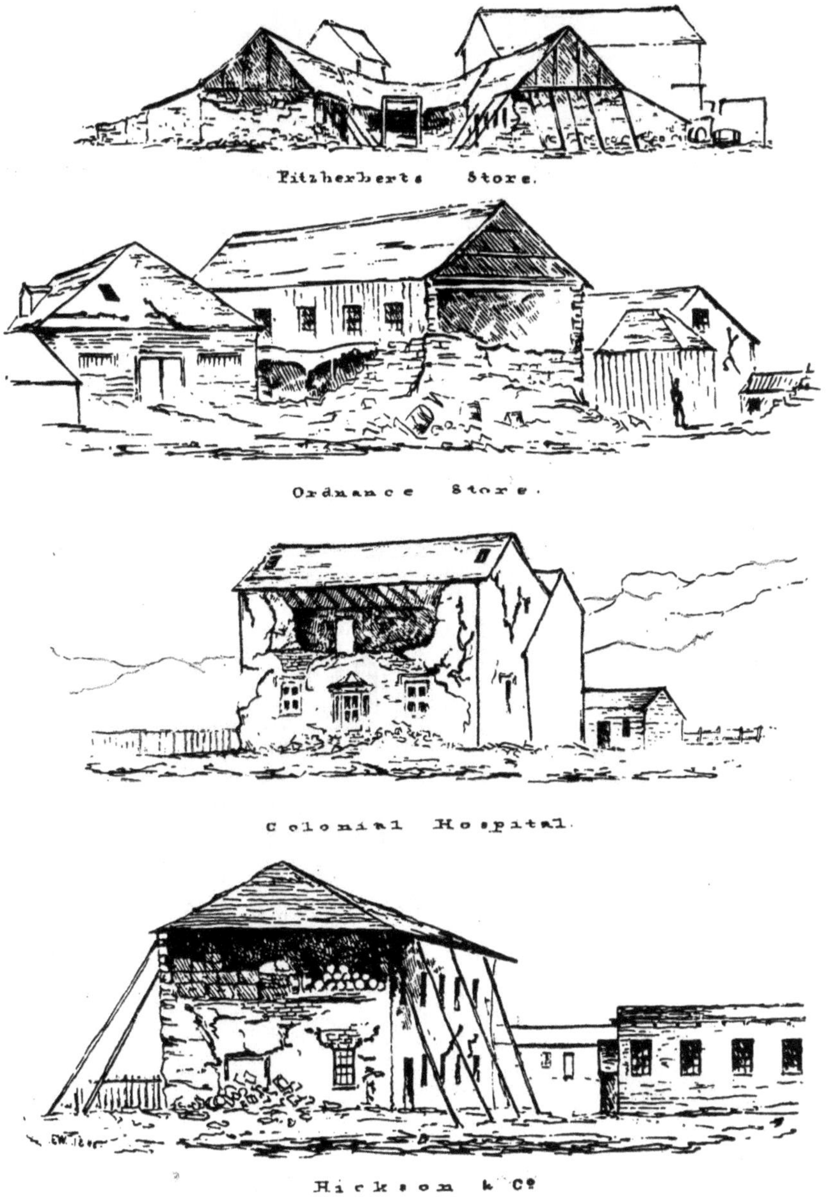

Fig. 2.57 The Marlborough earthquake in 1848

As with the Americas, the history of earthquakes in New Zealand is dependent on the advancement of European settlement. From around the 1790s, Europeans were making regular landings in New Zealand in connection with whaling voyages, or for trading with the Maori people. From about 1840, the first serious attempts were made to colonize the country, and it was very shortly after this, on 16 October 1848, that British colonists discovered that their new home was subject to earthquakes of a type never experienced in their homeland. The epicenter of the 1848 earthquake was in the Marlborough region, at the northern end of the South Island, and the magnitude is estimated to have been 7.5. Damage to recently built houses was extensive, especially in the Wairau, Waihopai, and Awatere Valleys, and in Wellington. The effects on the landscape were considerable, with numerous landslides over a wide area. One can compare it to the Kaikoura earthquake of 14 November 2016 (magnitude 7.8), in a similar part of the country, which again caused many landslides and, in one case, rather spectacularly threw a section of railway track onto the adjacent road.

The sketches reproduced here of damaged brick buildings near Wellington were made by Robert Park (1812–1870) and appeared in a booklet describing the earthquake, which was published in Sydney the same year (Fig. 2.57). Top to bottom, the buildings are Fitzherbert's Store, the Ordnance Store, the Colonial Hospital, and Hickson & Co.

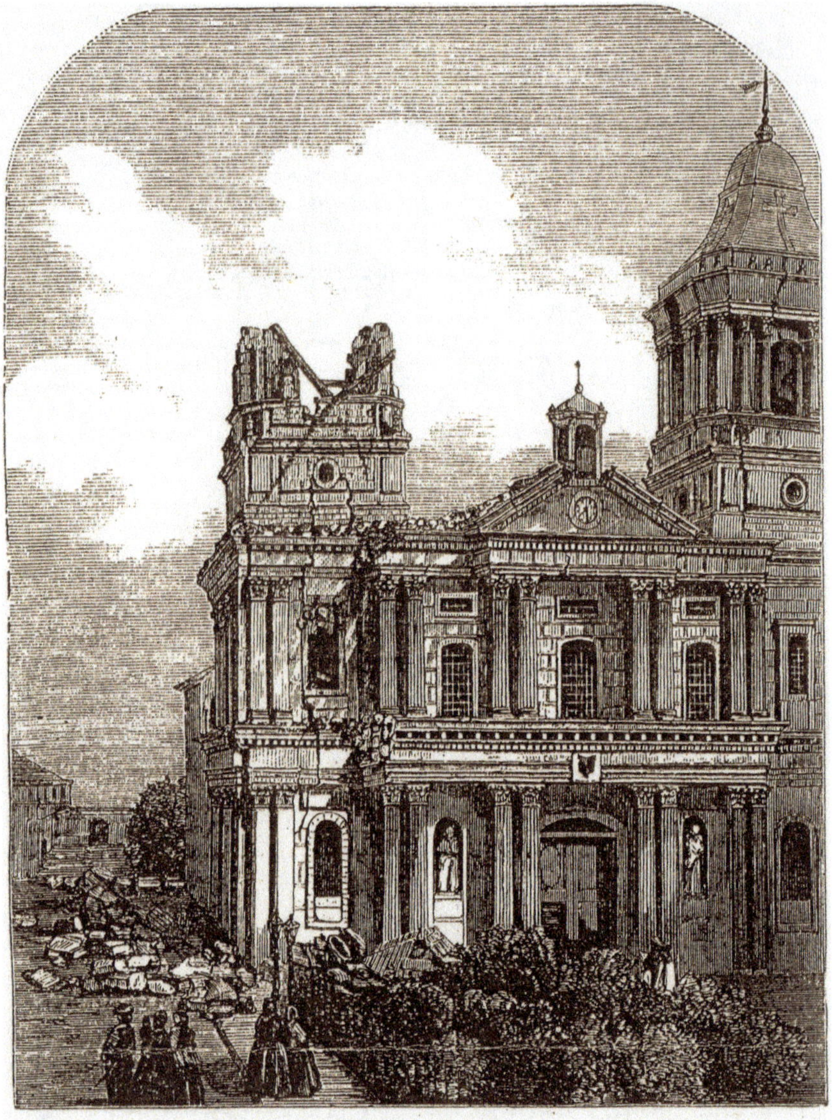

Fig. 2.58 An earthquake in Manila in 1863

The islands of the Philippines, as revealed by twentieth-century research, are not luckily situated from a geological point of view, being in an area of complex tectonic interaction between different plates. The islands suffer repeatedly from strong earthquakes and volcanic eruptions. The earthquake of 3 June 1863—with a magnitude estimated at 6.5—was not one of the largest, but it was a direct hit on the capital city, Manila, and caused much destruction, especially to monumental edifices. The cathedral, the governor's palace, the city hall, the barracks, and all of the public houses were ruined. Four hundred people were killed, and more than 2000 were injured. The material loss was assessed at US$8 million at 1863 values. Some 46 public and 570 private buildings were found to have been totally destroyed; all of the houses left standing in Manila were more or less damaged as well.

The destruction of the three aisles of the cathedral came at an hour when a religious service was under way, with the earthquake occurring at a time when preparations for Corpus Christi were being made. All of the persons seated on the west side of the choir were crushed under falling timbers and debris from the cupola, ceilings, and roofs, while those sitting on the east side managed to escape. The picture here (Fig. 2.58), though, is of the church of Santo Domingo; one tower fell, the other was so damaged it had to be taken down, and the whole structure can be seen to be leaning ominously. This wood engraving was published in the *Illustrated London News* and seems to be based on a watercolor by the English artist Charles W. Andrews (died 1869), who was active in the Philippines at the time.

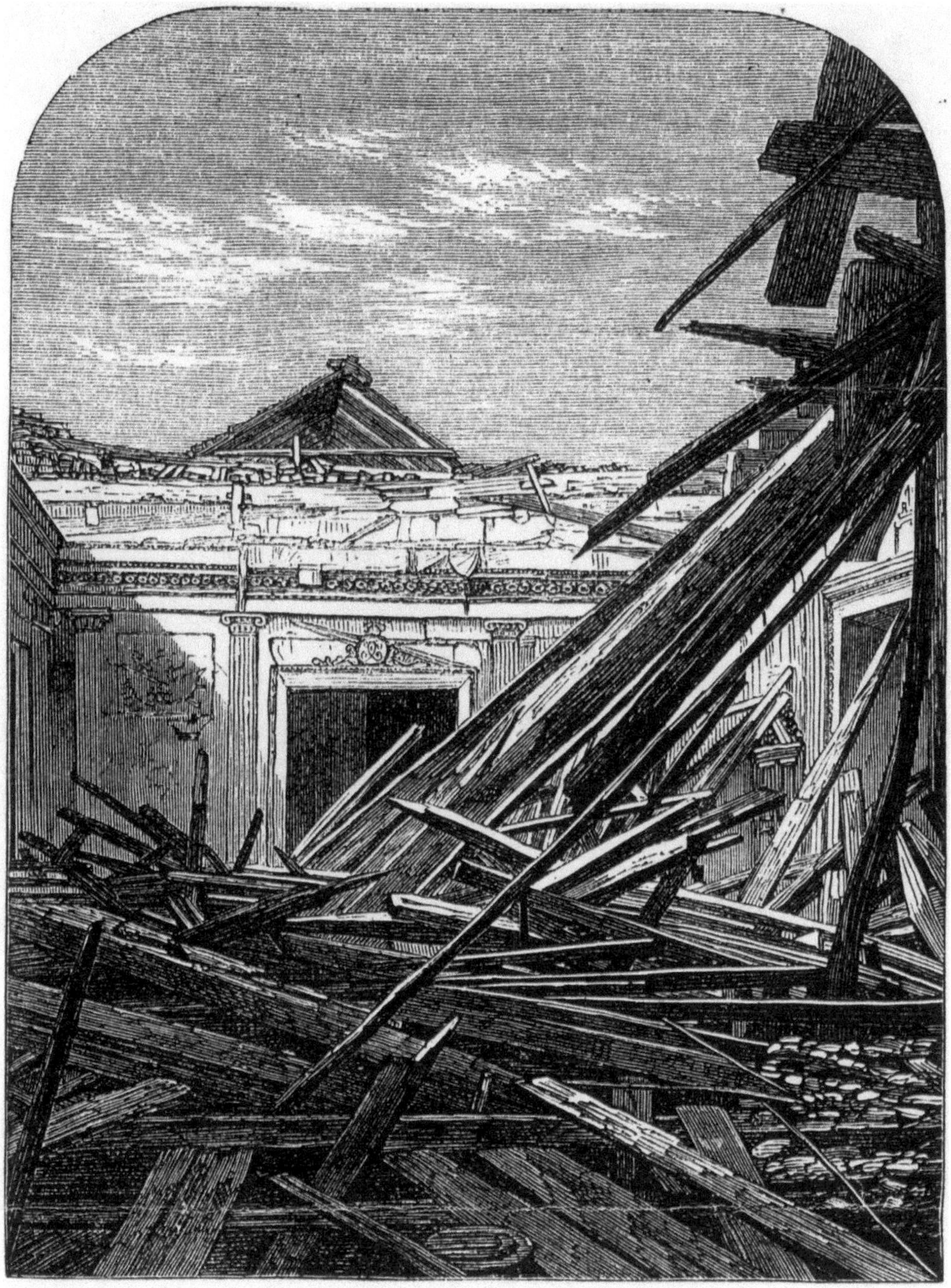

INTERIOR OF THE GOVERNOR'S PALACE, MANILLA.

Fig. 2.59 Damage to the governor's palace in Manila

This second image of the 1863 Manila earthquake shows the ruined interior of the governor's palace (Fig. 2.59). The source of the engraving is again the *Illustrated London News*, but this one is based on a photograph. The image gives a slightly misleading impression of the damage; the palace was a long building consisting of a central bay and two wings of eight bays each. The central bay and its tower collapsed, while the two wings were relatively unharmed. The image shows the interior of the central part, under the collapsed tower.

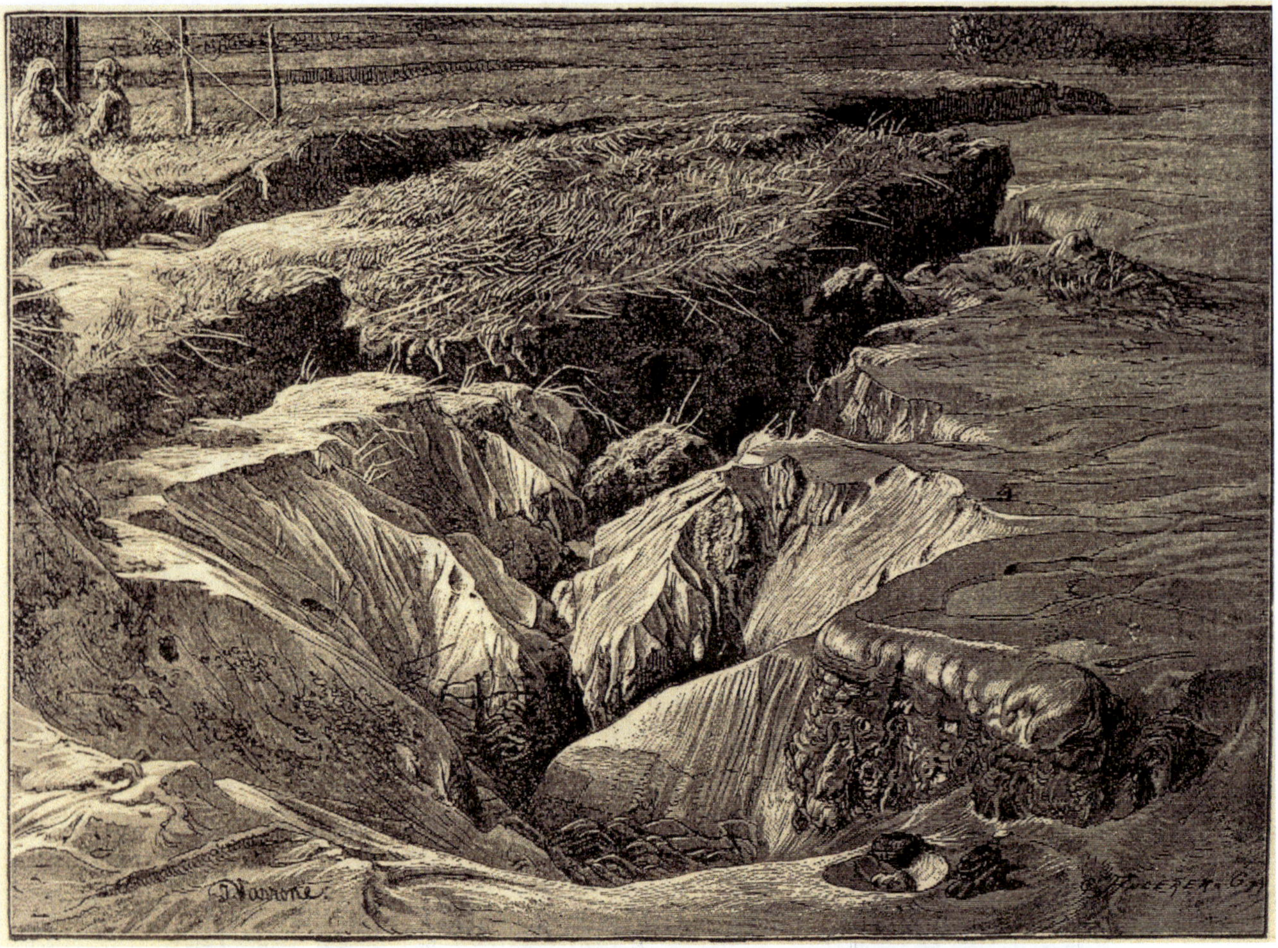

Fig. 2.60 The Cachar earthquake of 1869

If, as mentioned previously, the Alps are the result of a collision between the African and Eurasian plates, the Himalayas have a similar origin in the collision between India and Eurasia. And, necessarily, this process is accompanied by earthquakes, particularly along the front of the mountain chain. At the eastern end of the mountain chain, on the Indian–Burmese (Myanmar) border, the tectonic situation becomes quite complex, and some notable great earthquakes have occurred, the best-known being the Shillong earthquake of 12 June 1897. The illustration here, though, is from an earlier event, the Cachar (or Silchar) earthquake of 10 January 1869 (Fig. 2.60). Magnitude estimates for this earthquake vary greatly; a value of 7.5 or even less is often cited, but the most recent study by Szeliga et al. (2010) gives a value of 8.3, putting it in the category of great earthquakes.

Yet, for a great earthquake, it did very little damage, and very few lives were lost (nine, in fact), the reason being that most of the houses were made of wood and bamboo, and thus were elastic enough to escape much damage. Those that were made of masonry or brickwork tended to have low, heavy walls, which supported each other against toppling. A church tower, a sawmill, and a two-storied palace were the chief casualties. The earthquake was investigated at the time by Thomas Oldham (1816–1878), an Irish geologist, who was appointed as the first director of the Geological Survey of India. His report was incomplete at the time of his death but was completed and published by his son Richard Oldham (1858–1936), who is an important figure in the early development of seismology (Oldham 1882). The earthquake was notable for its effect on the ground surface; in many places where soft sand was overlain by heavy clays, differential movement occurred, with slumping or cracking seen at the surface. This is illustrated by the picture here, taken from Oldham's report. For dimensions, the figures of the two children sitting in the upper left corner of the composition give a sense of scale. It has to be stressed that cracks such as these are purely due to settlement in soft ground and are nothing to do with earthquake faulting.

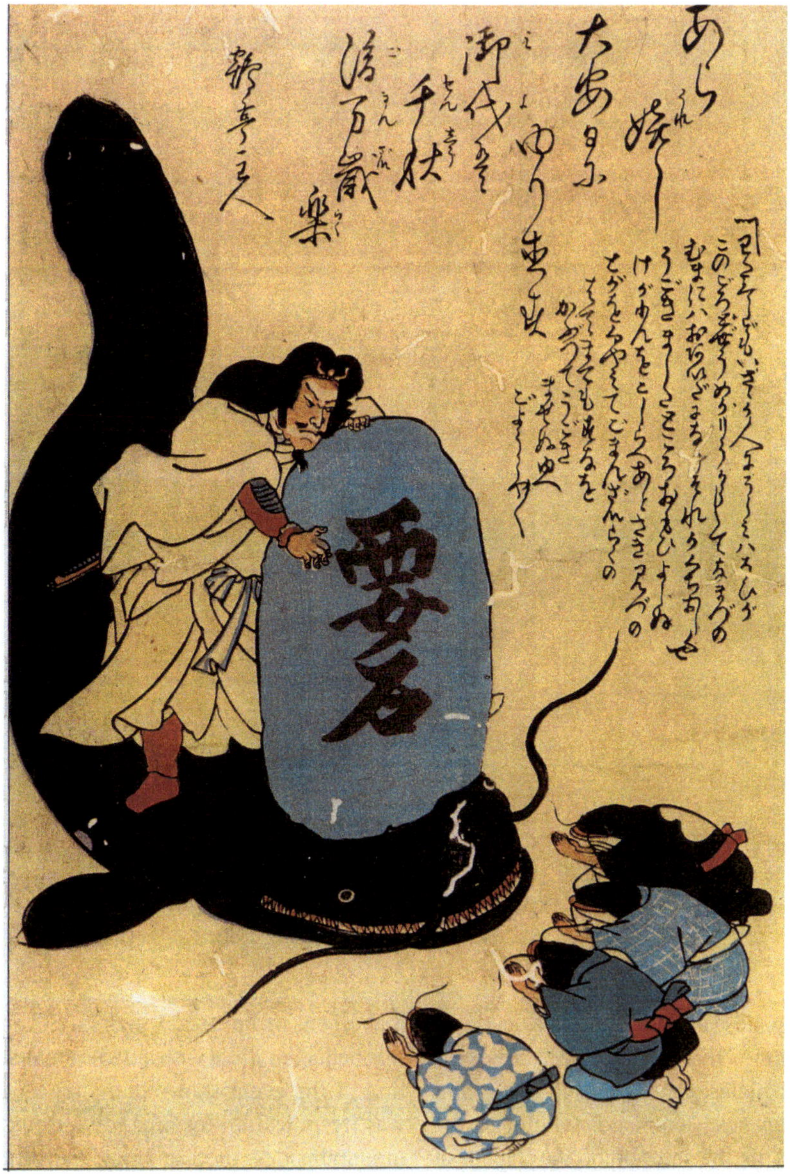

Fig. 2.61 The *namazu* (giant catfish) punished after the Edo earthquake of 1858

In pre-scientific times, different peoples devised their own explanations as to how earthquakes are caused. To the ancient Greeks, the god Poseidon was responsible. Elsewhere, various mythical creatures were invoked. In the Pacific Northwest of America, it was the giant thunderbird; in Japan, a giant catfish (*namazu*) lurking deep in the earth. Normally, the *namazu* is kept restrained by the god Kashima, who keeps a large stone on the catfish's head. But sometimes Kashima has to attend meetings with other gods (or he dozes off), and the *namazu* frees himself from the stone, thrashes his tail, and causes an earthquake. In one version of the myth, the *namazu* misbehaves in this way because he is jealous of other fish species that are considered more delicious in traditional Japanese cooking.

Since the fifteenth century, a tradition of artistic representations of the *namazu* in colored wood-block prints grew up, but it was only in the late eighteenth century that these images, known as *namazu-e*, became associated with earthquakes. Many were issued after the Great Ansei earthquake of 11 November 1855, which caused much damage in the Kanto region, Edo (now Tokyo), and its surroundings. Altogether, over 300 *namazu-e* are known today. The one reproduced here shows the giant *namazu* immobilized by Kashima, who presses him down with a great boulder (Fig. 2.61). Four little catfish gather round, representing other, smaller earthquakes, to be warned that this is their punishment if they don't behave themselves.

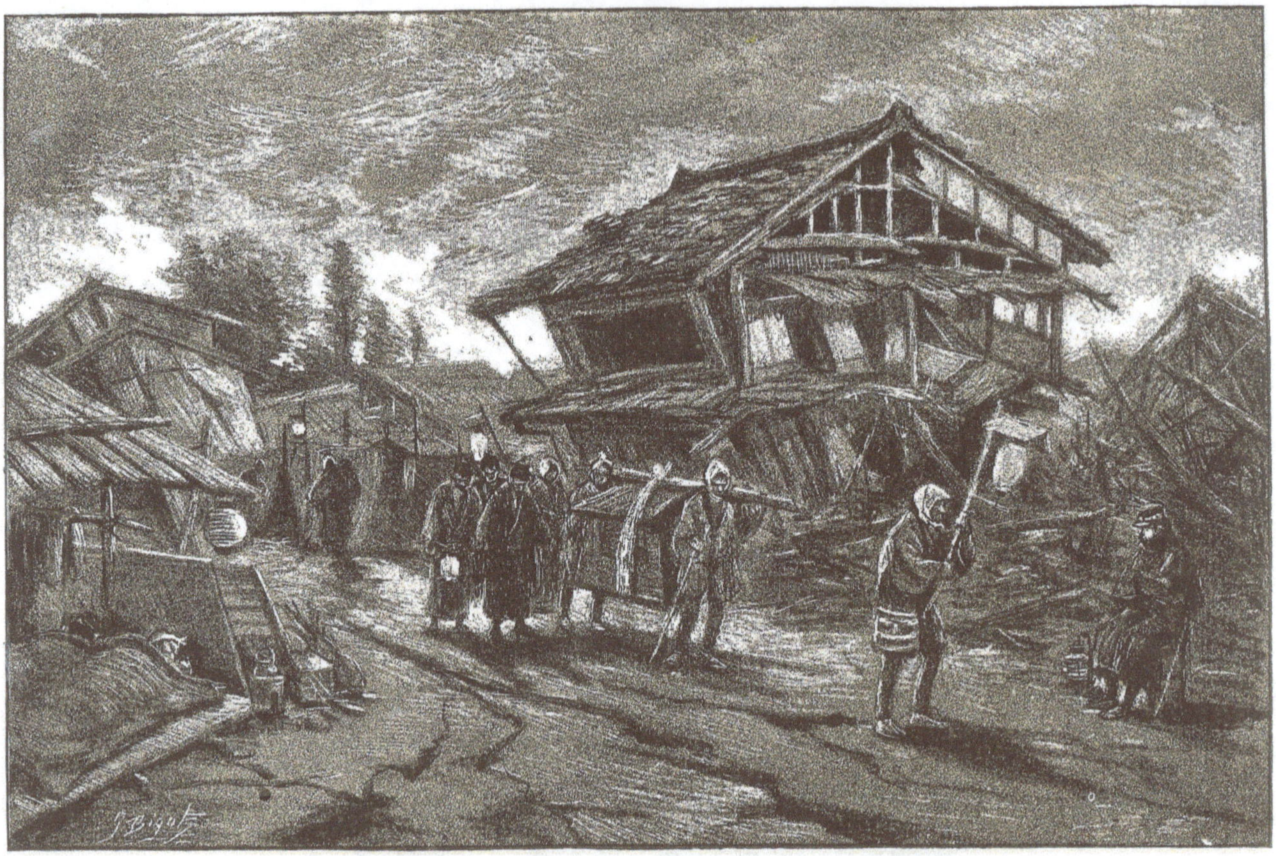

Fig. 2.62 The great Mino-Owari earthquake of 1891—a funeral procession

Most of the strongest Japanese earthquakes have occurred on or near the coast on either the east or south coasts of Honshu, where two oceanic plates, the Pacific and Philippine plates, collide with the Japanese mainland. Of the earthquakes that have occurred in the interior of the Japanese mainland, the largest is the great Mino-Owari earthquake of 28 October 1891, with a magnitude assessed by all recent authorities as 8. The epicenter was in the mountains near the town of Gifu, north of Nagoya in central Japan; the earthquake was felt widely, and in Tokyo it was strong enough to knock objects off shelves. The official death toll was 7273, with 17,175 being injured.

If it was the 1906 San Francisco earthquake that finally proved that earthquakes were caused by fault displacement, the Mino-Owari earthquake (also called the great Nobi earthquake) made it fairly obvious this was the case. The surface ground displacements were numerous, and large and deep scarps occurred; the vertical fault displacements reached up to 8 meters in some places, with horizontal movement of up to 2–3 meters. Seven of the surface faults reached a length of over 80 kilometers. Modern re-examination of the event shows that slippage on three major faults was responsible for this disastrous earthquake. On the mountainsides, thousands of landslides occurred and trees were uprooted. Fires, especially in Gifu, added to the destruction.

In the course of the 14 months that followed the main shock, more than 300 weaker aftershocks were recorded in Gifu. The loss is marked on the 28th day of every month in a Buddhist memorial service at the Earthquake Memorial Hall in Gifu. The Prefecture of Gifu marks the anniversary of the tragedy as Prefectural Earthquake Disaster Prevention Day, a day dedicated to earthquake preparedness against future disasters.

We show three European illustrations of the effects of the earthquake, all taken from *The Graphic* of 12 December 1891. This first (Fig. 2.62) depicts a funeral procession for one of the victims in Gifu.

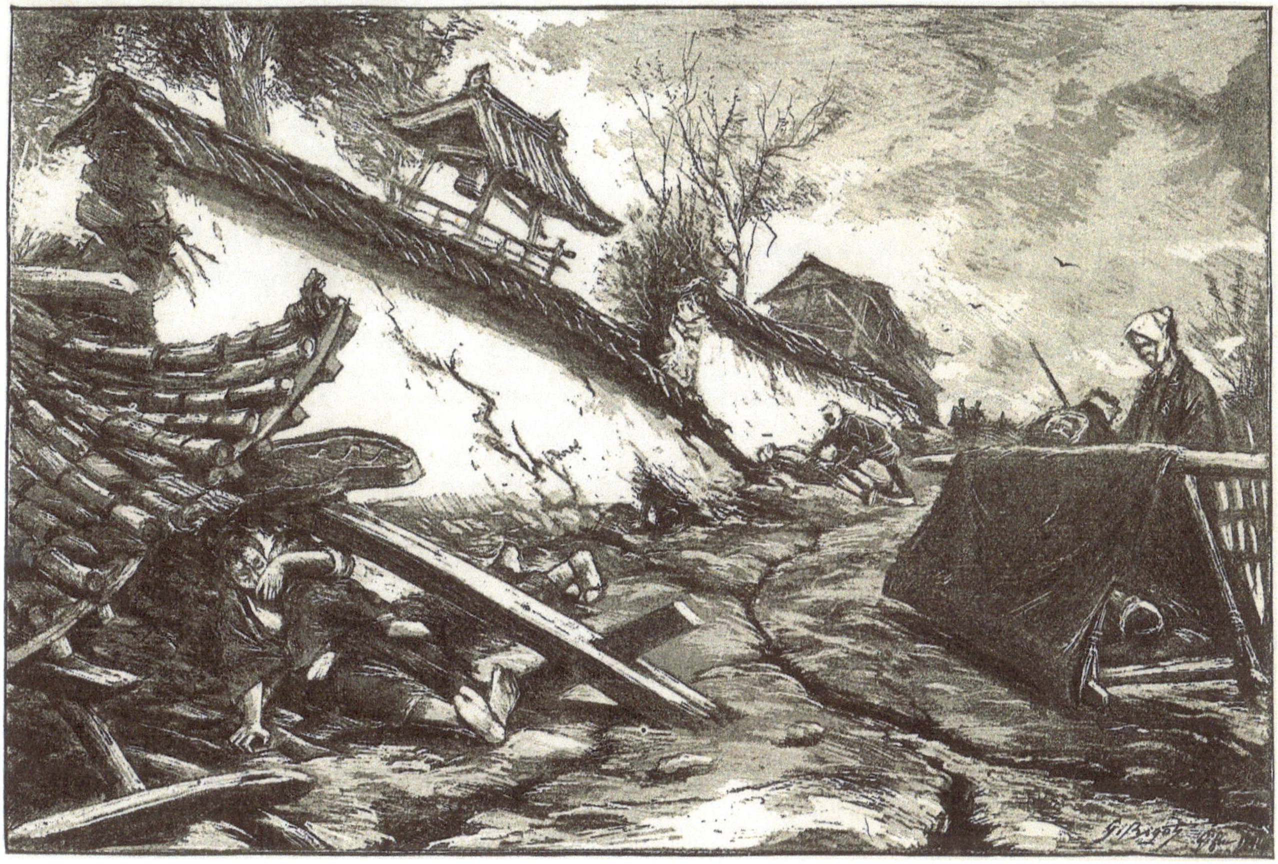

Fig. 2.63 Casualties in Gifu in 1891

The earthquake occurred at a time when Japan was modernizing rapidly. As part of this process, from the late 1870s onward the government had been inviting foreign experts to come and take up university positions to pass on Western technological developments. When British and American engineers were exposed to Japanese earthquakes, they rapidly developed an interest in seismology; as a result, important advances in seismograph design were made. The Mino-Owari earthquake was recorded on Gray–Milne–Ewing seismographs at weather observation stations in Gifu, Nagoya, Osaka, and Tokyo, as well as at a station housed at the Imperial University of Tokyo. This is therefore the earliest of the earthquakes featured in this book to be recorded by seismic instruments. (The only other one is the 1906 San Francisco earthquake; the 1884 Colchester earthquake was recorded on a barograph).

The image here (Fig. 2.63) is titled *Mother and Child—Killed by the Fall of a Gateway of the Outside Wall of a Temple near the Railway Station, Gifu*; the source is the same as that of the previous image.

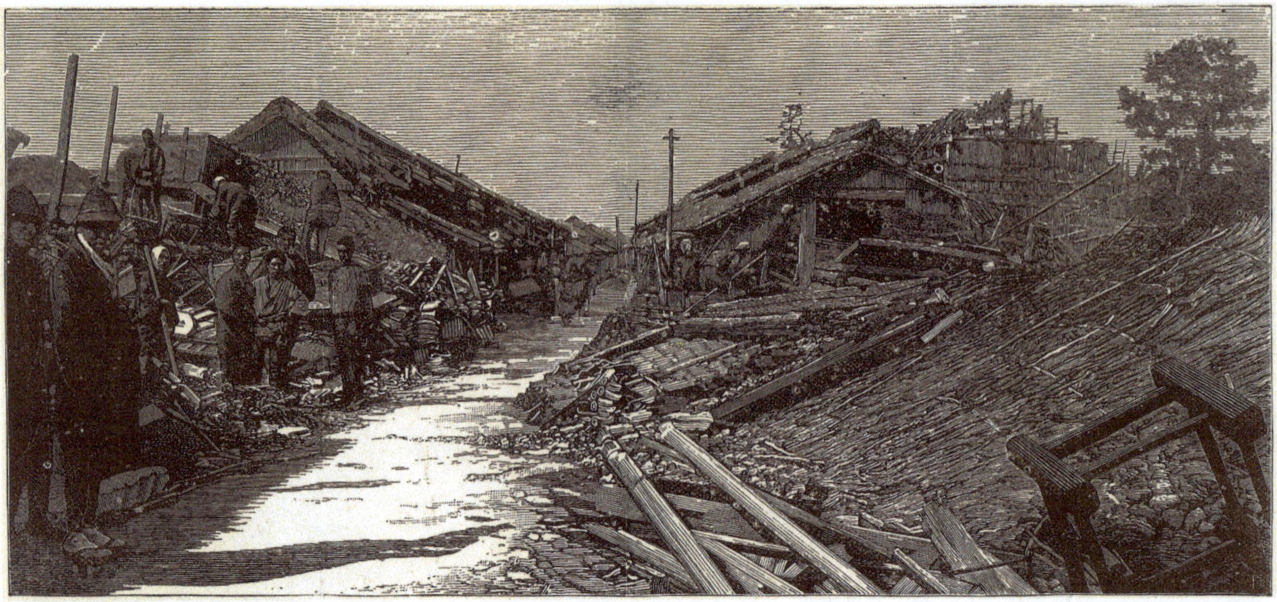

Fig. 2.64 Damage in Gifu

John Milne, often referred to as the "Father of Modern Seismology," was also a keen photographer and, with his friend William Burton, founded the first photographic society in Japan. After the Mino-Owari earthquake, Milne and Burton toured the damaged area and produced a book about the earthquake, illustrated with 29 photographic plates. The engraving here of a damaged street in Gifu (Fig. 2.64) was, according to *The Graphic*, made from a photograph supplied by "Messrs. Birch & Co, Sea View, Isle of Wight." Considering that Milne was to retire to the Isle of Wight four years after the earthquake, it might suggest that this image was based on one of Milne's photographs. If so, it is not one that appeared in the book he produced with Burton. Part of the fascination of that book is how it shows what life was like in provincial Japan in that period. Something of it can be glimpsed in this scene, where the houses appear as low, heavy-looking timber structures, not the elegant, light constructions that the European thinks of as typical Japanese houses.

2.5 Landslides and Other Gravitational Effects

As mentioned before, although everyone now knows that earthquakes are due to movements in the rocks of the Earth's crust, this was not always apparent. It was argued in the seventeenth century, for instance, that since earthquake effects are stronger on the upper floors of buildings, the vibration must come from above—from the air. It was even believed for a while in the mid-eighteenth century that earthquakes were electrical in nature. But there is surely no more direct way to appreciate that the element Earth is a hazard than having several thousand tons of it land on one's head.

So, in this last part of the chapter, we will show some images of notable landslides and other forms of gravitational collapse on steeply sloping ground. Landslides and snow avalanches can be triggered by earthquakes, as in the case of the numerous landslips caused by the Mino-Owari earthquake in 1891, but they can also occur because of heavy rain or just for no apparent reason. A technical distinction can be made between landslides (or landslips) and rockfalls, the former being something that happens when a cohesive mass of soil comes into motion, whereas in a rockfall, rocks, stones, gravel, and sand particles move freely. Rather than try and adjudicate each historical case for which we have illustrations, we present them together, but avalanches (which, consisting of snow, perhaps belong under Chapter 3) are grouped together at the end.

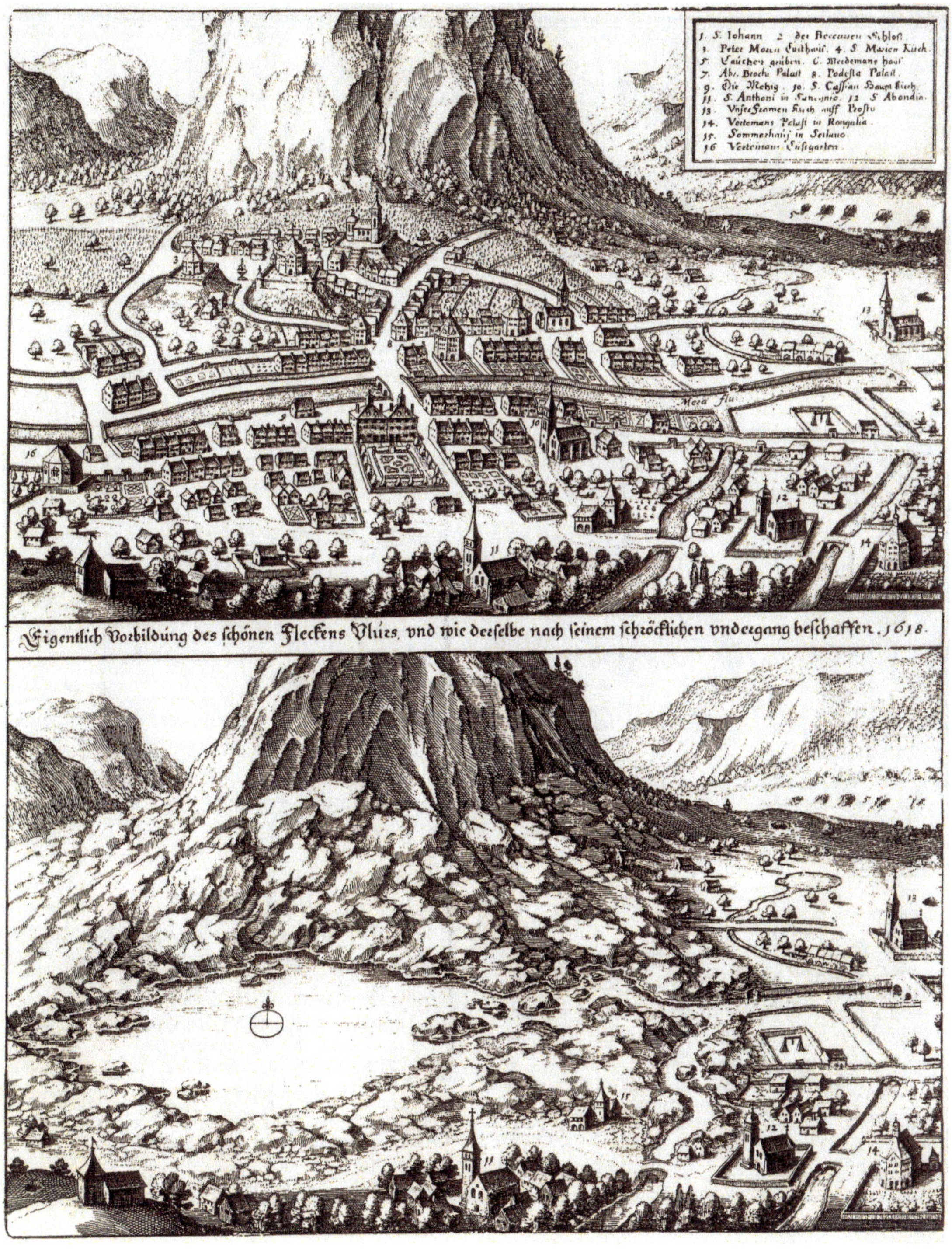

Fig. 2.65 Piuro before and after the landslide in 1618

Landslides are common phenomena and are not usually fatal, but there are occasions when a major landslide hits a village and buries it entirely, with heavy casualties. Such an occasion was the great Plurs landslide of 4 September 1618. The village of Plurs at that time was in Switzerland; it is now in Italy and is known as Piuro. The village's wealth came from mining talc, which was plentiful in the hill overlooking the village. However, the mining was carried out carelessly, with a lack of props or backfilling of old workings. After five days of heavy rain, the weakened hillside collapsed onto the village, wiping out the entire population (estimated at between 1000 and 2500), except for six people who were elsewhere at the time. The old village is still largely buried. This before-and-after representation was executed by Mattheus Merian (1592–1650), a Swiss engraver, for the *Theatrum Europaeum*, a work published in Frankfurt in 1635 (Fig. 2.65).

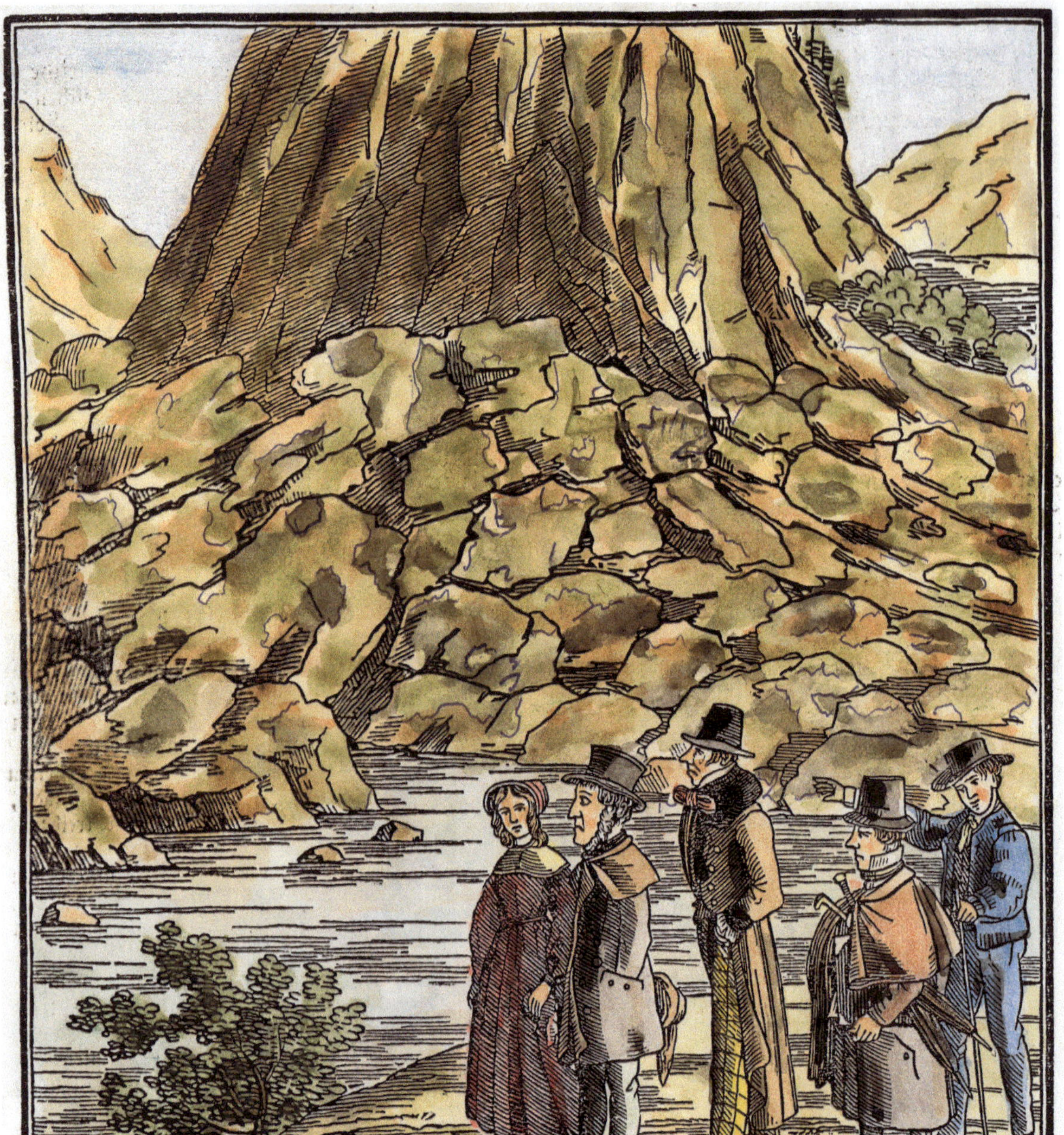

Fig. 2.66 Visitors to the scene of the Piuro landslide

This second image (Fig. 2.66) of the 1618 Piuro landslip is not at all well known. It is an anonymous, hand-colored copper engraving showing visitors to the scene of the Piuro disaster, who observe the locality of the tragedy some 229 years later; the image was composed as an illustration for a Swiss calendar in 1847. The image was discovered and reproduced in a book commemorating the 270th anniversary of the Piuro disaster by Scaramellini et al. (1988), and is included here by courtesy of the authors.

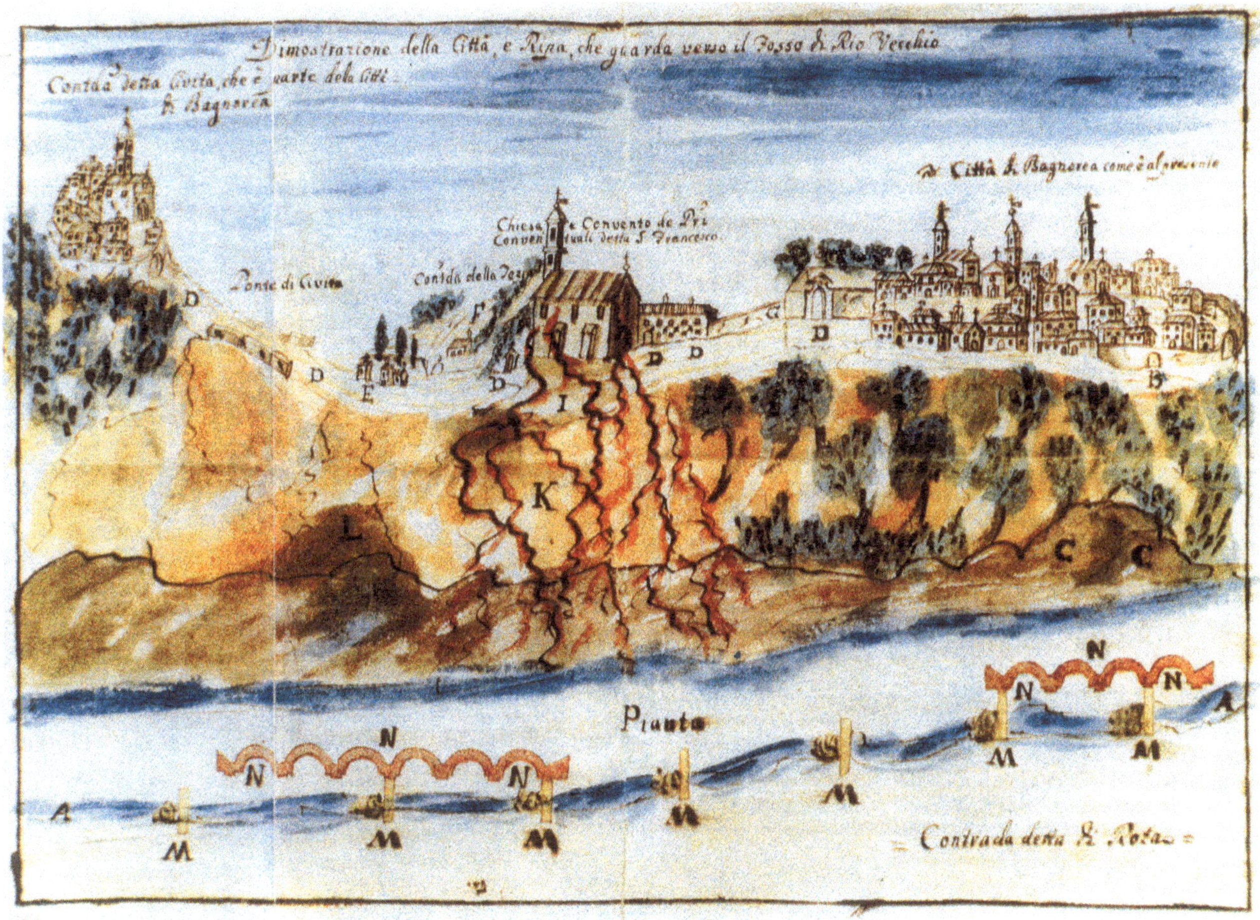

Fig. 2.67 A landslide at Civita di Bagnoregio in 1764

One of the worst geological situations, from a landslide hazard point of view, occurs where stiffer rock deposits are underlain by softer clays. A notorious case is the little Italian village of Civita di Bagnoregio in the province of Viterbo in central Italy, about 120 kilometers from Rome. The settlement was founded thousands of years ago in Etruscan times; hilltop sites were often selected for health reasons in order to avoid the malarial lowlands. In the case of Civita di Bagnoregio, the site is perched on the top of a very steep hill, with exactly the poor geological conditions that lead to landslides, as was soon discovered, as the village became plagued with landslides throughout its history. A large earthquake on 11 June 1695, triggered a major landslide, and after this the village became almost deserted.

In March 1764 a landslide occurred, which is the subject of this unusual contemporary watercolor image (Fig. 2.67). The bridge used for access to the village collapsed, together with part of the convent of San Francesco; the ground movement started at the convent church of San Francesco and continued down the hill, toward and over the River Pianto.

Today the population of Civita di Bagnoregio is about ten. One only has to glance at satellite imagery to see the scars left by past landslides. So, unusually, the village has turned landslide hazard into a tourist industry. The village hosts the world's only landslide museum; the Geological and Landslide Museum in the Palazzo Alemanni in the center of the village opened in April 2012, with exhibits that unsurprisingly focus strongly on the nature and impact of landslides in the Civita di Bagnoregio locality (Margottini and Di Buduo 2017).

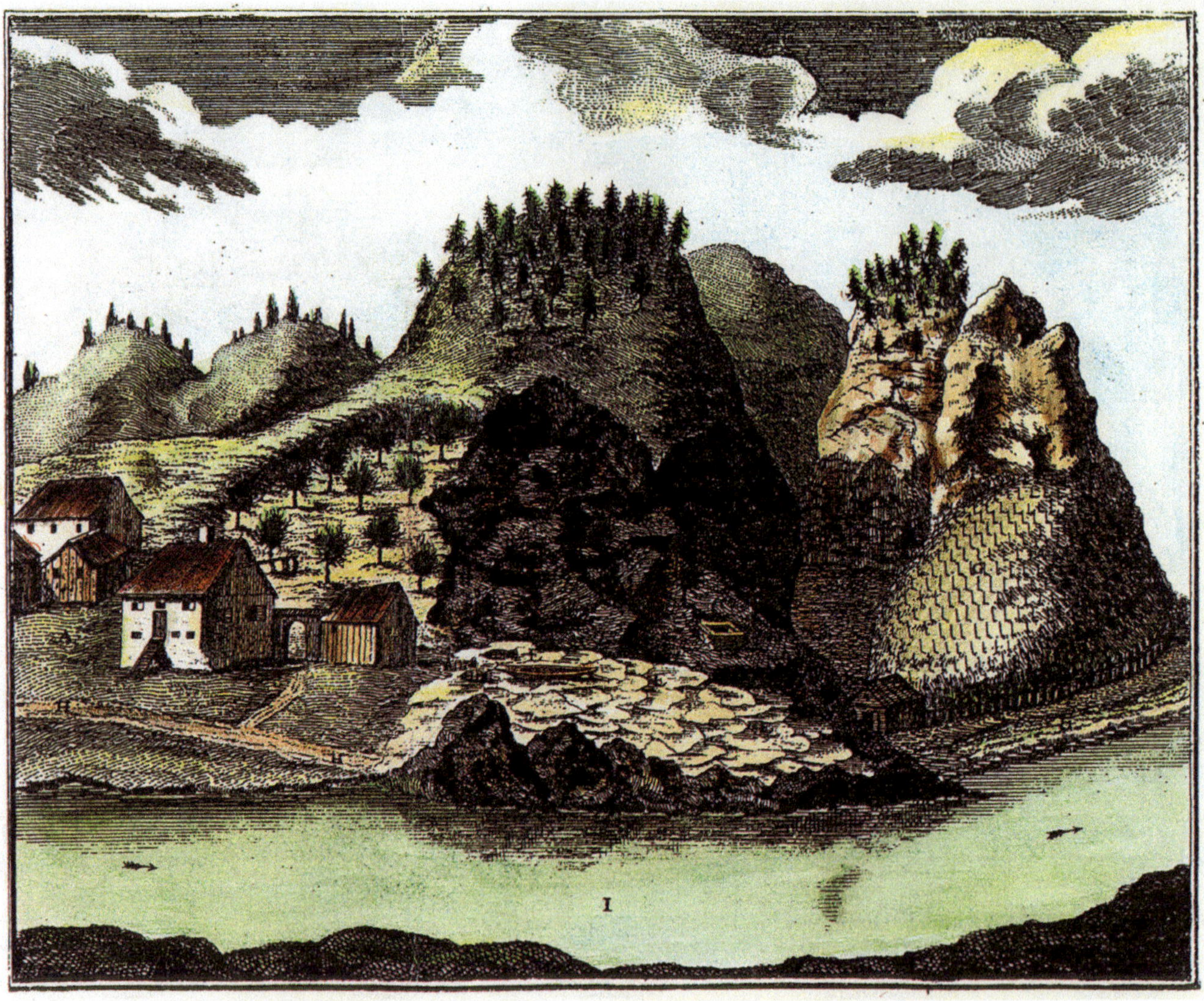

Fig. 2.68 The North Bohemian landslide in 1770

The image here, taken from a German broadsheet, is of the landslide on 5 January 1770, near the village of Veselí, North Bohemia (Fig. 2.68). As the illustration shows, the local landslide effects were not huge; modern examination suggests dimensions of about 700 meters in length and 200 meters. And it did not cause serious damage or loss of life. The reason why this negligible event appeared as important news in a leaflet published at the time can be found in the fact that the soil and rock avalanche at Veselí completely blocked the important communication route from Vienna and Prague to Dresden and Berlin (along the Elbe River, through the Bohemian Massif). Even today, the principal international railway line between Dresden and Prague follows the same route.

The Elbe River flows out of the Bohemian Basin through the middle part of the volcanic heights of the mid-Bohemian Massif. The steep slopes of the Elbe Valley suffer from mechanical instability; thus, landslides and rockfalls have occurred relatively often in this region. This is especially true in recent times when the continuous traffic of heavily loaded lorries and trains raises the level of local vibration. Today the most endangered segments of the railway line in the vicinity of Veselí are protected against rockfalls by well-fixed metal guards.

The most important triggering factor for earthquakes in Czech territory is heavy rain. In the case of the 1770 event in Veselí, heavy rain in June–July and September–December 1769 saturated the stratified volcanic rocks of the Kozí vrch Hill, making it unstable. Altogether, some 14 landslides occurred in northwest Bohemia in the year 1770 (Raška et al. 2016).

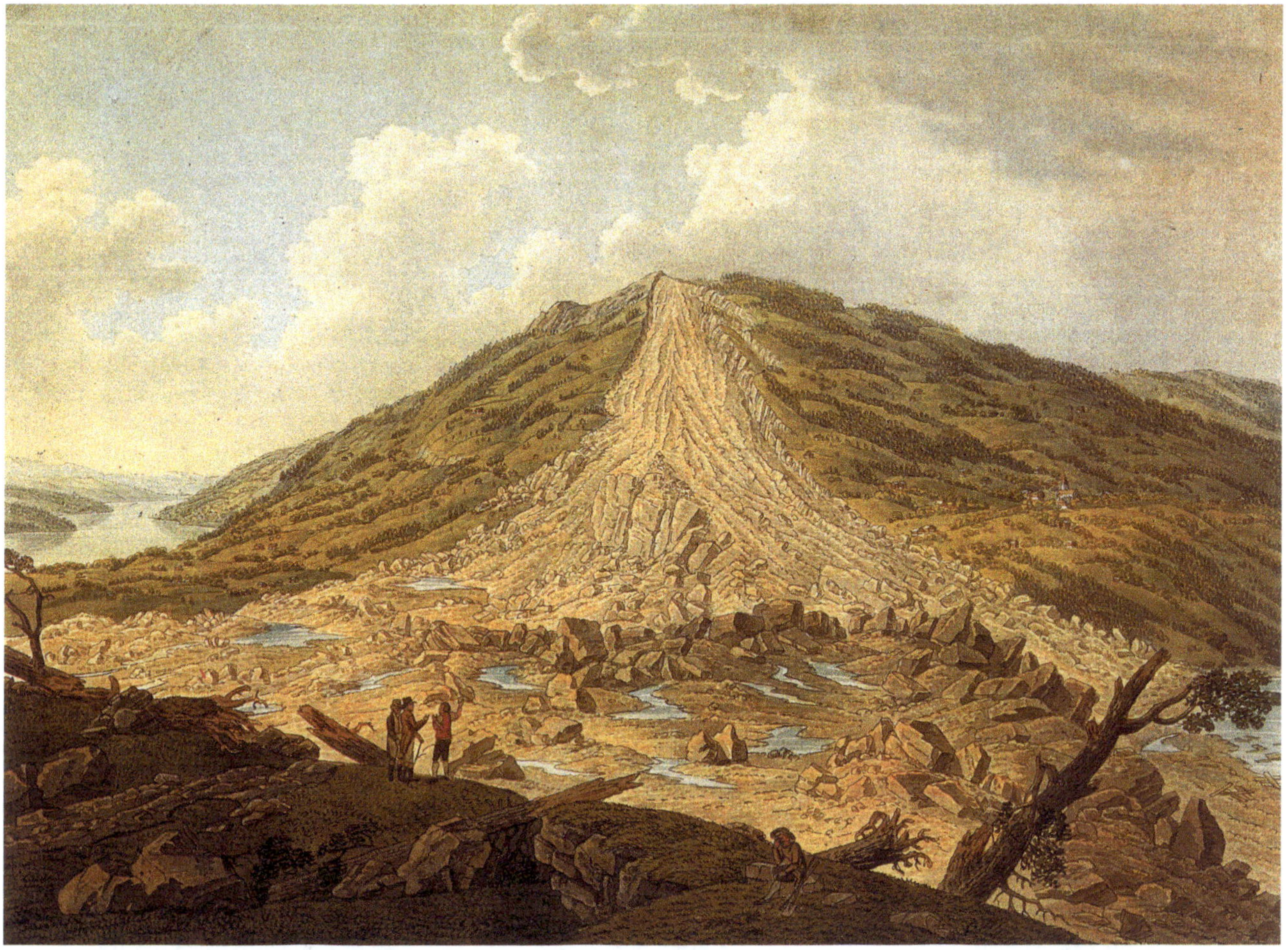

Fig. 2.69 The Rossberg after the 1806 Goldau landslide

Probably the first scientific investigation of a landslide was the case of the Goldau disaster of 2 September 1806. The cause of the landslide was partly geological and partly hydrological. The Rossberg—the hill to the north of the Swiss village of Goldau—is made up of layers of mudstone, sandstone, and conglomerate inclined at an angle to the horizontal. The rocks were destabilized after much wet weather in the summer of 1806 and, eventually, after some premonitory fissuring, gave way. A huge lump of the hill detached itself and slid down, breaking up into a wave, which hurtled down into the valley and partway up the slope of the hill on the other side. The technical term for this is a debris avalanche, or *Trümmerstrom*, to give it its German name. The village of Goldau was wiped out, with 457 people killed. The slide also impacted the waters of Lake Lowerzer, causing a tsunami that altered the shorelines. According to local legend, this was not the first such calamity on the Rossberg; supposedly, earlier events had happened 2000 years earlier and again in 1354, damming a river and eliminating the ancient village of Röthen.

The event on 2 September 1806, was witnessed by a Swiss physician, Dr. Carl Zay, who published an account of it the following year. The attractive image here is a painting by Xaver Triner (1767–1824), commissioned by the canton authorities (Fig. 2.69).

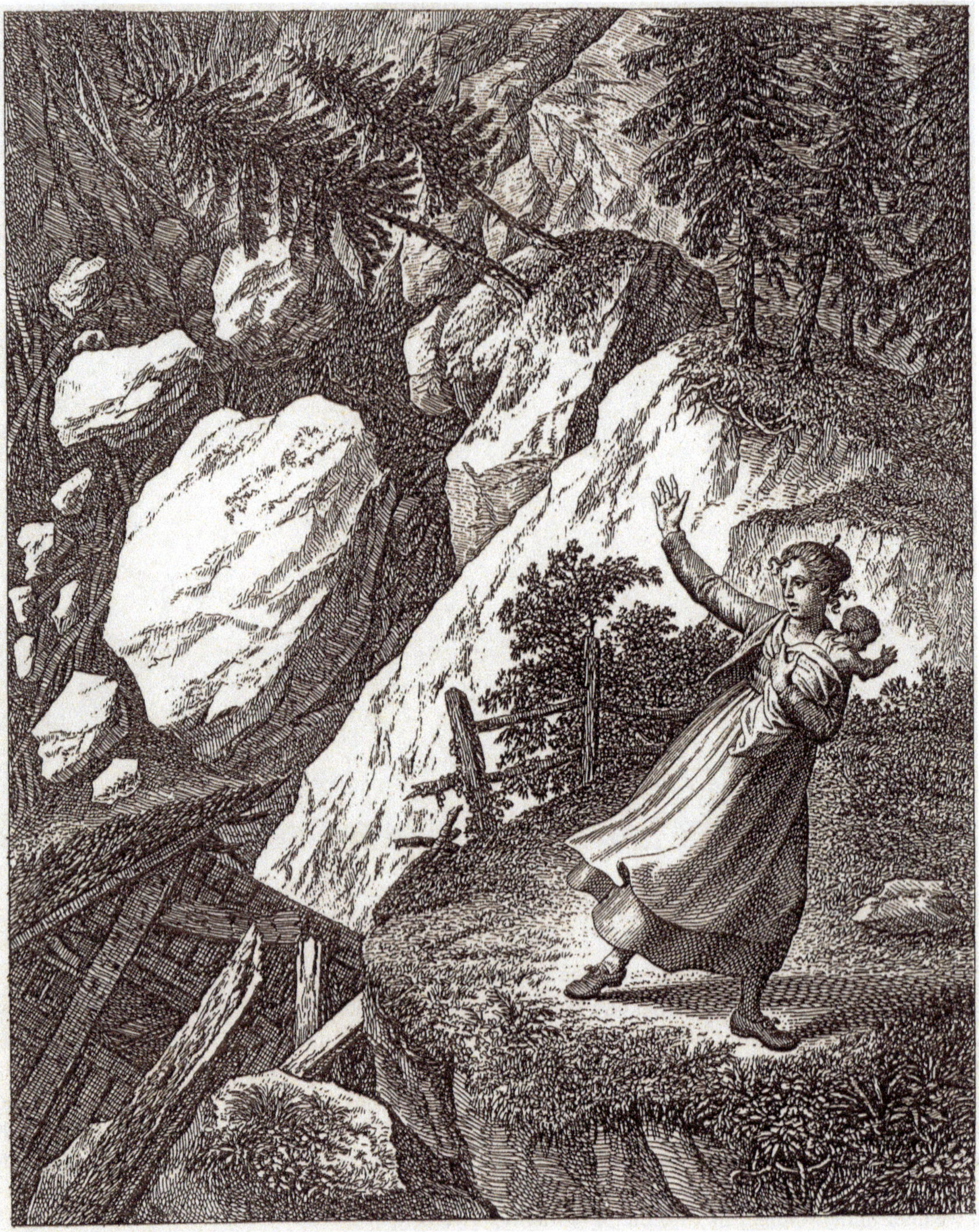

Fig. 2.70 Agatha Mettler saves herself and her child in 1806

The 1806 Goldau disaster sparked a remarkable public and cultural response. Help and money were sent to the survivors, and special arrangements had to be made for tourists who came to witness the spectacle of the buried village. These included no less a person than Goethe himself, who knew the area and made several drawings. Astonishingly, the event was even celebrated in music; the Austrian composer Joseph Weigl (1766–1846) wrote a three-act Singspiel (a light melodrama) on the subject, titled *Der Bergsturz.* Within two weeks of the disaster, the Zurich artist Johann Caspar Jahn had made four paintings of the event, converted them into copper engravings, and put them on sale as souvenirs. The highly dramatic engraving reproduced here is intended to depict Agatha Mettler, a villager who managed to save herself and her young child at the last moment (Fig. 2.70).

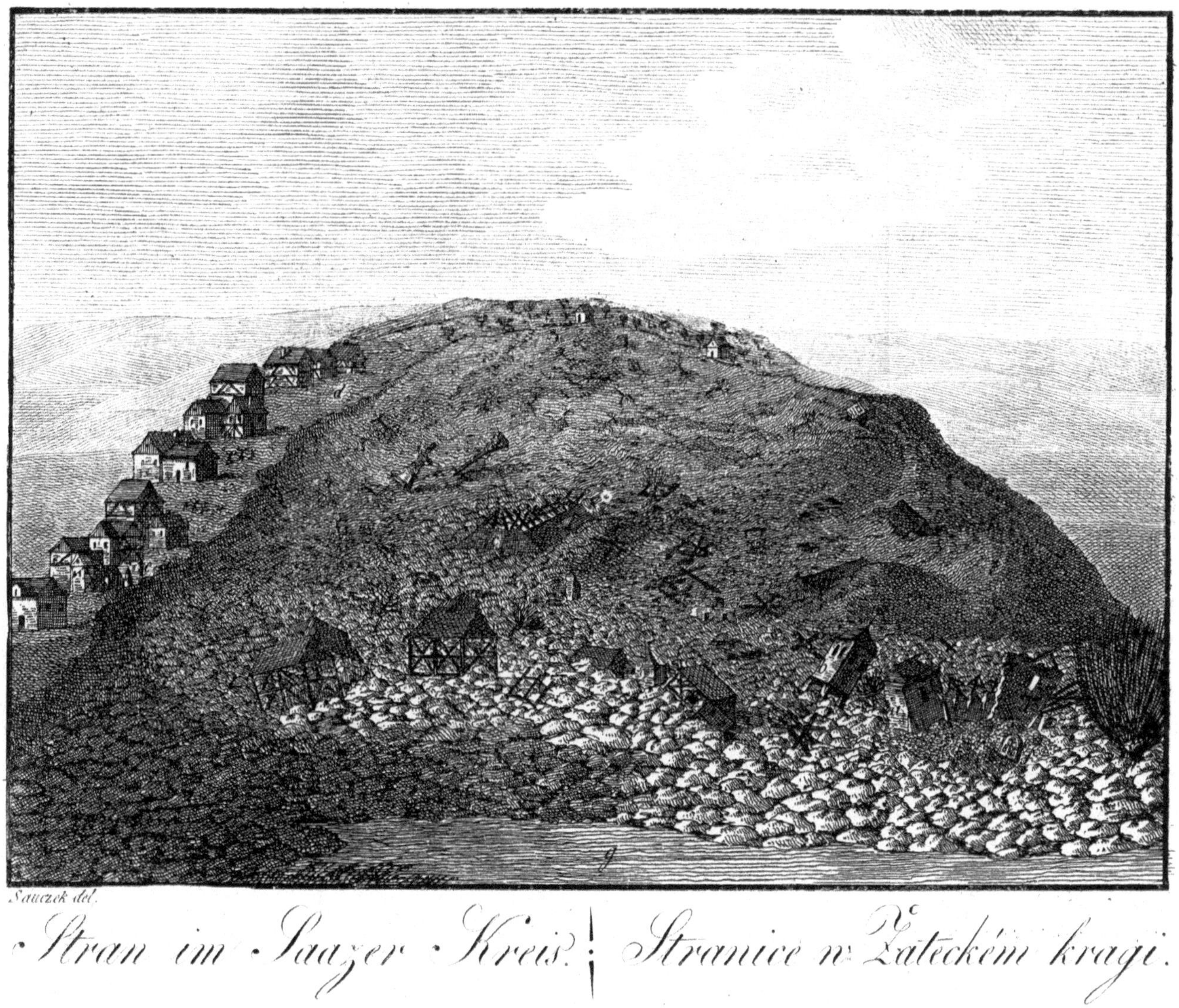

Fig. 2.71 The Stranná landslide in 1818

In this copper engraving, a small landslide near the village of Stranná in Žatec County in North Bohemia in 1818 is shown (Fig. 2.71). The landslide, which appeared on the hill slope above a meander of the Ohře (Eger) River, smashed a dozen shacks and log cabins erected on the hillside. In this part of Bohemia, massive strata of brown coal (lignite) often protrude at the very surface of the Earth. At the top of the Ohře riverbank, erosion due to the river revealed the protruding lignite deposits. This was an open invitation for many locals—mostly poor, unemployed people of the region—to come and fill their coal scuttles for personal use. In this way, the whole of the slope was undermined; thus, when the first strong rain came, saturating the hillside and lubricating the strata, the slope failed and a landslide resulted. Luckily, in this case, the damage was minimal and there were no casualties.

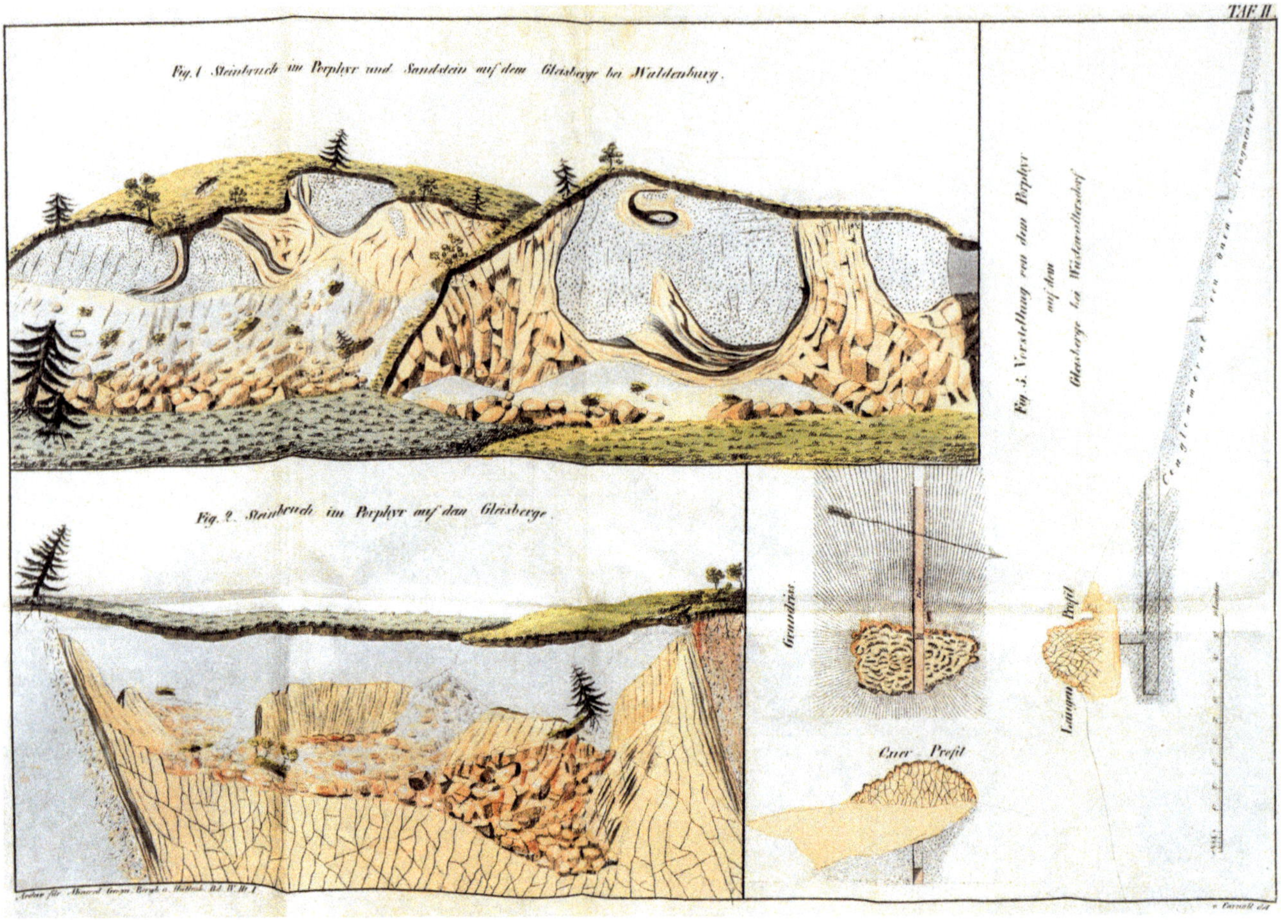

Fig. 2.72 Rockbursts at Wałbrzych in the 1830s

Surface slumping occurs as a simple collapse of the Earth's surface, when the subsoil material can no longer sustain the weight of the overlying rock. Collapses of this type occur often in karst (limestone) districts containing underground cavities, and also in the case of old abandoned mines, in which empty corridors and other excavated rooms may collapse because of a lack of support, often resulting in subsidence and sinkholes at the surface.

This particular situation is shown in the illustration here (Fig. 2.72), showing a surface collapse (German *Steinbruch*, or "rock burst") above old bitumen mines at GleisBerge near Waldenburg (now Wałbrzych) in Lower Silesia, where natural bitumen was mined in the early nineteenth century. This hand-colored copper engraving was published in a lengthy paper by two geologists working in Silesia in the 1830s: a certain J. von Zobel, about whom almost nothing is known, and Rudolf von Carnall, after whom the mineral carnallite was subsequently named. Their paper *Geognostische Beschreibung von einem Theile des Nieder-Schlesischen, Glätzischen and Böhmischen Gebirges* was published in the *Archiv für Mineralogie, Geognosie, Bergbau und Hüttenwesen* (1832, volume 4, pages 3–172), and the engraving appeared as a foldout at the back of the volume.

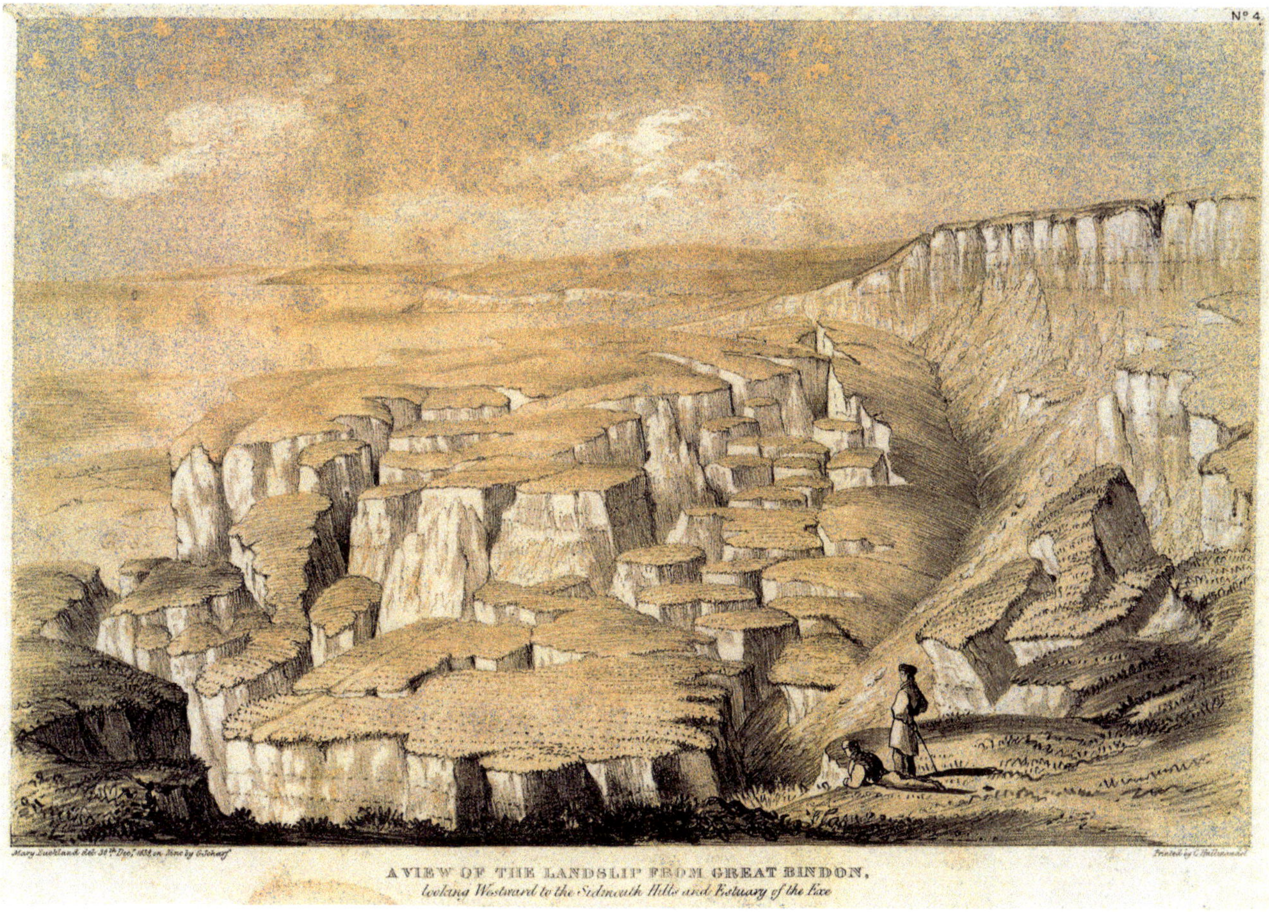

Fig. 2.73 A massive cliff collapse near Lyme Regis in 1839

This rather attractive tinted lithograph by George Scharf is done from a pencil sketch made on the spot by Mary Buckland, the wife of the celebrated English geologist William Buckland (Fig. 2.73). It shows a massive landslip that occurred on 26 December 1839, near Lyme Regis on the south coast of England (Buckland and Conybeare 1840). The fall of coastal cliffs is a common form of landslip when marine erosion weakens the base of the cliff, especially when the local geology is of weak rock such as chalk (as was the case here) or clay. It is not often, though, that such collapses occur on such a massive scale as here. Marine erosion typically takes the form of progressive slumps of cliffs over time, especially in storms, when significant amounts of land can be lost. Further east along the south English coast, in medieval times, was the important ecclesiastical site of Selsey Abbey. The location of the abbey buildings is now several kilometers out to sea.

An interesting feature of the 1839 landslip is that in contemporary news reports it was described as an earthquake. It is true that the sudden collapse of so great a volume of rock produced a seismic shock wave that seemed like an earthquake to villagers nearby, but actually, this is a late use of the word "earthquake" in the sense of *terra motus*—land moving. Ironically, if one examines any British book on earthquakes from the eighteenth century or earlier, one reads that the "greatest British earthquake" happened on 17 February 1575, near the village of Much Marcle in Herefordshire. This was another landslide, the trace of which is still visible today over 400 years later; however, the contemporary descriptions of it are all hugely exaggerated; it was only about 100 meters long by 40 meters across. It serves as an example of how accounts found in pre-eighteenth century "wonder literature" may be false reports spun from some modest incident. There is evidence that this was even recognized at the time, given the sideswipe made by the Tudor poet John Skelton (1463–1529) at "strange tidings from Wales."

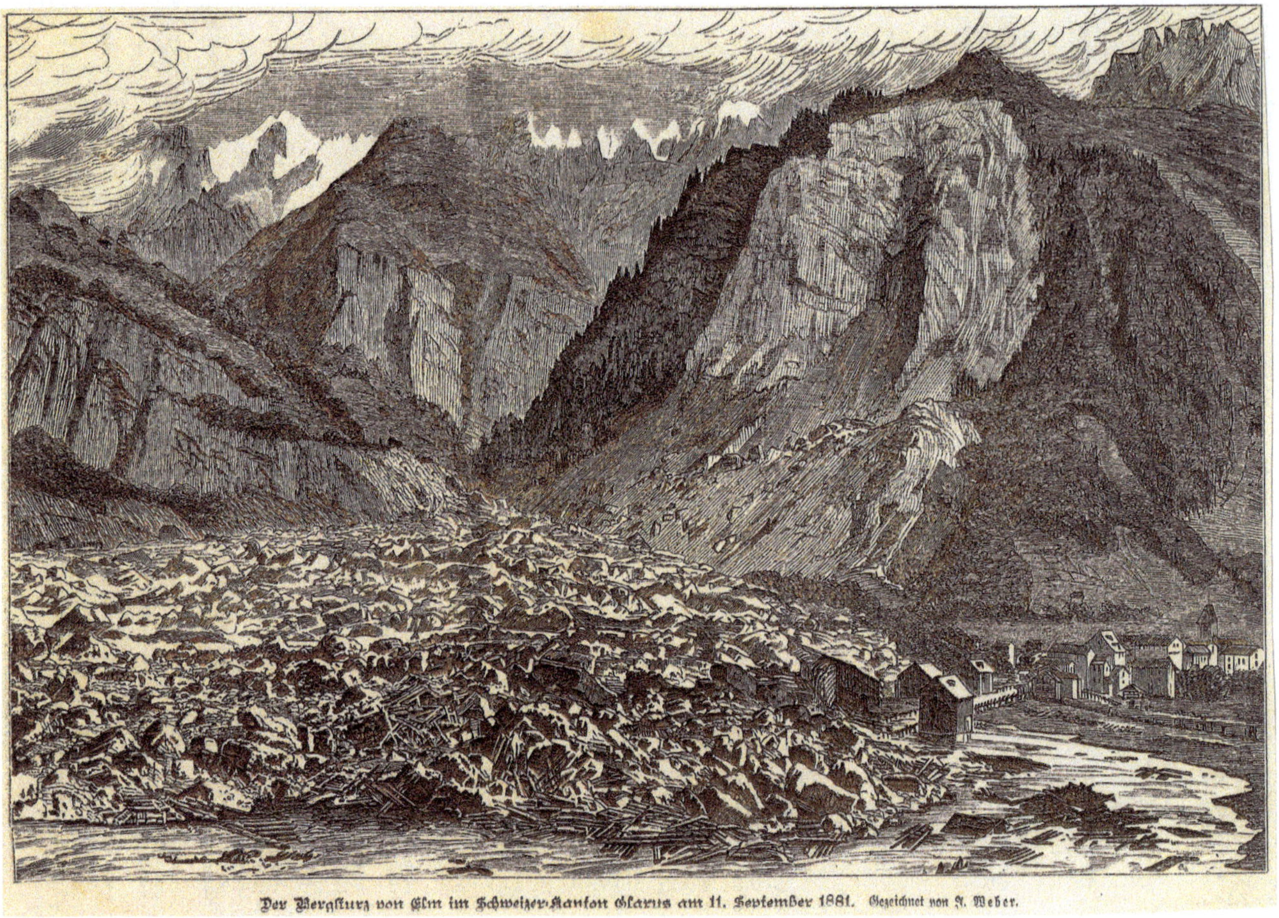

Fig. 2.74 The Elm landslide of 1881

Sometimes major landslides can occur as a result of what can only be described as human stupidity. Such a case was the devastating Elm landslide of 11 September 1881, shown here in a German engraving based on a photograph (Fig. 2.74). The mountain to the right of the image, the Tschingelberg, was a source of schist, which had been mined for centuries. In the 1870s, the demand for schist increased because of a demand for blackboards in schools. Mining activity increased, causing a few minor rockfalls. But what should have been cause for alarm was the opening of a large fissure above the quarry in 1876. Although this fissure was perceptibly growing, quarrying—with dynamite, of all inappropriate techniques—continued. Then, in August 1881, heavy rain weakened the mountainside further; new fissures appeared, and finally the local authorities realized that things were not safe. The quarry was closed, and the houses next to it were evacuated.

The scale of the danger was not appreciated, though. On 11 September, at 4 p.m., the mountainside began slowly to move. Some 75 minutes later, the first of three massive failures occurred, the other two following in quick succession. Huge volumes of rock seemed to flow into the valley at terrific speed, burying the river Tschingel and even surging up the valley slope opposite. The bulk of the village of Elm had a narrow escape; it can be seen on the extreme right of the illustration, right on the edge of the rock flow. Still, 115 people lost their lives, and most of the bodies were never recovered. Only 13 victims could be identified.

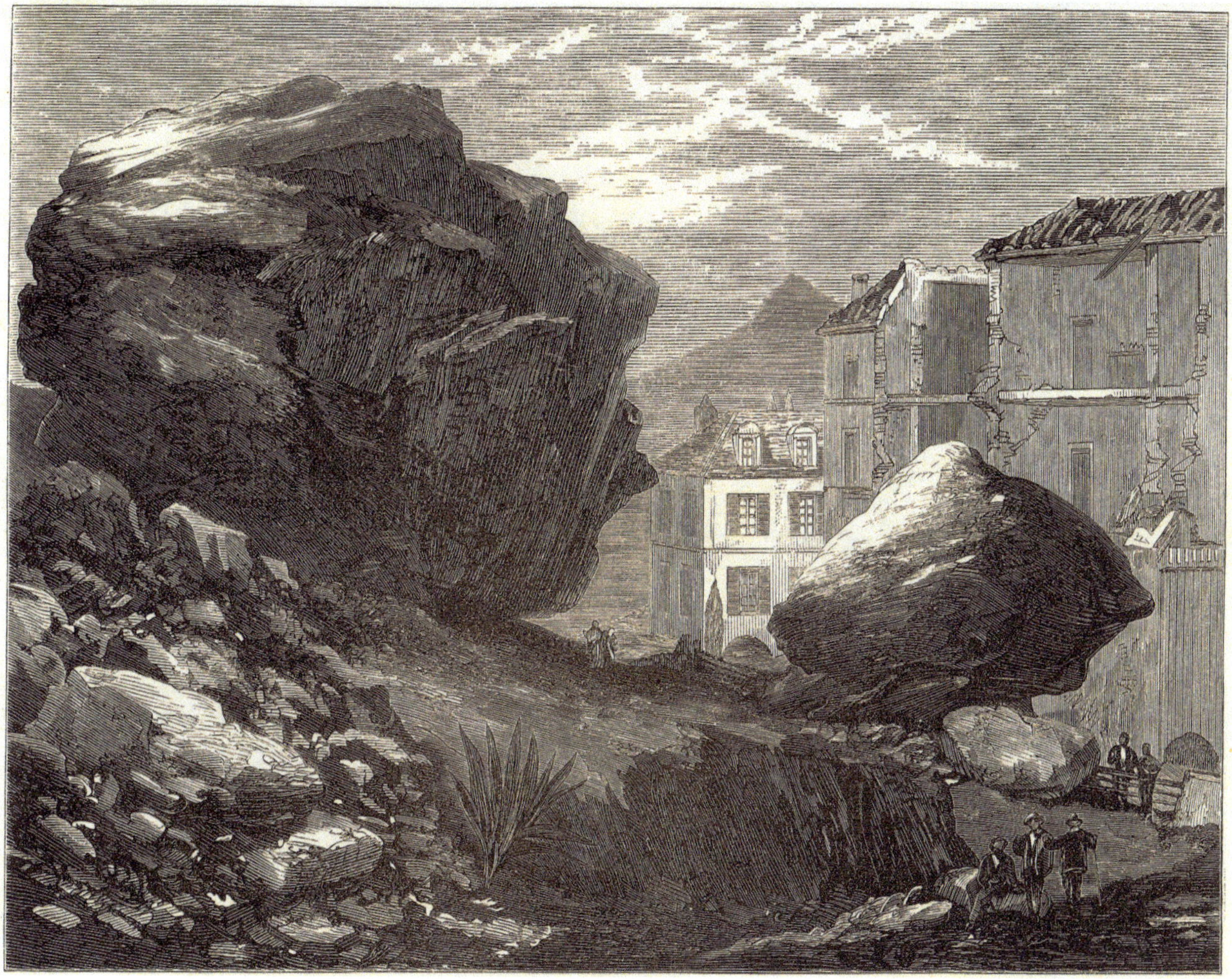

Fig. 2.75 A rockfall near Nice in 1887

In areas with rocky cliffs—one can mention as European examples the Alps and the Norwegian fjords—there is always a hazard that unstable stone blocks will detach themselves and come hurtling down. On reaching the base of the slope, a boulder can continue to roll some distance (and possibly hit a house). High mountain relief and steep slopes constitute a very favorable setting for failures of this sort in the French Riviera in southeastern France. The engraving here from a Czech publication (Fig. 2.75) shows a massive boulder that has rolled down toward the town of Nice, fortunately stopping before it hit any houses. The date of the image is 1887, and one can see damage to the buildings to the right, occasioned by the earthquake depicted previously (Figs. 2.32, 2.33, and 2.34). Reported falls of boulders in the hinterlands of Menton at the time of the earthquake actually owed more to heavy rains weakening the slopes than to the seismic shock (Vogt 2008).

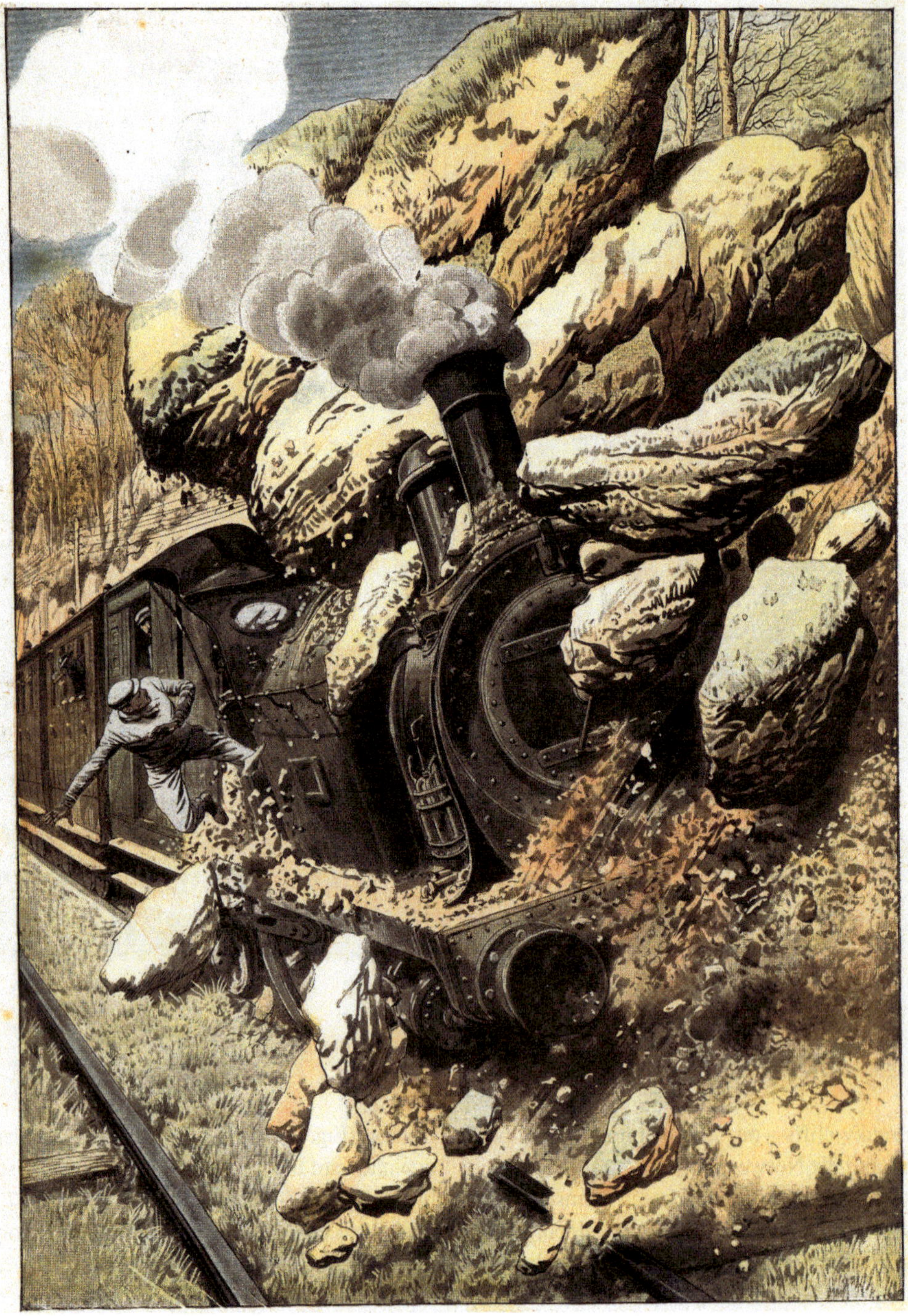

Fig. 2.76 A train hit by a rockfall near Vence in 1911

Rockfalls and landslides frequently block communication routes; narrow valleys or roads along the seashore beneath cliffs are frequently the only practical route but are also vulnerable to falls. Usually, this means traffic having to stop when a recent fall is encountered up ahead; more rarely, the rockfall actually hits the traffic—in this case, a train (Fig. 2.76). This incident happened near Vence on the route between Draguignan and Nice in the south of France in 1911. The locomotive was wrecked, and the engine driver was killed. This image was published in the illustrated supplement to *Le Petit Journal* on 10 December 1911.

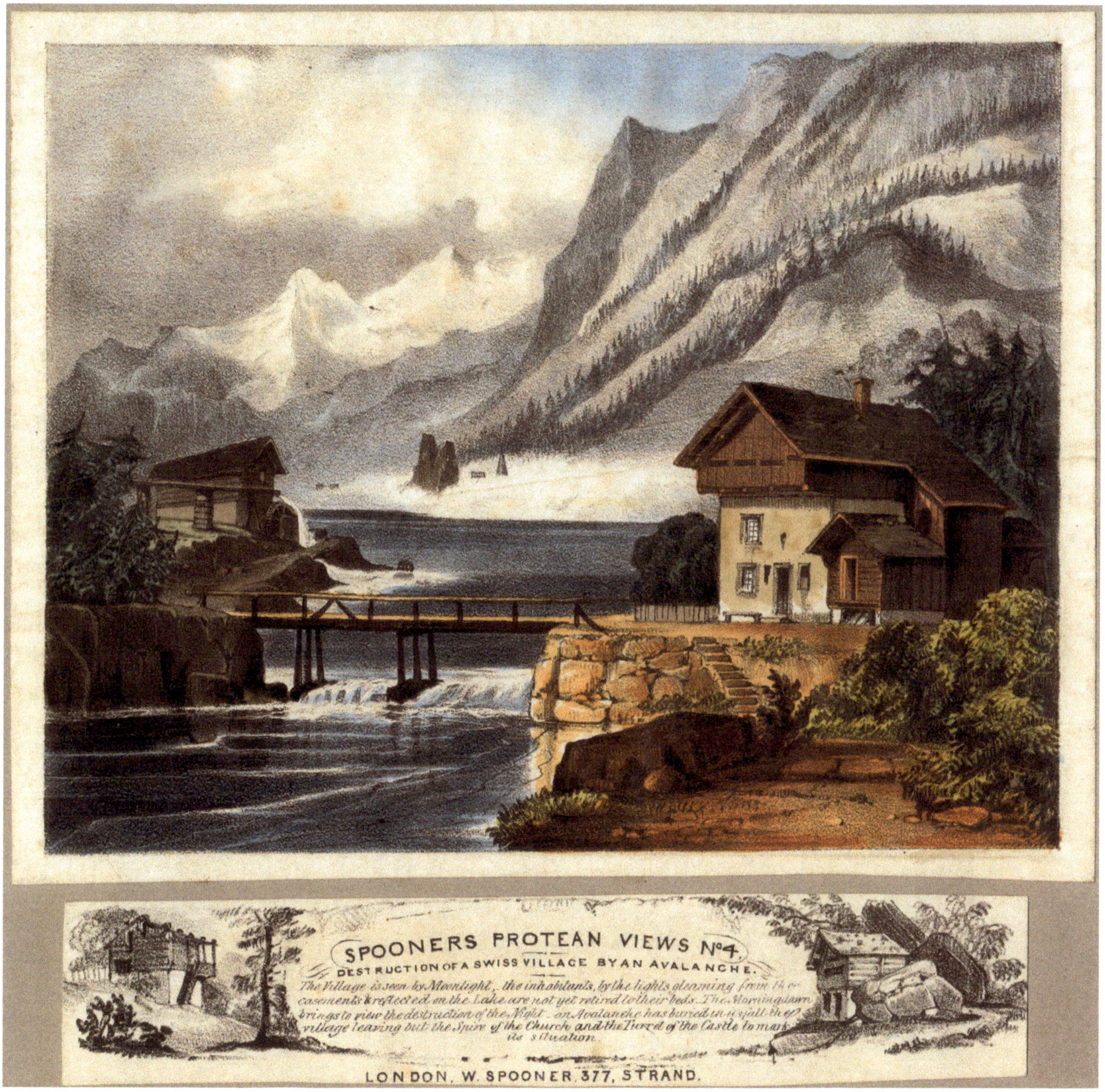

Fig. 2.77 A "Protean view" of a Swiss avalanche

Snow avalanches occur very commonly on snow-covered mountain slopes, when unstable masses of snow may slip down because of suitable atmospheric conditions or, often, ill-judged actions of skiers or mountaineers. Snow avalanches usually do not change the Earth's relief, but they can kill people and animals.

The first example shown here is an image of a very curious type (Fig. 2.77). The name of William Spooner is famous in connection with a Victorian academic who had an unfortunate habit of transposing consonants, and so he once referred to "the dear old queen" as "the queer old dean" (this effect is now referred to as a spoonerism). But there was another William Spooner of interest: a publisher who, in the 1830s, produced a series of what he termed "Protean views." These prints consisted of a scene pasted onto a single window mount, with another print adhered to the back. When held up to the light, the first picture would fade, revealing the second print. These often consisted of before-and-after scenes, where the initial image shows a tranquil scene, but when it is held up to the light, the scene after the disaster is revealed. In all, about 90 were produced, of which this is number 4. It is titled *Destruction of a Swiss Village by an Avalanche*. Exact details are not given.

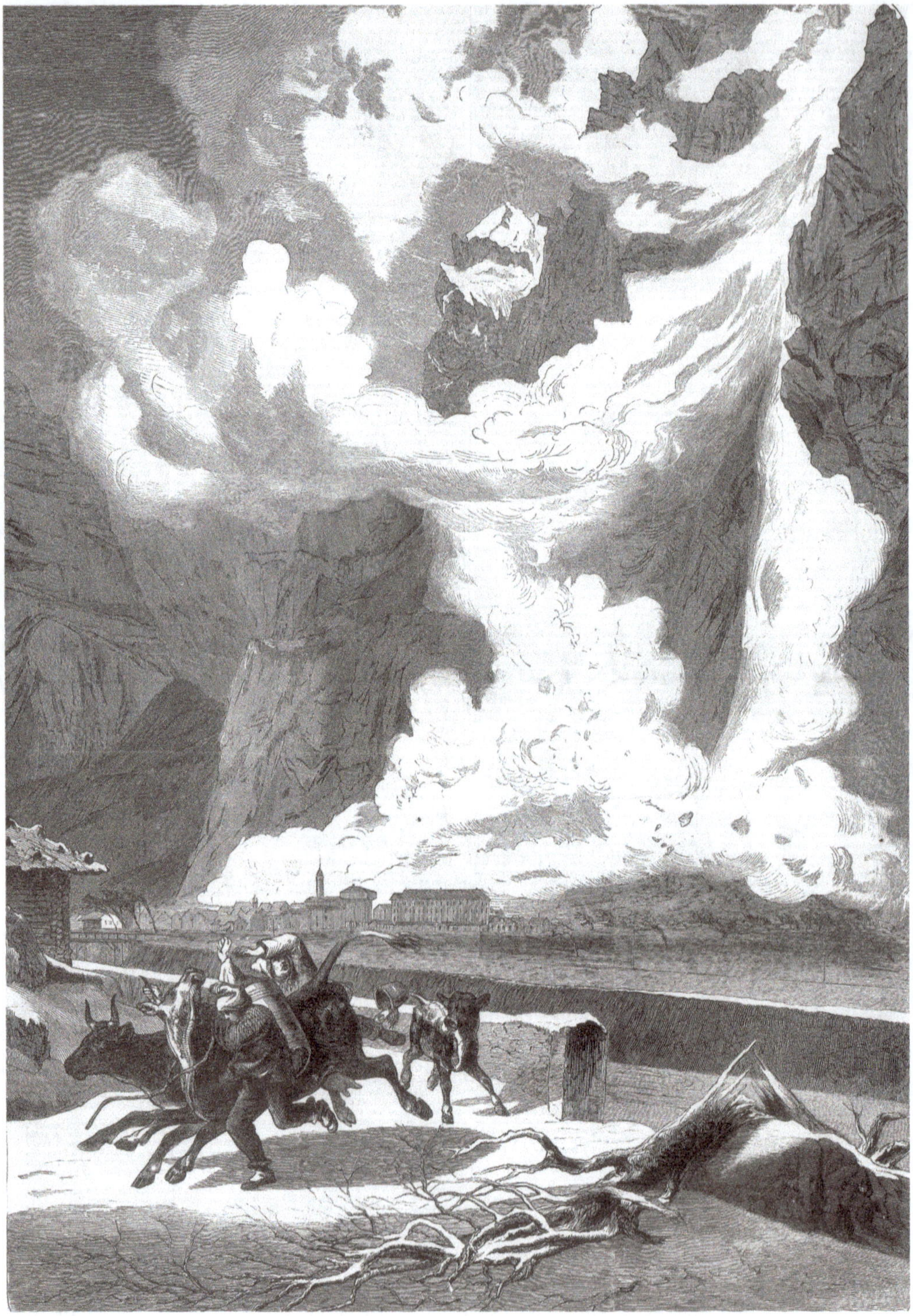

Fig. 2.78 An avalanche from Monte Rosa in 1871

This image (Fig. 2.78) comes from *Frank Leslie's Illustrated Newspaper* of 20 May 1871, and shows a dramatic avalanche from the east face of Monte Rosa Massif, the tallest wall in the Alps, 2600 meters high and four kilometers wide. The village is Macugnaga at the head of the Valle Anzasca. Note the small dimensions of the town church and houses compared with the size of the avalanche. Avalanches here are hardly uncommon; ten years after the incident depicted here, another huge avalanche killed one of the pioneers of Italian mountaineering, Damiano Marinelli, and his three companions.

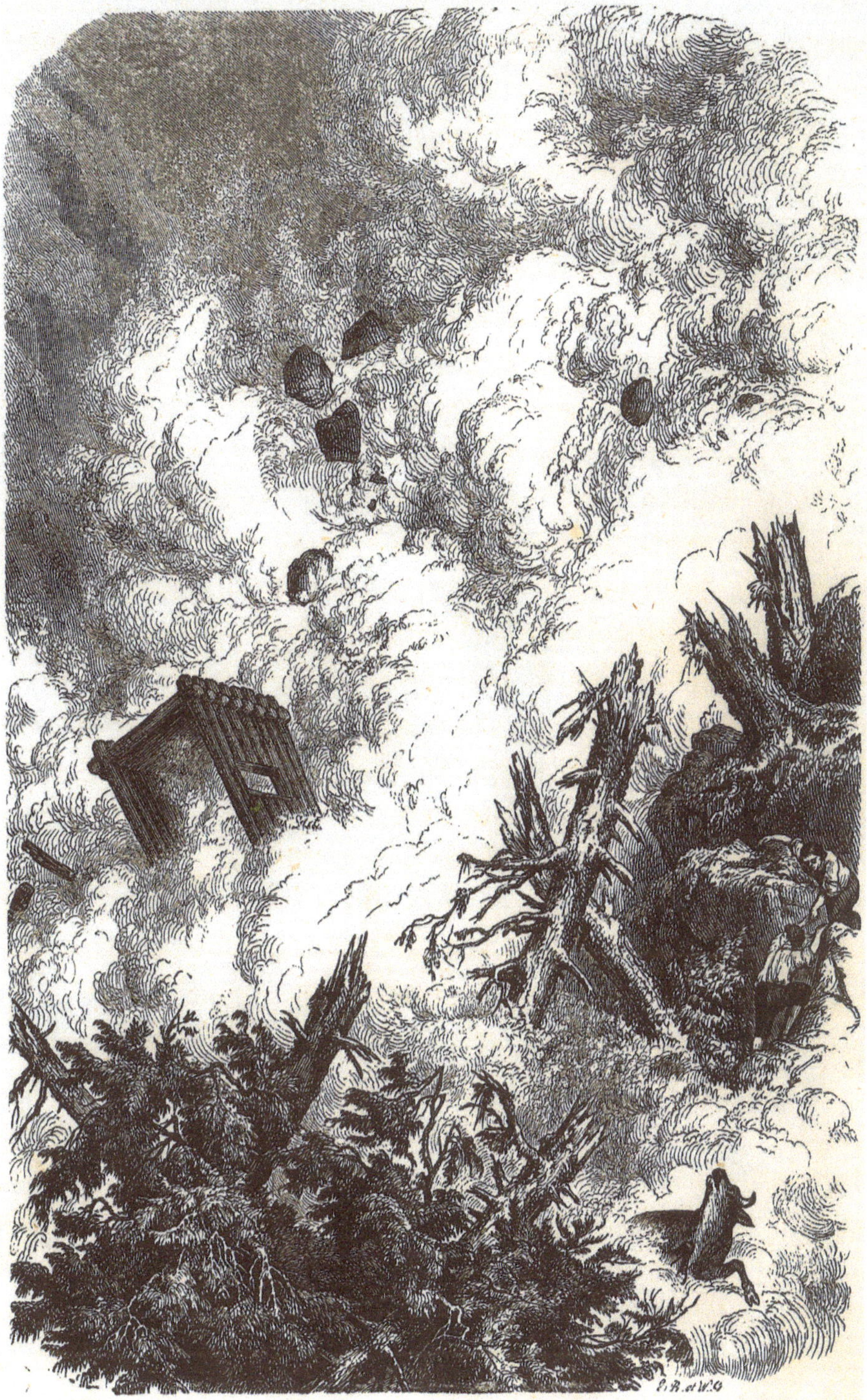

Fig. 2.79 An imaginative view of an avalanche

This dramatic image of a combined snow and rock avalanche (Fig. 2.79) is taken from an unidentified Russian book from the end of the nineteenth century. Presumably, the location of the event is in the Caucasus or the Urals, unless it is simply an imaginative image that could be anywhere.

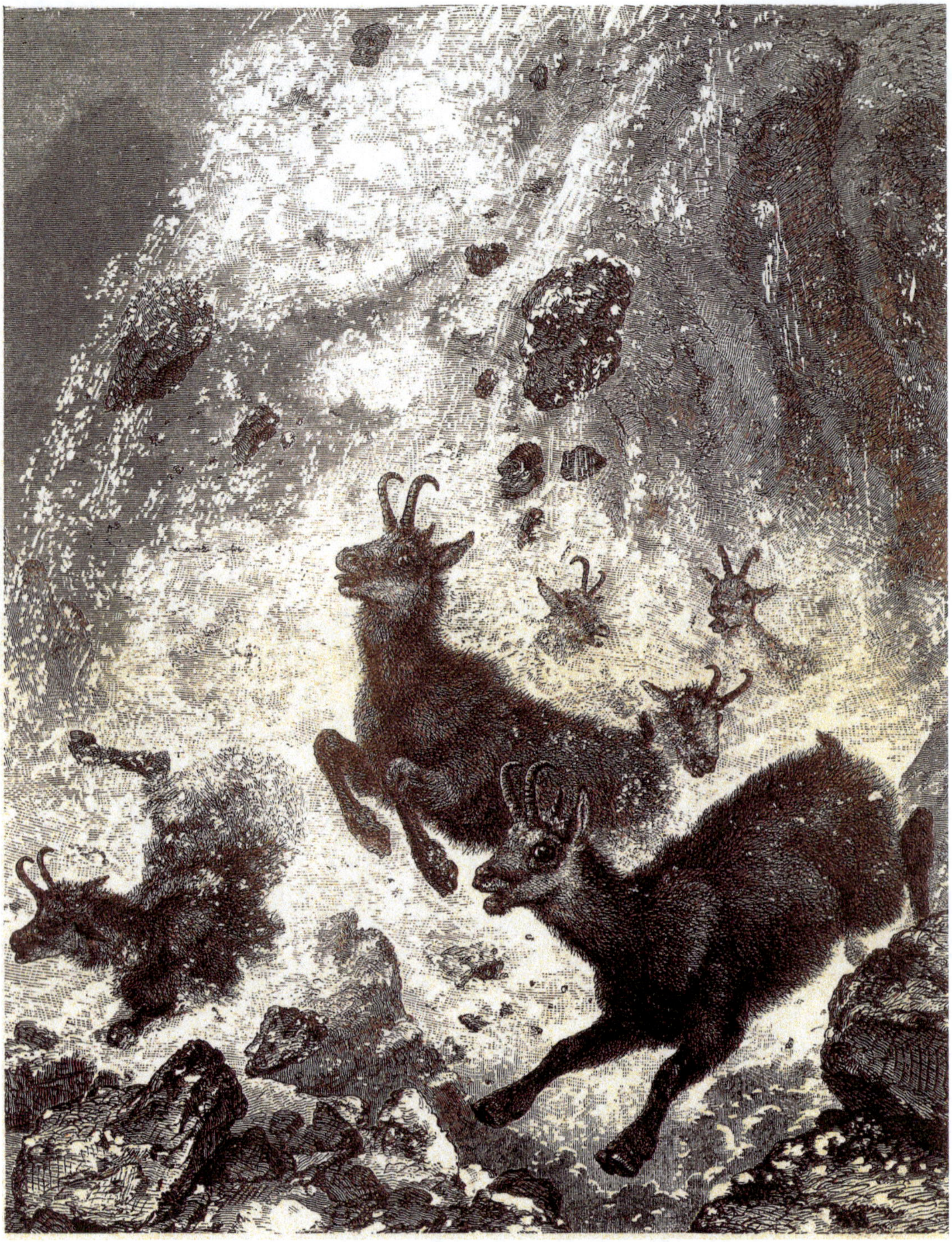

Fig. 2.80 Goats caught in an avalanche

In the case of powder snow avalanches, speeds of up to 300 kilometers per hour can be achieved. This imaginative picture of alpine goats caught in an avalanche (Fig. 2.80) comes at the end (somewhat irrelevantly) of a slim work titled *V krajích věčného ledu* (*In Regions of Eternal Ice*) by J. Kafka, published in Prague in 1901.

3 Water

3.1 Introductory Remarks

Water. H_2O, is probably the best-known chemical formula there is. It is also known as hydrogen monoxide or oxygen dihydride. It is one of the most abundant substances on the planet; much of the Earth's surface is covered with it. A view of the Earth from space, centered on the Pacific Ocean, shows almost no land at all, and the view showing the most land possible still has slightly more sea than land visible.

Water is vital for life; humans, animals, and plants are mostly water when one considers their chemical composition. In studying rocky planets in distant solar systems, the first criterion for their possibly hosting life is whether liquid water is possible. This is a combination of the star's temperature and the distance of the planet from its sun. Too close, and water boils away; too far, and the world is frozen. Our planet, of course, has optimal conditions: the occurrence of salt and fresh water in enormous quantities (possibly arriving from comets); the optimum distance of the planet from the sun (not too near, not too distant); an atmosphere regulating day and night temperatures and enabling the transport of water as vapor or precipitation over land; a suitable planetary mass whose gravity keeps atmospheric gases from floating away; and so on.

Also, while one might think of seas as barriers that isolate different land masses from one another, in fact, from ancient times onward, seas have been lanes of communication and transport, from Polynesians hopping across the Pacific from one island to another, to modern trade routes in which cargos are carried long distances, even from one side of the planet to another. At the same time, the sea has always been a vital source of food from fishing.

And yet Water is dangerous when you have too much of it in the wrong places. Flooding is a frequent cause of devastation of wide areas. Very heavy rain causes rivers to burst their banks, and the rushing water has terrific destructive force. Storms and high tides can produce marine incursions, destroying crops and leaving people homeless. It is significant that in the biblical story of Noah (and its counterparts in some other folk traditions), when God decides to wipe out mankind, Water is the element He chooses—a great flood, not a rain of fire, nor yet winds or earthquakes.

Watery disasters have links to the other elements too, of course. Storms link Water and Air, as high winds sweep the sea over coastal defenses and dykes. Wind and water together also produce that spectacular (but not actually very dangerous) phenomenon, the waterspout. And earthquakes cause the terrible tsunami, which can kill over huge distances. "Tsunami"—a Japanese word that literally means "harbor wave"—was a technical term unknown to many until the Sumatran earthquake and tsunami of 26 December 2004. Deaths from the earthquake were mostly confined to the northern part of Sumatra itself, but deaths from the tsunami occurred especially in Thailand, India, Sri Lanka, and even as far away as eastern Africa. Also, Western tourists were killed, which is a rarity in seismic disasters. In the case of the Tohoku (Japan) earthquake in 2011, the ratio of tsunami deaths to earthquake deaths was roughly 150:1.

J. Kozák, R. M. W. Musson, *The Illustrated History of the Elements*, https://doi.org/10.1007/978-3-030-21426-5_3

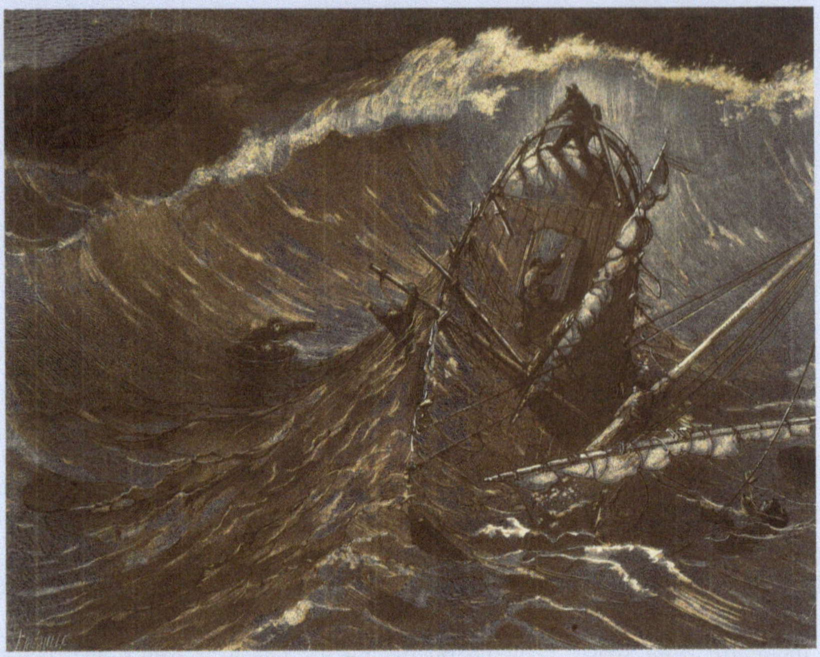

Given that we live on a planet most of the surface of which is covered by the element Water, it is no wonder that the sea has played such a large role in man's history and consciousness. On the positive side, the sea is a source of food and communication; on the negative side, when angered, the sea is a frightening peril, as this engraving (source unknown) demonstrates. The phrase "mountainous seas" is highly appropriate here. The small human figure, perched on the stern of the boat as it pitches alarmingly, seems to be weighing up his chances of survival and finding them poor

Included here are also pictures of shipwrecks—something that used to be a common source of anxiety to travelers in the days of sailing ships, which could be blown onto rocks, hit hidden reefs, or be struck by an iceberg, as the *RMS Titanic* famously was in 1912. Today, in the era of modern navigational tools, one seldom hears of shipwrecks, and when they do occur, fatalities are few.

We cannot live without Water, but its dangerous manifestations are well shown in the following pages. Above is a generic image of the element in its most frightening manifestation to set the tone.

3.2 Floods

Mankind is not an aquatic animal. We can take to the sea in ships and we can swim in water for recreation, but we are not conditioned to live other than on dry land. So when dry land ceases to be dry, we have a problem, as the following images show.

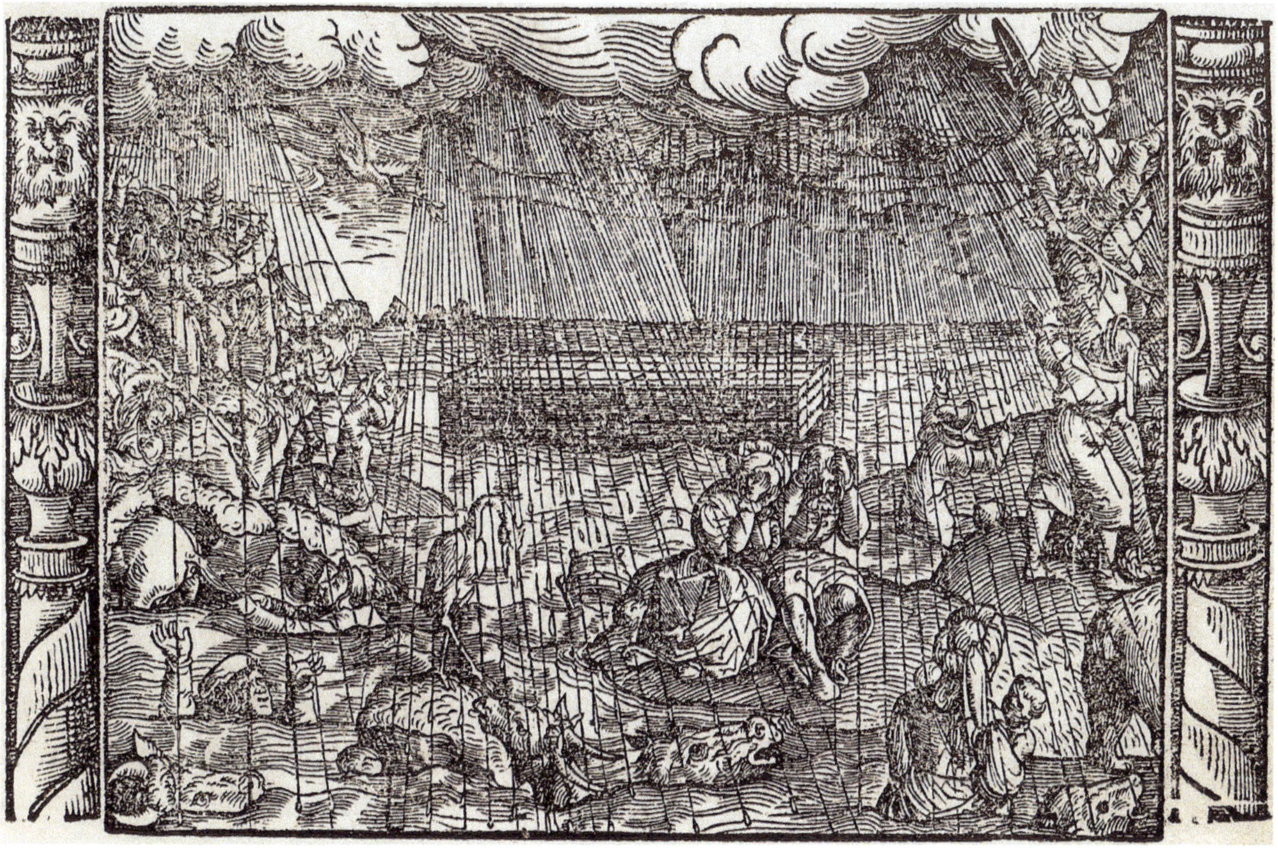

Fig. 3.1 The Biblical flood—a sixteenth-century view

Significantly, in the Bible, when God decides to eliminate mankind, he chooses Water as the element to do it with, sending a flood over the world to drown everyone—except, of course, for the godly Noah and his family, and the breeding stock of animals that he takes with him into the ark. One rather wonders how trees and plants survived, and what Noah and his family did for crops when the waters subsided. In this woodcut from an early Czech Bible (Fig. 3.1) published by Jiří Melantrich in 1557, the rain is seen falling heavily, and people and animals are already beginning to drown. It is interesting to note that while that the shape of Noah's ark that is familiar to modern readers is ship-like with a pitched roof, early depictions, like this one, show the ark as a gigantic box. Given the appearance of similar flood narratives in other mythologies around the Middle East, it has been suggested that these are vague memories of some actual calamity, possibly the drowning of a land basin to form the Black Sea.

It should be pointed out that most of the wood blocks from which Melantrich printed the illustrations for his 1549 and 1557 Bibles were not created in Prague; they were made between 1532 and 1534 in Nuremberg by German artists and were used to illustrate Luther's Bible in 1535. Shortly afterward, these blocks were sold to Czech printers in Prague, who used them to illustrate their own editions of the Bible published in 1537, 1549, and 1557.

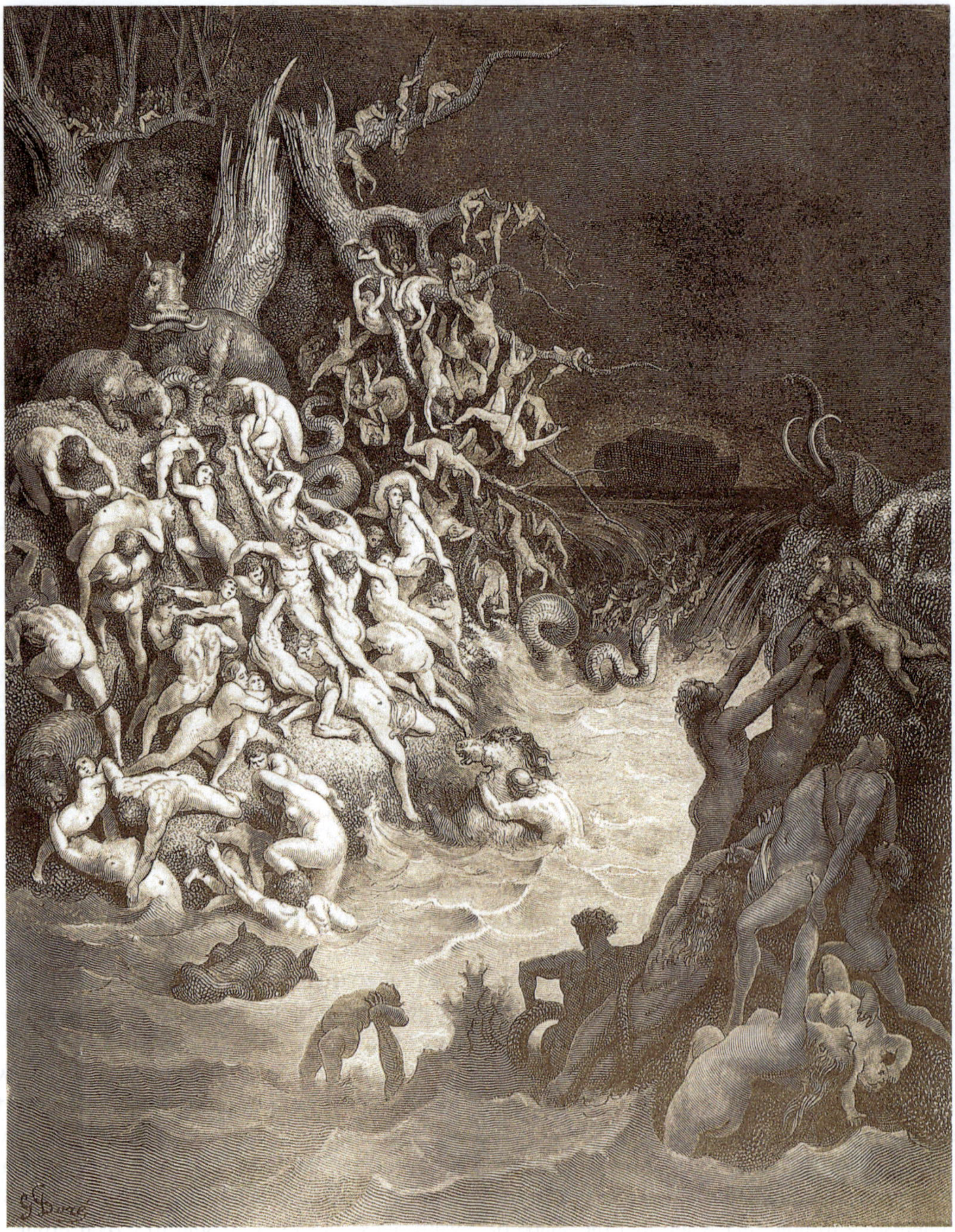

Fig. 3.2 The Biblical flood—a nineteenth-century view

This is a nineteenth-century view of the Deluge, a biblical illustration by Gustav Doré (Fig. 3.2). Noah's ark has a more familiar shape, but it is sailing off into the distance, barely noticeable. The image focuses instead on the sinners trying to cling to what remains of high ground. For some unaccountable reason, all of their clothes seem to have fallen off.

Leaving aside divine anger, large floods generally occur for one of four reasons: the melting of winter snow in spring, short torrential rains, very long intensive precipitation, and storm surges, when strong cyclonic weather conditions in conjunction with high tides drive sea waters inland. We can start with some cases of marine inundation and then move on to the subject of river flooding.

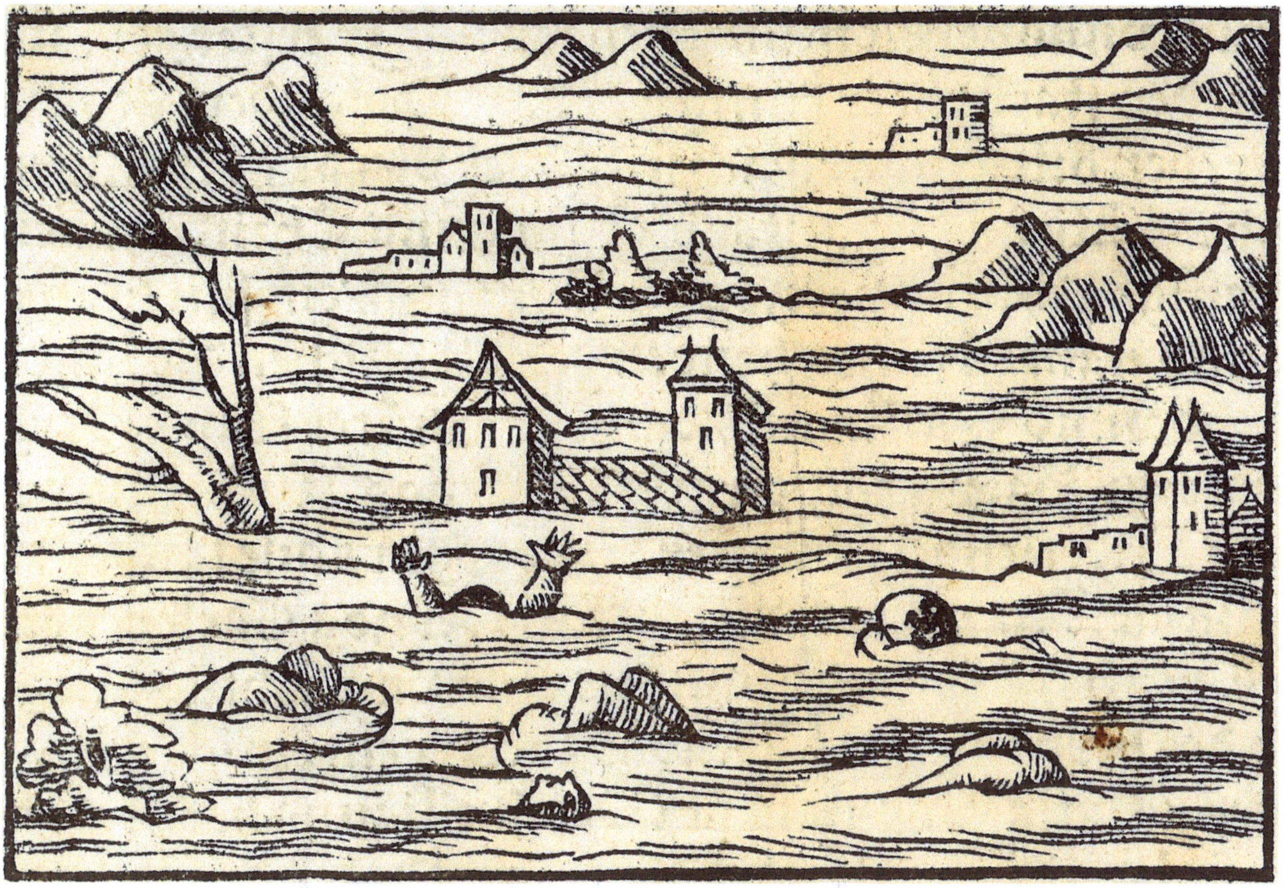

Fig. 3.3 The 1421 St. Elizabeth's flood

This woodcut from the *Nuremberg Chronicle* of 1493 (Fig. 3.3) supposedly depicts a great flood in Dordrecht in 1421, but, like so many images from this source, it is a stock "stamp" image, bearing no relation to any actual place, and certainly not Dordrecht, given the mountains sticking up out of the floodwaters. This was the devastating St. Elizabeth's flood of November 1421, considered to be one of the worst floods in history. A combination of a storm in the North Sea, a spring tide, a tidal surge, and strong north winds caused river dykes to burst and the land to flood. The text of the chronicle misdates the event to 1420. The little man with his hands in the air, as he is overwhelmed by the water, cuts a particularly pathetic figure.

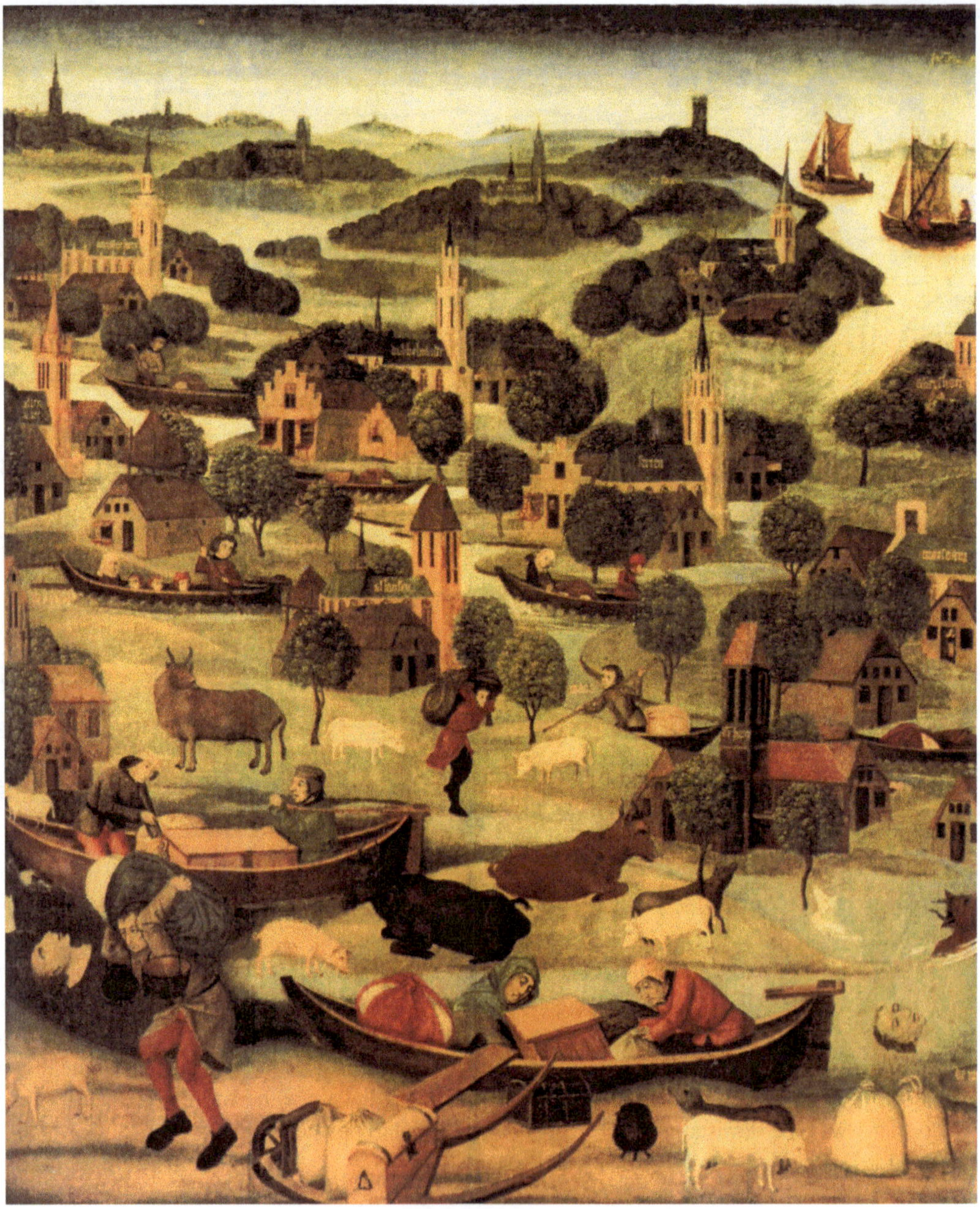

Fig. 3.4 The 1375 St. Catherine's flood

This is a rather more accurate depiction of flooding in the Netherlands (Fig. 3.4). This painting shows the St. Catherine's flood of 1375. The Dutch village of St. Catherine (now called Oostburg) was a small settlement drowned by the disaster. The painting (in the Rijksmuseum, Amsterdam) is by a Dutch artist of the late fifteenth century, known as the Master of the St. Elizabeth Panels. In the foreground, people in boats are attempting to carry their goods to safety. Note the churches perched on small hills in the background. Unlike the rocky mountains in the previous figure, these are man-made mounds known as terps. Especially in the northern Netherlands, churches would be built on terps so that they would remain above the floodwaters, as shown here, thus providing places of refuge.

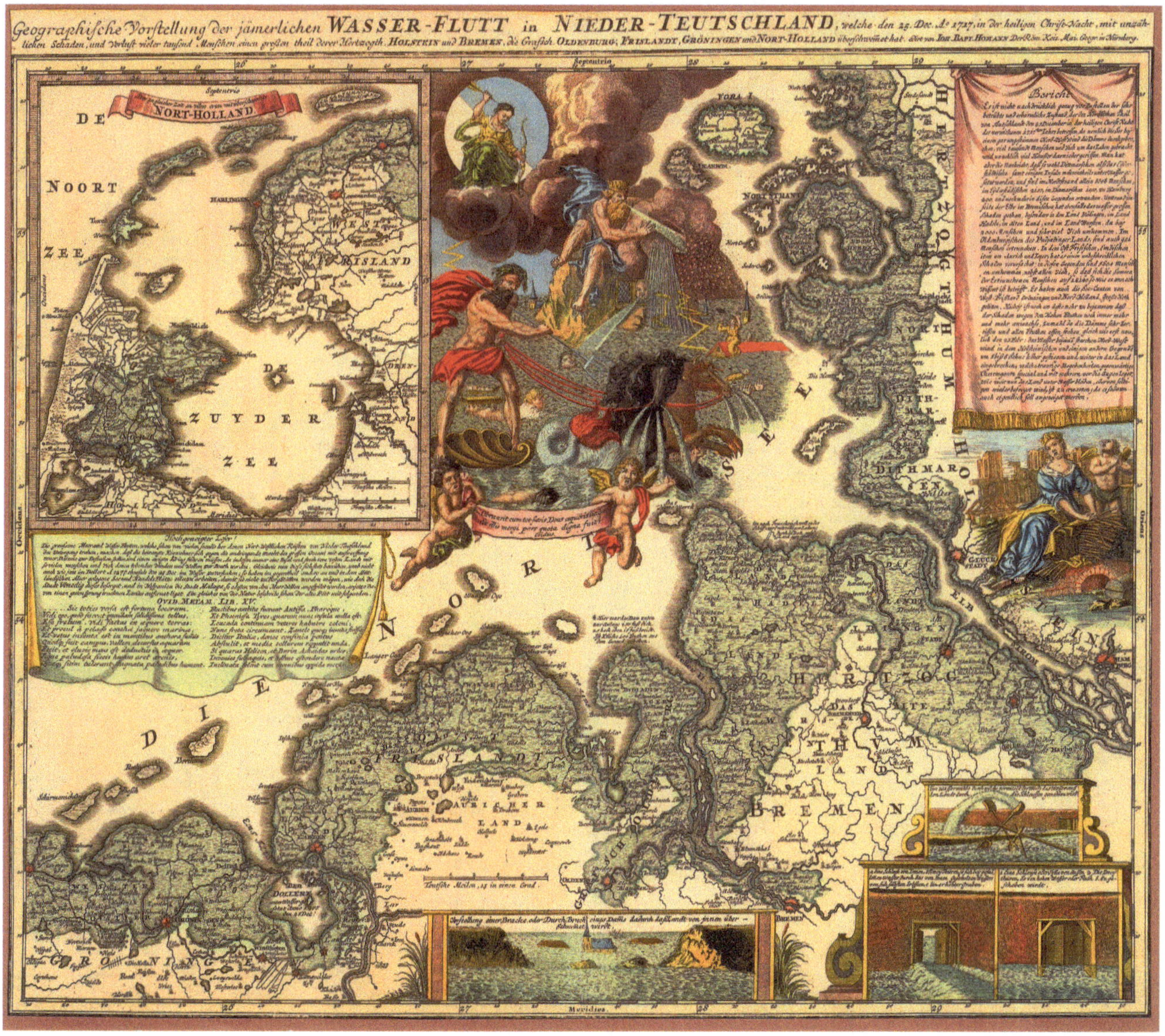

Fig. 3.5 A map of the Christmas flood of 1717

The continental coasts of the North Sea, from the flat coastal area of western Denmark and Lower Saxony down to the Netherlands, have been often visited by great natural floods in history. Violent storms accompanied by heavy rain and spring tides were the cause of these watery disasters. In the low-lying districts, the effects of the flood were multiplied by the failure of numerous artificial polder and river dams, which often did not bear the onrushing surge of floodwaters. As one such low-lying country, the Netherlands has been often subject to flooding throughout its history. The last really great flood occurred at Christmastime, 1717, and is thus often referred to as the St. Nicholas flood or the Christmas flood. A northwesterly storm in the North Sea hit not just the Netherlands but also the northwest coast of Germany and parts of Denmark as well, causing immense damage. About 14,000 people, and many cattle, were drowned. This elaborate map shows the areas affected and was published the following year by the accomplished German cartographer Johann Baptist Homann in Nuremberg (Fig. 3.5).

Homann's map is decorated by three small pictures at the bottom and by a special inset showing "North Holland." There are also rich verbal data engraved inside the map, and the whole should be appreciated as one of few geographic maps related directly to a natural disaster—a flood.

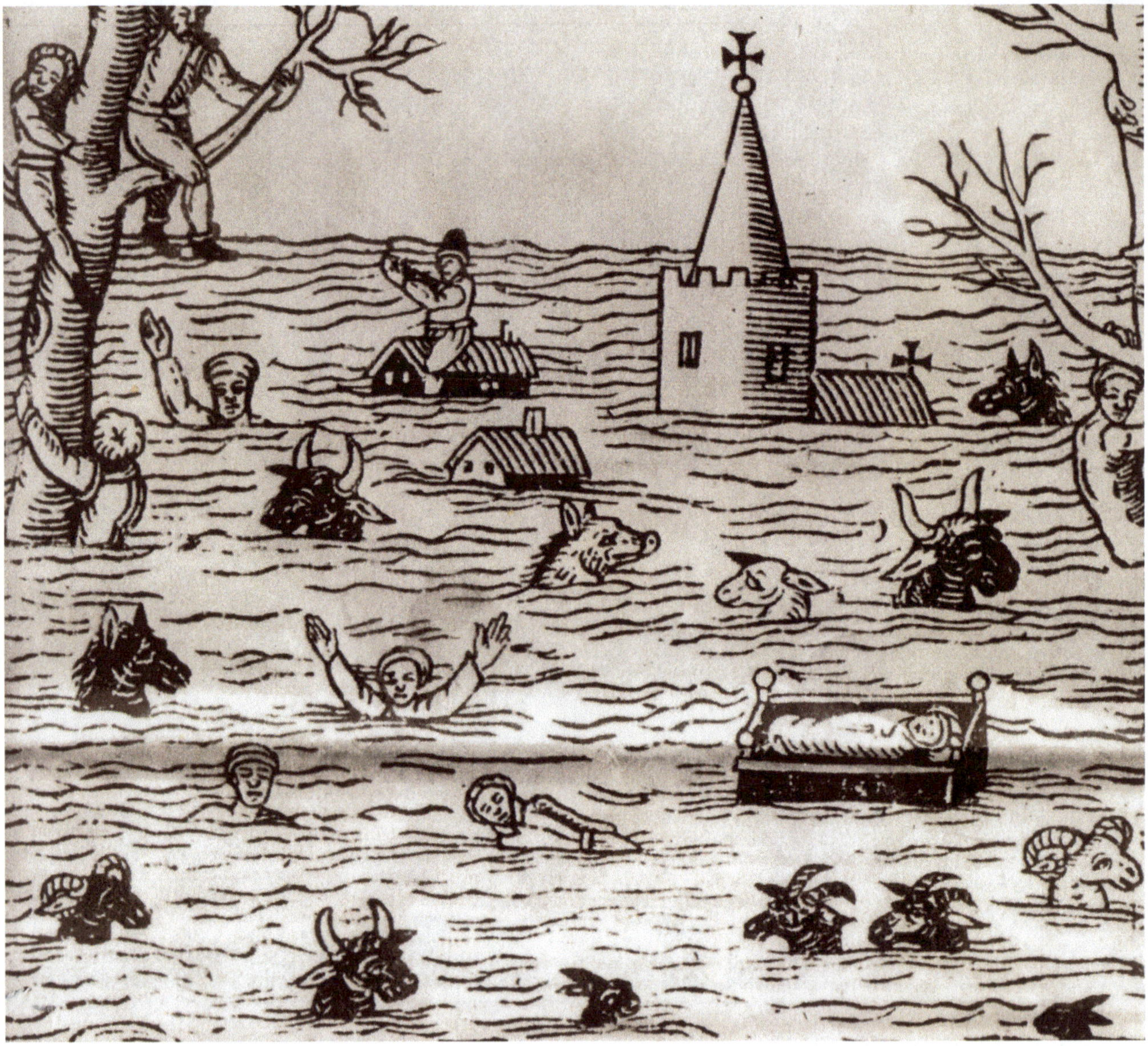

Fig. 3.6 Flooding in the Bristol Channel in 1607

Storm surges have also affected parts of Britain, most notably in 1953, when much of East Anglia was affected. This woodcut comes from the cover of a pamphlet describing flooding in Somerset in 1607 (Fig. 3.6). In this case, the flood came as a surprise, since no stormy weather was observed at the time. However, a combination of a high spring tide and strong winds from the southwest were shown to be quite sufficient to produce the reported flooding, which also affected Monmouthshire (Wales) on the opposite side of the Bristol Channel. A second pamphlet describing the flooding in Wales carried on its front page a very similar but not quite identical woodcut to that shown here, again demonstrating the use of stock images.

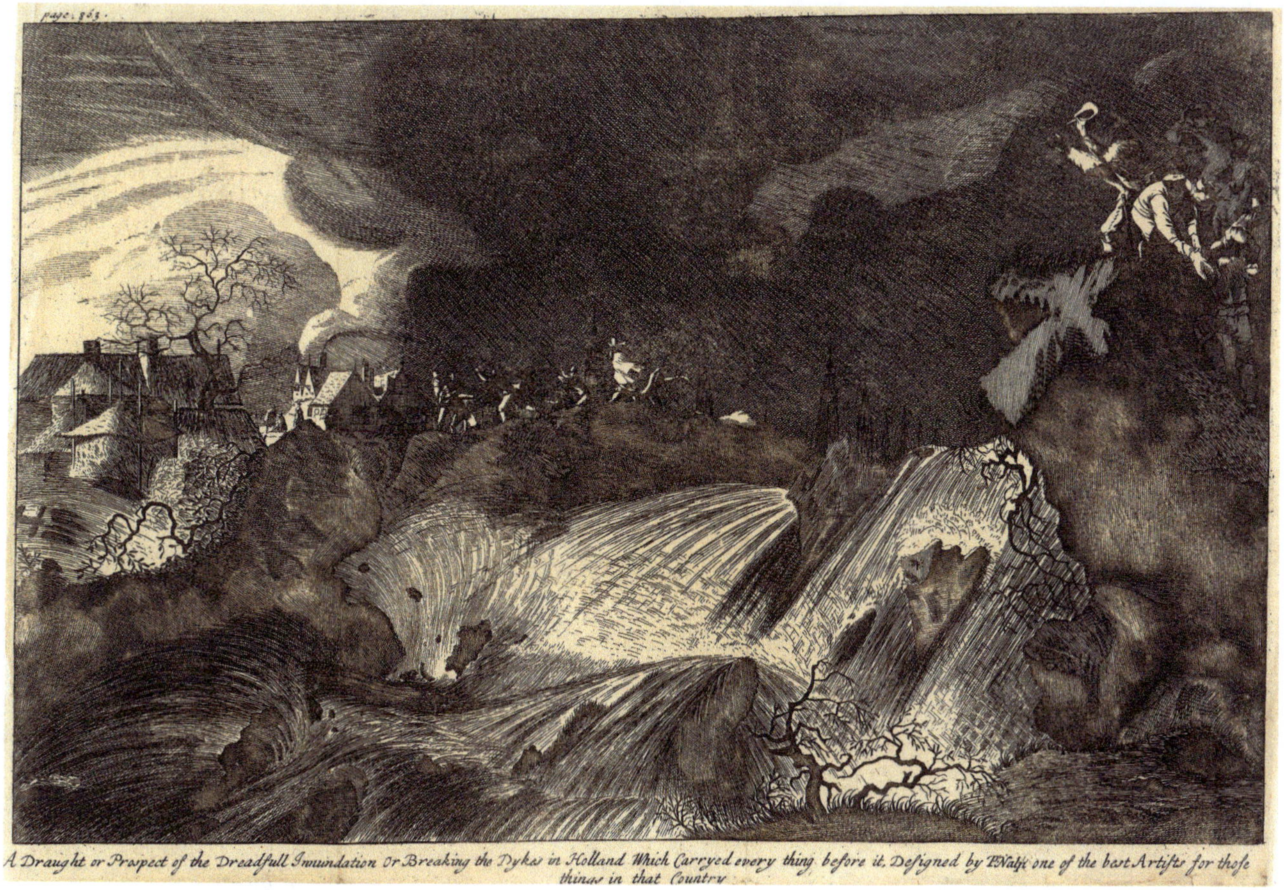

Fig. 3.7 Dykes breaking in the Netherlands in 1675

In the Netherlands, much of the territory lies below sea level. This means that in the "Low Countries" there is an additional factor in the causation of floods. This is the breaking of the dykes that line the streams of mighty rivers flowing through the Low Countries into the North Sea; they also line the coast itself. Quite clearly, any dyke collapse results in inundation of the surrounding land. It follows that the Netherlands has to suffer not only from heavy rains on its own territory: floods can also be initiated by heavy rains or mass snow-melting in distant countries, in which the big rivers such as the Rhine rise. This was a historical risk, but today technological developments have reduced the risk of breaking dykes to a minimum.

The picture here shows another flood in the Netherlands (Fig. 3.7), probably that of 3 November 1675, which inundated most of the southern parts of the country, including Amsterdam and Utrecht, together with parts of the north, around Hoorn. The artist here imaginatively captures the moment of the dykes breaking, waterlogged and weakened after the extended heavy rains. The caption of the image (source so far unidentified) reads, in English, "A Draught or Prospect of the Dreadfull Inundation or Breaking the Dykes in Holland Which Carryed every thing before it. Designed by P. Nalp [?] one of the best Artists for those things in that Country." (The name of the artist is unclear and does not seem to correspond to any known Dutch artist of the period.)

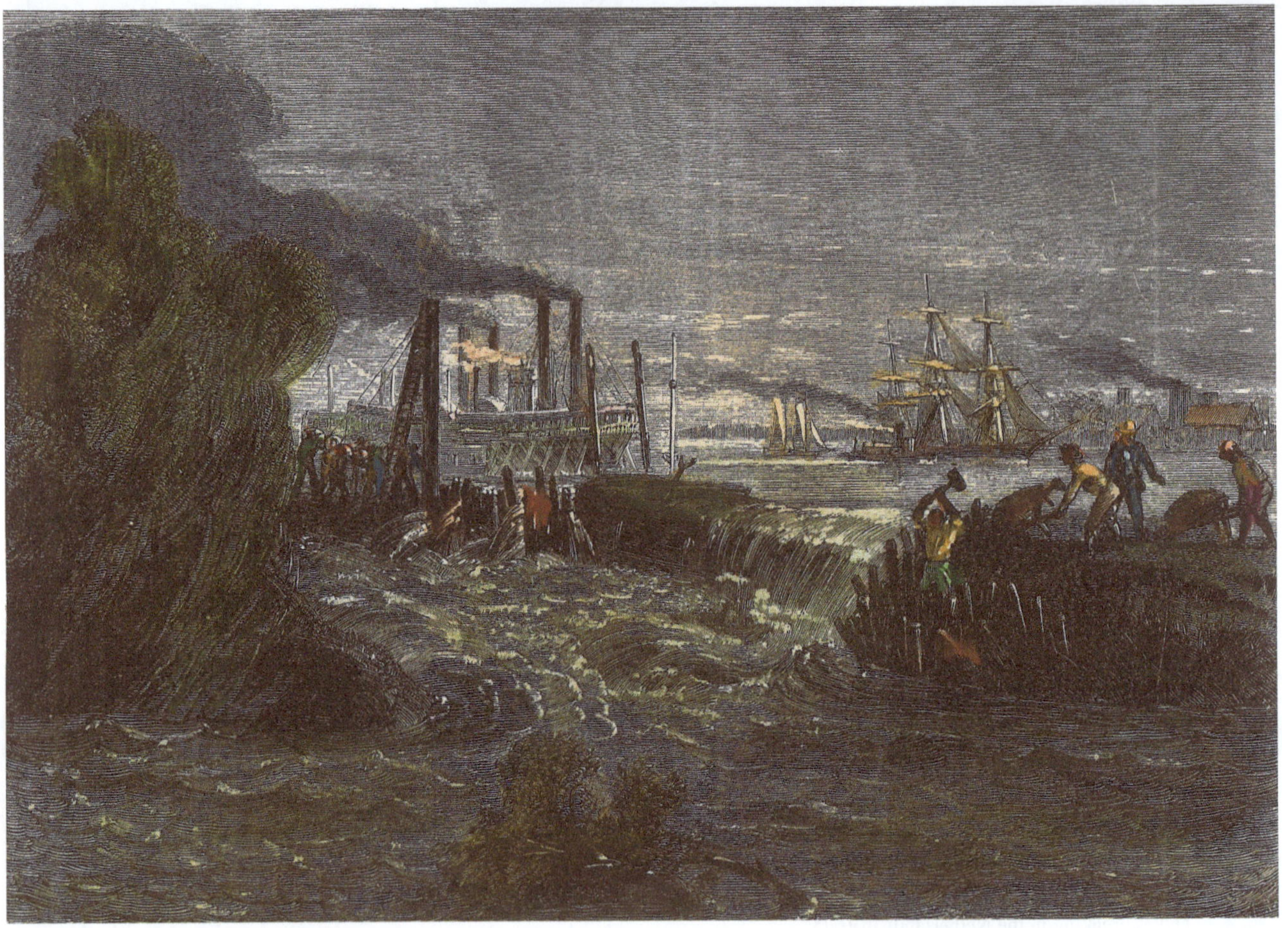

Fig. 3.8 Mississippi levees breaking

On the other side of the Atlantic, the same situation arises. Dykes called levees protect low-lying land from river flooding. In places, the Mississippi River actually runs some meters above the surrounding fields. Inevitably, sometimes the levees are breached and flooding results, as shown here. This nineteenth-century engraving, published in Germany, has been rather crudely colored in by hand (Fig. 3.8).

This type of river flooding is restricted to those few cases where river courses have been confined by man into narrow channels bounded by dykes or levees; that a river should flow at a level actually higher than the surrounding countryside is obviously not something that can occur naturally. But even when a river is flowing in its natural channel, it is commonplace for it to burst its banks after very heavy rain. Not for nothing is the land either side of a river known as its floodplain. The next group of images show more conventional river flooding.

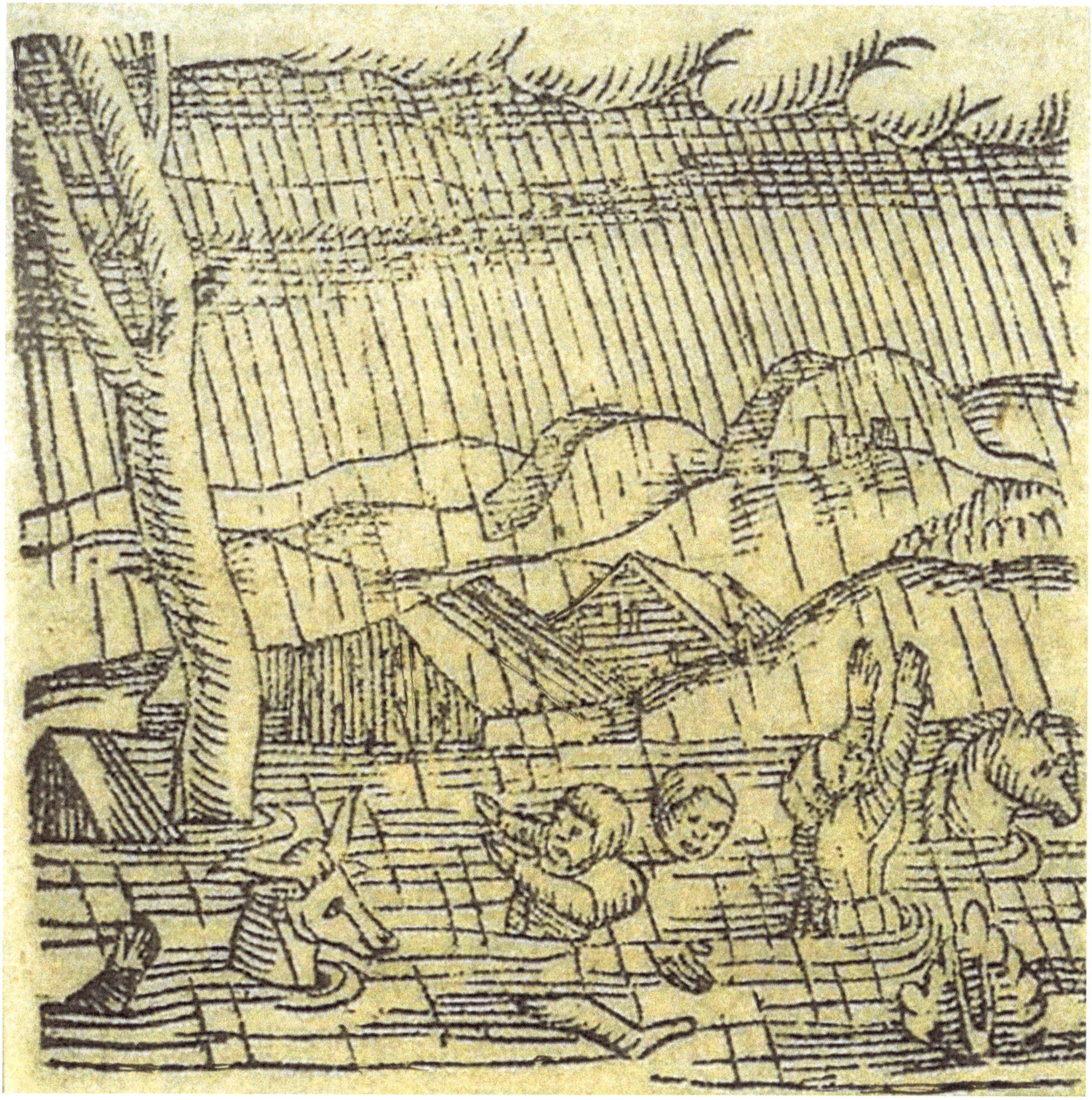

Fig. 3.9 A sixteenth-century picture of flooding

This small, unidentified woodcut (Fig. 3.9) shows an early scene of flooding, probably generic. People, cattle, and horses seem to be doing their best to stay afloat in the rising waters. This anonymous image of calamity caused by the bursting of riverbanks in heavy rains comes, most probably, from the sixteenth century.

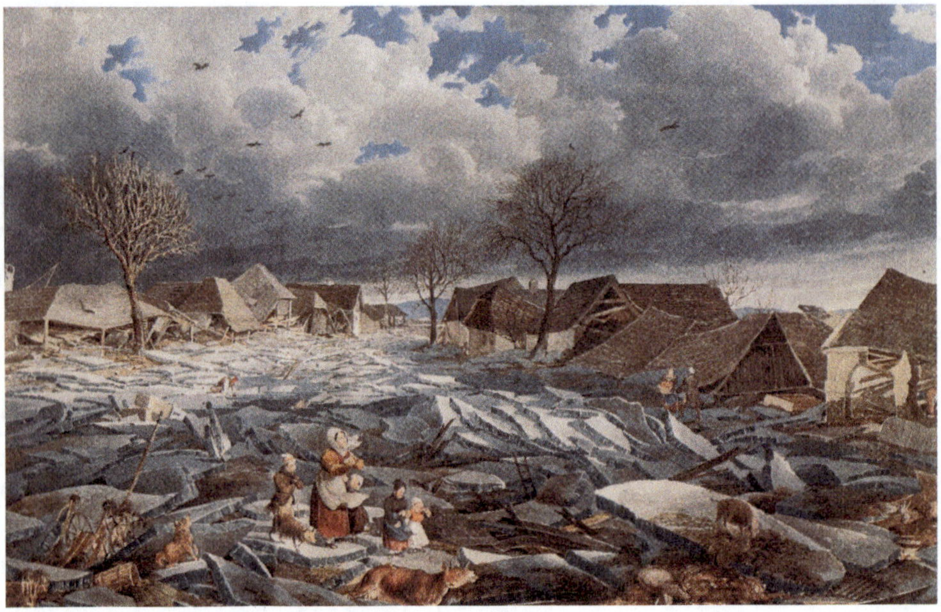

Fig. 3.10 The 1830 flooding of Vienna—the Leopoldau district

Since major cities tend to be built next to rivers, when a river bursts its banks, it can lead to serious damage. Our next two illustrations depict a river flood—the 1830 flood in the River Danube basin, in this case due not so much to rainfall as to ice.

The winter of 1829–1830 was extremely cold; the thickness of ice covering rivers and lakes reached 30 centimeters, and the Danube was actually dammed by the ice. On 27 February 1830, the ice dam on the Danube broke, sending a wall of water down the river. The flood hit Vienna between 28 February and 1 March, when the lower parts of the town were under water (for instance, the large Augarten Park in the suburbs was reported to be flooded to a depth of 1.75 meters). This painting (Fig. 3.10) by the contemporary artist Eduard Gurk shows the effect in the Leopoldau district of Vienna, where the flood carried great ice floes with it. A recent study of Danube floods has been published by Blöschl et al. (2015).

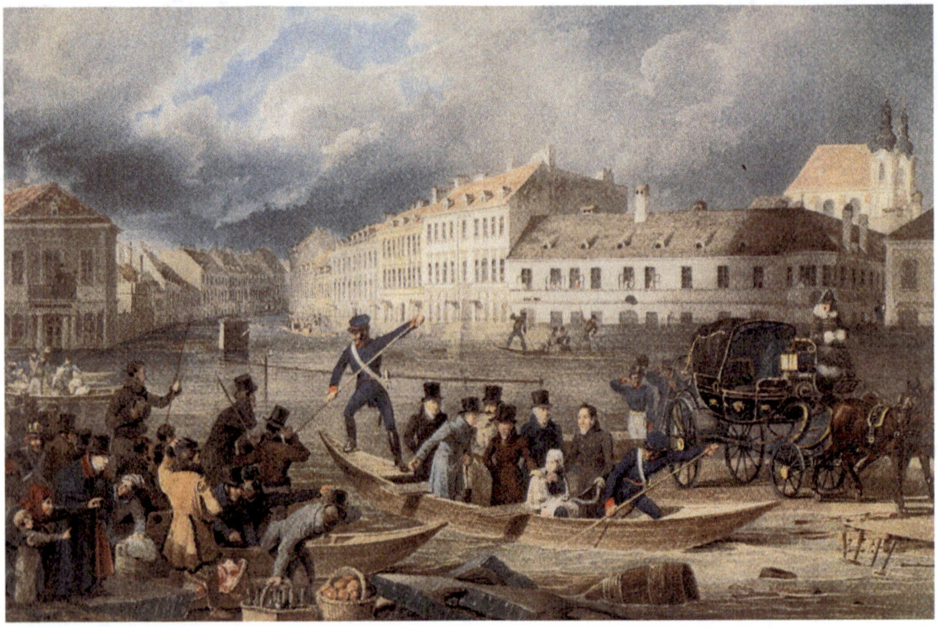

Fig. 3.11 The Roßau district of Vienna in 1830

Another of Gurk's paintings, this shows a scene in Schmidgasse, in the Roßau district of Vienna, on 2 March 1830 (Fig. 3.11). The military are helping people evacuate their flooded homes.

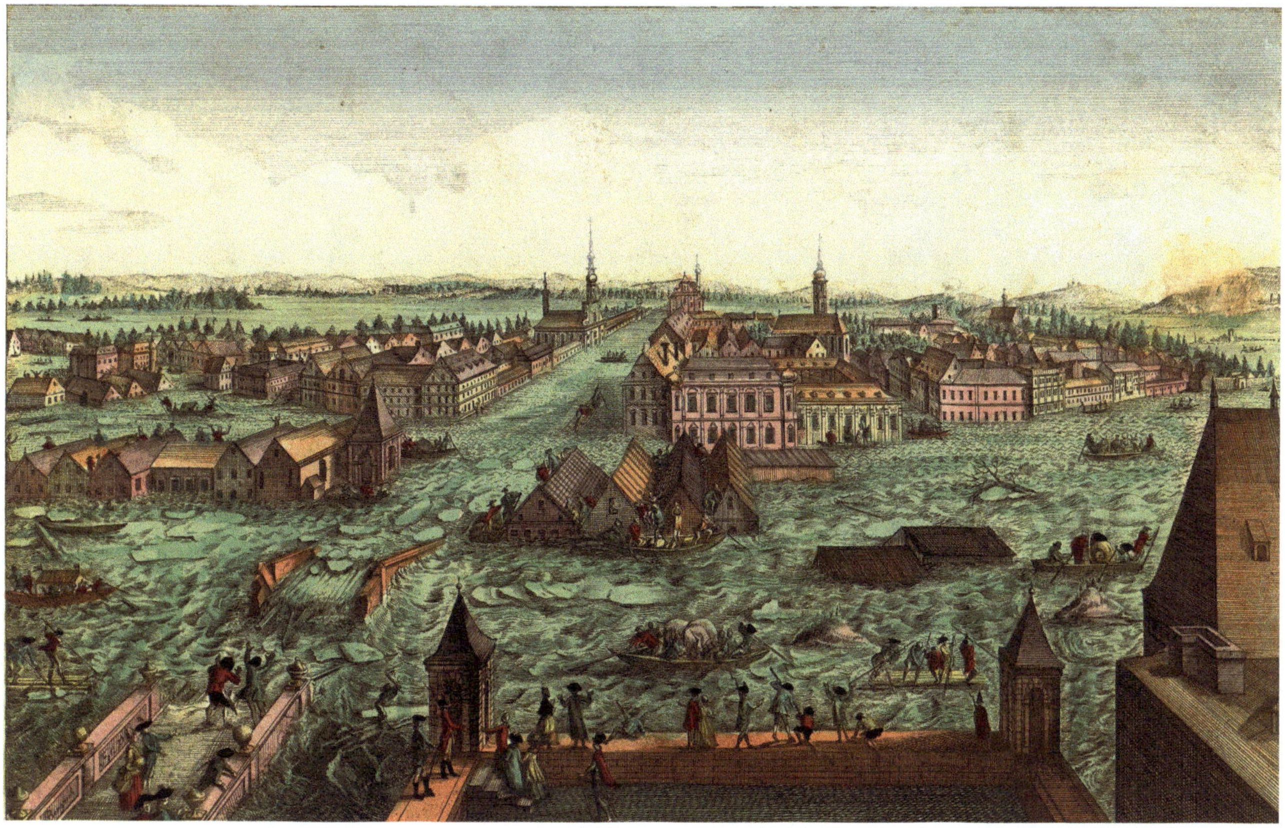

Fig. 3.12 Flooding in Vienna in 1784

The winter of 1783–1784 was harsh and long-lasting over much of Europe because of the widespread climatic disruption caused by ash and gas expelled by the great fissure eruption of the volcano Laki in Iceland. The long-lasting winter resulted in thick snow cover in central Europe combined with extreme thickness of river and lake ice. Then, in late February 1784, a warm southerly wind arrived, which quickly thawed the frozen rivers, and the result was flooding from Belgium and France through to Bohemia, Slovakia, parts of Germany, and Poland.

This scene shows the flooding in Vienna on 28 February 1784 (Fig. 3.12). The view is of Leopoldstadt, and the image is what is known as a *vue d'optique*—a print designed to give an illusion of depth when viewed through a zograscope (a tabletop viewer featuring a large magnifying glass and sometimes an angled mirror), which enhances the sense of perspective in images of this type (Fig. 2.18 is another example). Typical characteristics of such images are bright hand-coloring and strong diagonal lines, as shown here. Also, the text of the title will be mirrored. The mirror-text title of this image, when reversed, reads *Ravage causé par le debordement du Danube a Vienne, residence de l'empereur dans le fauxburg Leopold.*

The zograscope was developed in England in the 1740s, and many perspective views were printed for it, generally featuring urban prospects, where the strong diagonals of the street pattern provided an extra visual cue to depth.

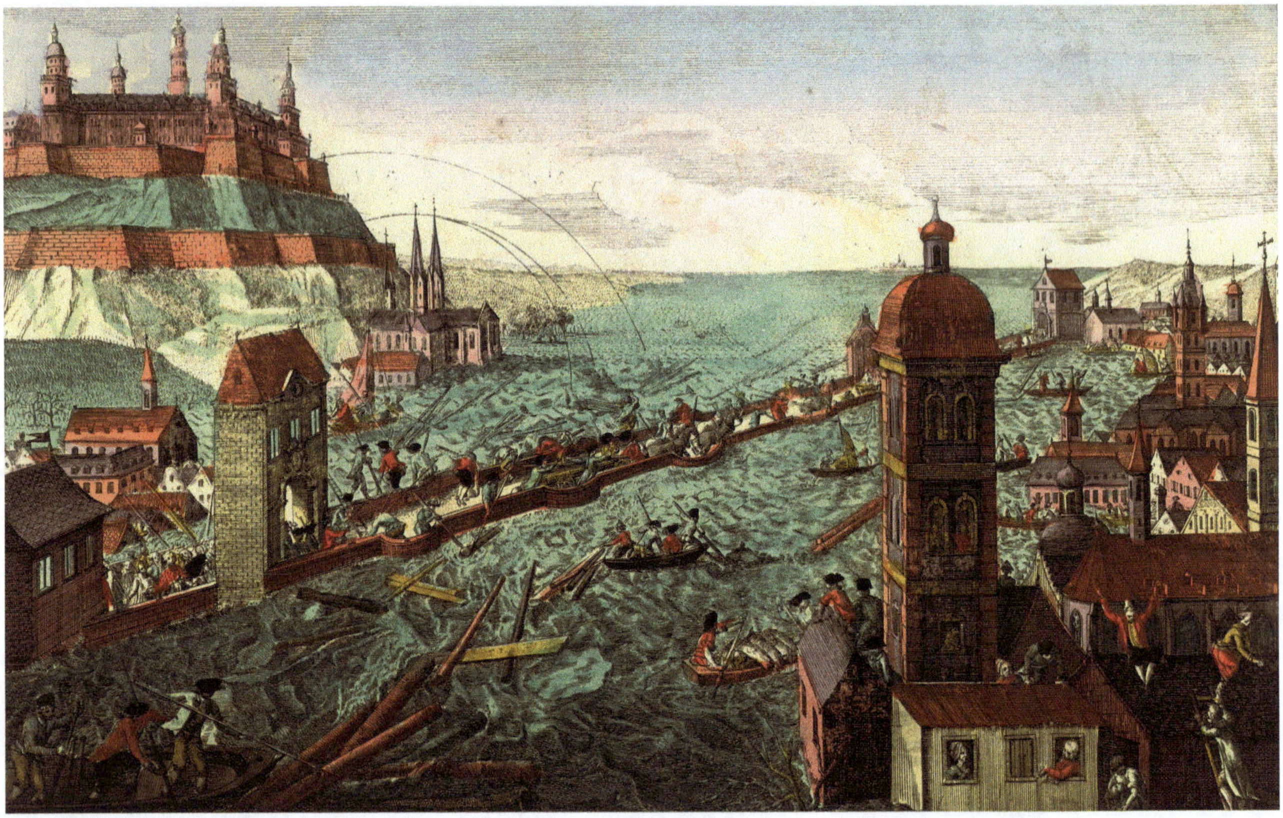

Fig. 3.13 Flooding in Wurzburg in 1784

The city of Wurzburg, on the River Main, was affected in a similar way to Vienna. Shown here (Fig. 3.13) is another *vue d'optique* from the same series as the previous figure (Fig. 3.12), documenting the effects of the February 1784 flood. The title of this one is *La Grande Desolation Arrivée a Wurzbourg par le Debordement du Mayn*. The accompanying text for both images is presented in both German and French. Given that Wurzburg is on the River Main and Vienna is on the River Danube (and Prague on the Vltava and Cologne on the Rhine were also affected), one gets an idea of the extent of this great flood.

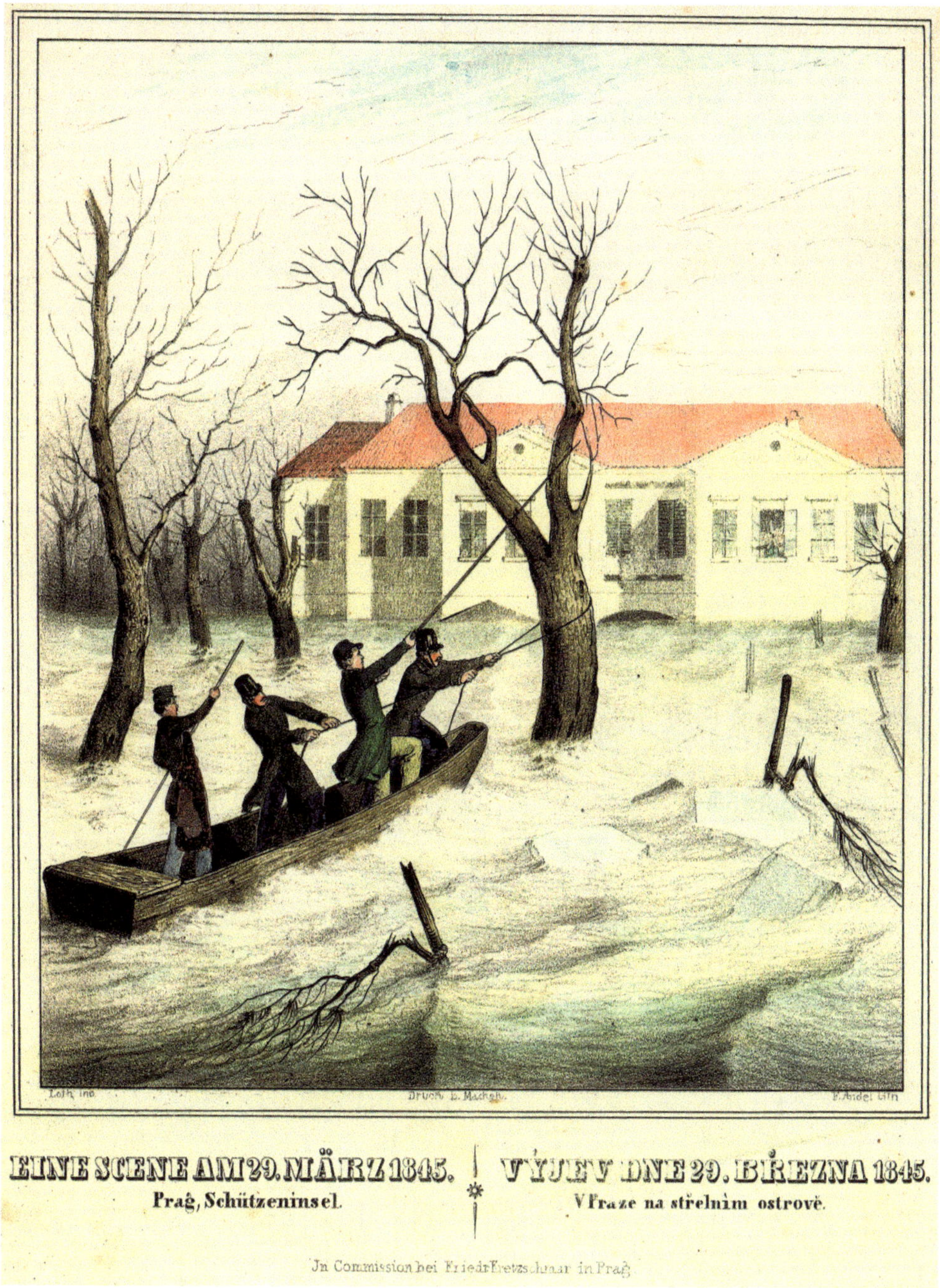

Fig. 3.14 Floods in Bohemia in 1845

After the floods of 1830, the next similar, and even stronger, event was the flood that affected central Europe in March 1845, known as the Spring Flood. In some places, such as Dresden, this was even stronger than the 1784 flood. Again, the sudden melting of ice was the cause. A sudden increase in temperature in central Europe while it was still covered by a blanket of snow, in combination with intensive rainfall in west Bohemia, resulted in a giant flood. What started as a local calamity quickly grew into a large mid-European disaster. The Elbe River valley was hit particularly hard, since the Elbe gathers together all of the outflow from the rivers in Bohemia. In this image from Prague, chunks of ice can clearly be seen bobbing about, while four survivors use ropes to try and keep their boat steady in the rough water (Fig. 3.14). This image is one of a series of prints issued by the publisher Friedrich Kretzschmar in Prague.

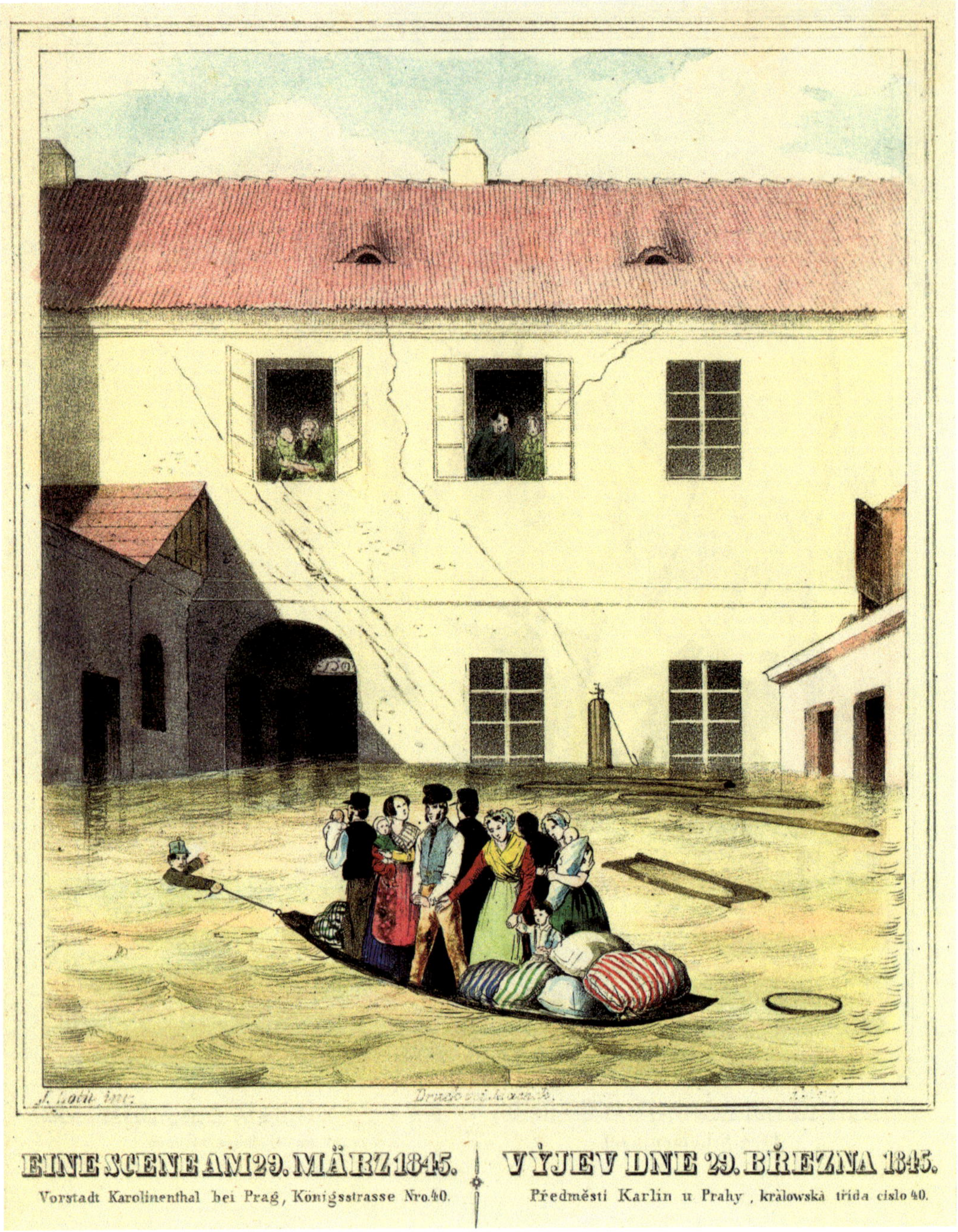

Fig. 3.15 Prague flooded in 1845

This print (Fig. 3.15) is another of the series published by Friedrich Kretzschmar. Again, the scene is Prague, and the caption (in both German and Czech) identifies the scene as occurring on 29 March 1845. This time, the scene is set in an enclosed courtyard, where the water is not so choppy. A group of residents, with some of their belongings, escape on what appears to be a very overloaded little boat. They are being pulled to safety by a man wading in water that comes up to his chest, while others watch from an upper floor.

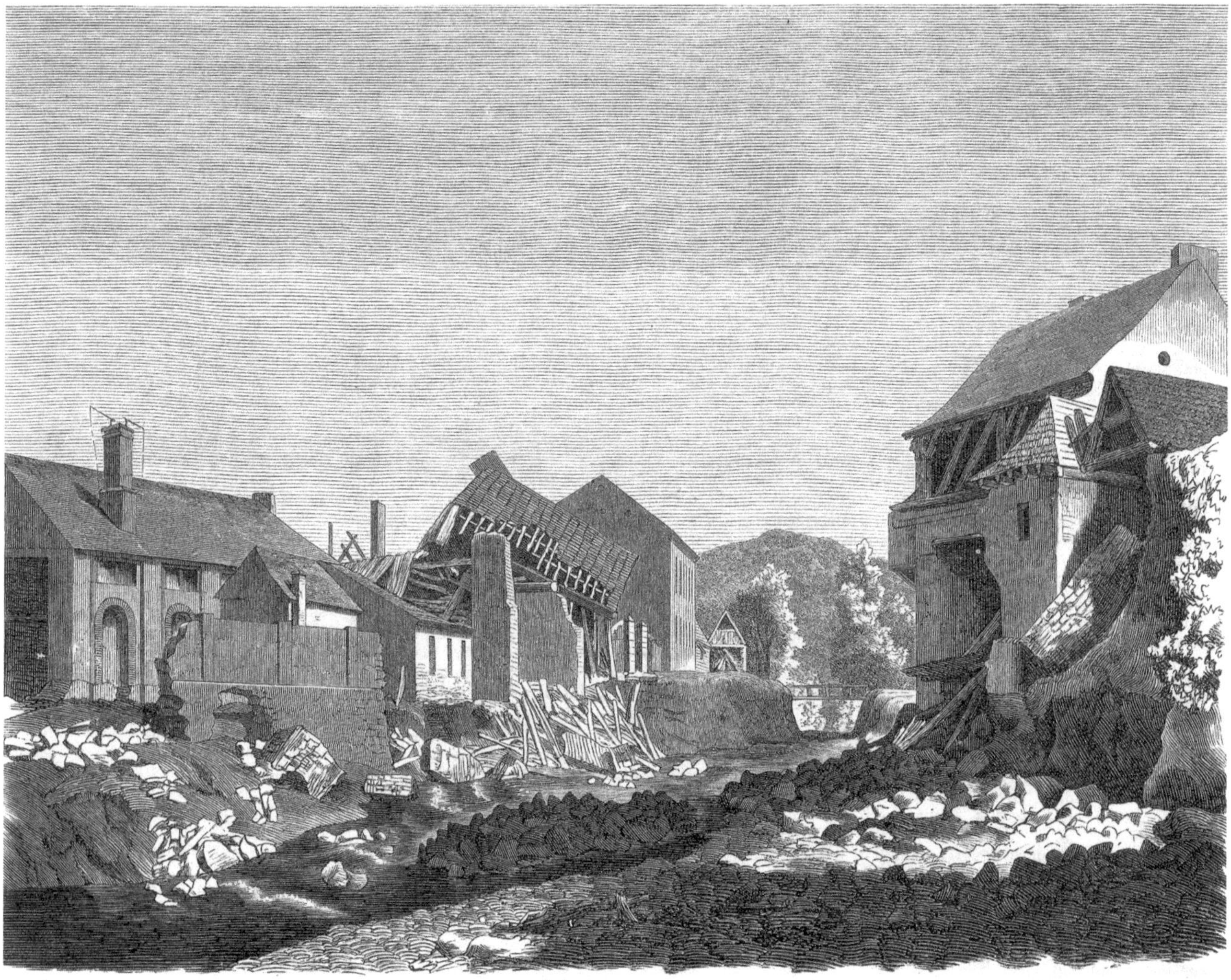

Povodeň v Čechách. 42. Ves Komárov na Hořovicku. (Dle skutečnosti kreslil Ed. Herold.)

Fig. 3.16 Flash flooding in Bohemia in 1872

The destructive power of floods is not only evident in continent-wide calamities such as the floods of 1784 and 1845. In May 1872, severe convective storms struck an area west of Prague in what was then the Bohemian province of the Austro-Hungarian Empire. Intense rain beat down, with hail and even tornadoes. The event was one of the worst natural disasters to have affected Czech territory; some 240 people lost their lives. Near the village of Odlezly, in the wooded countryside west of Prague and north of Pilsen, the storm triggered a landslide, which dammed a local stream, forming a lake.

When the dam eventually burst, muddy water deluged down the valley, wreaking destruction (Fig. 3.16). The damage was depicted in a series of sketches by Edward Herold and other artists, published in the illustrated magazine *Světozor*.

This and the next three images are part of a long series of these depictions showing the effects of the extensive flood of 25–27 May 1872, in central and western Bohemia. Steady rainfall was supplemented by numerous local rainstorms, which dramatically increased the water level in already swollen streams. The population of the basins of the Vltava and Berounka Rivers suffered most of the heavy damage and casualties. Bohemia topographically is like a giant bowl, combining the basins of the Vltava and Elbe Rivers; all of the extensive flat plain west of Melnik town, where the Vltava meets the Elbe, temporarily became one big lake, and people had to use boats to travel.

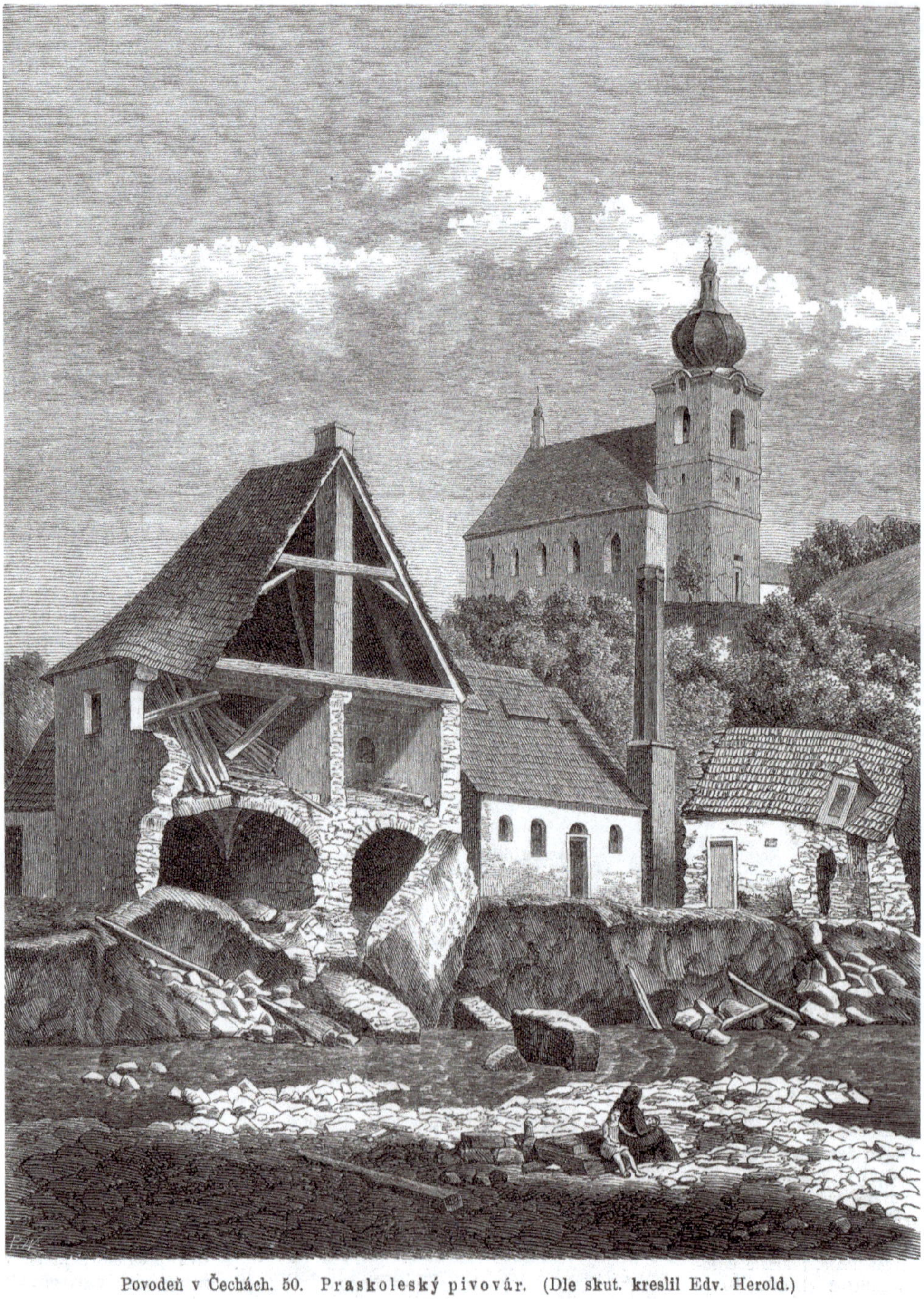

Fig. 3.17 The 1872 flood at Praskolesy

This print (Fig. 3.17) is another of Edward Herold's illustrations (drawn in situ) of damage from the flash flood caused by the dam collapse in May 1872. Here, the subject is the destruction of the brewery in the village of Praskolesy. The force of the water has ripped off a wall of the building and much of the riverbank.

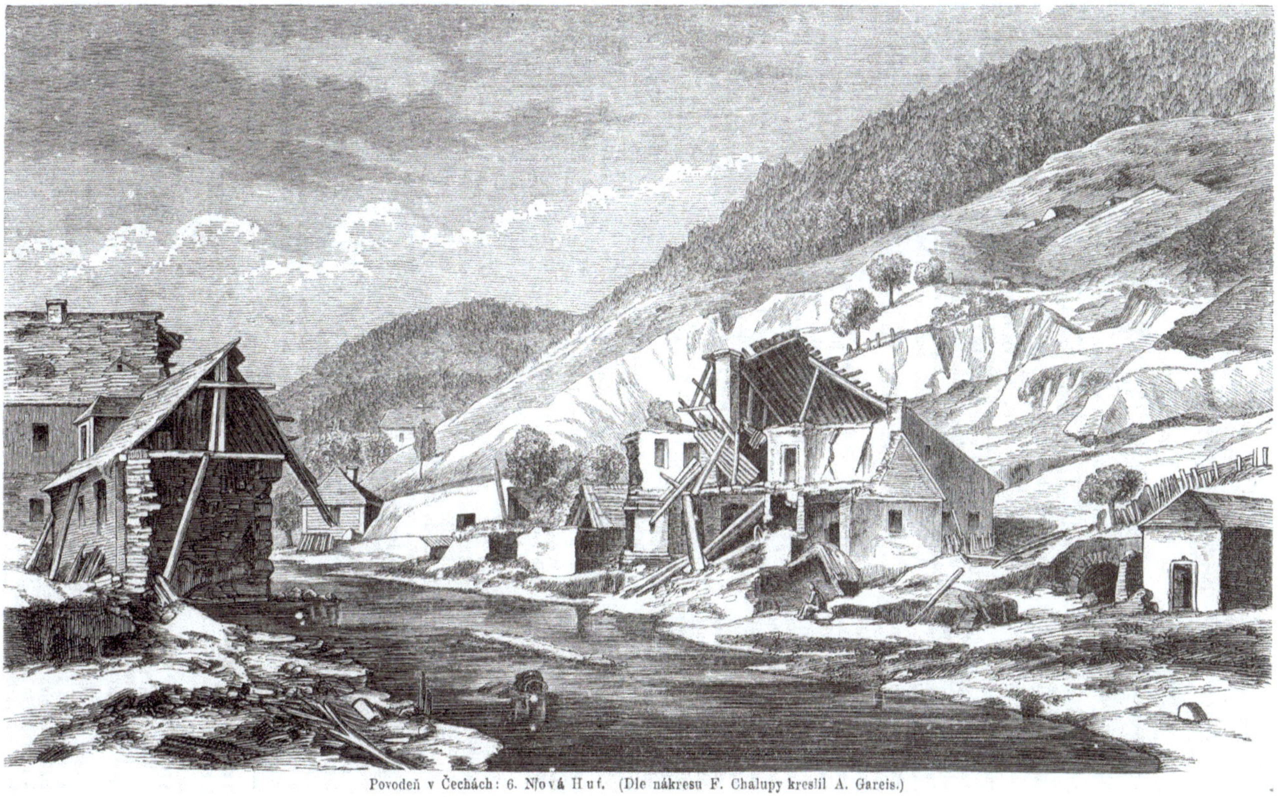

Fig. 3.18 The 1872 flood at Nová Huť

The capacity of flowing water to move objects increases as the fifth power of the increase in speed. In other words, water moving twice as fast as normal can shift rocks 32 times as much. In the flood disaster in central Europe in 2002, videos showed floodwaters rushing down streets and picking up and tossing container lorries about as if they were made of balsa wood. In 1872 there were no container lorries, but the effect of a wave of water, mud, stones, and debris was quite damaging enough to any buildings close to the course of the river, as shown here (Fig. 3.18) in the village of Nová Huť, in another of the series of illustrations from *Světozor.*

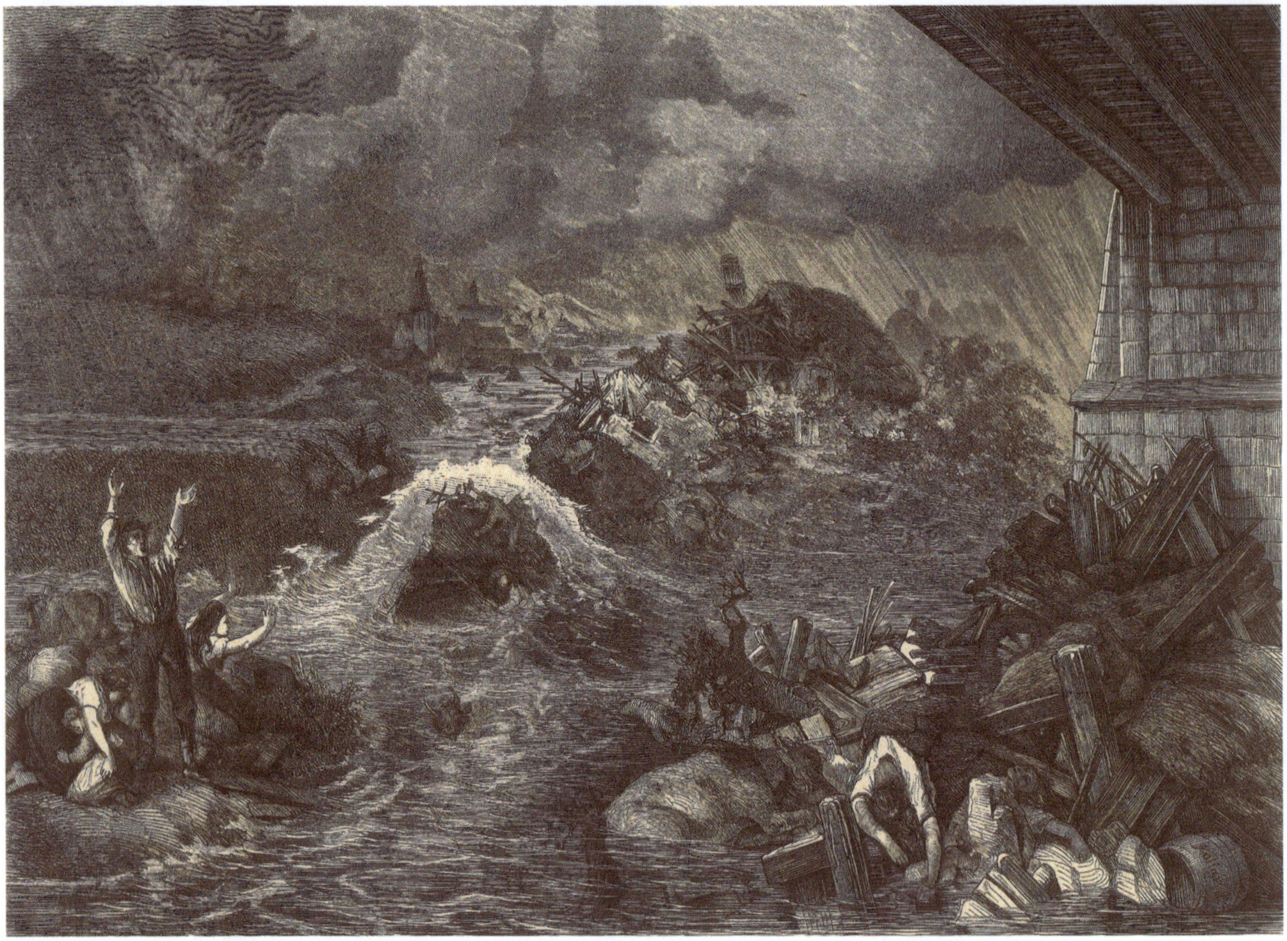

Fig. 3.19 An imaginative view of the 1872 flood

This, the last of four images of the 1872 flash flood in western Bohemia (Fig. 3.19), owes more to imagination than reportage. Unlike the previous images, which resulted from the artist visiting the damaged locations after the event, this purports to show the floodwaters actually surging downstream, while local villagers look on in dismay, and one poor woman appears to be trapped in the rubble. Three more bodies appear in the center of the picture, and one can also make out the head of a bull, which is being carried away by the flood. Obviously, the artist was not there at the time! This engraving comes from the German popular magazine *Über Land und Meer*; the artist is identified as the Czech illustrator Franz Kollarz (Kolář in Czech; 1825–1894).

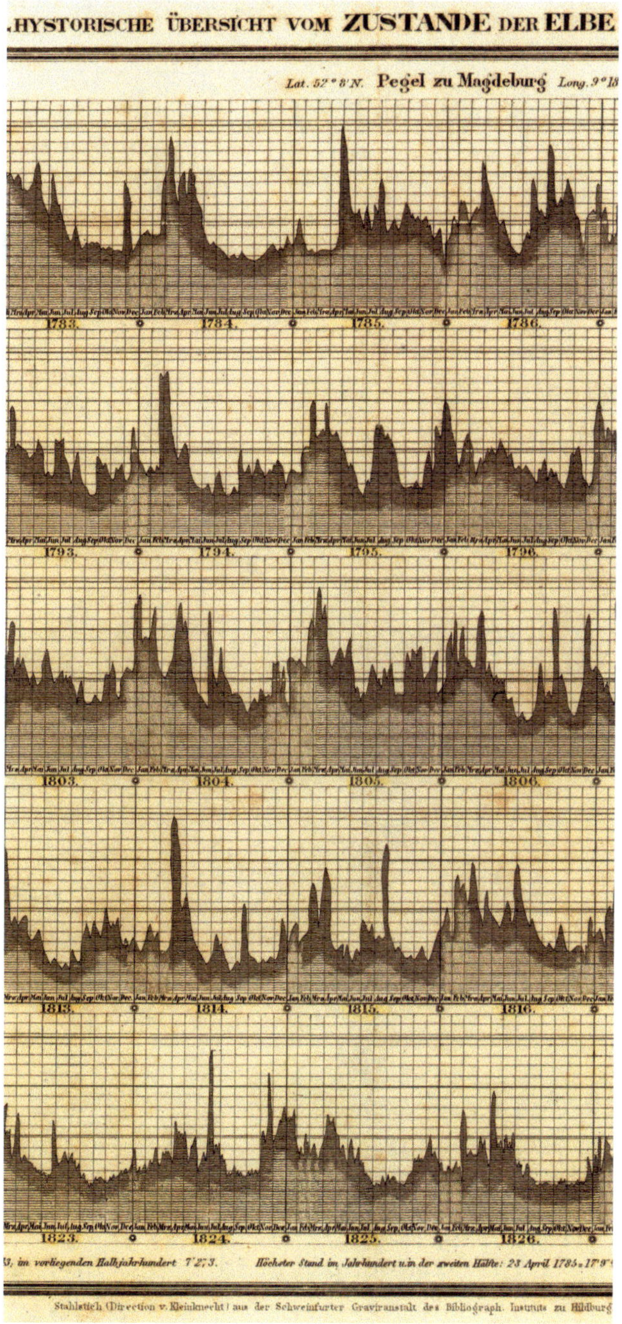

Fig. 3.20 A chart of flood heights on the River Elbe

The eighteenth century is really when modern science was born. Regular measurement was found to be the optimum way of providing objective data, whose collection, analysis, and interpretation was seen as the correct approach to deciphering what had previously seemed to be incomprehensible natural phenomena. In this period, culminating in the second half of the eighteenth century—sometimes called the Enlightenment—many physical disciplines were founded and pertinent means of measuring developed. The first results obtained from both short- and long-term observations started to appear in newly established scientific periodicals. In the first half of the nineteenth century, this effort culminated in the publication of the first physical atlases, in which all of these scientific achievements were gathered together.

Recording of flood heights requires no sophisticated instruments. It is enough to erect a graded stick in some suitable spot in a town and have someone examine it whenever the river overflows its banks. Sometimes one sees a scale of heights simply painted on the side of a building, with notes about the height reached by the floodwaters on particular historical occasions. This chart (Fig. 3.20), first published in 1837, shows monthly flood heights of the River Elbe in the city of Magdeburg between 1781 and 1830, and is taken here from the first *Physical Atlas* (1849) by Heinrich Berghaus. The image is a cropped version of the original chart; some years are missing to both the left and the right.

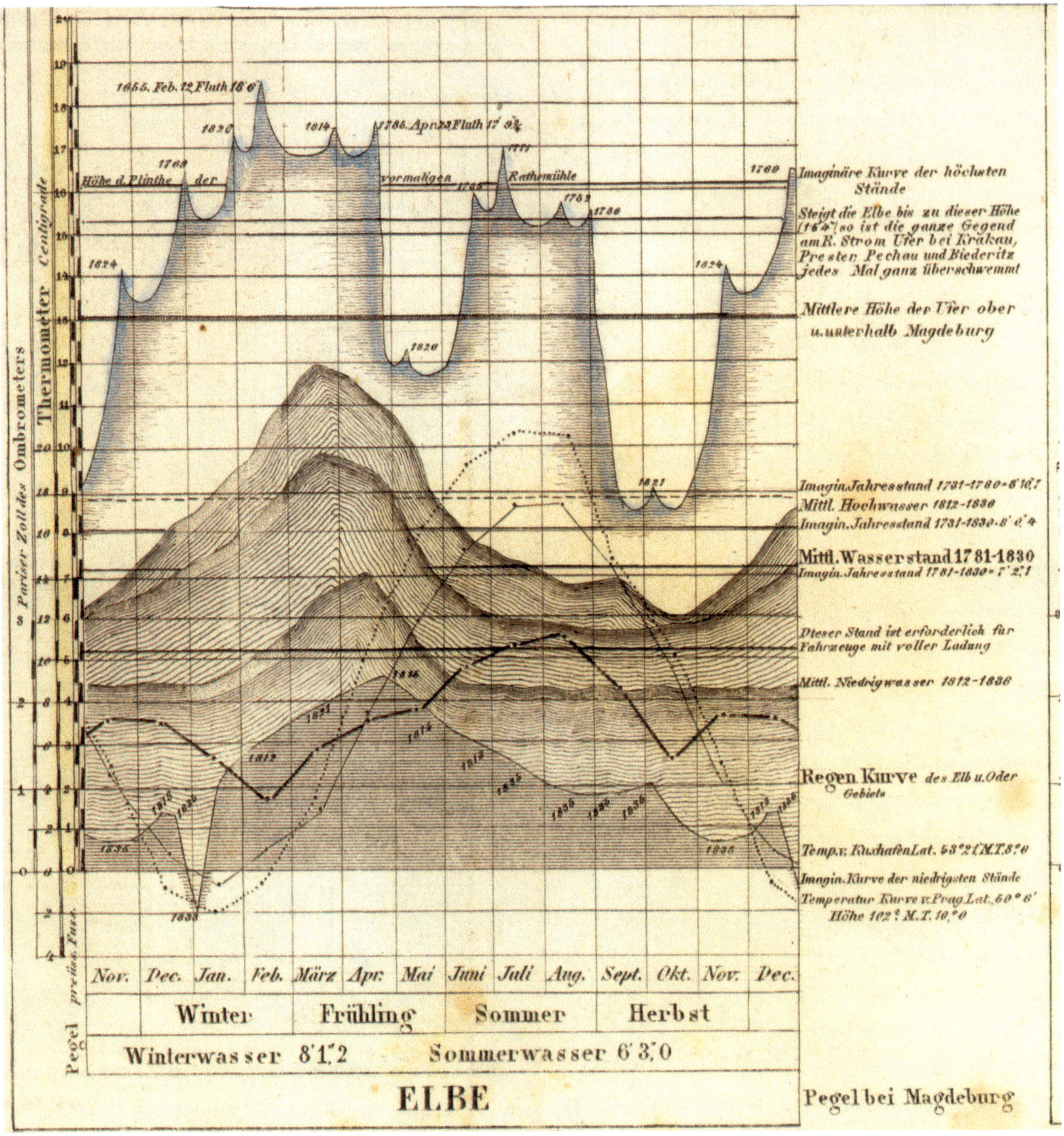

Fig. 3.21 Seasonal flood heights on the River Elbe

This chart (Fig. 3.21) comes from the same study as the previous image (Fig. 3.20); this time the data are arranged by month and season, so one can see that (unsurprisingly) winter is the worst season, and the best months are September and October. One notes also that the worst flood on record at the time (5.8 meters) was the flood of 22 February 1655—another ice-related flood. This record was broken in 2002, when the water height rose to 6.72 meters, and again in 2013, when the floodwaters exceeded 7 meters.

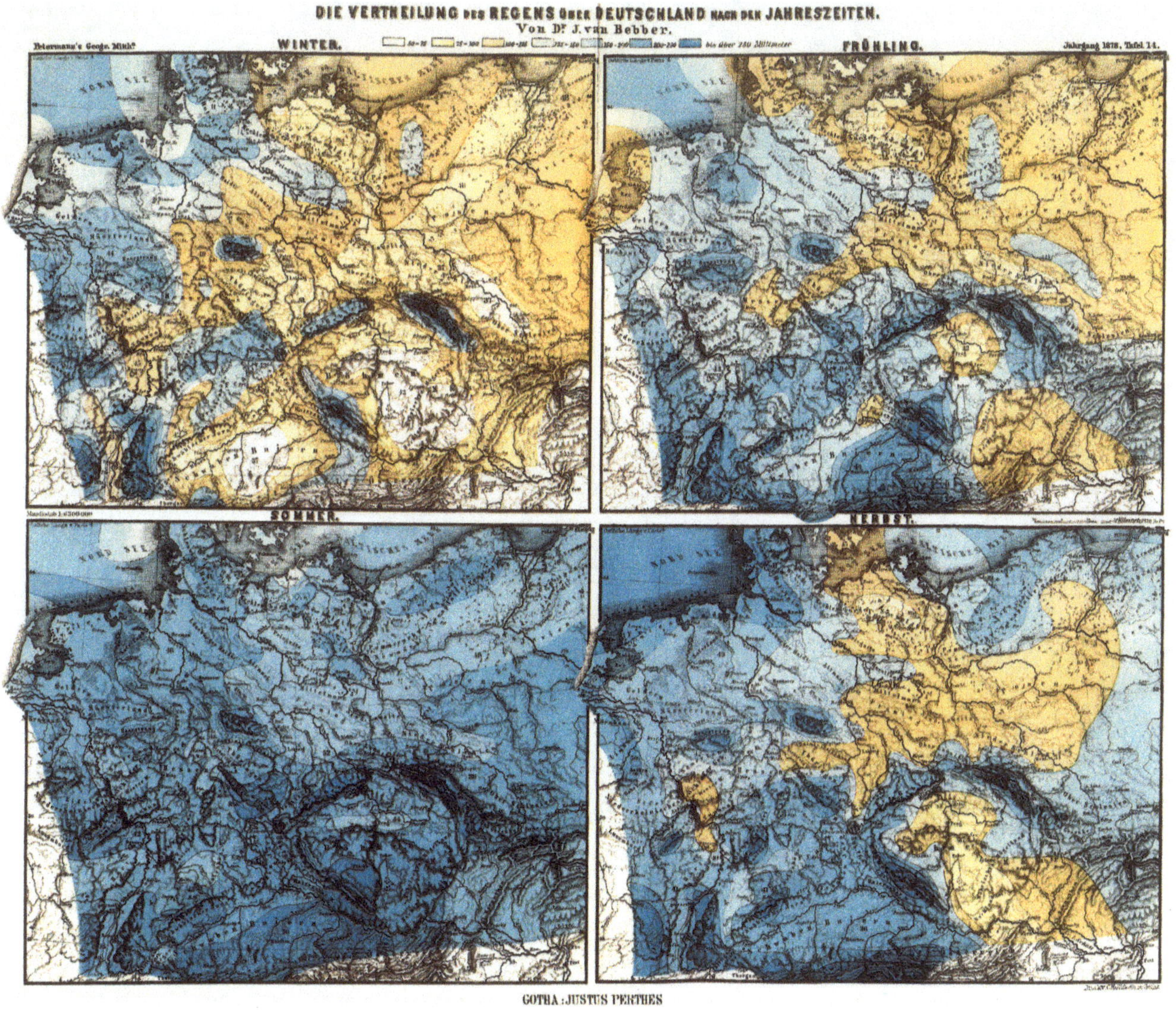

Fig. 3.22 An 1878 rainfall map of Germany

While the previous two figures presented data chronologically, here (Fig. 3.22) we see a study of the spatial distribution of rain over Germany, published in 1878. This was the work of the German meteorologist Wilhelm Jakob Van Bebber (1841–1909), who was one the first people to study the movement of cyclonic weather systems across Europe. The map is a colored lithograph and shows the different patterns of rainfall in each season. It appeared in *Petermanns Geographische Mitteilungen*, a magazine published by Justus Perthes in Gotha.

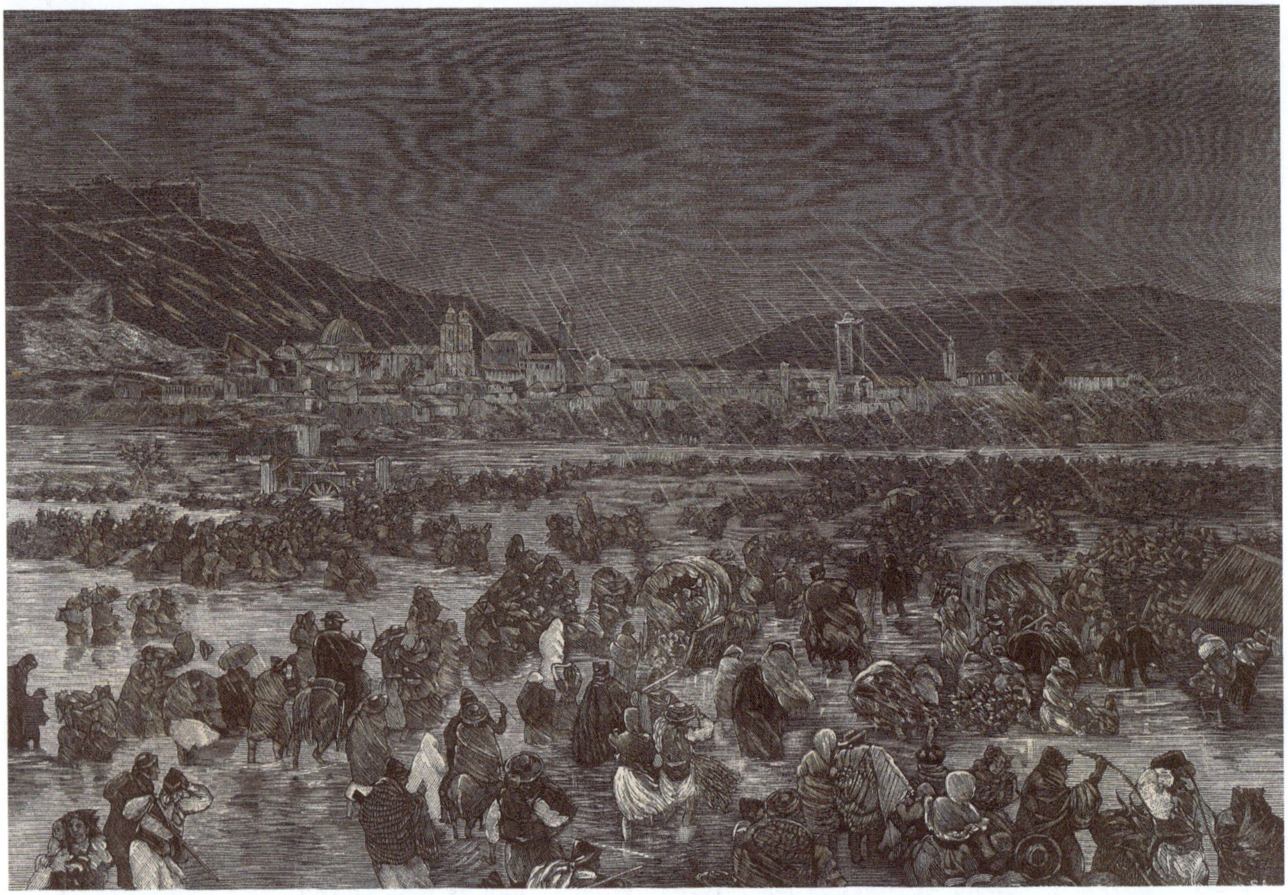

Fig. 3.23 The Santa Teresa flood in Murcia in 1879

While most of the images so far have been drawn from central Europe, flooding can be hazardous in many parts of the world, some unexpectedly so. The engraving here (Fig. 3.23) shows the Santa Teresa flood in Murcia, Spain, on 15 October 1879. After a long drought, lasting three years, people at first welcomed the rain, but this was no ordinary shower. It is estimated that perhaps 60 centimeters of rain fell in just an hour throughout Murcia, Almeria, and Alicante. The Guadalentín River became a raging torrent, wrecking bridges and dams, and washing away irrigation systems and houses. The Segura, Mula, Mundo, and other rivers similarly burst their banks. Over 770 people were killed and over 22,000 animals drowned. Thousands of homes were destroyed. The disaster was widely reported in the press of the day, with a number of engravings appearing that showed the damage. This particular image is taken from the Polish magazine *Kłosy*.

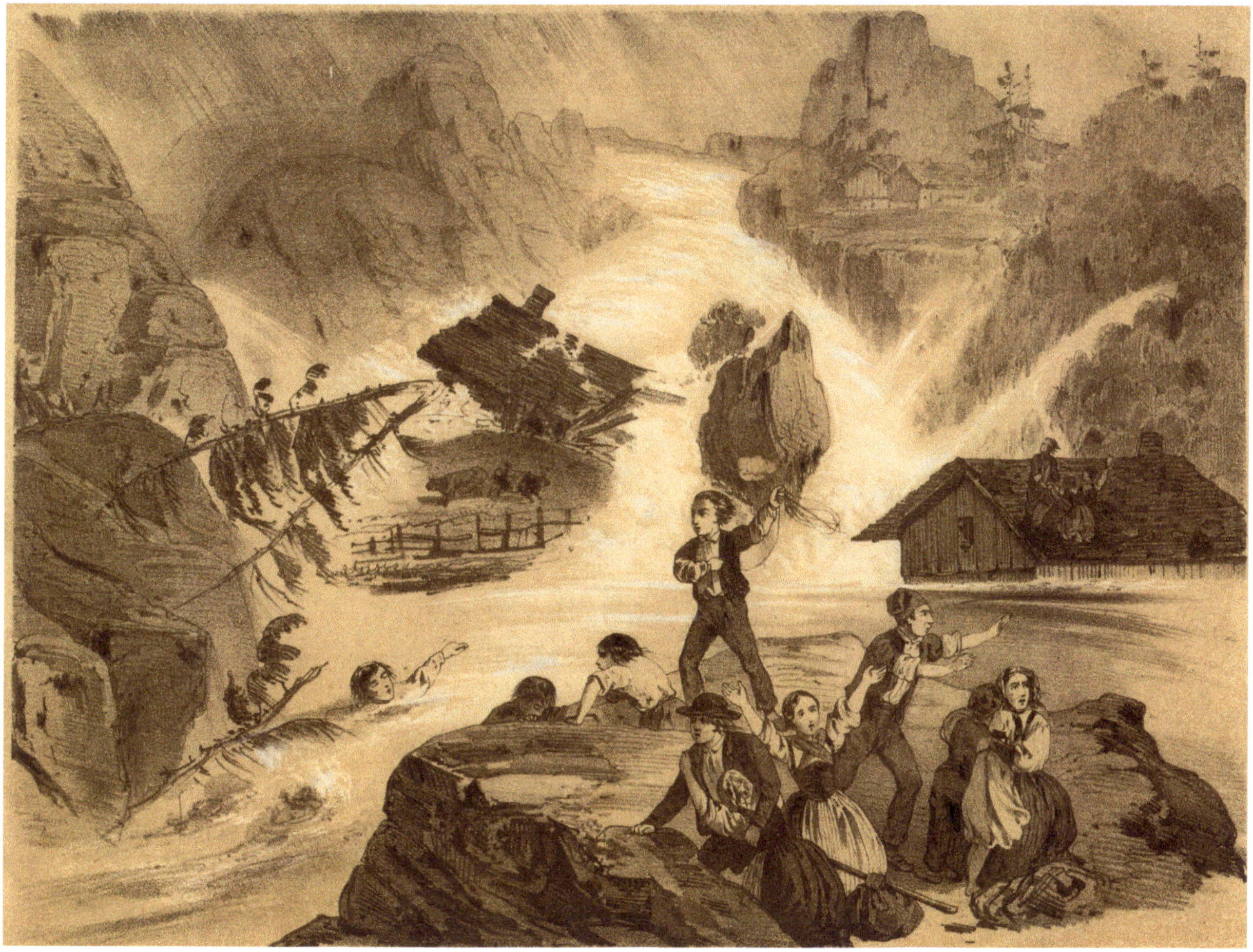

Fig. 3.24 *Hochwasser* (*High Water*)

Our final image of flooding is this unidentified and unsigned print (Fig. 3.24), titled *Hochwasser* (*High Water*). The scene looks like it might well be Swiss, but this is clearly a work of imagination rather than a specific event. Water streams down the mountainside, destroying houses and uprooting trees. A group of survivors huddle miserably on a rock above the foaming water, while one less lucky figure attempts to join them; from the apparent speed of the current, her chances do not look good. The picture demonstrates well that floods may also visit high mountainous areas, especially in deep and narrow valleys and canyons.

3.3 Storms and Shipwrecks

When assigning storms and hurricanes to the four classical elements, they have to be considered to come under both Water and Air, depending on whether the winds sweep across the land or the sea. Today we worry chiefly about storms on land, which can bring flooding, damage to property, landslips, and the uprooting of trees. Historically, storms at sea were greatly to be feared on account of the great loss of life from the sinking of sailing ships with large crews.

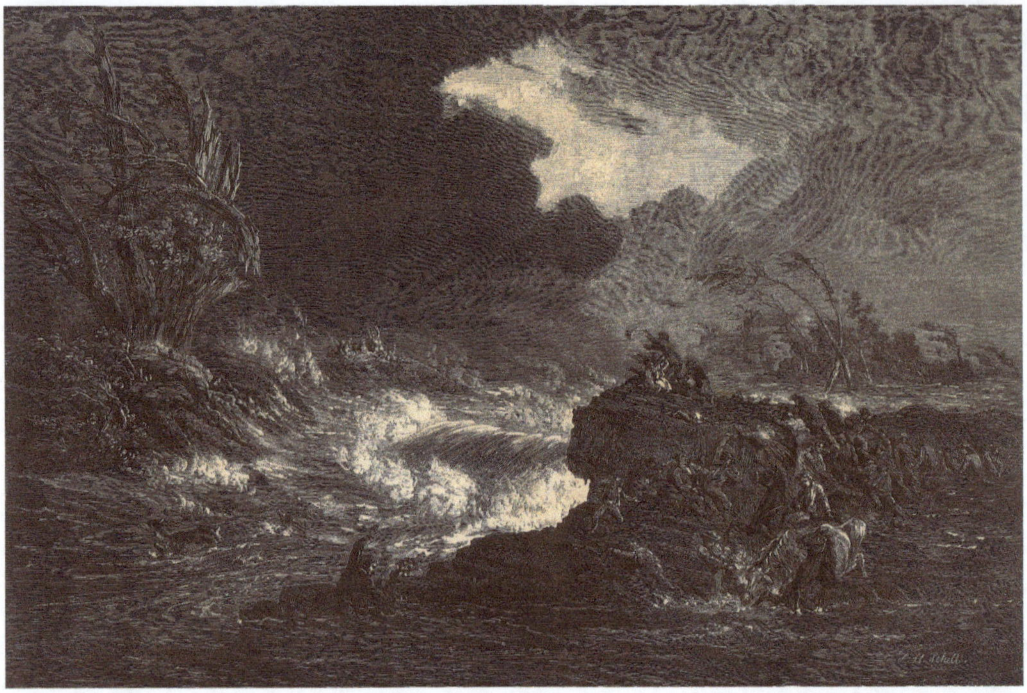

Fig. 3.25 A hurricane in 1867

An image of the great hurricane of 1867 in the Caribbean appears in Chap. 4 in this book (Fig. 4.1); this (Fig. 3.25) is another picture from the same source, *Frank Leslie's Illustrated Newspaper* of 7 December 1867. Unlike Fig. 4.1, this picture focuses on the pounding of the sea waves against the shore.

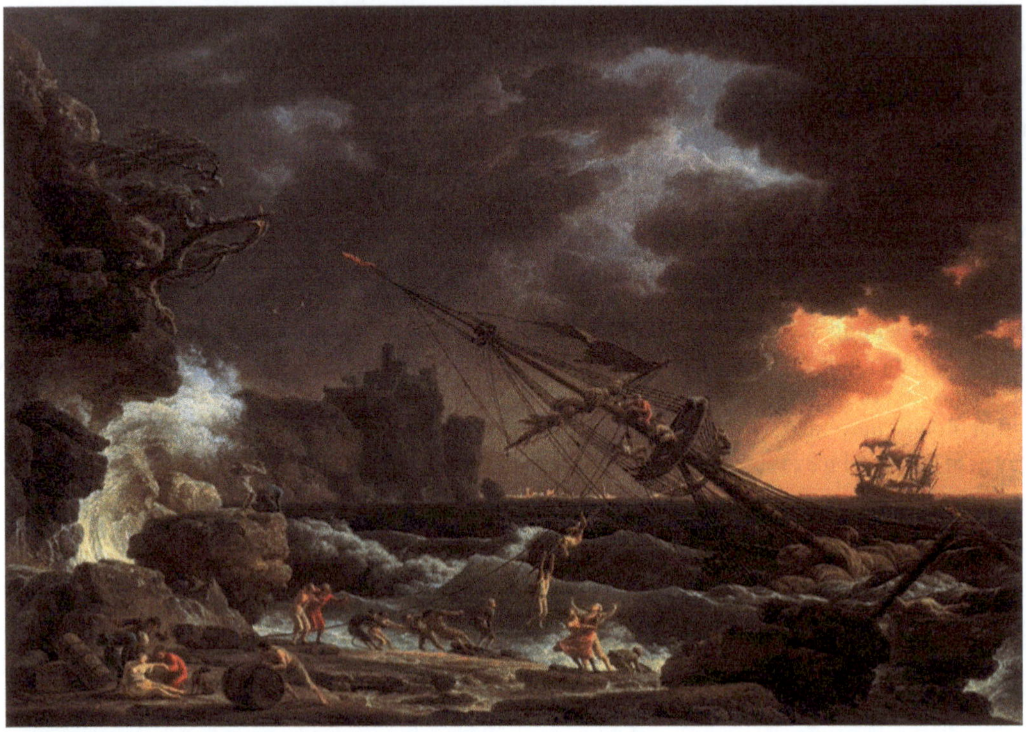

Fig. 3.26 A painting of a shipwreck

Claude Joseph Vernet (1714–1789) was a French painter who particularly specialized in pictures of Nature, especially in its more dramatic aspects, such as storm scenes like this one. These paintings found a ready market among British aristocrats visiting the artist in Rome, where he lived for much of his life. This particular example (Fig. 3.26) was executed in 1772 and is now in the possession of the National Gallery of Art, Washington DC. It is purely a decorative work and does not correspond to any actual event, although scenes like this were all too common in the period.

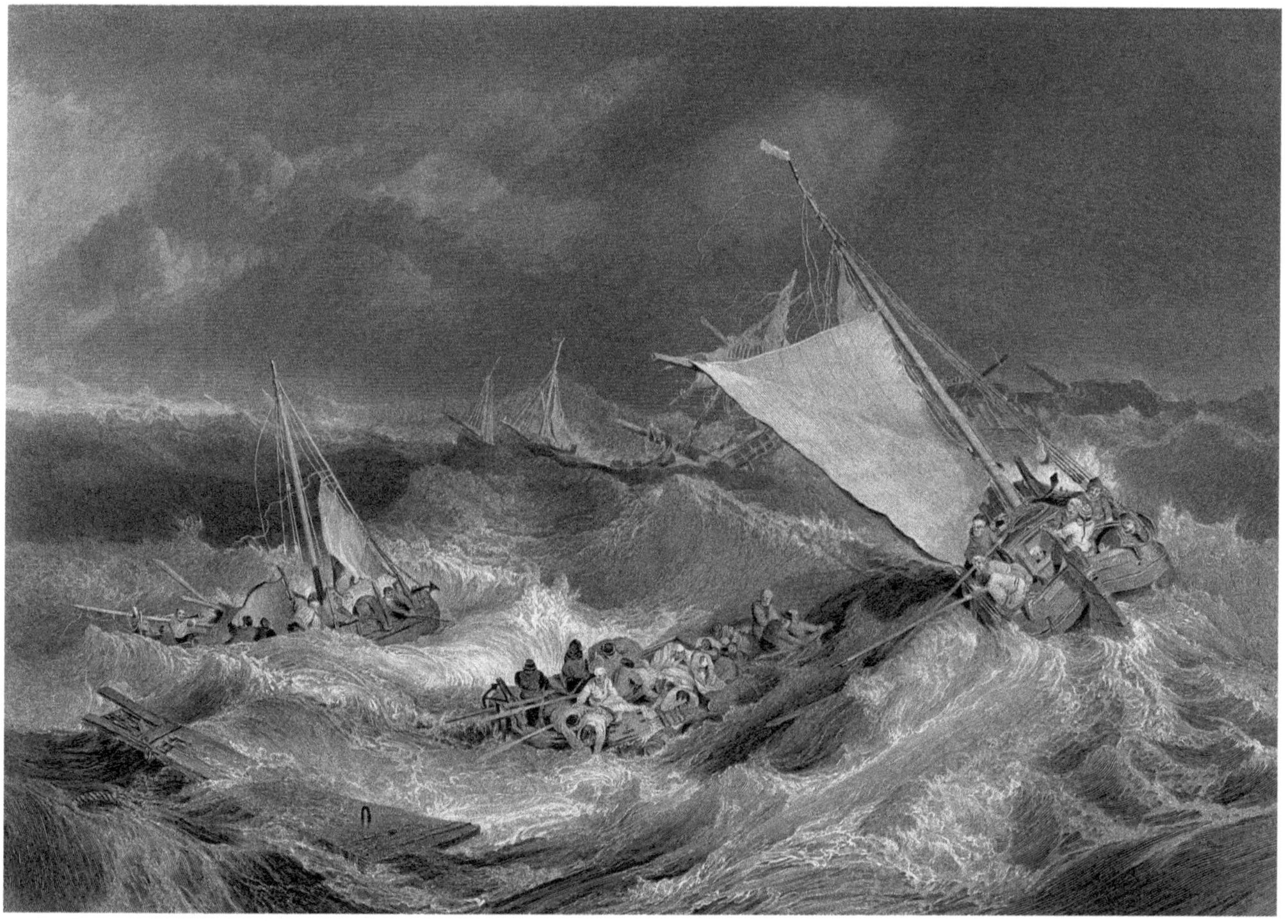

Fig. 3.27 Rough seas in 1875

Eternal Father, strong to save,
Whose arm hath bound the restless wave,
Who bidd'st the mighty ocean deep
Its own appointed limits keep;
Oh, hear us when we cry to Thee, For those in peril on the sea!

These are the words of a popular hymn written by William Whiting in 1860. In a period where sea travel was both common and hazardous, these were heartfelt words. Reports of loss of life from shipwrecks were commonplace among news reports in the days of sailing ships. The numbers killed were far greater than those in air crashes today, and modern shipwrecks, when they very occasionally occur, usually result in no loss of life from lightly crewed freighters. The image here (Fig. 3.27) is an engraving by William Miller, after a painting by Turner, one of a series published in 1875. The risks of contemporary sea travel are well evident.

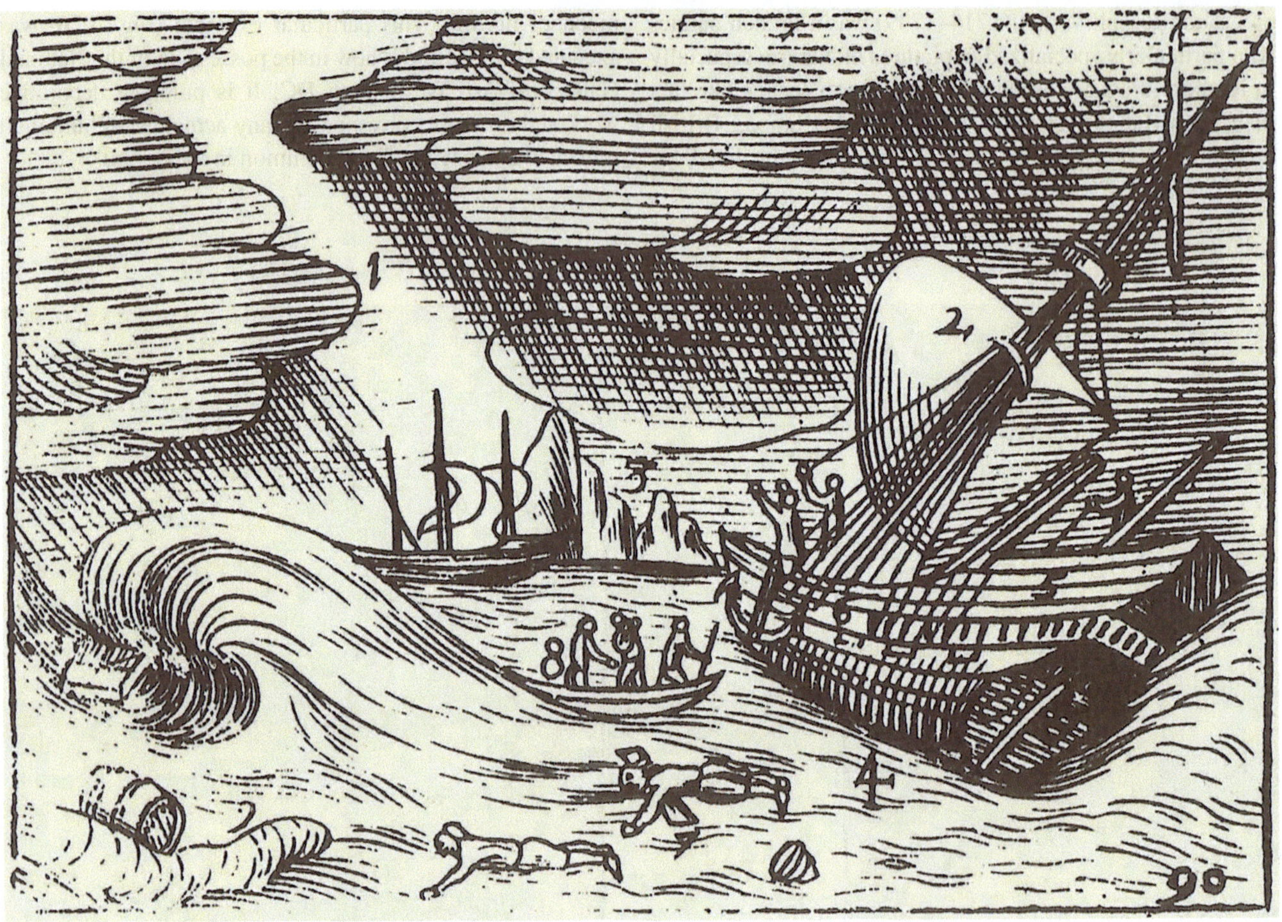

Fig. 3.28 A shipwreck depicted for children in 1658

As a further example of how the fear of shipwreck and the hazards of sea travel were a part of the European mind-set in the days of sailing ships, we present this image (Fig. 3.28) from the pictorial textbook *Orbis Sensualium Pictus* (1658) by Jan Amos Comenius, a famous European educationist of the seventeenth century. His aim in this book was to instruct children in "basic general knowledge" by means of annotated pictures. Since shipwreck was a basic fact of life at this time, it was something children should be instructed about. This cartoon-like and very schematic illustration shows a violent storm (1), a foundering ship (2), a rocky shore (3), and the poor mariners trying to swim for safety or using a lifeboat (4). In the far left corner, one can see some of the cargo washing up. For some rural communities in places where shipwrecks were common, retrieval of washed-up cargo could be quite lucrative.

3.4 Tsunamis

The word "tsunami" was not widely known to the general public in Europe until the great Sumatran earthquake of 26 December 2004. After that, everyone wanted to know about tsunamis. The cause is simply explained; when a great undersea earthquake suddenly displaces the seabed, a huge mass of water is likewise displaced and comes roaring in to shore. Out at sea, the wave moves rapidly in deep water. As it enters shallower coastal waters, the wave slows and piles up. The word "tsunami" translates from the Japanese as "harbor wave," precisely because it is a wave observed only in shallow waters. Hence, writing "tsunami wave" is akin to writing "harbor wave wave"!

In the years and decades before a great earthquake is caused, when an oceanic plate is pushed under a continental one, the seabed of the upper plate is gradually dragged down by the slow movement of the lower one. This continues until the force of friction that binds the two together is no longer strong enough to hold them together. Freed from the downgoing lower plate, the seabed flips upward, displacing a massive amount of water. Tsunamis can have other causes: underwater landslides, landslides falling into water, or even volcanoes on rare occasions, such as the 1883 explosion of Krakatoa. But the great tsunamis that can cause damage and destruction on the other side of an ocean are all produced by great earthquakes.

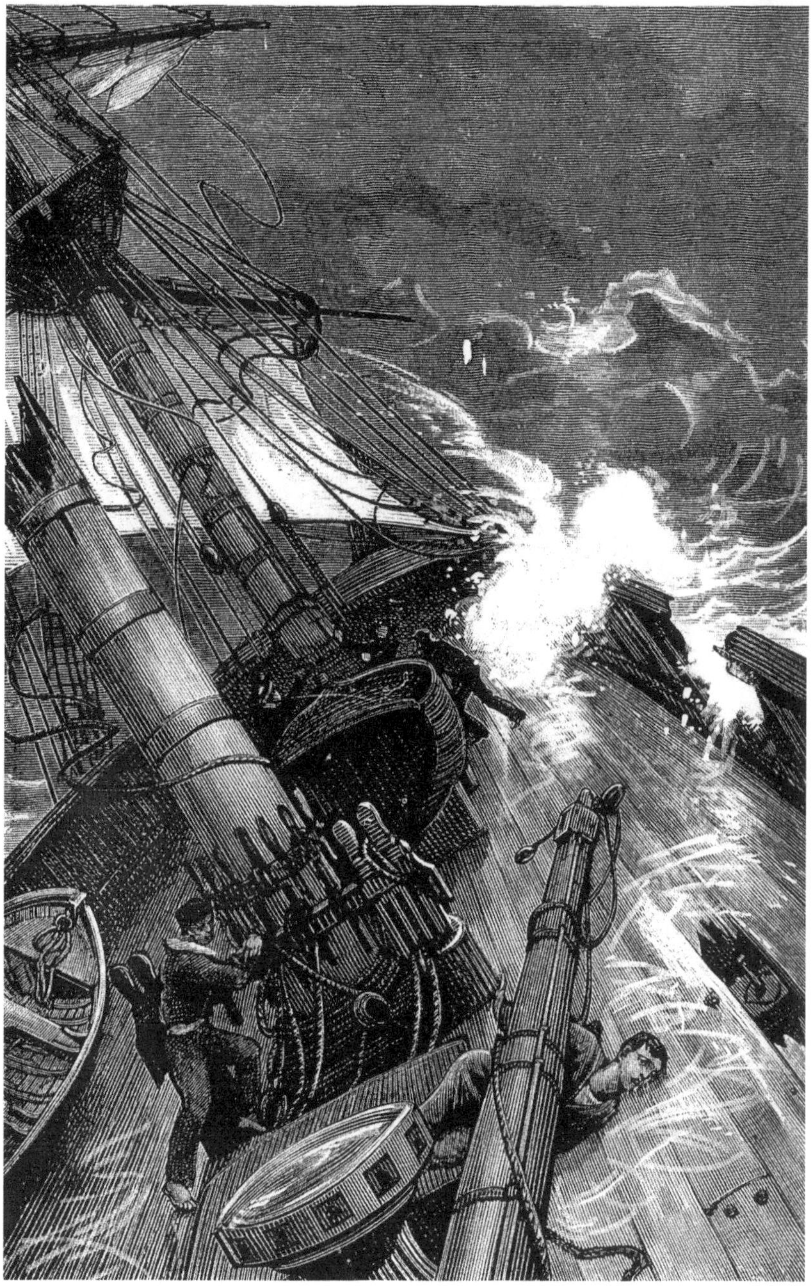

Fig. 3.29 A tsunami at St. Thomas in 1867

At first sight, this appears to be another storm-tossed boat, this time from the point of view of a deckhand. But, actually, this perilous sea is due not to a storm but to a tsunami. The capacity for damage from tsunamis is enormous. In the case of the largest ever recorded earthquake, which occurred in Chile in 1960, ships were picked up by the subsequent tsunami and hurled several kilometers inland.

This particular example (Fig. 3.29) is the tsunami that followed the earthquake—or rather the earthquakes (two shocks ten minutes apart)—that struck the US Virgin Islands on 18 November 1867. The tsunami was particularly damaging to the town of Charlotte Amalie on the south coast of the island of St. Thomas. The wharf was destroyed, along with many small boats anchored there, and the buildings along the waterfront were all flooded or destroyed. This imaginative depiction is an illustration from a popular book on earthquakes by Arnold Boscowitz (first English edition, 1890).

One of the unhappy sailors seems to have been crushed by a falling mast. Although the image owes more to imagination than to observation, it is actually possible to identify the probable identity of the ship that is shown, though Boscowitz does not name it. The largest ship in the harbor was the *USS De Soto*, a wooden-hulled sidewheel steamship that was probably the vessel intended to be shown in the picture. The tsunami swept the vessel off its moorings and onto the wharf, but the next wave lifted it up again and carried it back into deep water. It took the crew and carpenters ten days to repair the damage before the ship was seaworthy again.

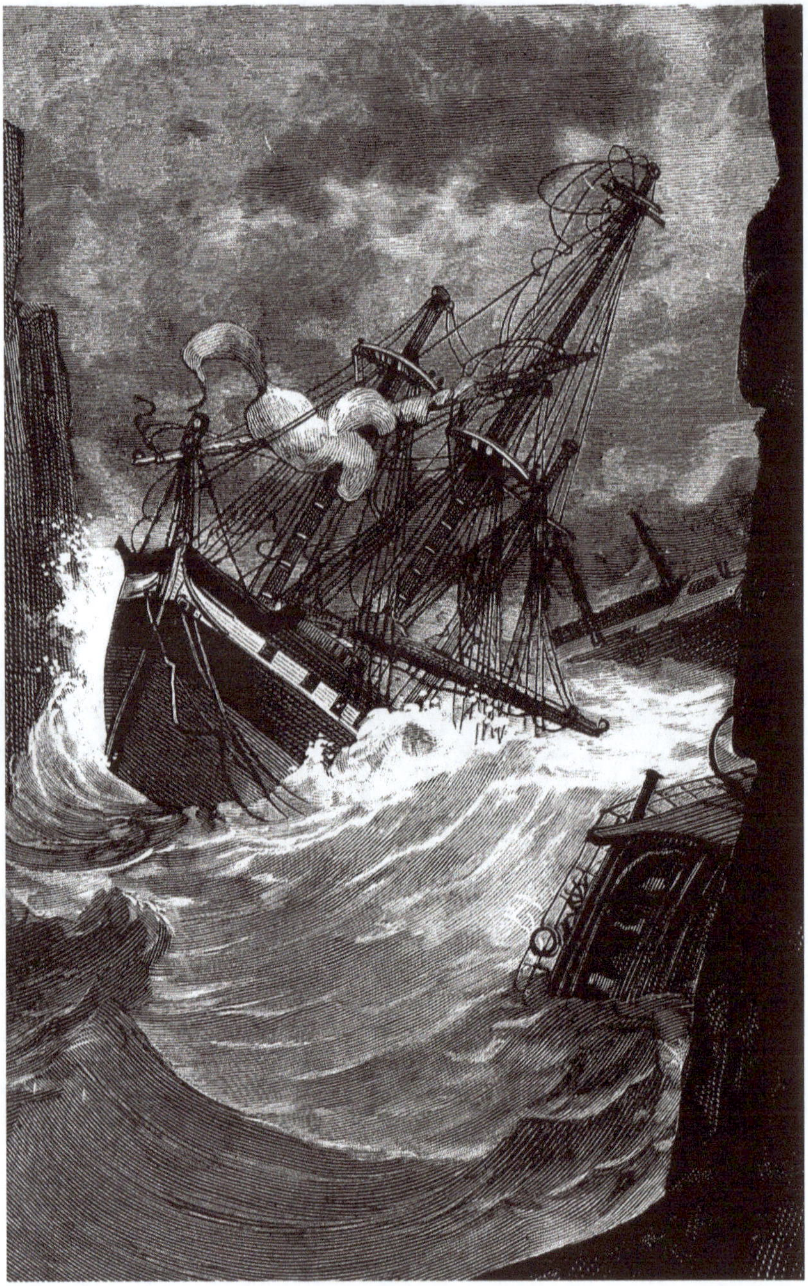

Fig. 3.30 A tsunami from the Ansei-Tōkai earthquake of 1854 (a European view)

This is another illustration from Boscowitz's book (Fig. 3.30). This time, the earthquake is a Japanese one: the Ansei-Tōkai earthquake of 23 December 1854. Shown here is the Russian frigate *Diana*, the flagship of the Russian explorer Yevfimy Putyatin. The ship was anchored in the harbor of Shimoda at the time of the earthquake; the water of the harbor was so convulsed with whirlpools that the ship was spun round 43 times in 30 minutes, so rapidly that the crew were giddy and fell down, and the ship's chains became twisted. The water fell and rose again, damaging the ship as it hit the bottom. There was no hope of getting the ship fixed in Shimoda, which was in ruins, so the *Diana* attempted to sail to Heda for repairs, but it sank en route in the bay of Miyajima-mura. Obviously, the picture here is unrealistic; the scene seems to be a rocky shoreline, whereas in fact the *Diana* was riding at anchor in Shimoda Harbor when the tsunami hit.

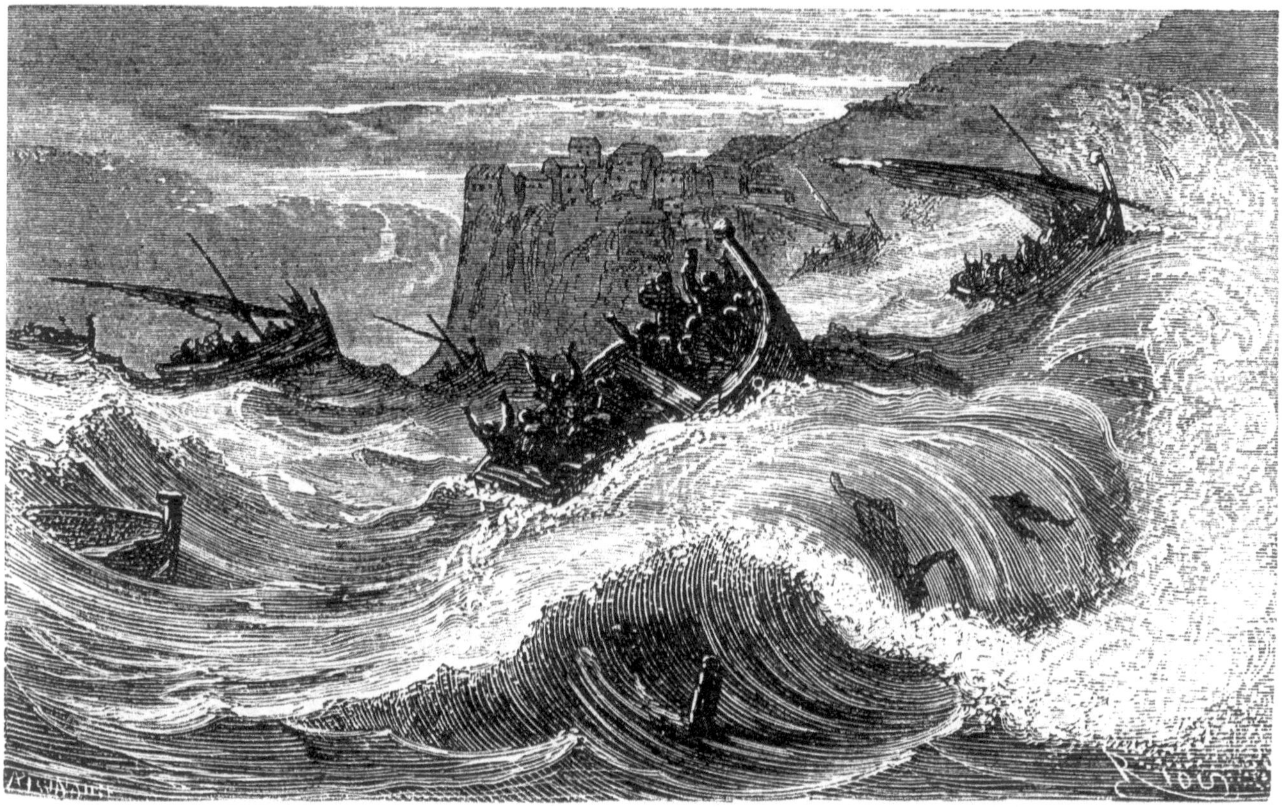

Fig. 3.31 The Sumatran tsunami in 1861

The direction a tsunami wave takes is perpendicular to the orientation of the fault causing the earthquake. In the earthquake on 26 December 2004, the fault was north–south and extended north of Sumatra itself, so the wave traveled west and east, striking India and Thailand. Earlier great Sumatran earthquakes occurred where the plate boundary fault was more northwest–southeast, resulting in a wave that, traveling southwest, would be lost in the expanses of the Southern Ocean. However, the northeast component could still wreak havoc on the coast of Sumatra itself, as shown in this picture (Fig. 3.31) of an earthquake on 16 February 1861 (magnitude about 8.5). The engraving is the work of the prolific French illustrator Édouard Riou (1833–1900), who is now largely remembered for his illustrations of the works of Jules Verne but was somewhat of a specialist when it came to the subjects of travel and adventure. It is taken from a book by Frédéric Zurcher (1816–1890), *Volcans et tremblements de terre*, published in 1868 (so the Sumatran earthquake was still topical).

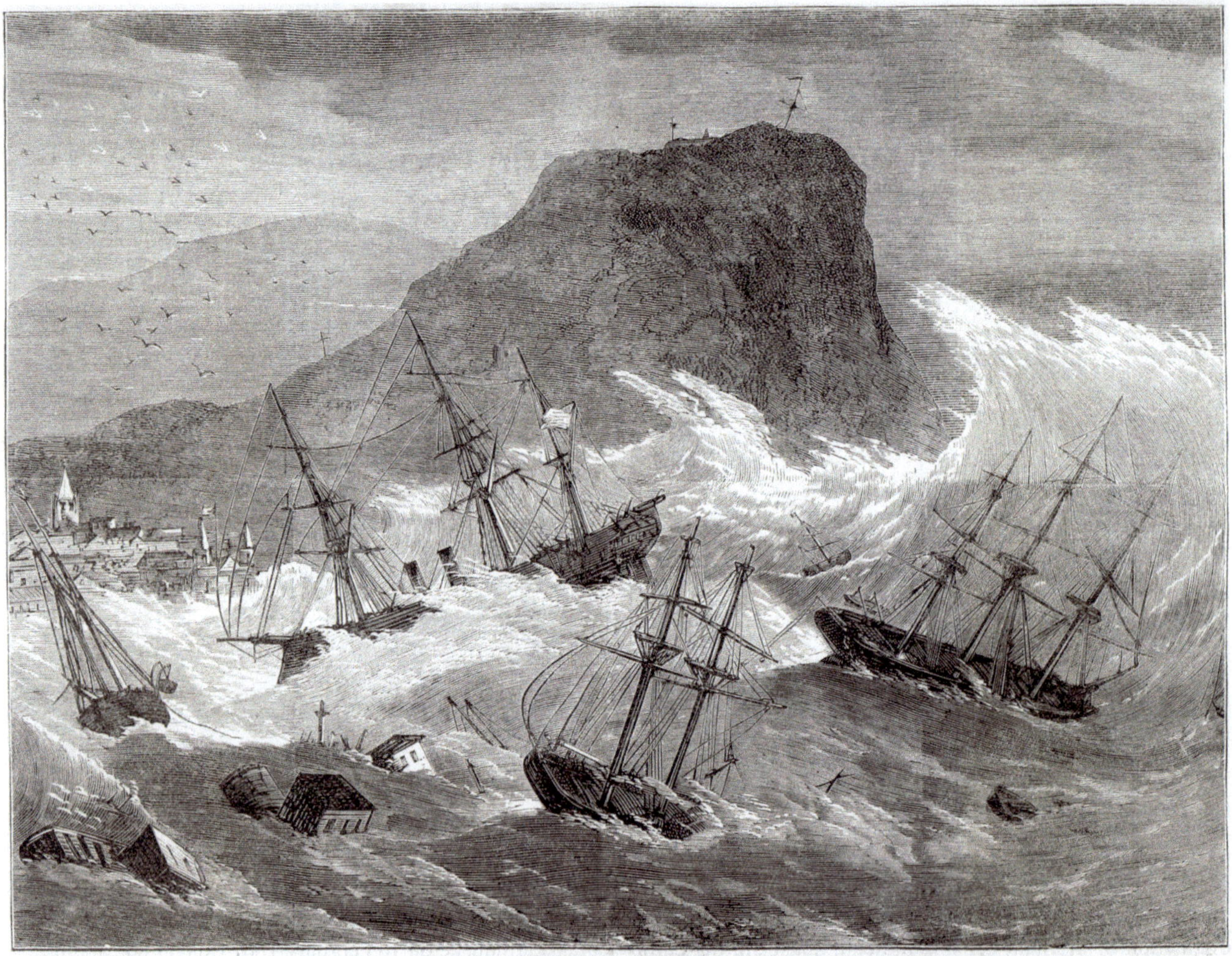

EARTHQUAKE IN PERU—SUBMERSION OF ARICA.—[SEE FIRST PAGE.]

Fig. 3.32 A tsunami at Arica in 1868

The earthquake of 13 August 1868 occurred near the Peruvian–Chilean border and has already been discussed in Chap. 2 of this book. The earthquake wiped out the city of Arica but is particularly known for the devastating tsunami it generated.

The tsunami is shown here in a much-reproduced contemporary wood engraving (Fig. 3.32). The notion of a tsunami as a towering wave is not usually what is observed. Observers reported that the sea seemed perfectly smooth. It just kept on coming onshore, reaching up to 14 meters above sea level, and then draining back out again. As it receded, conflicting currents swirled in all directions, carrying masses of wreckage, debris, and boats, all of which caused secondary damage as they were dragged over the ruins of the town. Extensive video of the tsunami from the 2011 Tohoku earthquake in Japan confirms that this is exactly what happens. However, at Arica, a receding wave happened to collide with a subsequent incoming one, and they merged into a "solid sea wall," which is perhaps what is shown here.

This earthquake was the subject of a pioneering study by the German geologist Ferdinand von Hochstetter, who studied the propagation of the wave across the Pacific; many hours after the earthquake, the wave reached New Zealand, where it did significant damage. This was not quite the first occasion on which an attempt was made to investigate the arrival times of a seismic sea wave; such a study of the arrival times associated with the Lisbon tsunamis of 1755 and 1761 was made by the Scottish naturalist David Milne in 1842 (Milne 1842).

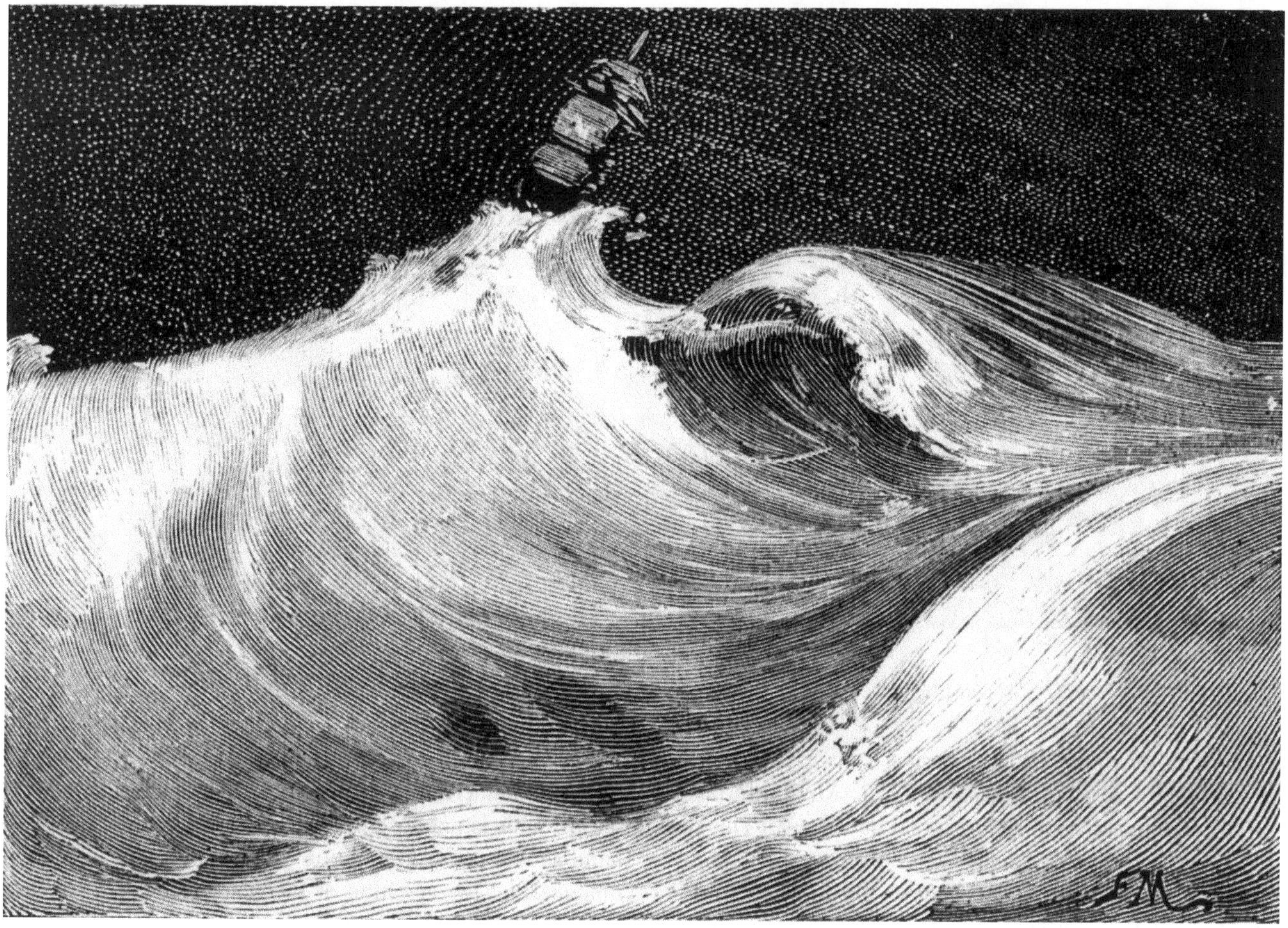

Fig. 3.33 A fantasy depiction of a tsunami

It is a widespread myth that a tsunami, when encountered out at sea, is seen as an immense wave, as shown here in another illustration from Boscowitz's book on earthquakes (Fig. 3.33). In deep water, a tsunami manifests itself as a barely perceptible swell, as evidenced by the fact that boats offshore from Sumatra in 2004 noticed nothing. In shallow water, the wave slows and the water piles up, but seldom does it create a giant wave towering over buildings. The usual trough-to-crest height of a tsunami as it comes onshore is a few meters, but it keeps on coming with great force.

When one reads of a "30-meter-high tsunami" this means the wave ran up to the 30-meter contour above sea level. The tsunami from the 2011 Tohoku earthquake in Japan was particularly well captured on video, and one can see clearly how it caused so much damage despite not being a "giant wave." Not only did the tsunami cause much damage coming in, but also when the water drew back again, it carried much debris with it, which had a scouring effect on anything still standing.

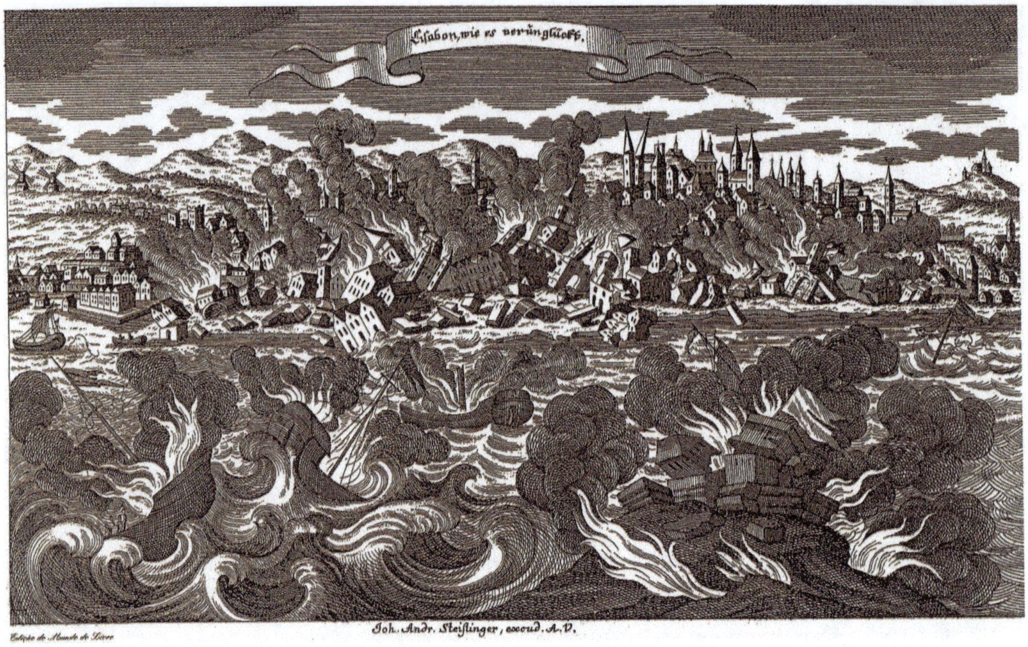

Fig. 3.34 The tsunami at Lisbon in 1755

Perhaps the most famous historical tsunami is that caused by the great Lisbon earthquake of 1755. (More information on the earthquake itself can be found in Chap. 2.) This print by the artist Johann Andreas Steisslinger (dates unknown; floruit mid-eighteenth century) shows the destruction of the city of Lisbon by earthquake and fire, and also the agitation of the waters of the River Tagus, but not any single sweeping wave (Fig. 3.34). In fact, a tsunami is usually a series of waves. It is interesting to note that the artist has shown ships on fire. This is not absurd; it seems ironic that a watery tsunami should cause destruction by fire, but it does. One can imagine that the agitation of the sea would upset cooking stoves on board ships, setting wooden vessels alight.

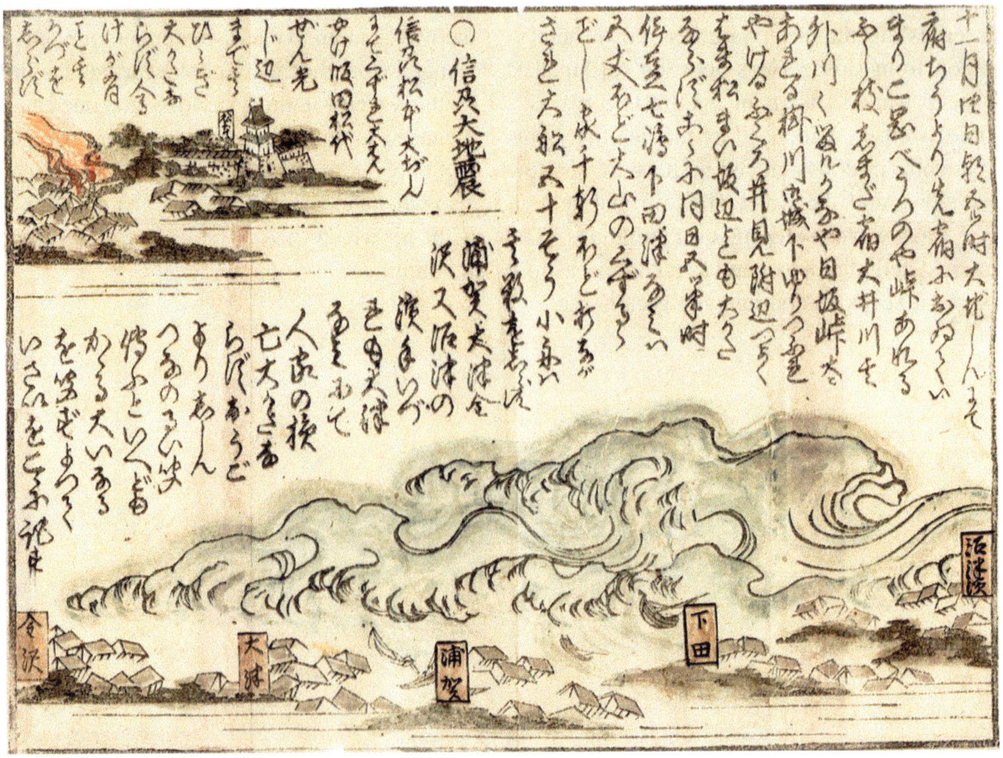

Fig. 3.35 A tsunami from the Ansei-Tōkai earthquake of 1854 (a Japanese view)

The most famous image associated with tsunamis is Katsushika Hokusai's *The Great Wave off Kanagawa*. For practically any publication or conference on tsunamis, Hokusai's iconic wood-block image is wheeled out, despite the fact that it does not depict a tsunami at all. Here, we skip it and show instead a very definite tsunami (Fig. 3.35) caused by the great Ansei-Tōkai earthquake (magnitude about 8.4); the same event was depicted in a previous image (Fig. 3.30) by a European artist (who surely never witnessed a tsunami for real). This wood-block image is reproduced with kind permission from the University of Tokyo Earthquake Research Laboratory.

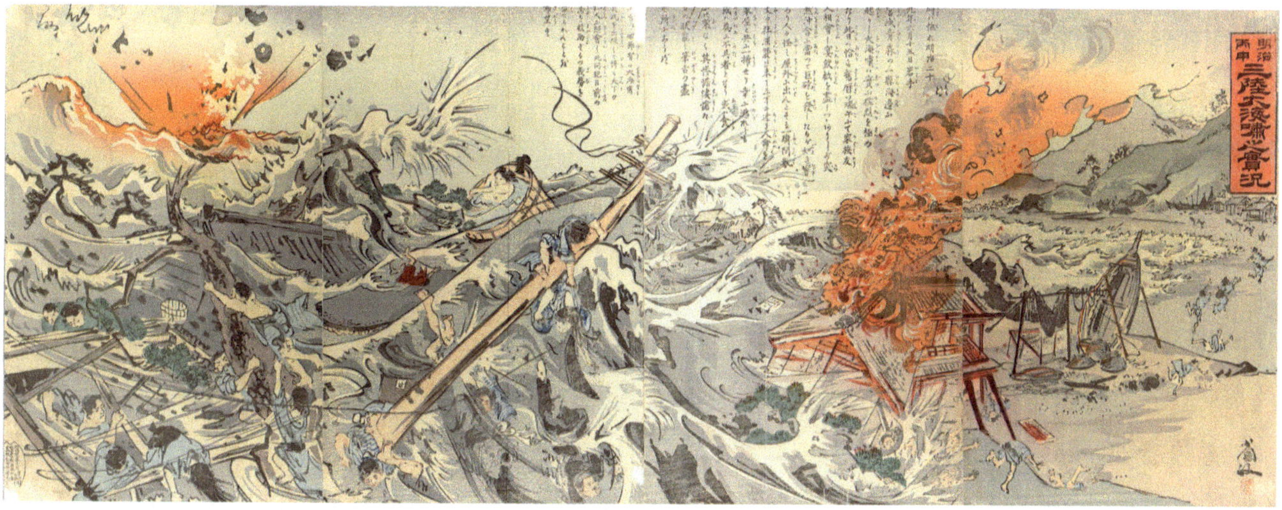

Fig. 3.36 A tsunami from the Sanriku earthquake of 1896

The Japanese wood-block print shown here (Fig. 3.36) is conspicuously more modern in style, almost approaching the character of modern manga illustrations. The tsunami depicted was caused by the Sanriku earthquake on 15 June 1896, one of the most destructive tsunamis in Japanese history. Some earthquakes cause much larger tsunamis than would be expected for their magnitude, because of the way in which they rupture the seabed. This is one such event; the magnitude was relatively modest, at about 7.2, but the tsunami had a run-up height of 25 meters. One can note especially how the tsunami is setting buildings alight as stoves are upset, scattering hot ashes. Again, this image comes from the collection of the University of Tokyo Earthquake Research Laboratory.

3.5 Waterspouts, Whirlpools, and Other Perils of Water

Finally, we present some other watery perils. Waterspouts and whirlpools are more dangerous in imaginary fiction than in reality, but having too little water can be as deadly as having too much. And while we think of water first in its liquid form, as a solid it also causes problems.

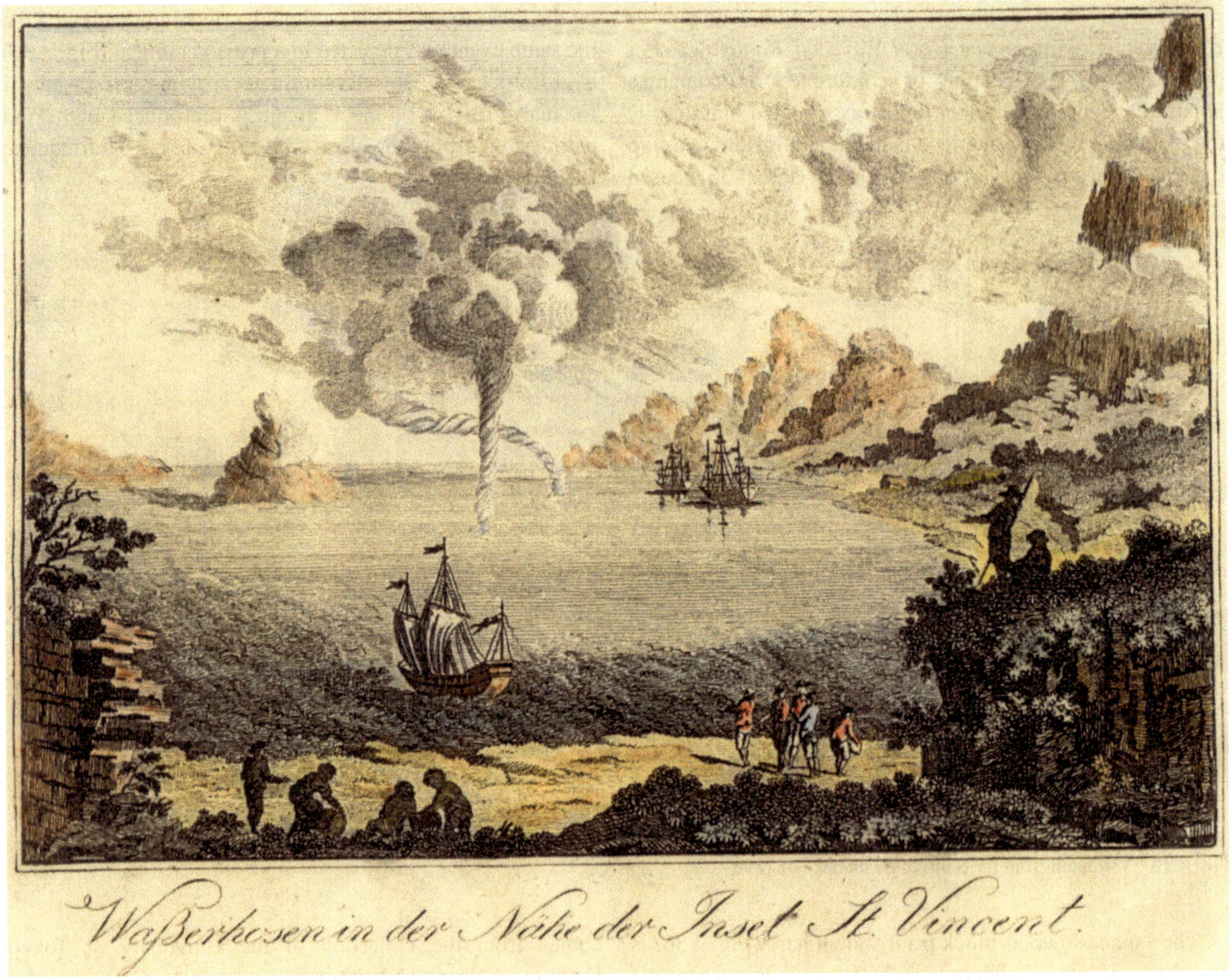

Fig. 3.37 Two waterspouts in the bay of St. Vincent

The marine equivalent of the tornado on land or the dust devil in the desert (see Chap. 4) is the waterspout. These are typically weaker than their land equivalents, and it is rare for them to cause serious damage. They do pose a threat to shipping but can usually be avoided. Waterspouts are mostly a tropical phenomenon, although they do occur in other parts of the world. This engraving (Fig. 3.37) shows two waterspouts in the bay of St. Vincent in the Caribbean. It comes from the *Magazin des Außerordentlichen in der Natur, der Kunst und im Menschenleben* by Johann Adam Bergk (1769–1834) and Friedrich Gotthelf Baumgärtner (1759–1848), published in Prague in 1815.

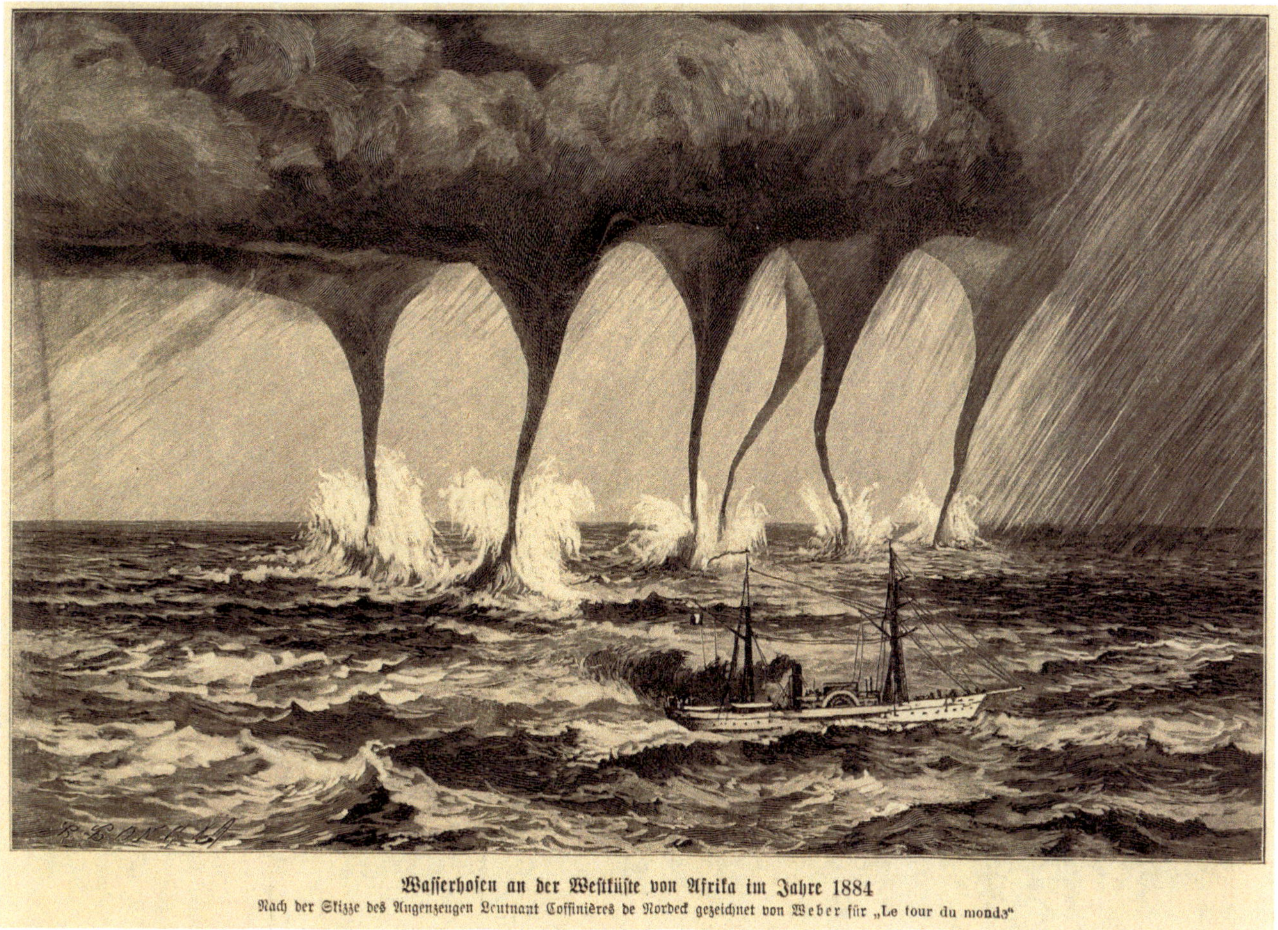

Fig. 3.38 Multiple waterspouts off the coast of Africa in 1884

In French, a waterspout is *une trombe marine*; the word *trombe* comes from the Italian word for a trumpet—rather appropriate, given the trumpet-like shape. This is seen clearly in the image shown here of multiple waterspouts, six in all, observed off the west coast of Africa in 1884 (Fig. 3.38). In German, the term is *wasserhosen* (which literally means "water trousers"), and it is perhaps not too fanciful in this picture to see three pairs of rather spindly legs.

Multiple waterspouts, as shown here, are not uncommon, and they make impressive images. This illustration is an engraving made for the magazine *Le Tour du Monde*, after a sketch by an eyewitness, Lieutenant Coffinières de Nordek (1811–1887), and was published in Paris in 1886. The record for the largest number of simultaneously observed waterspouts is nine, on Lake Michigan, USA, in 2012.

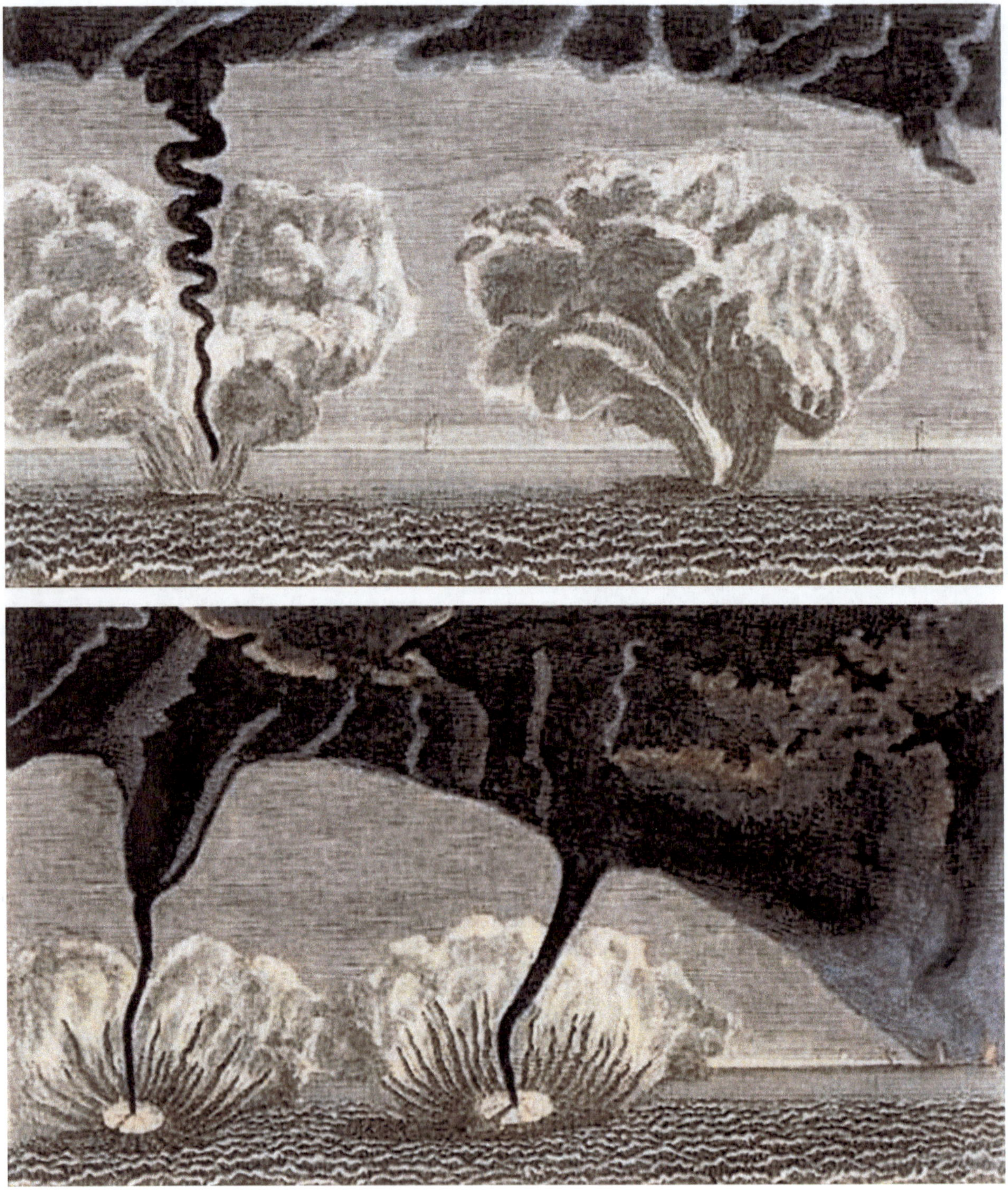

Fig. 3.39 Waterspouts near Nice in 1789

Most waterspouts are not actually tornadoes at all. They are known as "fair-weather waterspouts" and originate under dark, flat-bottomed, developing, convective cumulus towers. These are clearly shown here in these two prints (Fig. 3.39) from *Bilderbuch zum Nutzen und Vergnügen der Jugend* by Friedrich Justin Bertuch (1747–1822). The original engraving is French, from a paper by M. Michaud, titled *Observations sur les trombes de mer vues de Nice en 1789, le 6 Janvier et le 19 Mars*, published in the *Memoires de l'Academie de Turin* in 1801. The spray thrown up as the funnel impacts the sea surface can be seen clearly. Waterspouts such as this usually do not last long, and they move slowly, if at all, and are therefore not especially hazardous.

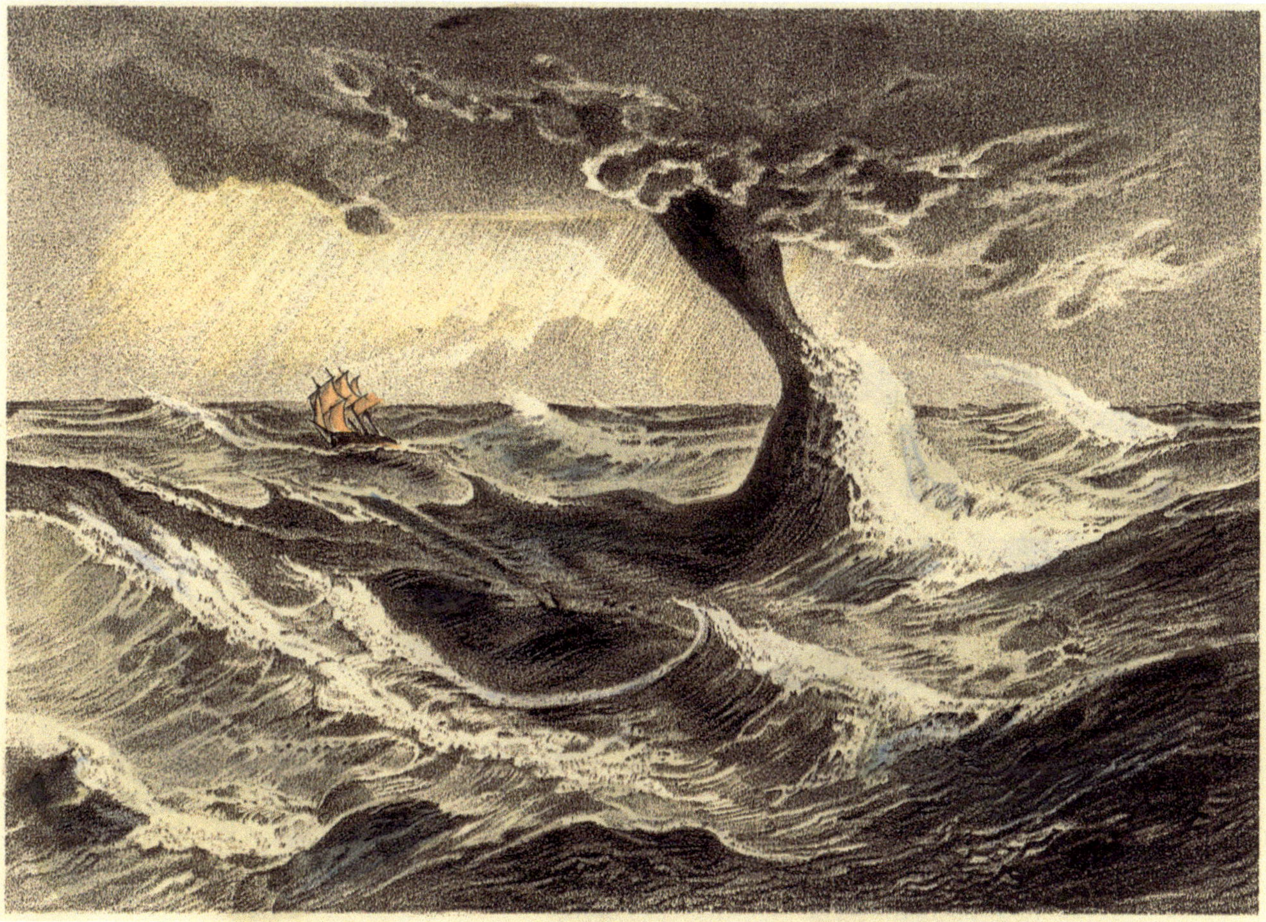

Fig. 3.40 An imaginary view of a waterspout

The French astronomer and author Camille Flammarion (1842–1925) has already been mentioned in Chap. 1. The illustration here is taken from his book *L'atmosphere: meteorologie populaire*, published in 1873 (Fig. 3.40). One cannot regard this as a very realistic image; the cloud base appears to be only a little above the tops of the waves. Still, it presents a dramatic scene, and one would rather not be on board the ship in the middle distance, despite the fact that the author could find no case of a ship ever being sunk by a waterspout, although he reports a case in the Ionian Sea in 1832 of a ship being struck by one, much to the alarm of the passengers, but without serious injury. Flammarion recommends that "generally the base of the column is severed by discharging a cannon into it." Apparently, this really was the practice; there is a record of Captain Patrick Napier doing exactly this in the West Indies in 1783.

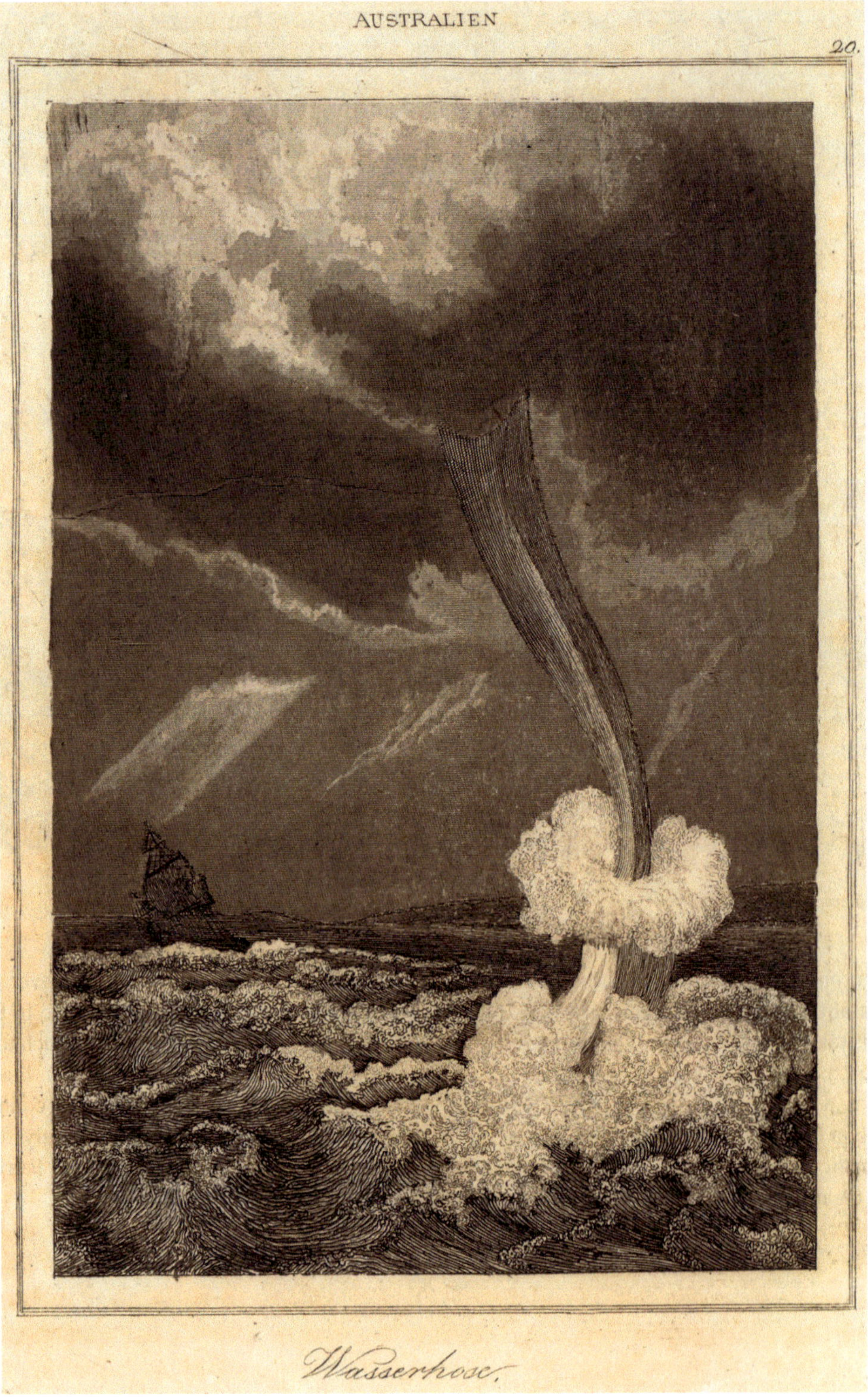

Fig. 3.41 A supposed waterspout off the coast of Australia

This print (Fig. 3.41) was published in 1836; it is a steel engraving, finished with watercolor, by the French landscape artist Victor Marie Félix Danvin (1802–1842). Supposedly, it represents a waterspout and a typhoon off the coast of Australia. Curiously, an extremely similar but not quite identical picture by the Strasbourg lithographer Emile Lemaître appears in a work from 1856, *Les Phénomènes et les Curiosités de la Nature*. It seems to have been a principle of artistic composition to show the waterspout on the right side of the image and a single sailing ship far left; other examples can be shown (including the previous picture, Fig. 3.40).

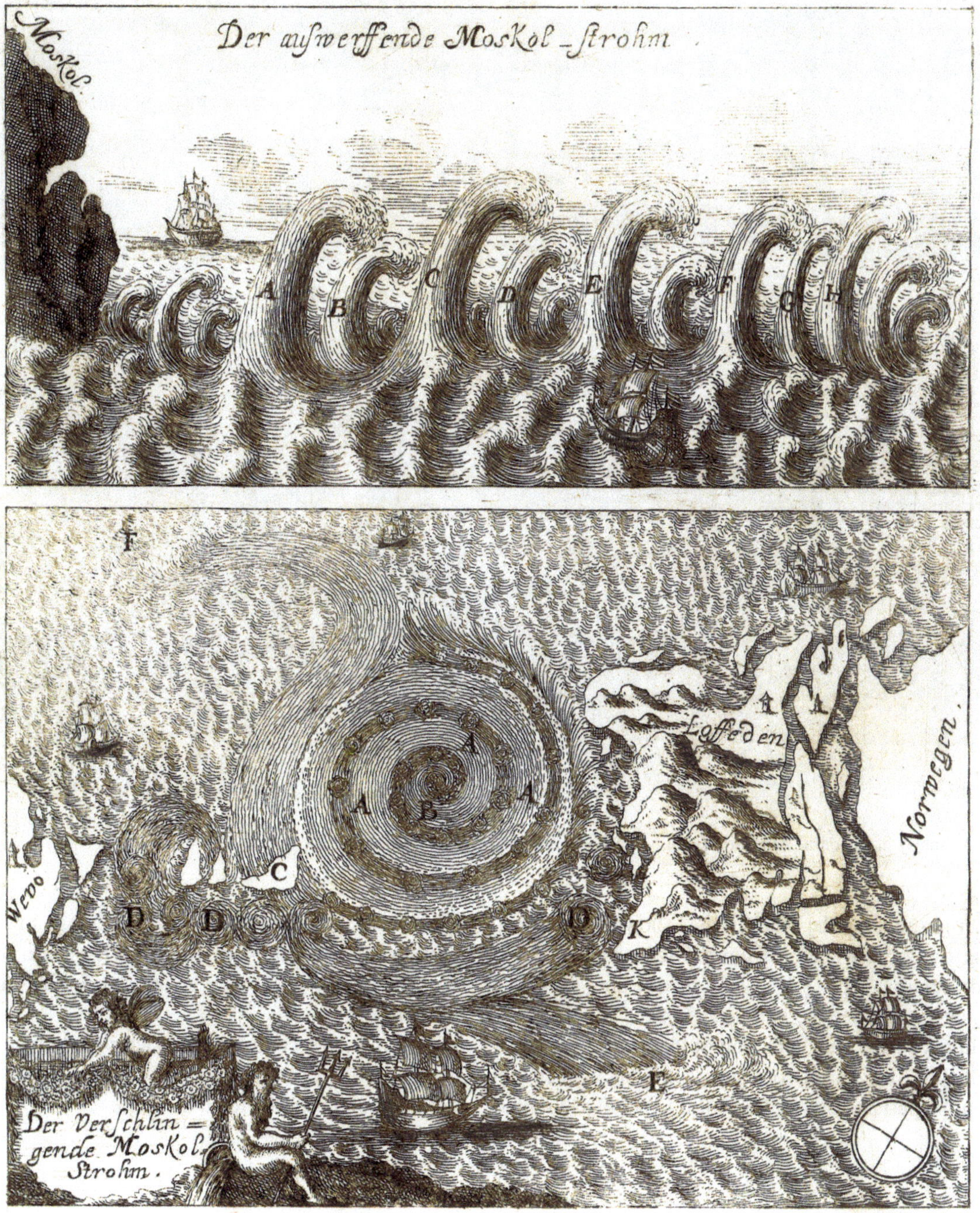

Fig. 3.42 The Maelstrom

We continue this group of pictures of rough, stormy sea waters by presenting two images of the legendary Maelstrom. On some fifteenth- and sixteenth-century maps of Norway and the North Atlantic, one spot is marked in which a gigantic whirlpool, surrounded by giant waves, gulps down whole vessels. This dangerous spot was located in shallow seas southwest of the southern tip of the rocky cliff of the island of Moskenesøya, one of the Lofoten Isles. The name Maelstrom was introduced into the English language by Edgar Allen Poe in a short story titled *A Descent into the Maelstrom*, published in 1841. Since then, the word "maelstrom" has often been used for any violently confused situation.

The two images of the Maelstrom reproduced here (Fig. 3.42) seem to be highly exaggerated, as the unrealistic dimensions of the sea waves confirm. Presumably, both pictorial compositions were based on some traveller's tales. A whirlpool is actually caused by powerful tides interacting with the unusual shape of the seabed, which amplifies the currents and introduces a vortex-like motion.

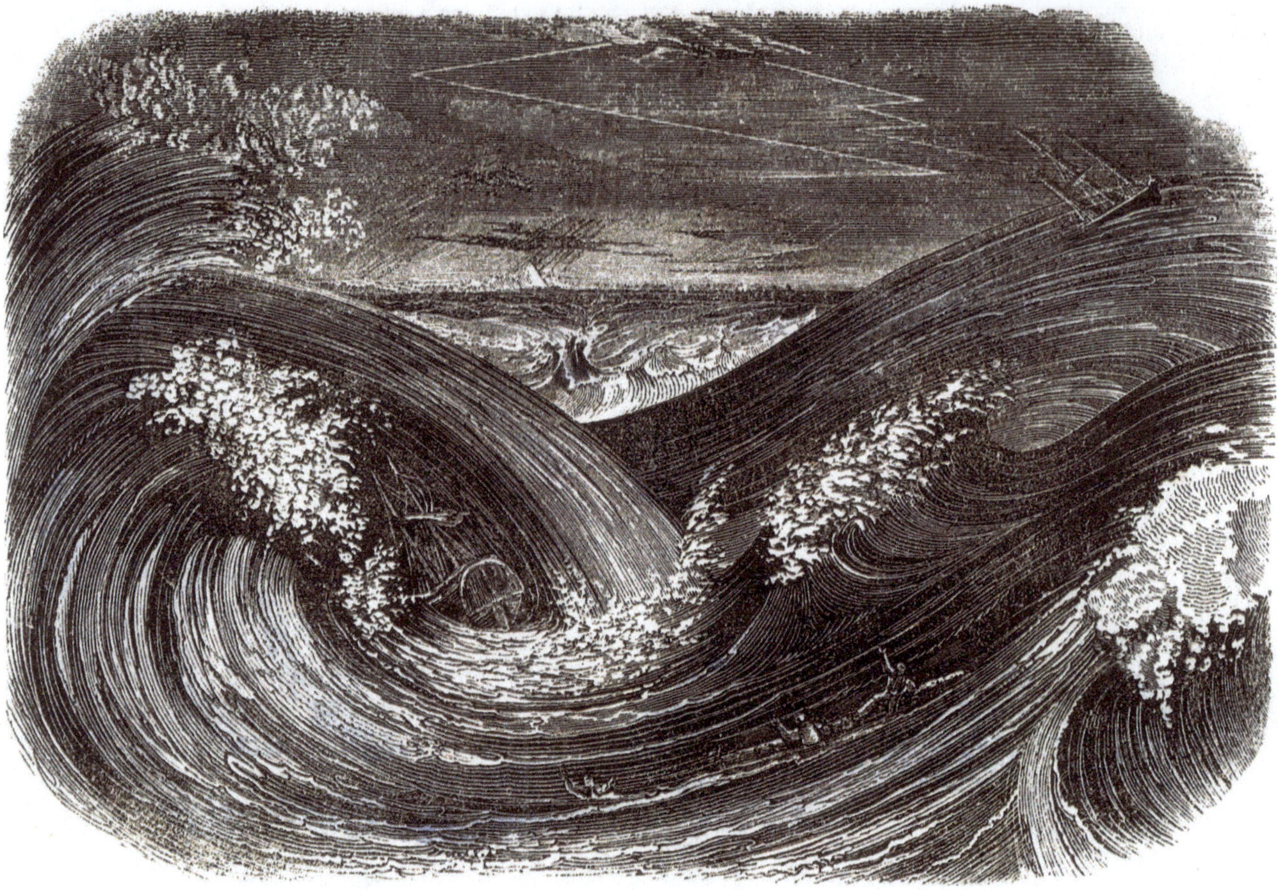

Fig. 3.43 A fantasy image of the Maelstrom

Another portrayal of the Maelstrom is shown in this steel engraving, a book illustration from the second half of the nineteenth century (Fig. 3.43). The size of the waves is greatly exaggerated; see the silhouettes of the two ships in the right upper corner and on the left of the picture's center.

Nevertheless, the Maelstrom was definitely something dreaded in the past. Actually, although it is popularly represented as a giant watery funnel reaching down into the depths, the Maelstrom—or, to give it its proper local name, Moskstraumen—is more a set of dangerous currents and cross-currents.

It is only the second-strongest whirlpool in the world; the strongest is also in northern Norway. Its name is Saltstraumen, and it is found near the town of Bodø.

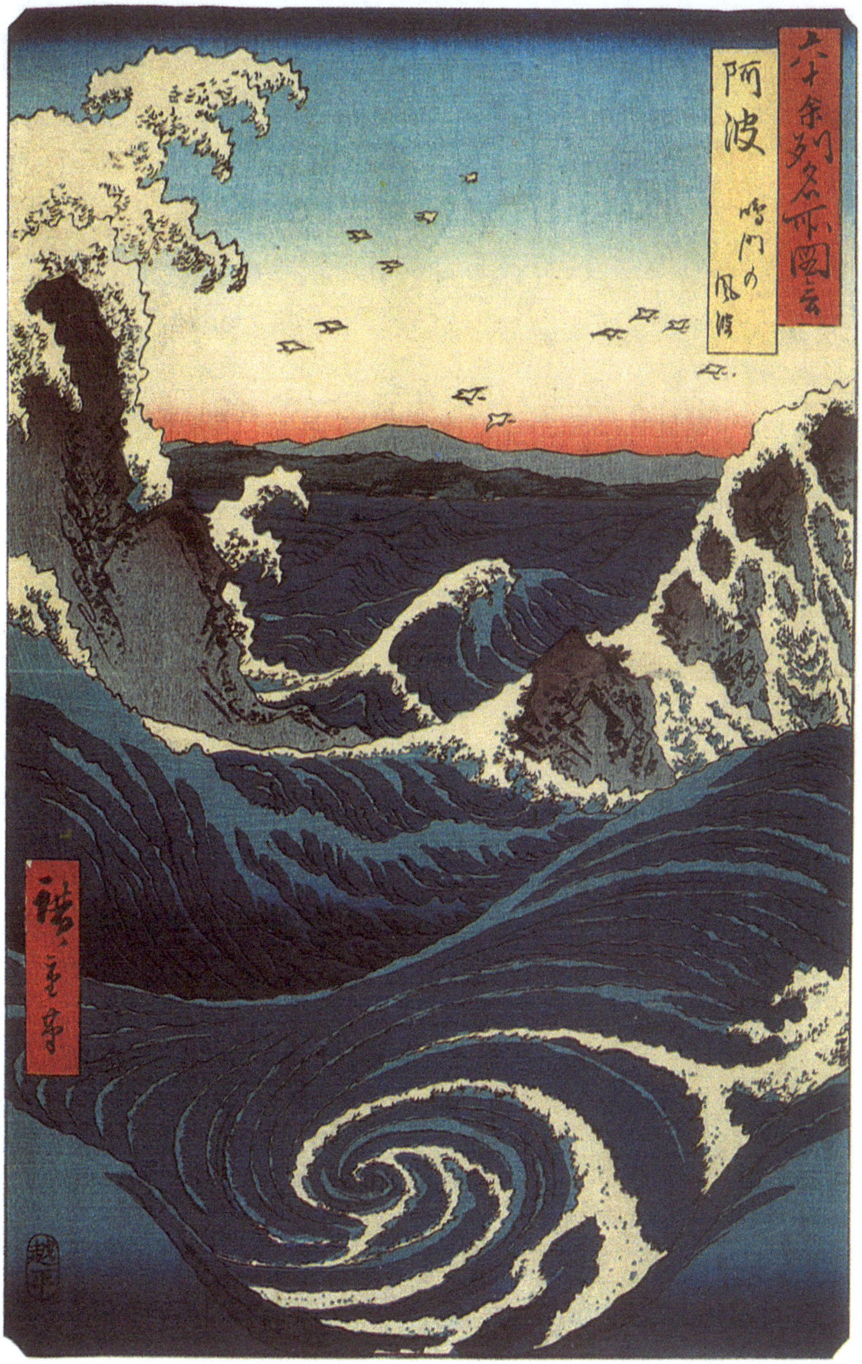

Fig. 3.44 The Naruto whirlpools

While Saltstraumen can reach water speeds of 40 kilometers per hour, and Moskstraumen 32 kilometers per hour, the Naruto whirlpools in Japan can generate currents of a respectable 26 kilometers per hour. They are found in the Naruto Strait, a channel between Naruto in Tokushima and Awaji Island in Hyōgo prefecture. This image (Fig. 3.44) is a color wood block of a painting by Utagawa Hiroshige (1797–1858). His style is typical of the *ukiyo-e* tradition in Japan, although his choice of subjects, mostly landscapes, is less so. One can trace here the influence of the older artist Katsushika Hokusai, who is well known for his *Great Wave off Kanagawa* in the collection *Thirty-six Views of Mount Fuji*. The picture here of the Naruto whirlpools is from the series *Famous Views of the Sixty-odd Provinces*, his last collection, which was incomplete at his death and issued posthumously. It is unusual among *ukiyo-e* works in adopting what we would now call a portrait format—an innovation of Hiroshige's for landscapes.

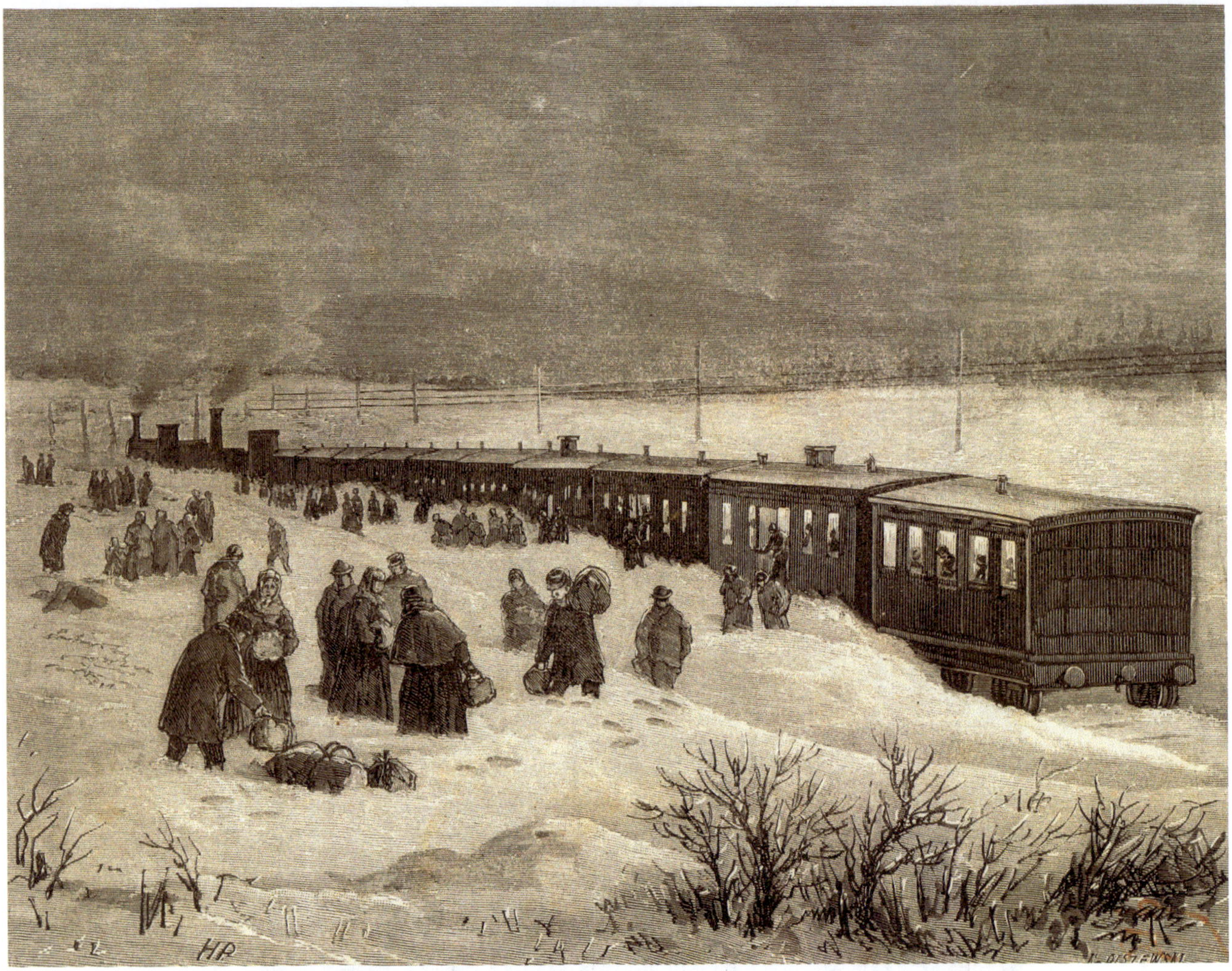

Fig. 3.45 A train stuck in snowdrifts in Galicia in 1871

Water is not only hazardous in its liquid form. Snow and ice are hazardous as well. Blizzards of snow block transport routes, and stranded travellers may be at risk of freezing to death. This engraving (Fig. 3.45) from the Polish magazine *Klozy* (1872, volume 16/244, page 76) depicts a train brought to a standstill by snowdrifts blocking the line; the place is somewhere on the Karola–Ludwika Railway in Galicia, and the date is December 1871. The image is signed by the engraver Kazimierz Olszewski and the artist Henryka Pillatiego. The train passengers, weighed down with their luggage, sink into deep snowdrifts. It can be imagined that they were hardly satisfactorily equipped for such an eventuality.

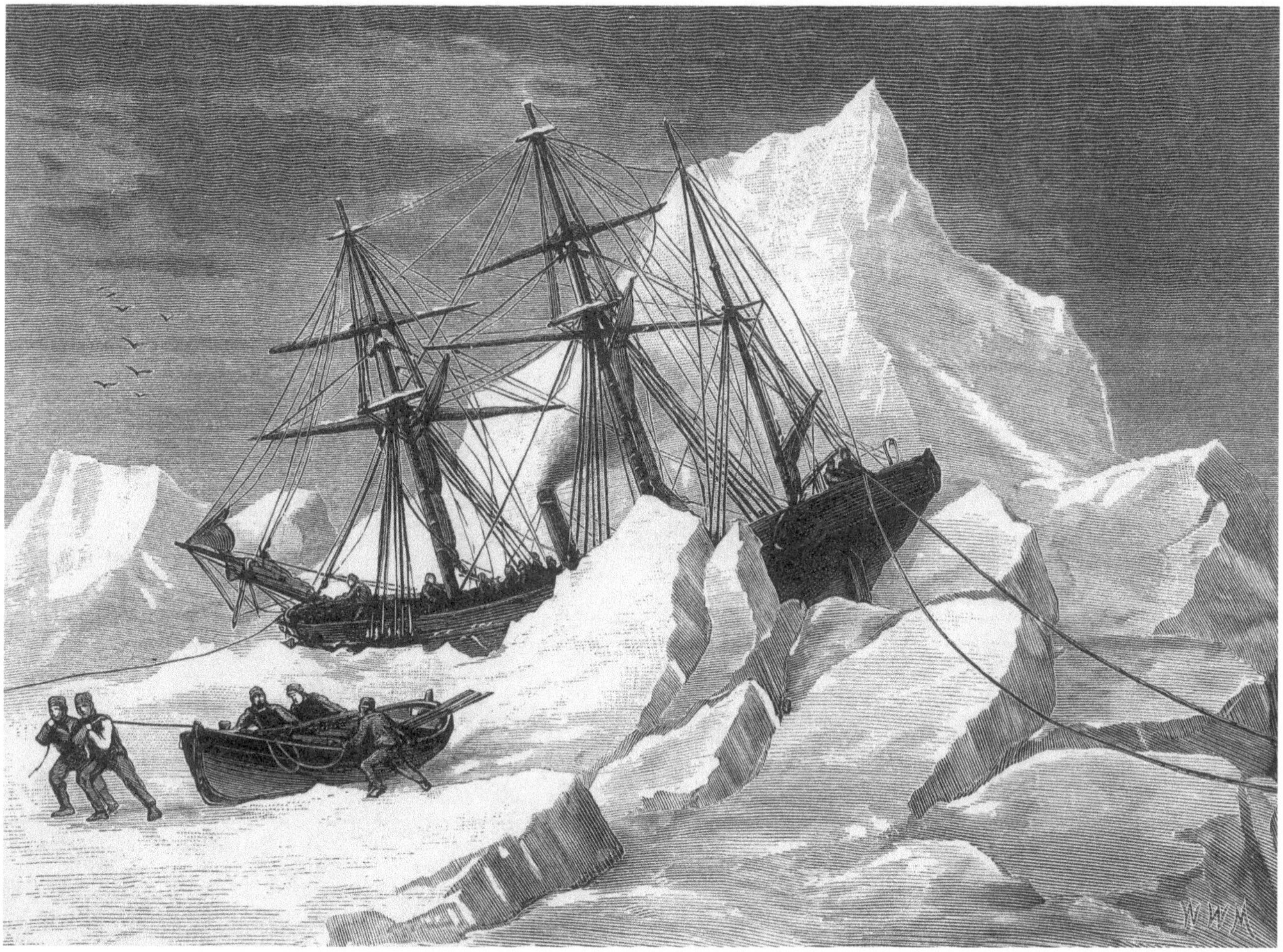

Fig. 3.46 *HMS Intrepid* stuck in ice

Ice is a hazard to shipping in two ways: collision with icebergs (such as the one that sank the "unsinkable" *RMS Titanic* in 1912) and the freezing up and trapping of ships in pack ice—a problem fortunately restricted to polar latitudes. One celebrated case was that of Sir John Franklin, whose expedition to try and find a navigable passage between the Atlantic and Pacific Oceans via the Canadian Arctic (the Northwest Passage) vanished in 1848. A number of expeditions were mounted in search of the missing explorer and his two ships, *Erebus* and *Terror*. The expedition of Admiral Sir Edward Belcher in 1852 ran into trouble itself. Of his five ships, four became trapped in ice (all except for his depot ship, *North Star*). The crews abandoned their vessels and continued to explore on foot; in this engraving, made from a sketch by one of the officers on the expedition, one can see *HMS Intrepid* trapped in the grip of ice (Fig. 3.46). By July 1854, Belcher had concluded that it was too dangerous to continue; he abandoned all of his ships and made his way back by sledge to the *North Star* and thence home.

As to Franklin's fate, *Erebus* and *Terror* were both abandoned when they became trapped, and the crews perished from inadequate supplies. The wreck of the *Erebus* was not discovered until 2014; the *Terror*, which had sunk, was found two years later.

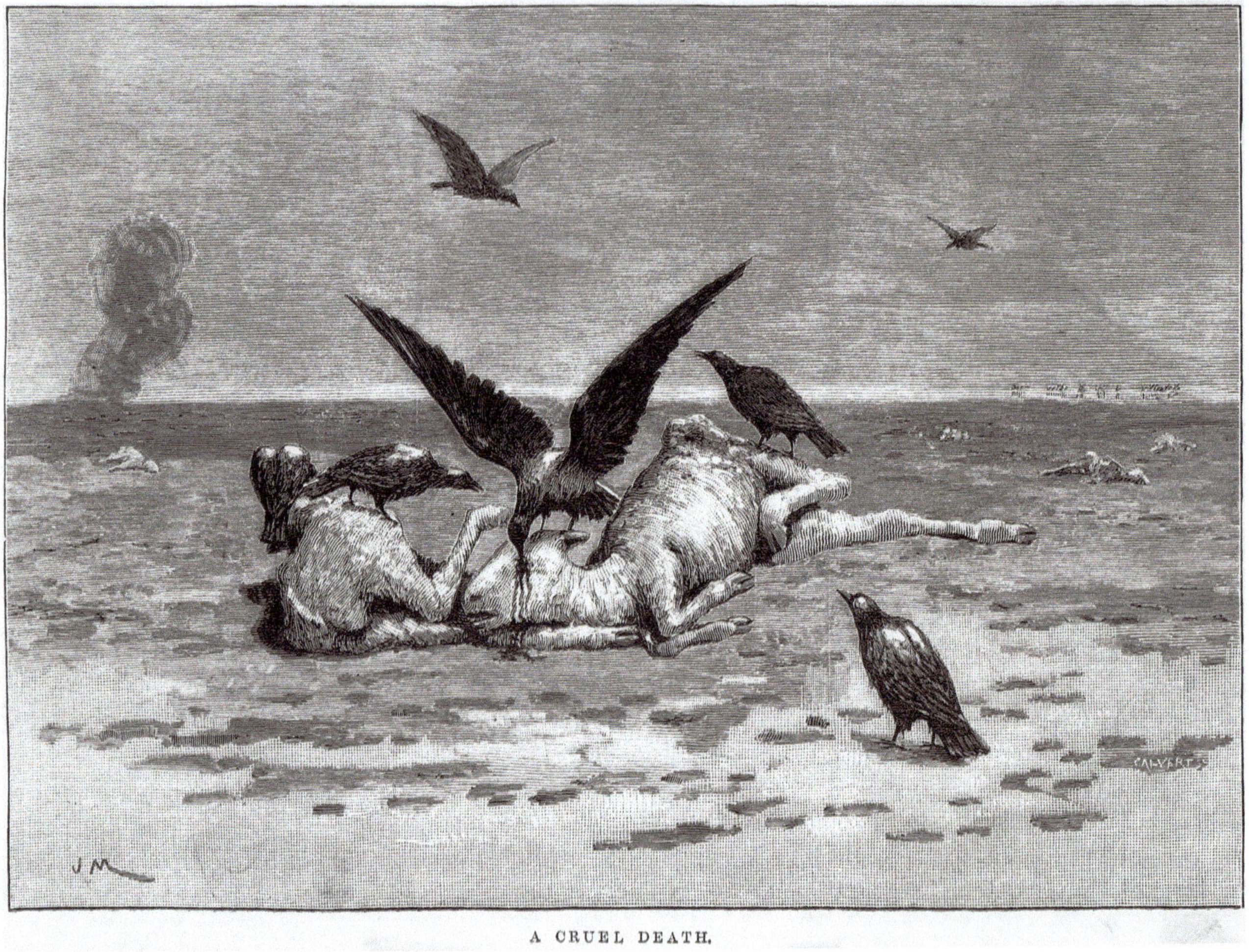

Fig. 3.47 A drought in Australia in 1889

Let us close our excursion around the extreme manifestations of the element Water with one warning example, which tends to be forgotten when floods and other excesses of water are discussed: drought. Drought is the absence of water—not too much, but too little. It is not to be ignored as a natural disaster. It kills more than earthquakes and volcanoes put together, but whereas an earthquake can kill hundreds of thousands of people in a matter of seconds, a drought can kill millions of people, but over a longer period of time, so much so that it rarely grabs the headlines in the way that earthquakes do. This engraving, titled *A Cruel Death*, is by the artist and illustrator John Macfarlane and depicts a scene during the drought that struck New South Wales, Australia, in 1889 (Fig. 3.47). It was published in the *Illustrated Australian News* on 12 January 1889.

Air

4

4.1 Introductory Remarks

According to various dictionary definitions, air is that mixture of gases that surrounds our planet—in other words, the atmosphere. In ancient times, people were aware that there was something invisible that surrounds us, that we breathe every day, but they had no way to analyze it. It was considered some sort of invisible stuff, and "air" was the name given to it.

Today we know that Earth's atmosphere is made up of a mixture of gases, which, fortunately for us, is conducive to life. Oxygen is the gas we need for breathing (about 20% of air by volume), but the largest part of the atmosphere is nitrogen (nearly 80%), and the remainder is made up of argon, carbon dioxide, water vapor, and a few other gases in trace quantities. Besides giving us something to breathe, the atmosphere also absorbs ultraviolet solar radiation, reduces the extremes of temperature between day and night, and makes weather systems possible, all of which make our planet pleasant to live on.

Other planets have their own atmospheres—or not, as the case may be. Our moon has no atmosphere. Venus has an atmosphere consisting mostly of carbon dioxide, with some nitrogen. It is an atmosphere, but could one ever call it air? No one could ever breathe it. But if we could visit a planet orbiting some distant star and found there an atmosphere similar to that on Earth, we might well think of it as air. "I feel air from another planet," wrote the poet Stefan George. So perhaps a simpler definition would be "that which we breathe."

Certainly, to pre-scientific writers, that would have made sense. Air exists; it does not seem to be made of Fire, Earth, or Water; therefore, it must be a fundamental element. For the artist, it presented a problem. How do you portray something that is invisible? Air can only be represented by its visible manifestations or effects. On a still day, the air around us is not very noticeable. On a windy day, one sees the branches of the trees blowing about and one feels the wind on one's face—a tactile reminder that air is all around us. So pictures have to show things blowing about or clouds racing across the sky.

In fact, weather in general—and anything else in the sky—could be considered the domain of Air, including meteors, auroras, and the like. Most atmospheric phenomena are not very disastrous in the normal run thing of things, but some could be—for instance, if a comet should actually impact the Earth rather than passing by and providing an interesting spectacle. One should also reflect on changing vulnerability. An auroral display that was perfectly harmless in historical times might today be associated with a geomagnetic storm capable of harming electrical infrastructure.

Thinking about the four elements, there are connections between Air and the three others: rain is Water, meteorites are Earth, and comets (at least superficially) resemble Fire, as do meteors; it was common in medieval times for those watching a blazing meteor at night to mistake it for the fiery breath of a passing dragon overhead.

Another connection with Fire—and Earth too, for that matter—relates to volcanic eruptions. We think of people being killed by fiery lava or falling rocks (volcanic bombs) or by rushing clouds of superheated ash (*nuées ardentes*), as in the catastrophic eruption of Vesuvius in AD 79, which killed the inhabitants of Pompeii and Herculaneum, or, more recently, the eruption of Mt. Pelée (Martinique) in 1902, which wiped out the town of St. Pierre. But volcanic eruptions can change the weather as ash is injected into the upper atmosphere. The massive eruption of Tambora in 1815 was followed by the "year without a summer" of 1816, and people died of starvation.

In this context, we should also mention the fissure eruption of Laki (Iceland) in 1783, one of the greatest gas disasters. It released vast amounts of sulfur dioxide, which then drifted over Europe, causing acid rain (sulfuric acid), respiratory problems, and heavy mortality. Unfortunately, we lack a good illustration of the event.

So, in our pictorial excursion below, we will encompass a wide variety of phenomena, starting with the severest tropical storms, taking in European windstorms, whirlwinds, lightning, hail, comets, meteorites, and even a nod to volcanoes and earthquakes. We shall also take in pandemics, as disease was historically thought to be due to bad air, even if we know better today.

J. Kozák, R. M. W. Musson, *The Illustrated History of the Elements*, https://doi.org/10.1007/978-3-030-21426-5_4

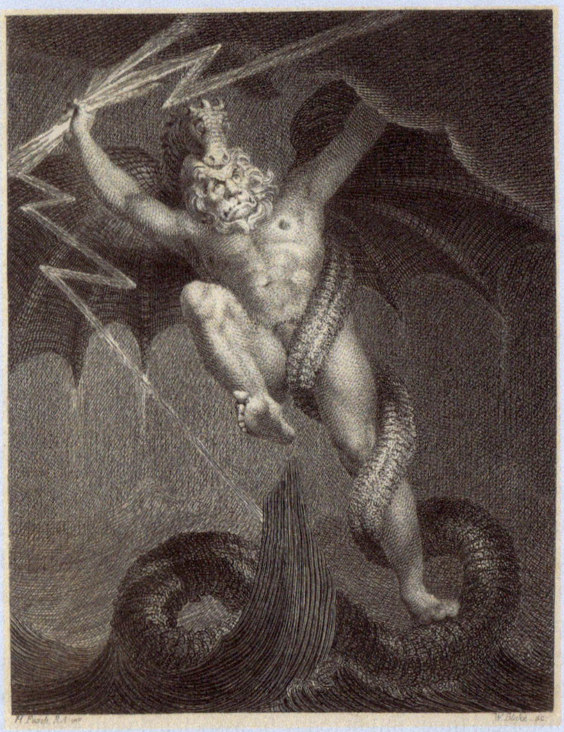

This engraving by William Blake, after a design by Henry Fuseli, is an artistic interpretation of a tornado, although it does not make much of a literal depiction. The tornado is the god Zeus, clutching a handful of lightning bolts as he battles the sea serpent Typhon. The story is found in Hesiod: "Zeus thundered hard and mightily: and the Earth around resounded terribly and the wide heaven above, and the sea and Ocean's streams and the nether parts of the Earth… . And through the two of them heat took hold on the dark-blue sea, through the thunder and lightning, and through the fire from the monster, and the scorching winds and blazing thunderbolt. The whole Earth seethed, and sky and sea: and the long waves raged along the beaches round and about at the rush of the deathless gods: and there arose an endless shaking." The image dates from 1795, and was prepared for a book The Botanic Garden by Erasmus Darwin. The similarity of sound between "Typhon" and "typhoon" is probably not a coincidence."

4.2 Wind

When it comes to disasters associated with the element Air, windstorms are what spring first to mind. Hurricanes, typhoons, and tornadoes have the capacity to deal much destruction. Loss of life was typically greater in historical times than for similar modern events, because of the effect of such storms on sailing ships, with their large crews. One might reasonably ask what the difference is between a hurricane and a typhoon; it is really a matter of geography. A fierce cyclonic storm in the Caribbean is referred to as a hurricane, one in east Asia is called a typhoon, and one in the Indian Ocean is called a cyclone. Tornadoes are much more localized and sudden. One might have days of warning of the approach of a hurricane. For those unlucky enough to be in the path of a tornado, the warning is likely to be a matter of minutes.

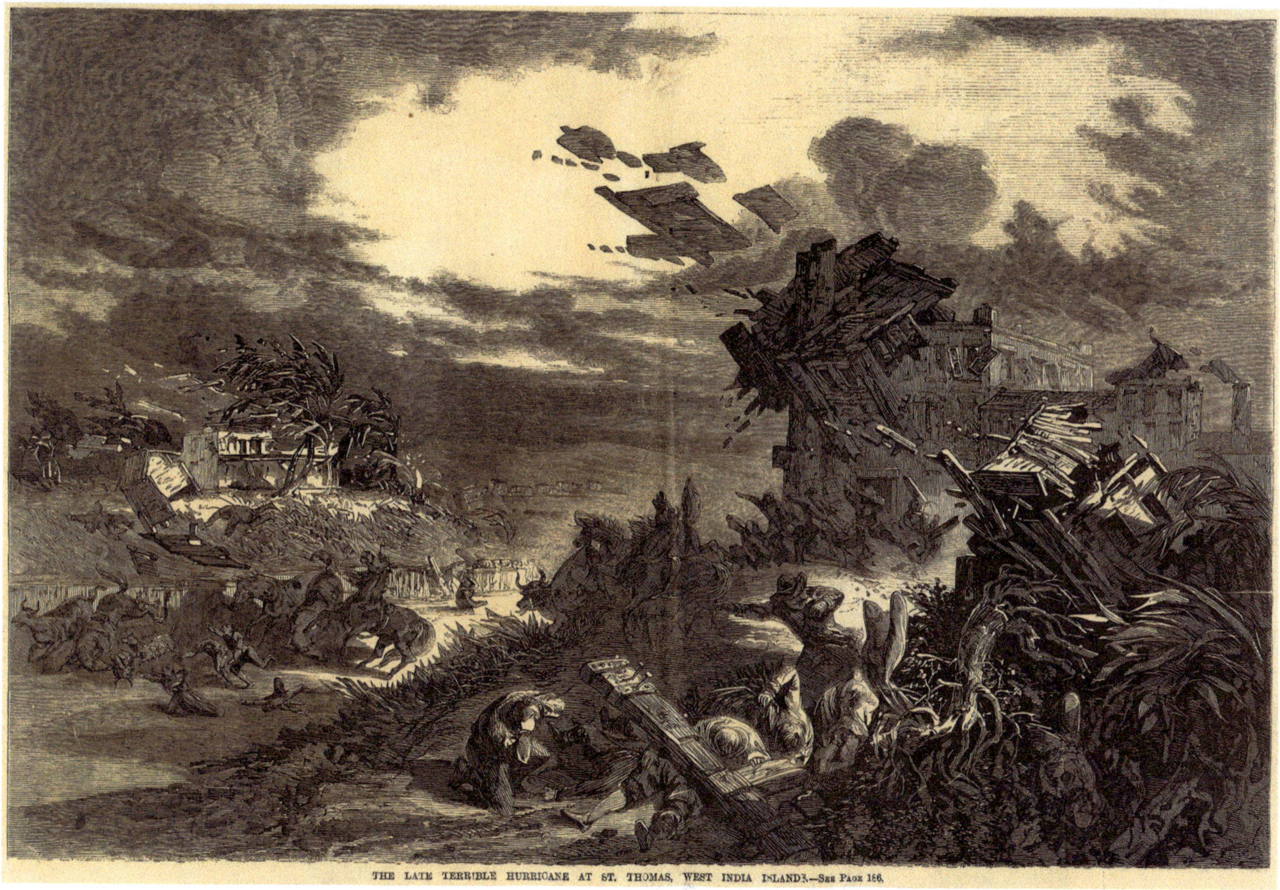

Fig. 4.1 The San Narciso hurricane of 1867

This image shows a severe hurricane striking the island of St. Thomas in the Caribbean (Fig. 4.1). The storm hit the island on 29 October 1867, sinking 80 ships and killing around 600. It then moved on to Puerto Rico, wreaking even more death and destruction. The hurricane season in 1867 in the Caribbean was particularly severe. Between 21 June and 31 October, a total of nine tropical storm systems developed, of which the one shown here—the ninth and last, referred to as the San Narciso hurricane—was the strongest. This engraving was published in *Frank Leslie's Illustrated Newspaper* on 7 December 1867.

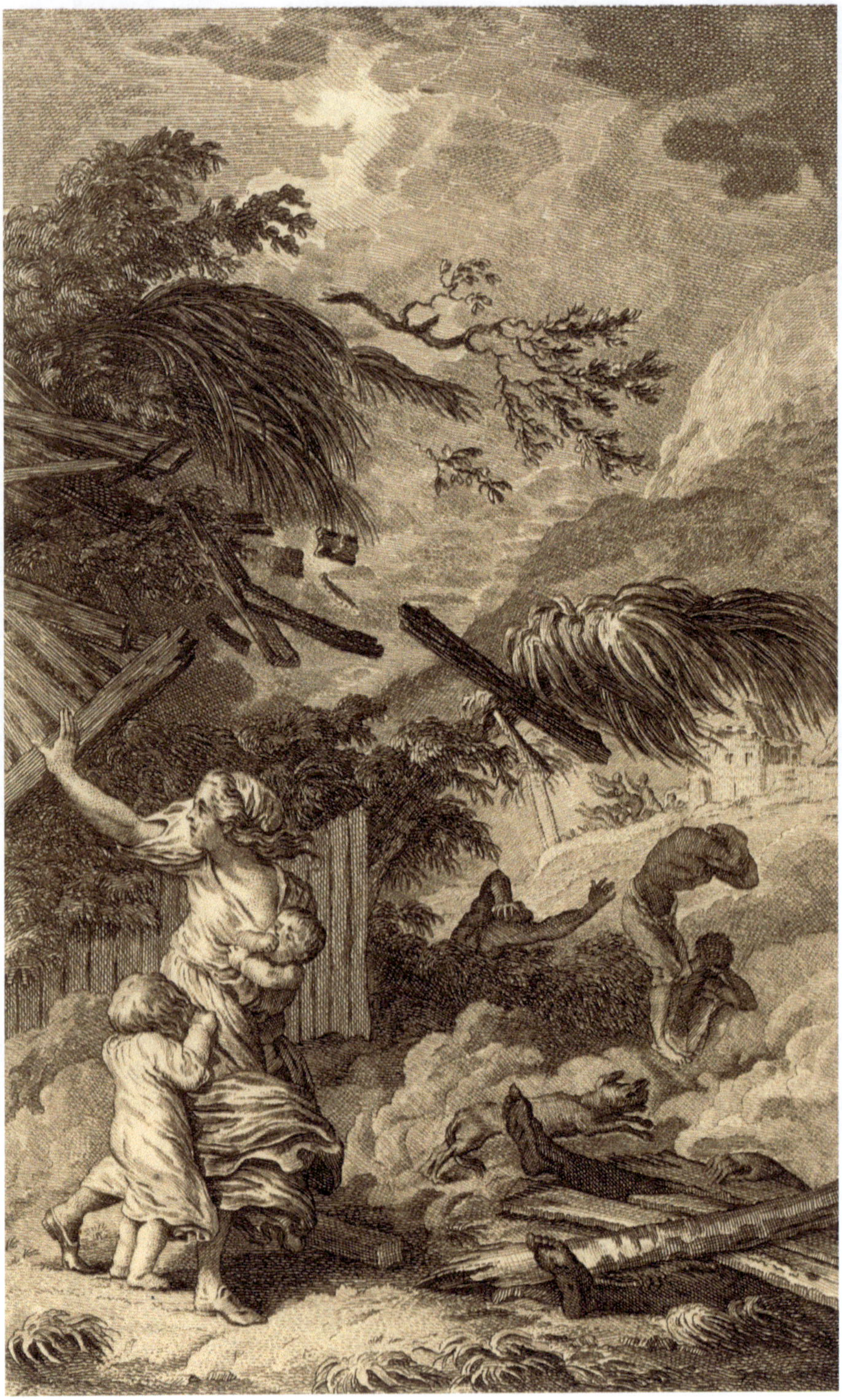

Fig. 4.2 The San Mateo hurricane of 1819

The Caribbean and Gulf of Mexico is the part of the world most at peril from hurricanes. Tropical cyclones can also be very damaging in the Philippines and the west Pacific, but these storms are less rich in iconography. The image here (Fig. 4.2) is a copperplate engraving of uncertain but Dutch origin, probably showing the San Mateo hurricane of September 1819, which struck the Virgin Islands and Puerto Rico, sinking over 100 ships, with great loss of life.

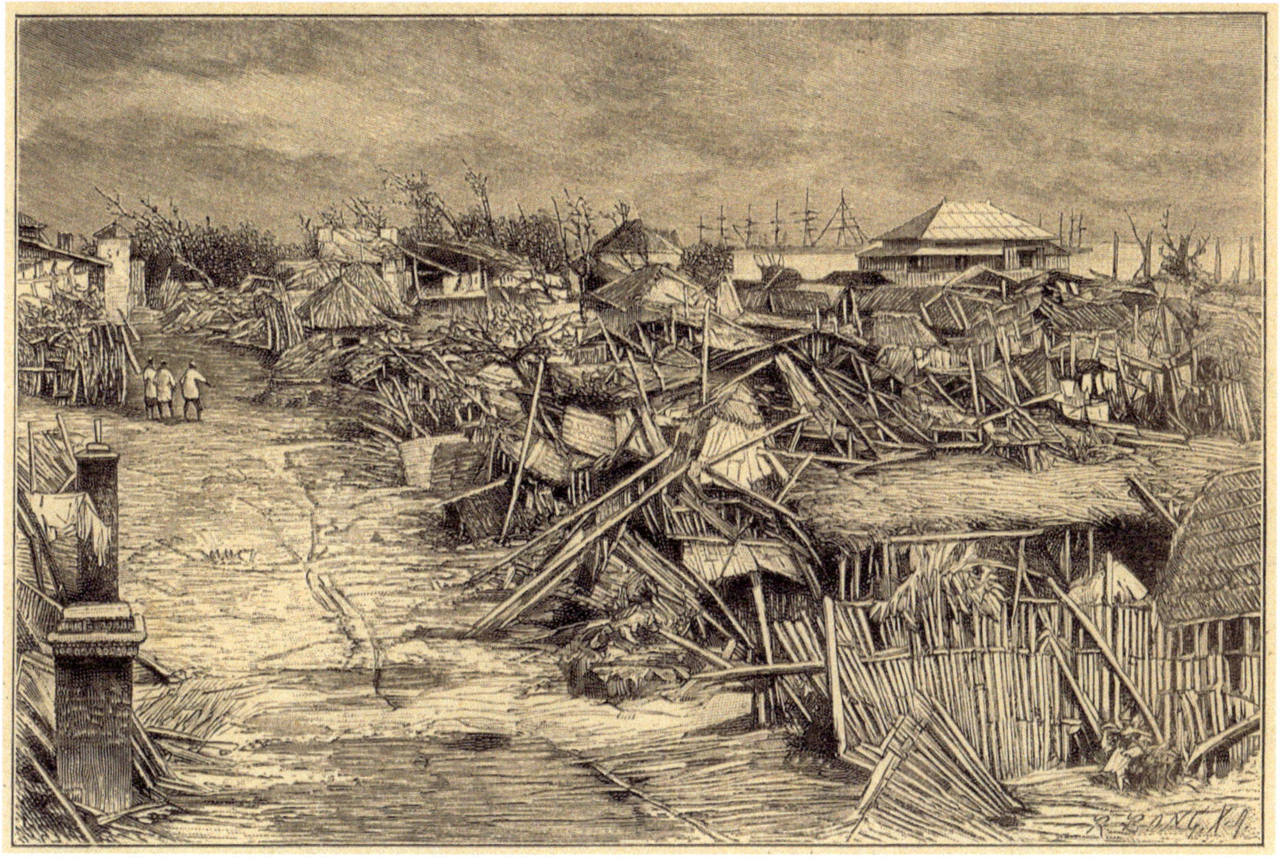

Fig. 4.3 Typhoon damage at Luzon in 1867

Images of Pacific typhoons from the prephotographic era are not, however, entirely lacking. This engraving (Fig. 4.3) from the French travel magazine *Le Tour du Monde, Nouveau Journal des Voyages* shows the destruction wreaked by the typhoon of 20 September 1867, on Luzon Island. The deadliest of all pre-1900 typhoons in the Philippines, it killed about 1800 people, not counting sailors lost in shipwrecks. A history of typhoons that affected the Philippines was compiled by a Jesuit, Miguel Selga (1879–1956), who was the last Spanish director of the Manila Observatory.

One can see in this image that most buildings in this region were of wood (in contrast to the magnificent stone palaces of the very wealthy). Even many of the port facilities were of wooden construction. These wooden structures had little resistance to the force of a typhoon, and, as the picture shows, most collapsed totally.

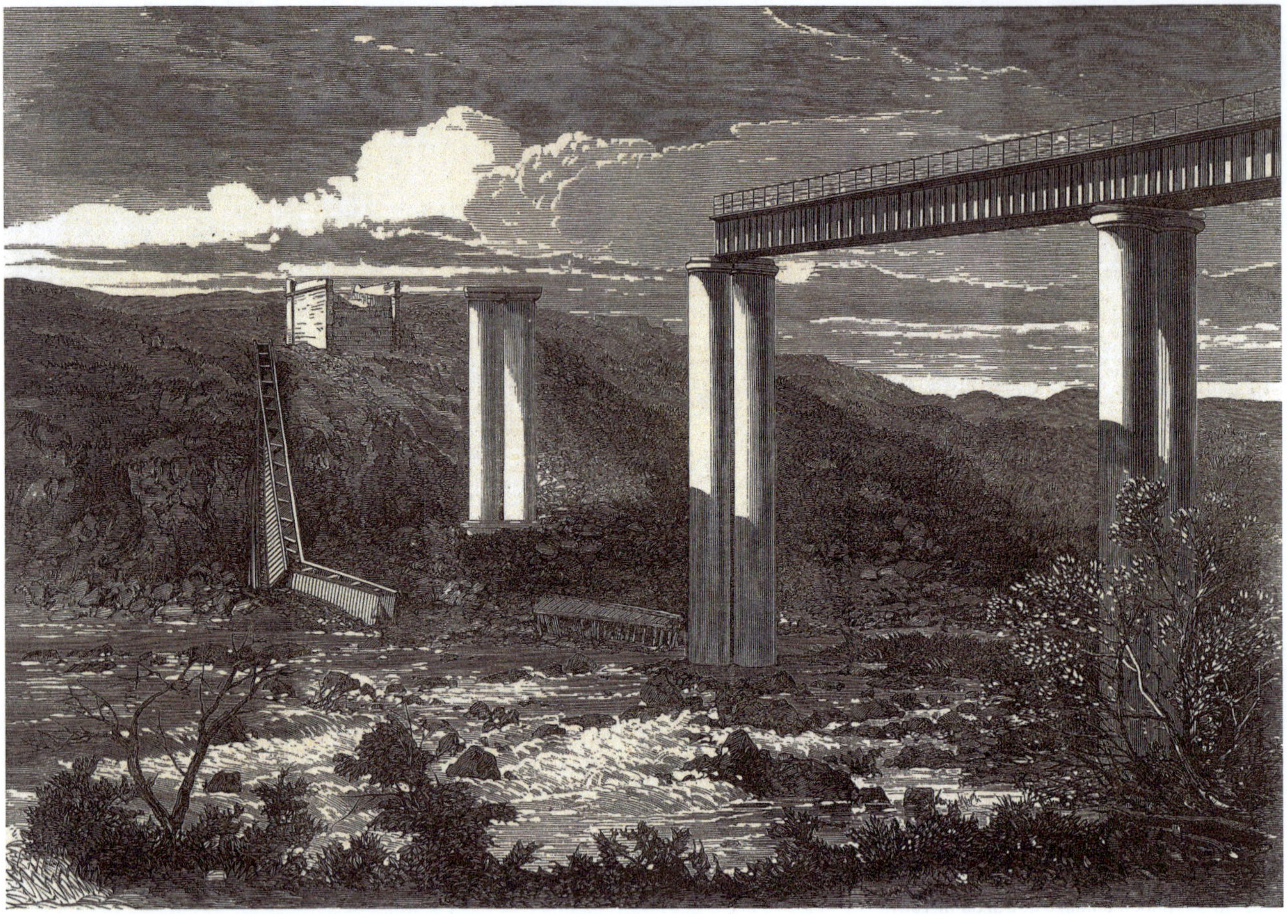

Fig. 4.4 Cyclone damage in Mauritius in 1892

The island of Mauritius in the Indian Ocean is located east of Madagascar; it is relatively small (about 2000 square kilometers), and its rich variety of natural beauty gives it the reputation of a paradise island. Yet, every year between January and April comes the "cyclone season"; a period of rains and storms visits the island, always accompanied by one or two strong cyclones. In January 2018, Cyclone Berguitta swept past the island, causing flooding and damage, and sending residents to seek refuge in shelters.

One of the most devastating cyclones to hit the island was that of 29 April 1892, during which—within two hours—a great deal of the island's constructions were smashed, trees were broken or uprooted, and 1000 people were killed. The picture here, though, is of an earlier event. On 11–12 March 1868, extensive damage was caused by a cyclone that was described at the time as one of the most violent experienced. Huge destruction was visited upon the island's sugar plantations, and the Grande Rivière bridge on the island's Central Railway was partly destroyed, as shown in this engraving (Fig. 4.4) from *the Illustrated London News* of 18 June 1868 (page 585). Part of the bridge was actually uplifted bodily by the wind and dropped into the ravine. Possibly the bridge's designers in 1865 thought mostly of the need to bear the vertical load of heavy passing trains and underestimated the need for lateral resistance against the strong horizontal impact of a mighty tropical storm.

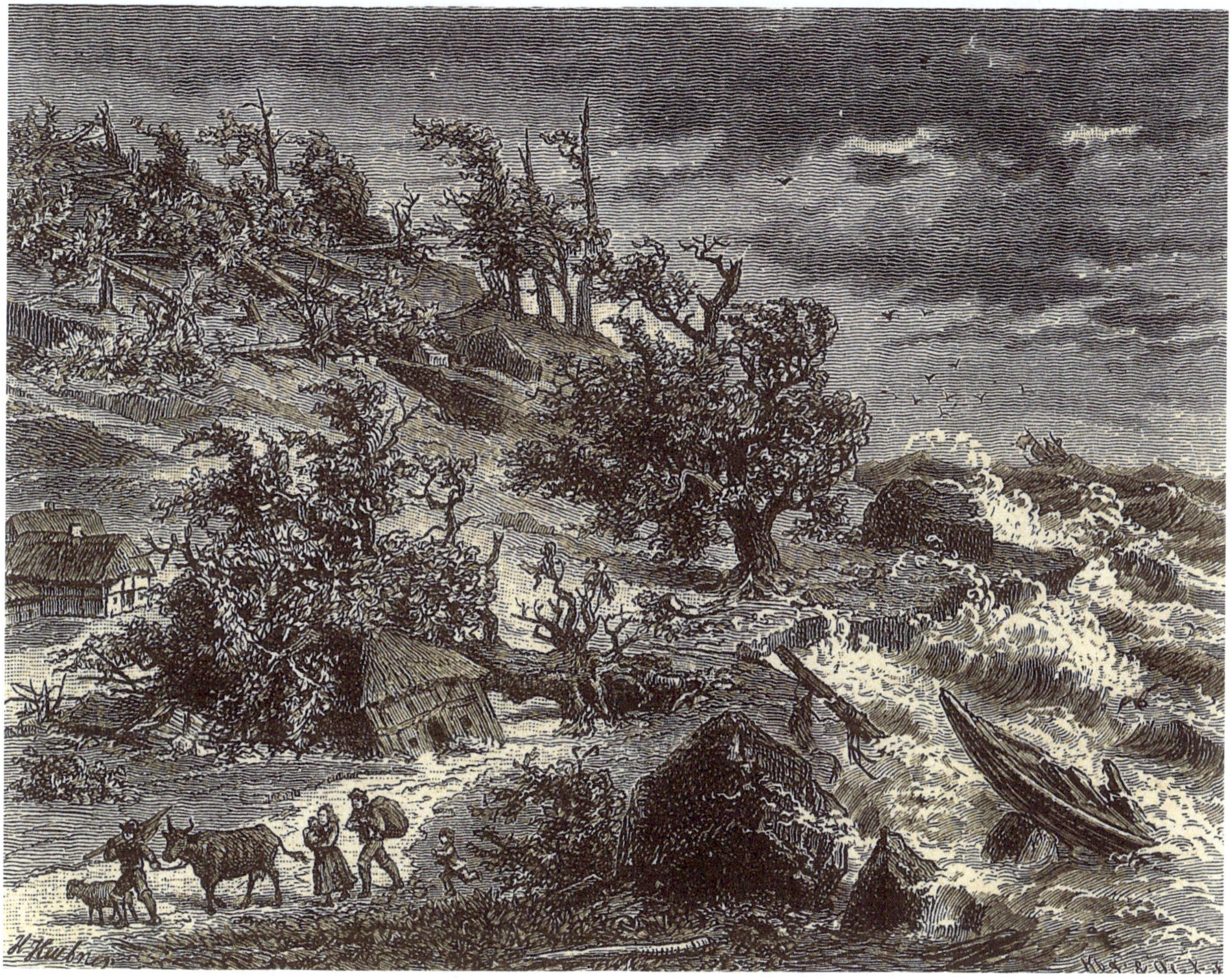

Fig. 4.5 Stormy weather in Silesia

Windstorms can be devastating outside the tropical zones as well. 1881 was a particularly stormy year in central Europe, and this image, captioned *"devastating effect of wind"*, shows something of it. Storms and floods affected many areas; the scene here, probably with some artistic license, depicts the effects in Silesia. Trees are blown down, a house is destroyed, and a boat is wrecked on the stormy waters of a lake. This depiction (Fig. 4.5) comes from a collection of engravings by Ferdinand Hirt, published in Breslau (now Wroclaw) in 1881 as *Ferdinand Hirt's Geographische Bildertafeln* by Alwin Oppel and Arnold Ludwig. This scene, in which strong winds and rough waters fight each other, is titled *Kampf der Elemente* (*Struggle of the Elements*).

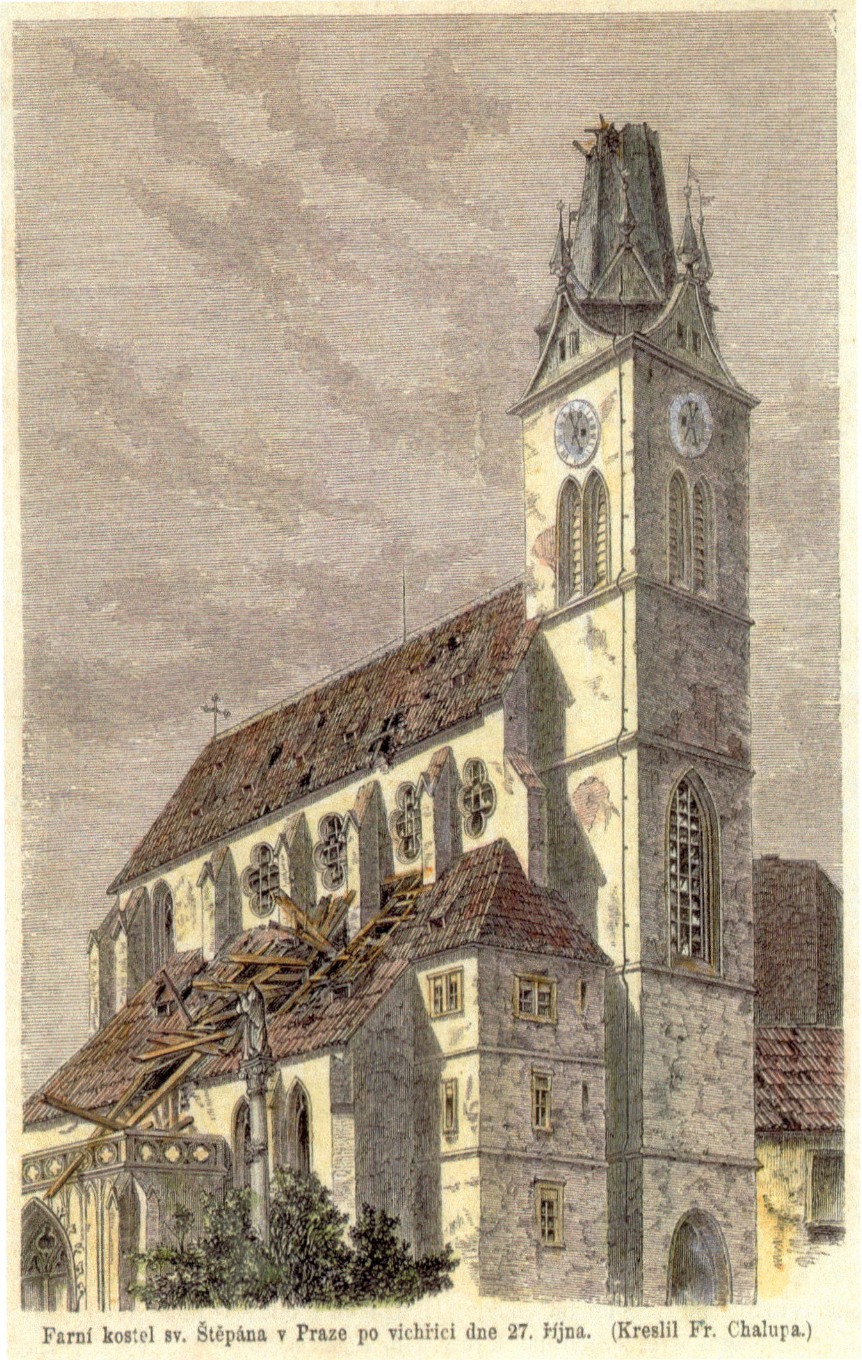

Fig. 4.6 Storm damage in Prague in 1870

This is a Czech example of the destructive power of the wind. The picture (Fig. 4.6) shows the Church of St. Stephen in Prague after one of the severest windstorms to affect the city. It occurred on the night of 26–27 October 1870, and threw down half of the spire of the church. The windstorm originated in the transalpine region of Bavaria and lower Austria, and entered Bohemia by advancing northward. It then turned to the east, toward the town of Brno, and up to the east Beskydy Mountain ridge. Heavy damage was reported in Prague and in other Czech and Moravian towns, in which the storm pulled down roofs and church steeples.

During the storm's culmination, which started at 11 p.m. on 26 October and lasted four hours, the windstorm was accompanied by repeated hailstorms. Damage from winter storms is common across northern and central Europe, especially damage to chimneys. It is not so common for churches to be damaged as badly as this. The damaged steeple shown here dated from 1605 and was rebuilt to a design by the architect Joseph Mocker in 1875, this being the structure one sees today.

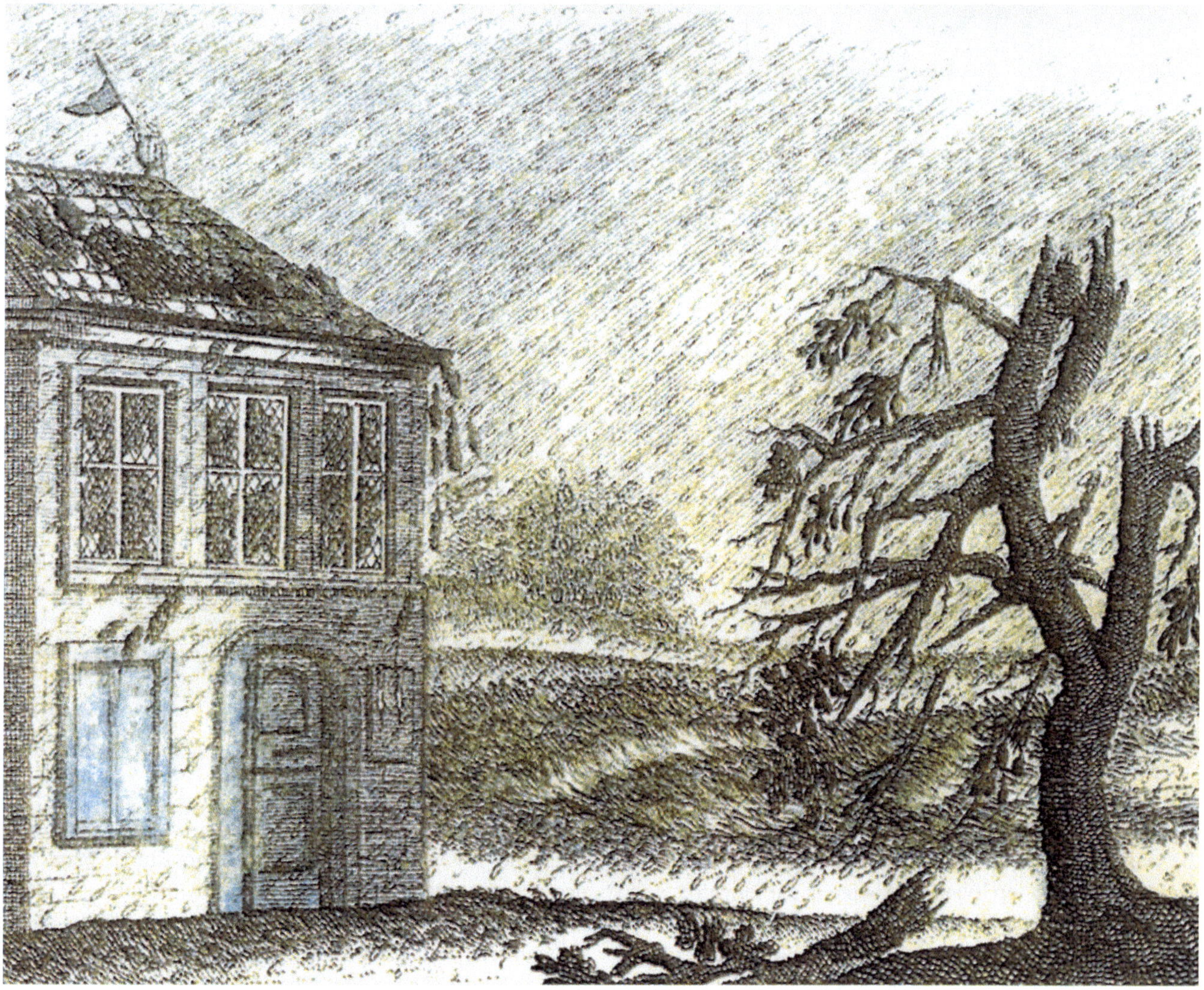

Fig. 4.7 Miserable weather, circa 1750

This unidentified image (Fig. 4.7), probably from central Europe, circa 1750, is another example of the misery that winter storms inflict on home owners. Gusts of wind have snapped off branches of the tree and blown slates off the roof of the house; to make matters worse, heavy rain is pouring down. Nowadays, fatalities in such storms usually result from people being hit by falling branches or parts of roofs. Gone are the days when high fatality rates were reported on account of shipwrecks from even modest winter storms.

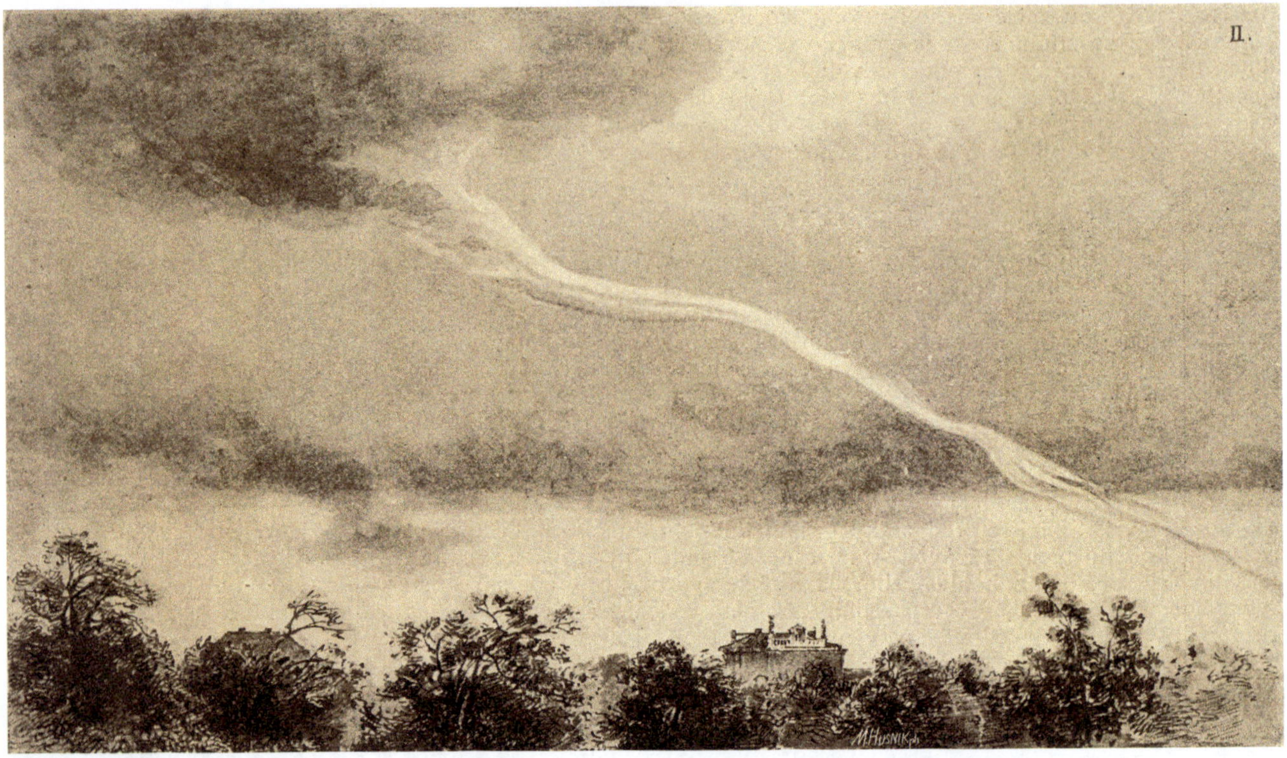

Fig. 4.8 The Teplice tornado of 1887

One particular type of windstorm is particularly visual: the whirlwind. This is a vortex of wind rotating at speed. The greatest and most terrifying whirlwind is the tornado.

Tornadoes are typically thought of as affecting the central region of the USA, but in fact they can occur in many places. Even in a country such as the UK, where people imagine that tornadoes are unknown, they are actually quite common, only much less powerful than their American cousins. This image shows a Czech example—a tornado that struck Teplice on 16 May 1887. The engraving was published in the illustrated weekly journal *Zlatá Praha* (Fig. 4.8).

A tornado in Bohemia—this really is an extraordinary spectacle. It is worth mentioning that this strange phenomenon was described and commented on by the famous meteorologist and polar explorer Alfred Wegener (1880–1930), remembered today especially for his ideas on continental drift. In his 1917 book *Wind- und Wasserhosen in Europa*, he discussed this instance. He stressed that on the day before the tornado, a frontal system passed by; the complex stormy weather finally formed an almost horizontal tornado funnel about 8–10 kilometers in length. Repeated hailstorms accompanied the spectacle.

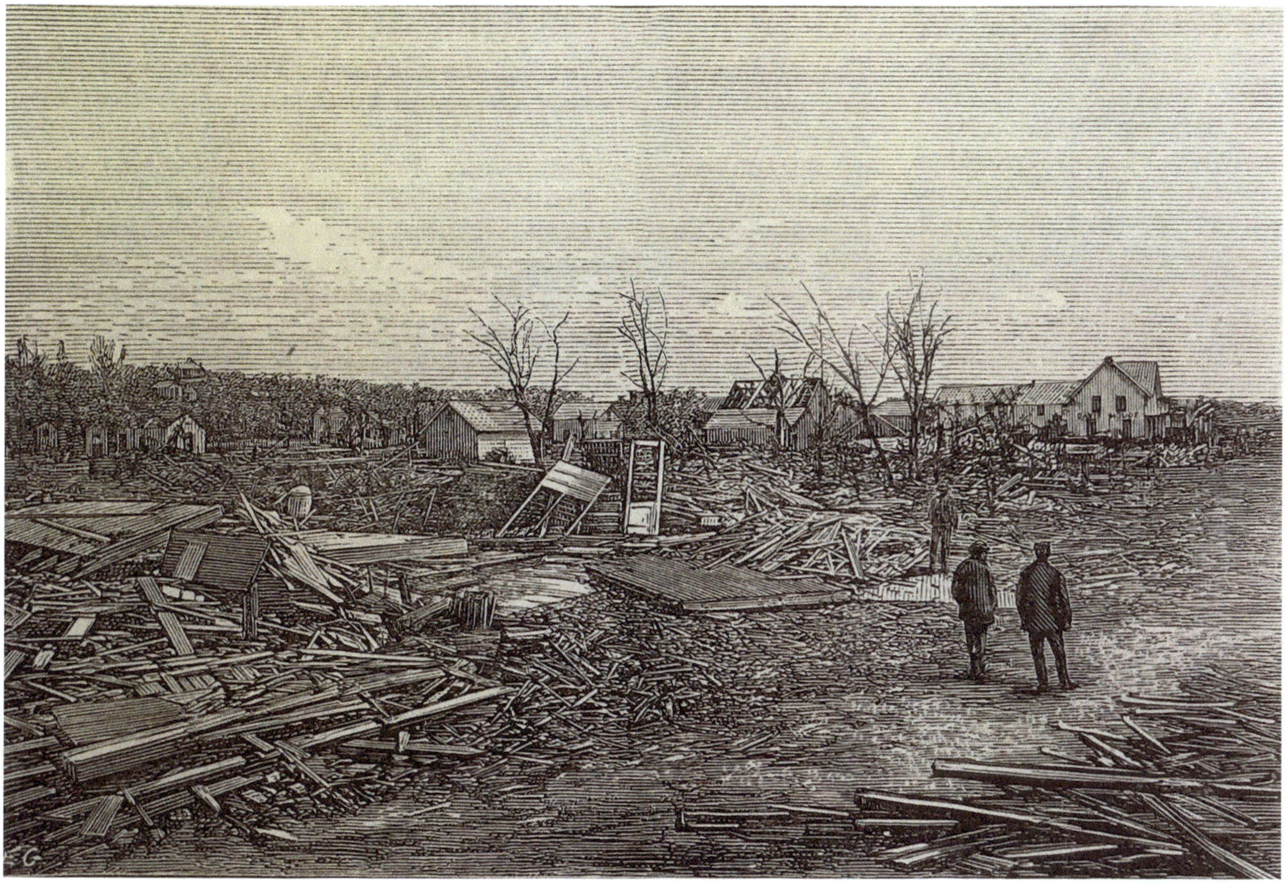

Fig. 4.9 The Sauk Rapids tornado of 1886

For an American midwest example of a historic tornado, one can turn to one of the deadliest tornadoes ever to affect that area: the one that struck on 14 April 1886 (Fig. 4.9). The cities of Sauk Rapids, St. Cloud, and Rice in Minnesota were the settlements worst affected. This contemporary engraving, from the *Illustrated London News*, shows the devastation caused in Sauk Rapids. Most of the timber buildings have been flattened. Altogether, 72 people lost their lives; 200 more were injured. This disaster had a significant economic impact; Sauk Rapids had been an important commercial center in Minnesota, but the destruction of most of its commercial buildings led to a decline as trade shifted to other towns.

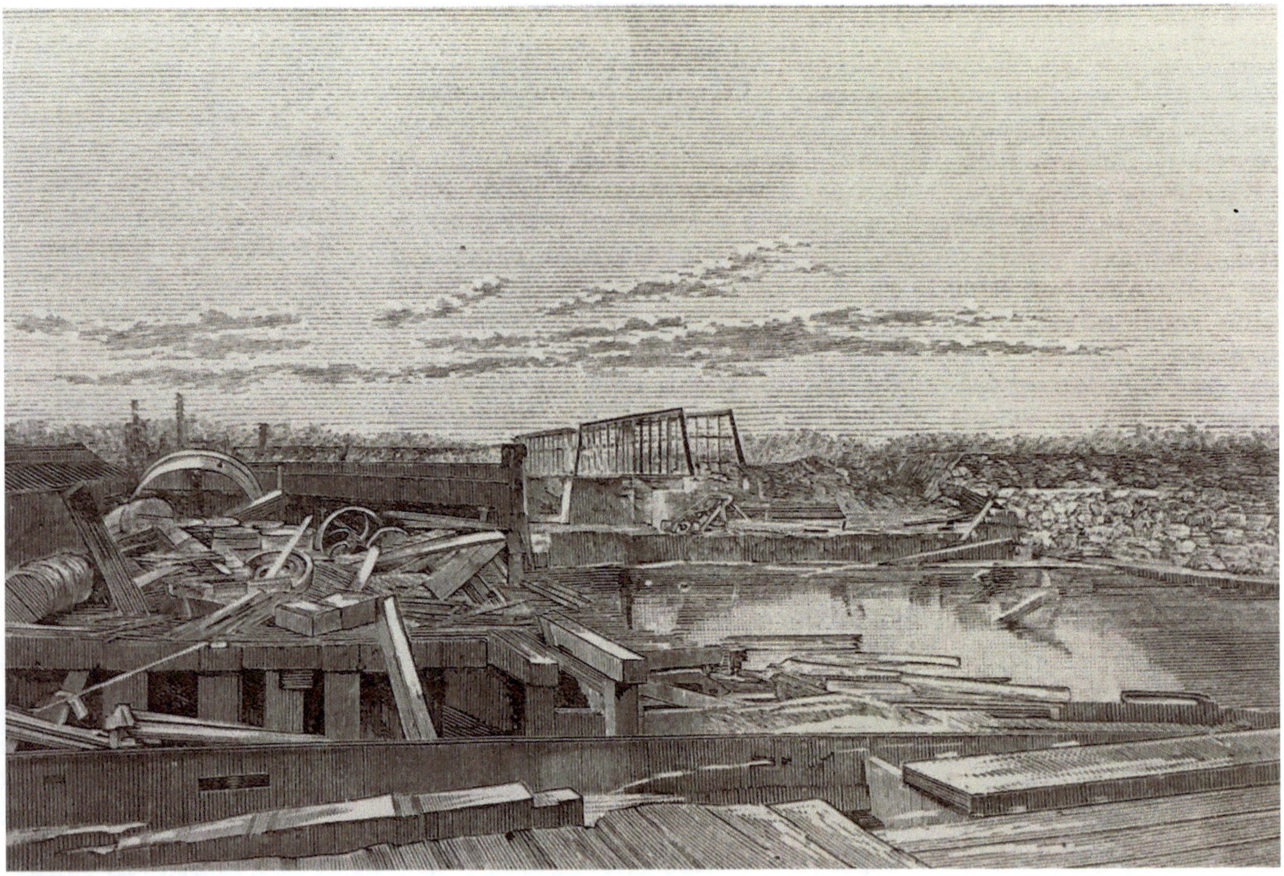

Fig. 4.10 Bridge damage at Sauk Rapids

The Sauk Rapids tornado of 1886 crossed the Mississippi River, sucking up so much water that the river is said to have run dry. At the same time, it hit an iron truss bridge across the river, wrecking it, as shown in this engraving (Fig. 4.10) from the same source as Fig. 4.9. Parts of the bridge were picked up by the twister and carried some distance before being dropped. People and cattle were also lifted into the air; clothes were even torn from people's bodies. The railway freight depot also lay in the path of the tornado; the freight house was destroyed, and sections of track were ripped up and tossed about like wires. When the tornado first appeared, it had the appearance of a very black, funnel-shaped cloud, but with clear skies all around it. Once it was in full swing, and afterward, rain fell heavily, and the air was so dark that it was hardly possible to see more than a few meters. All the time, the twister was accompanied by a roaring and crackling sound, almost deafeningly loud. The swathe of destruction, which ran right down the Main Street of Sauk Rapids, was about 100 meters wide and extended for some 30 kilometers.

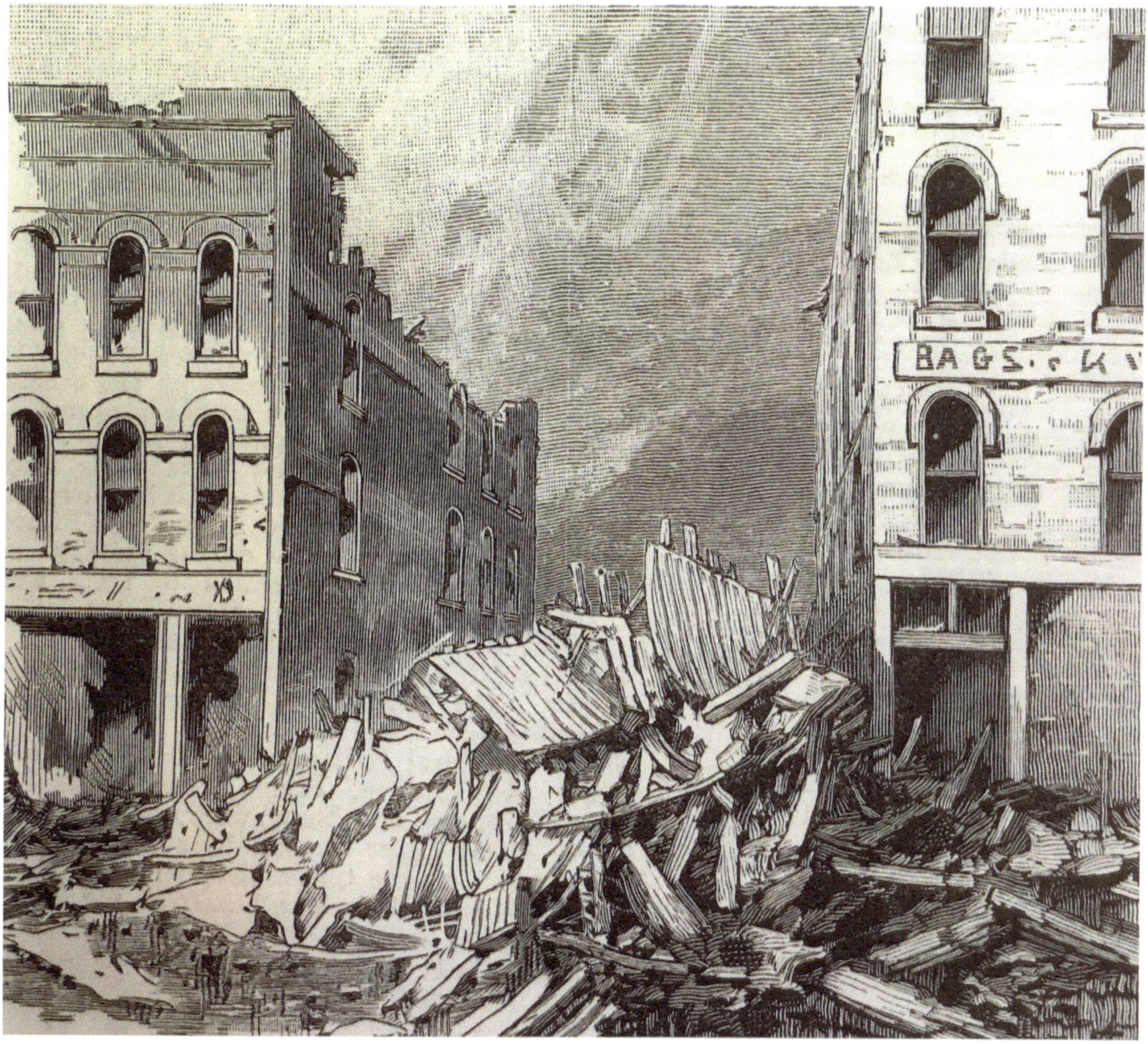

Fig. 4.11 Kansas City in 1886

Minnesota was not the only state to suffer on 14 April 1886. Tornadoes also hit Iowa, Texas, and Kansas. But the 1886 "tornado season" was not over yet—on 11 May another very destructive tornado struck Kansas City, destroying a number of important buildings, including the courthouse and, tragically, Lathrop School, where 15 children were buried in the ruins of the building (Fig. 4.11). It was described by a local newspaper as follows: "The space between the clouds and the Earth presented an almost perfectly white appearance made by the rain, and blown in every imaginable direction by the wind as the storm advanced. It swept every small object away from its path. These filled the air and fell to the ground only to be lifted up again and to be carried to a still greater distance. Many people in the city not aware of the seriousness of the storm, were caught in it and lifted bodily from their feet. Many were hurled fifteen or twenty feet away, receiving terrible bruises. In the suburbs of the city a large number of frame houses were blown down and the inmates barely escaped with their lives." About 25 people in all were killed.

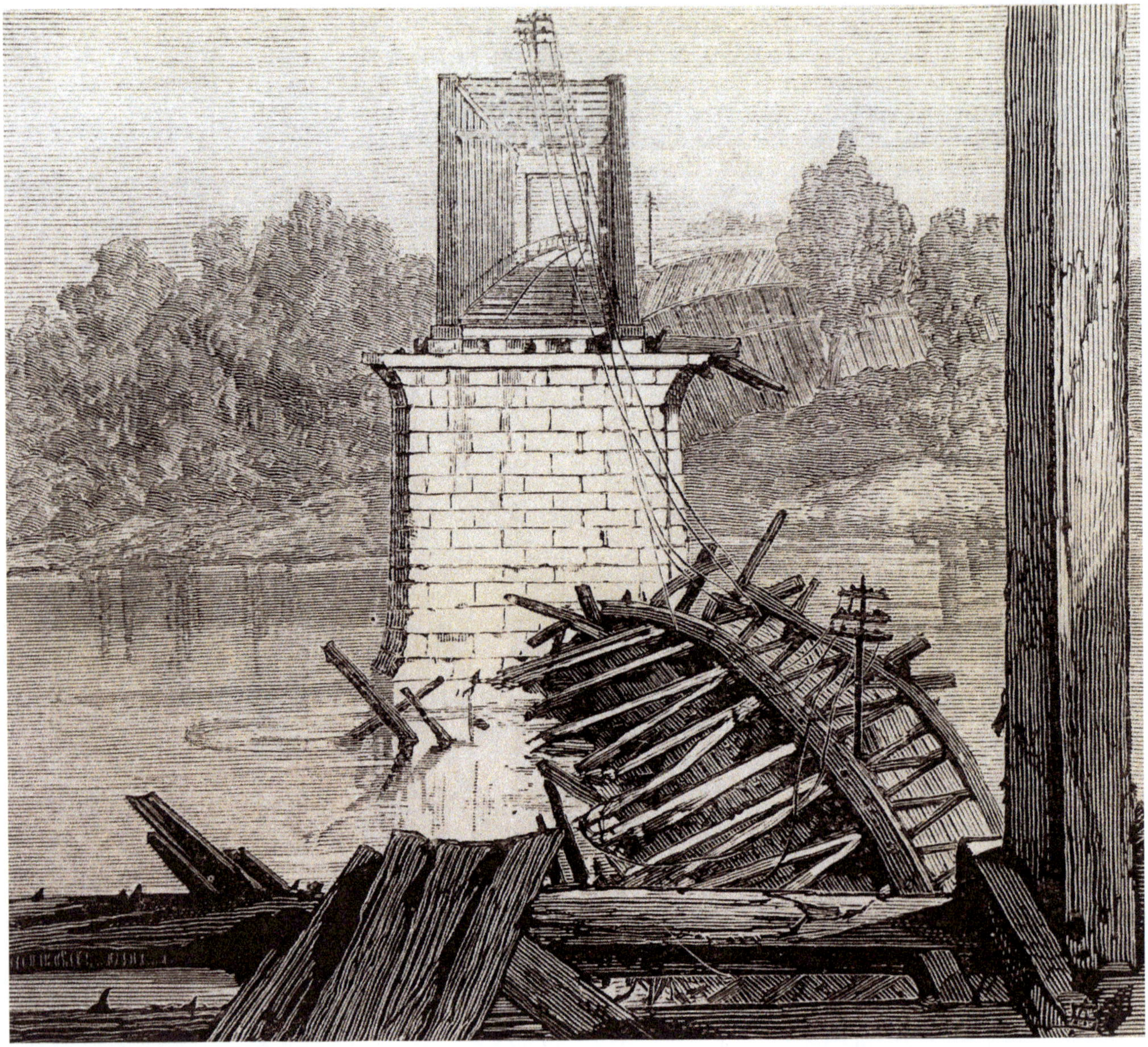

Fig. 4.12 A bridge over the Missouri River in 1886

The tornado on 11 May 1886, also took out a rail bridge; this is the Hannibal and St. Joseph Bridge over the Missouri River (Fig. 4.12). One span was destroyed, and all rail traffic had to be diverted through Leavenworth. This was the first rail bridge built over the Missouri (in 1869), and it contributed significantly to the growth of Kansas City as a commercial center. Again, this engraving is from the *Illustrated London News*.

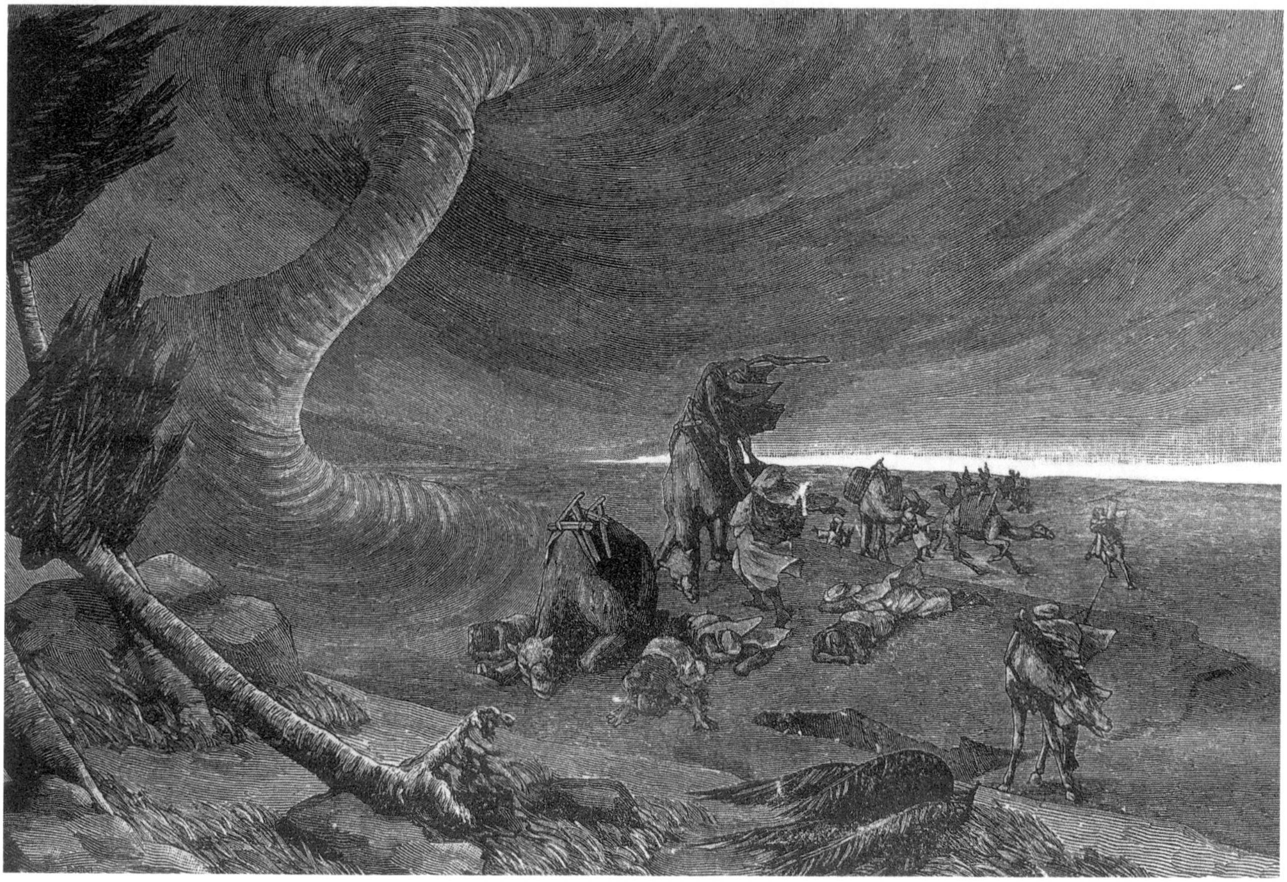

Fig. 4.13 A fantasy image of a dust devil

Here, a tornado in a desert landscape is creating a deadly-looking sand spout (Fig. 4.13). This phenomenon is known as a dust devil, and this is a rather large specimen. The origin of this nineteenth-century engraving is unknown; the signature is hard to read and appears to be "CPAN." The picture is surely a work of imagination rather than observation. If a caravan like this is traveling across a sandy desert and has to face a sandstorm, in most cases there is no way to avoid the peril. Probably the most effective measure is for the caravan members to lie on the ground and make some sort of simple shelter to assist breathing. The caravan's draft animals must also be attended to.

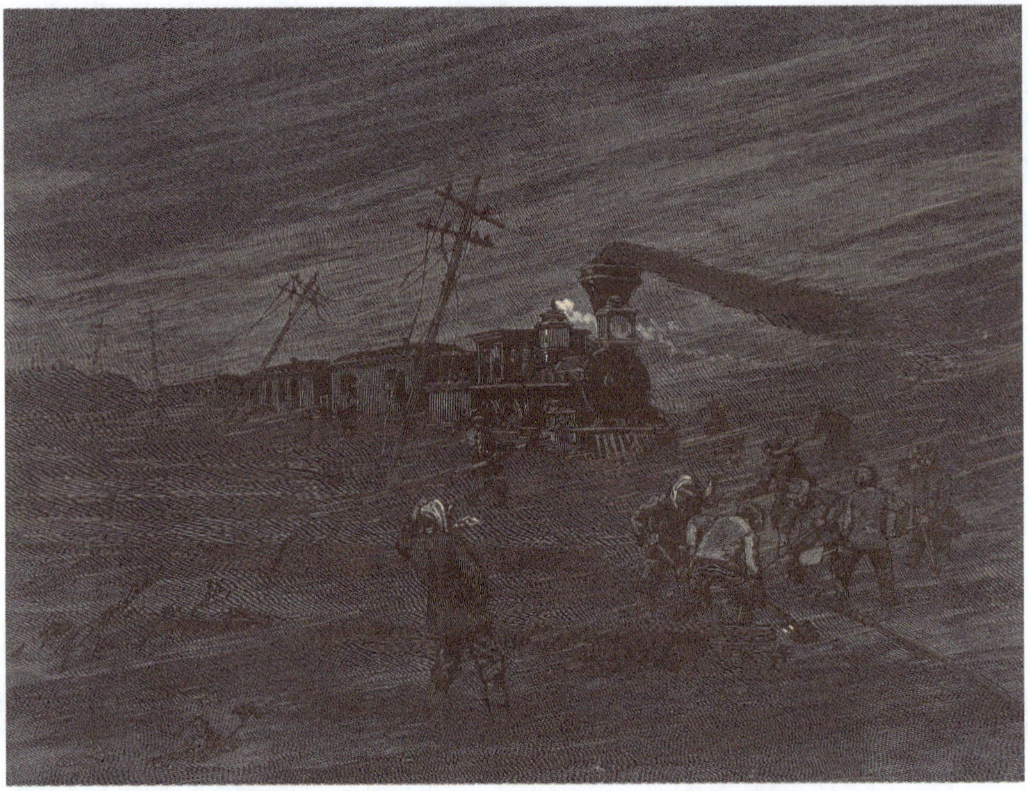

Fig. 4.14 A sandstorm in Oregon in 1887

In desert or arid areas, even without whirlwinds or dust devils, sandstorms are a considerable peril. Sand or dust is whipped up by the wind, reducing visibility, causing much discomfort, and, in this case, burying railway tracks, which have to be dug clear before the train can proceed any further. This fine and dramatic image of a sandstorm in eastern Oregon (Fig. 4.14) comes from a sketch made by the artist James Norman Biles (1865–1914). The sketch was elaborated into an engraving by Charles Graham and published in *Harper's Weekly* on 16 April 1887.

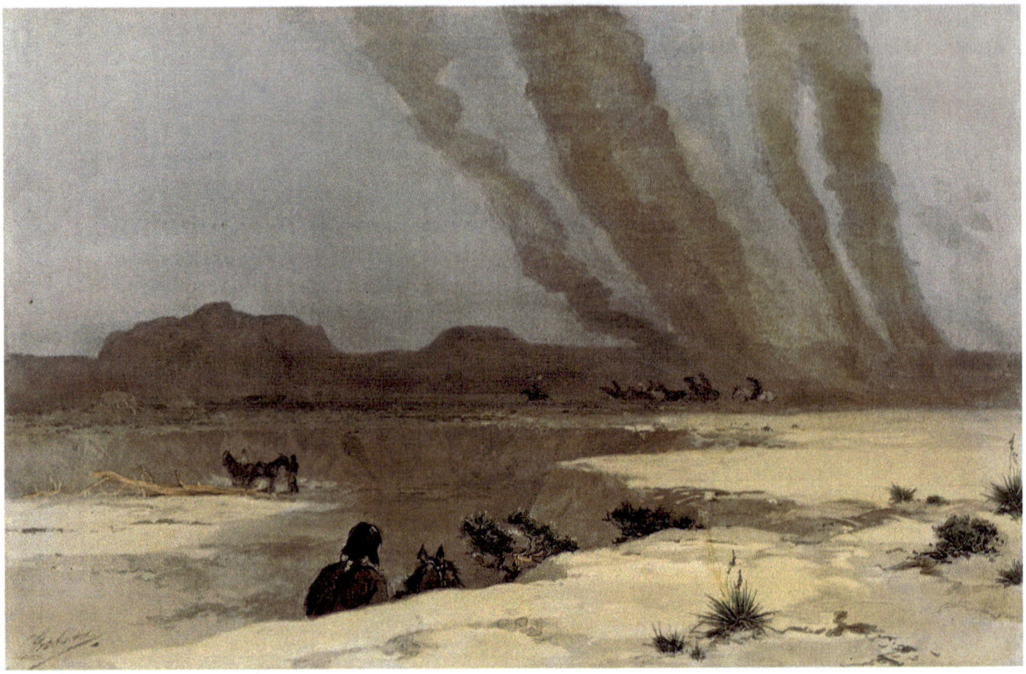

Fig. 4.15 A sandstorm in 1896

This colored engraving is again by the illustrator Charles Graham (1852–1911), showing another sandstorm, this time perhaps in a more southerly location in American desert territory (Fig. 4.15). In contrast to the previous image (Fig. 4.14), where the artist shows what it is like to be caught up in a sandstorm, here the viewpoint is from further away, showing how clouds of sand are swept high into the air. If they reach high enough into the atmosphere, the sand grains may be washed out of the air far away as colored rain, much to the puzzlement of observers. Depending on wind conditions, grains of sand from Saharan sandstorms may be swept high up into the atmosphere, eventually to fall as red-colored rain over parts of Europe, thousands of kilometers away. To the medieval mind, these were seen as "rains of blood"—clearly something ominous portending future disasters. The Native American in the foreground of this picture is watching the clouds anxiously and hoping they are not coming his way. Again, the source is *Harper's Weekly*, published on 10 October 1896.

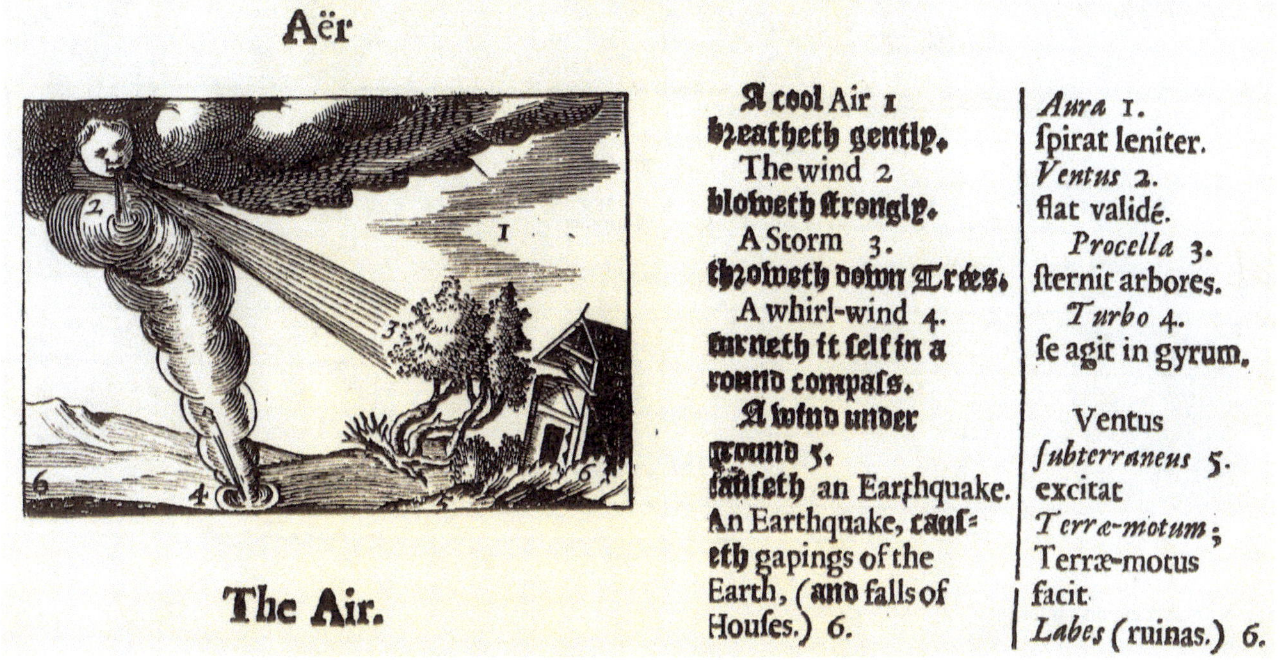
Aër

The Air.

A cool Air 1 breatheth gently.	*Aura* 1. ſpirat leniter.
The wind 2 bloweth ſtrongly.	*Ventus* 2. flat validè.
A Storm 3. throweth down Trees.	*Procella* 3. ſternit arbores.
A whirl-wind 4. turneth it ſelf in a round compaſs.	*Turbo* 4. ſe agit in gyrum.
A wind under ground 5. cauſeth an Earthquake.	Ventus *ſubterraneus* 5. excitat *Terræ-motum*;
An Earthquake, cauſeth gapings of the Earth, (and falls of Houſes.) 6.	Terræ-motus facit *Labes* (ruinas.) 6.

Fig. 4.16 Effects of air depicted for children in 1658

John Amos Comenius (1592–1670) was a Czech philosopher and educator, who conceived the idea of the picture book as an educational tool. His *Orbis Sensualium Pictus* aimed to provide pictures and annotations of everything a child might need to know about that could be expressed pictorially (his picture on the theme of a shipwreck was shown in Fig. 3.28). The book consists of a series of double-page spreads: on the left, the picture, with each salient feature numbered; on the right, the key to the numbers, both in the child's native language (according to the edition) and in Latin. It was first published in Nuremberg in 1658, and an English edition, translated by Charles Hoole, appeared the following year. As the preface states, the book may serve "to entice witty Children to it, that they may not conceit a torment to be in the School, but dainty-fare. For it is apparent, that Children (even from their Infancy almost) are delighted with Pictures, and willingly please their eyes with these sights. And it will be very well worth the pains to have once brought it to pass, that scarcrows may be taken away out of wisdomes Gardens." To give an idea of the book's character, the illustration here (Fig. 4.16) from the English edition includes the text. Air is shown as causing both a storm and a whirlwind, and, in true Aristotelean fashion, an earthquake as well.

Comenius (Komenský in Czech) is remembered for his great achievements in the fields of pedagogy and education. However, his purely scientific concept of physics, particularly concerning astral and planetary physics, did not follow the new ideas related to our universe that occurred in his period. For instance, he believed in a geocentric structure of our solar system. His main achievements in the field of education, however, allow us to appreciate him as a "teacher of nations."

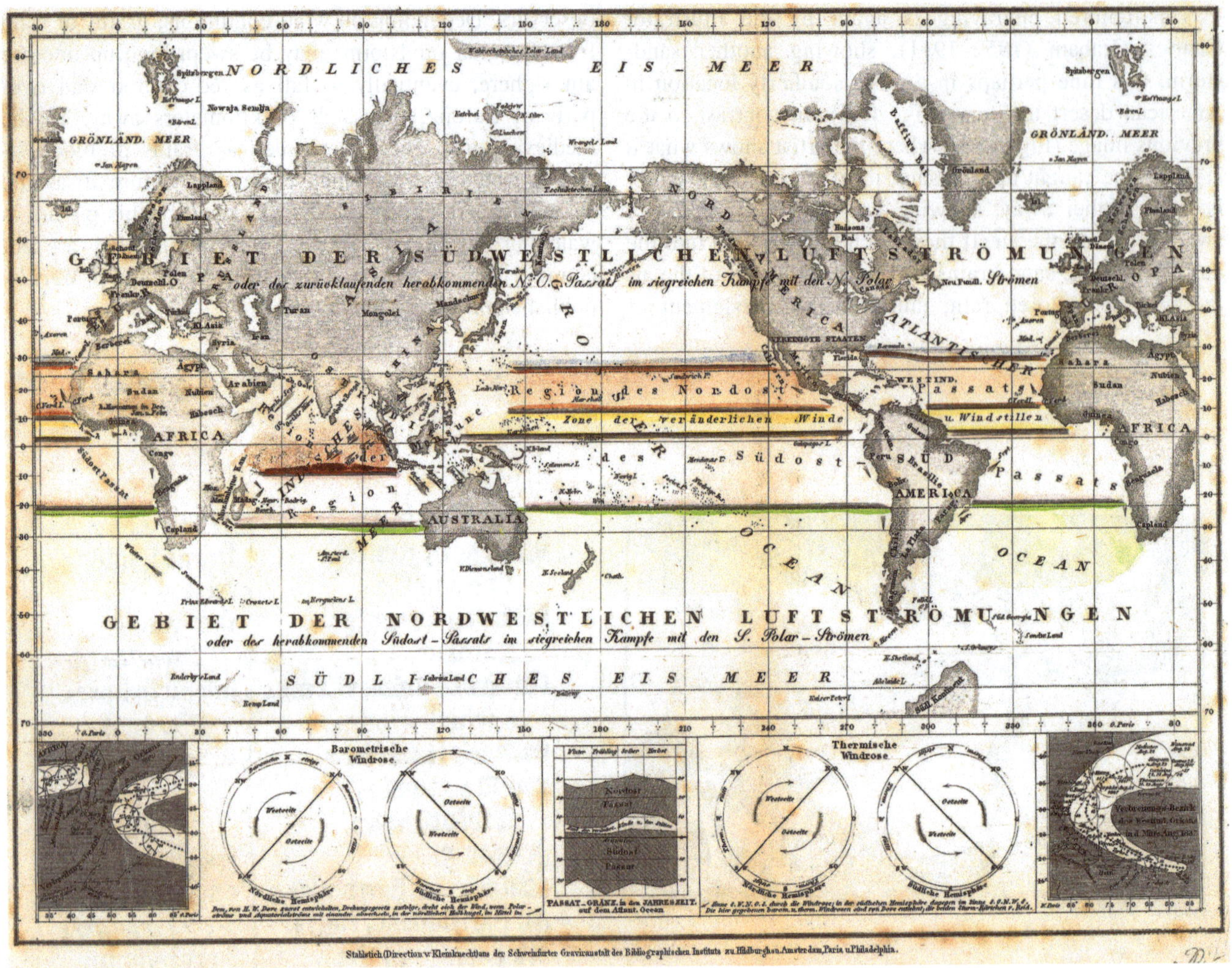

Fig. 4.17 A wind map from 1849

Wind hazard was becoming understood scientifically by the early nineteenth century. This German world map of winds, titled *Windkarte der Erde*, was first published in 1840 as a single map "drawn and engraved in the Geographische Kunstschule at Potsdam" and was published by Justus Pethes in Gotha, Saxony (Fig. 4.17). Subsequently, it was included as a plate in Heinrich Berghaus's *Physikalischer Atlas* of 1849 and later in publications by Joseph Meyer for the Hildburghausen Bibliographisches Institut.

Today we can say that the insufficiency of world meteorological data at the time of this map's compilation prevented the author from plotting a more accurate depiction of wind patterns. However, we must appreciate this schematic wind map as a pioneering meteorological attempt to chart global patterns of wind direction and power. Note also the inset (bottom right), showing typical hurricane tracks in the Caribbean and on the east coast of North America.

4.3 Other Atmospheric Phenomena

In historical times, it was accepted that the universe was centered around our Earth—and one could not even write "our planet Earth," for planets were celestial bodies that wandered around the sky, whereas the Earth was fixed in place. The sublunary atmosphere was a strange place, inhabited by comets, dragons, and who knows what. We know now from geological evidence that the impact of a large asteroid (for instance) would be a major disaster. Fortunately, mankind never had to cope in historical times with an impact event such as that believed to have wiped out the dinosaurs. Nevertheless, strange things seen in the sky were generally seen as ominous.

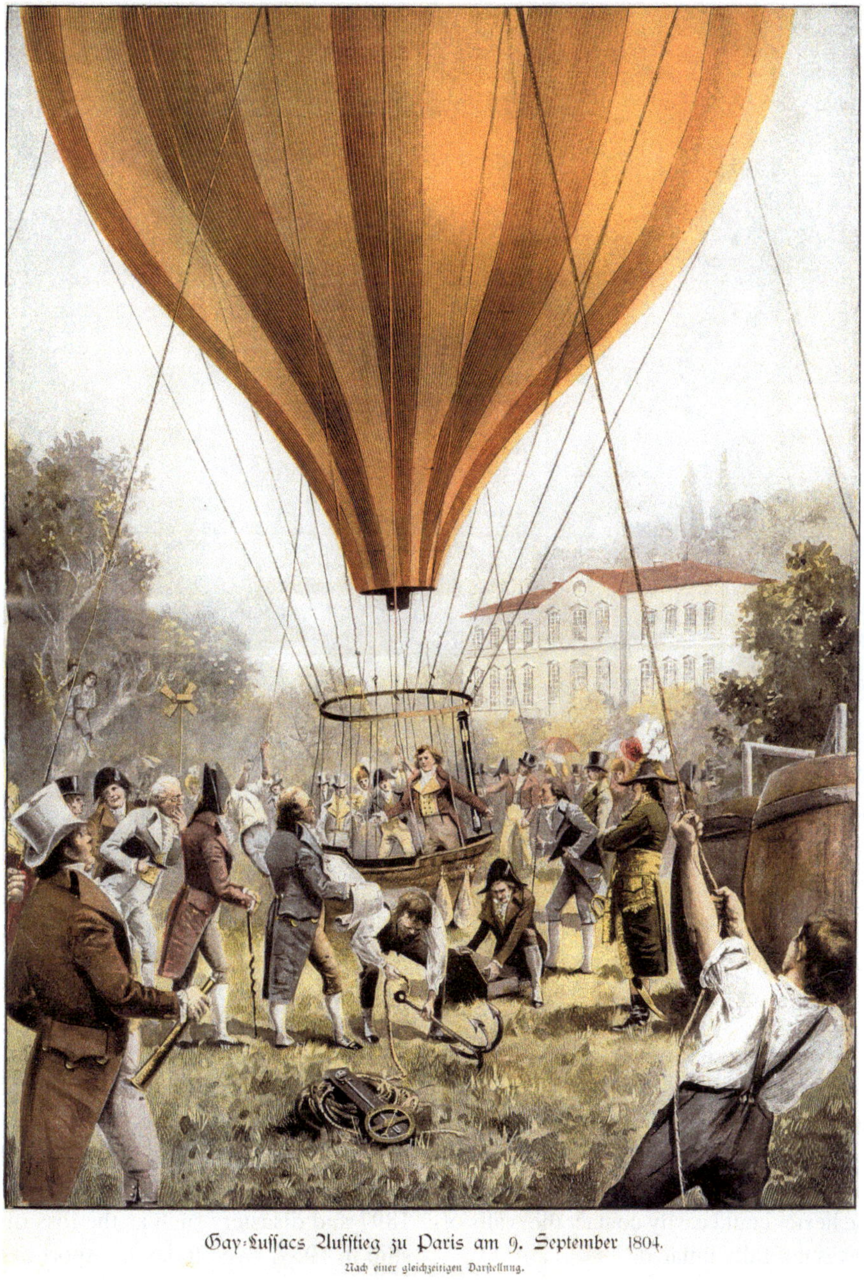

Fig. 4.18 Gay-Lussac about to ascend by balloon

Studying the atmosphere without modern equipment was difficult, if not dangerous. Here, the French chemist and physicist Joseph Louis Gay-Lussac (1778–1850) prepares to make a balloon ascent from Paris on 9 September 1804, together with Jean Baptiste Biot (1774–1864). This was the first balloon ascent made specifically for scientific purposes, reaching a height of 4000 meters. A second ascent was made by Gay-Lussac alone, on 16 September. With only one person aboard, the balloon weighed less, and a height of 7000 meters was reached. On both of these expeditions, Guy-Lussac took samples of air and proved that the air at different altitudes has the same composition.

Gay-Lussac belongs among the most notable French scientists of the first half of the nineteenth century. He made numerous fundamental discoveries in the fields of chemistry, physics (namely, thermodynamics), and the Earth sciences. During his balloon ascent to an altitude of over 7000 meters, he made the first high-altitude measurements of geoelectrical and thermometric parameters. He paid particular attention to the alteration of the Earth's magnetic field with increasing altitude as the balloon ascended.

The image shown here is much later, from a book by Joseph-Louis Lecornu, titled *La Navigation Aérienne: Histoire documentaire et anecdotique*, published in 1903 (Fig. 4.18).

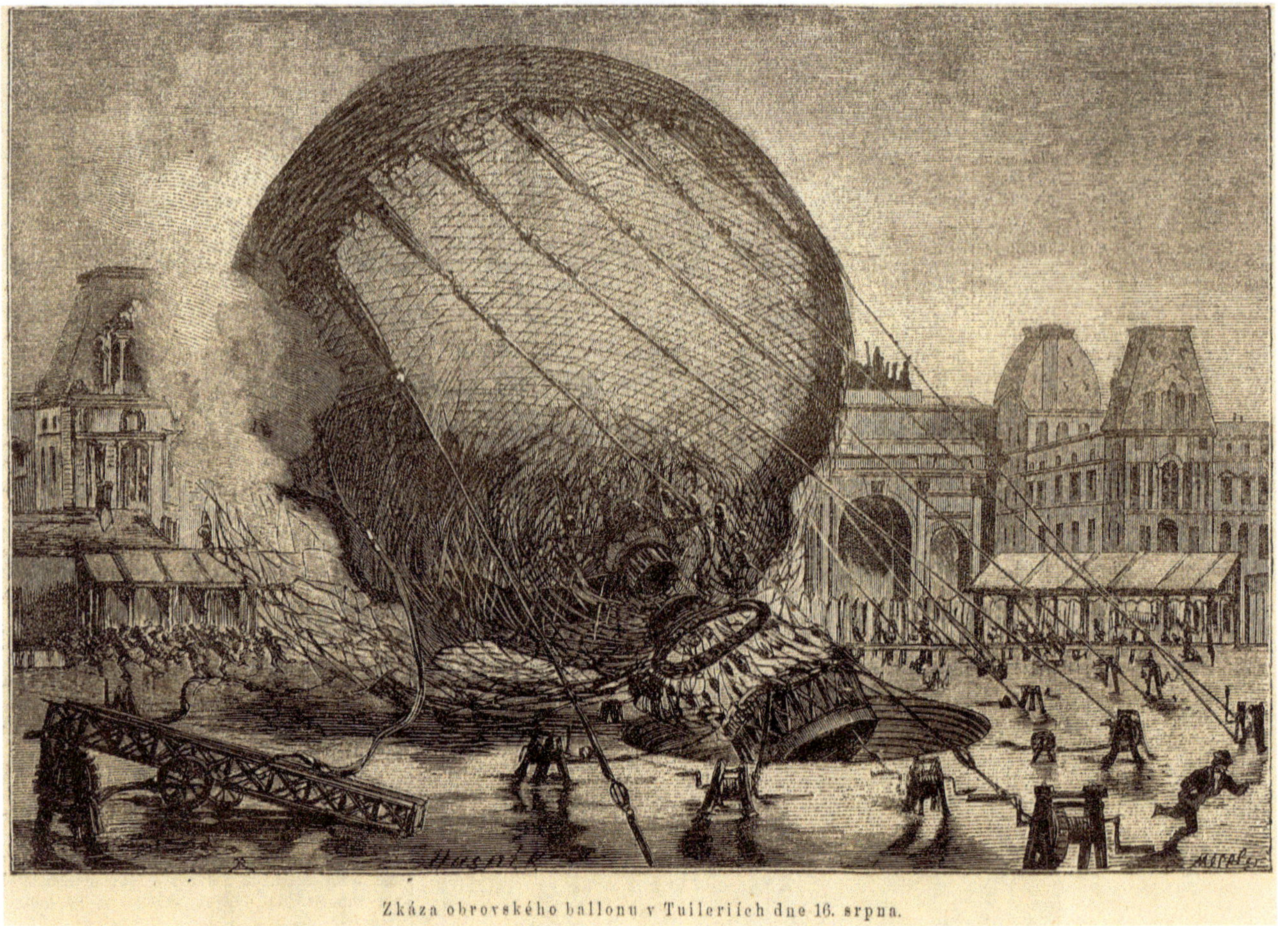

Fig. 4.19 A balloon crash in 1879

Balloon ascents, however, were far from being easy and reliable excursions. The first models of balloons used for such ascents were pulled upward by warm air heated by a flame in the balloon's nacelle. The travellers had to face a number of dangerous situations. During the filling of the balloon, sparks from the heater could easily contact the walls of the balloon while it was not fully inflated.

Once the balloon reached high elevations, it was at risk from impacts caused by strong winds, or lightning strikes, or even electrical sparks occurring in the electrical field of the higher atmospheric strata. The balloon's occupant had to suffer the effects of extremely low temperature. And, of course, there was a grave danger of the balloon being blown far away by winds, perhaps even out to sea, given the difficulty of maneuvering the craft horizontally.

These limitations and hazards of balloon ascent and lighter-than-air craft were persistent, even in the twentieth century, as the frequent collapse of this kind of expedition confirms. One can mention the fatal balloon crash in the failed attempt by S.A. Andrée to reach the North Pole in 1897 and disasters such as the loss of the *Hindenburg* airship in 1937. Even today the sport of hot-air ballooning is not without accidents. The illustration here (Fig. 4.19) shows the collapse of a hot-air balloon in the Tuilleries, Paris, on 16 August 1879, reproduced from the Czech illustrated weekly *Světozor*; the original engraving appeared in *L'Illustration: Journal Universel* on 23 August 1879.

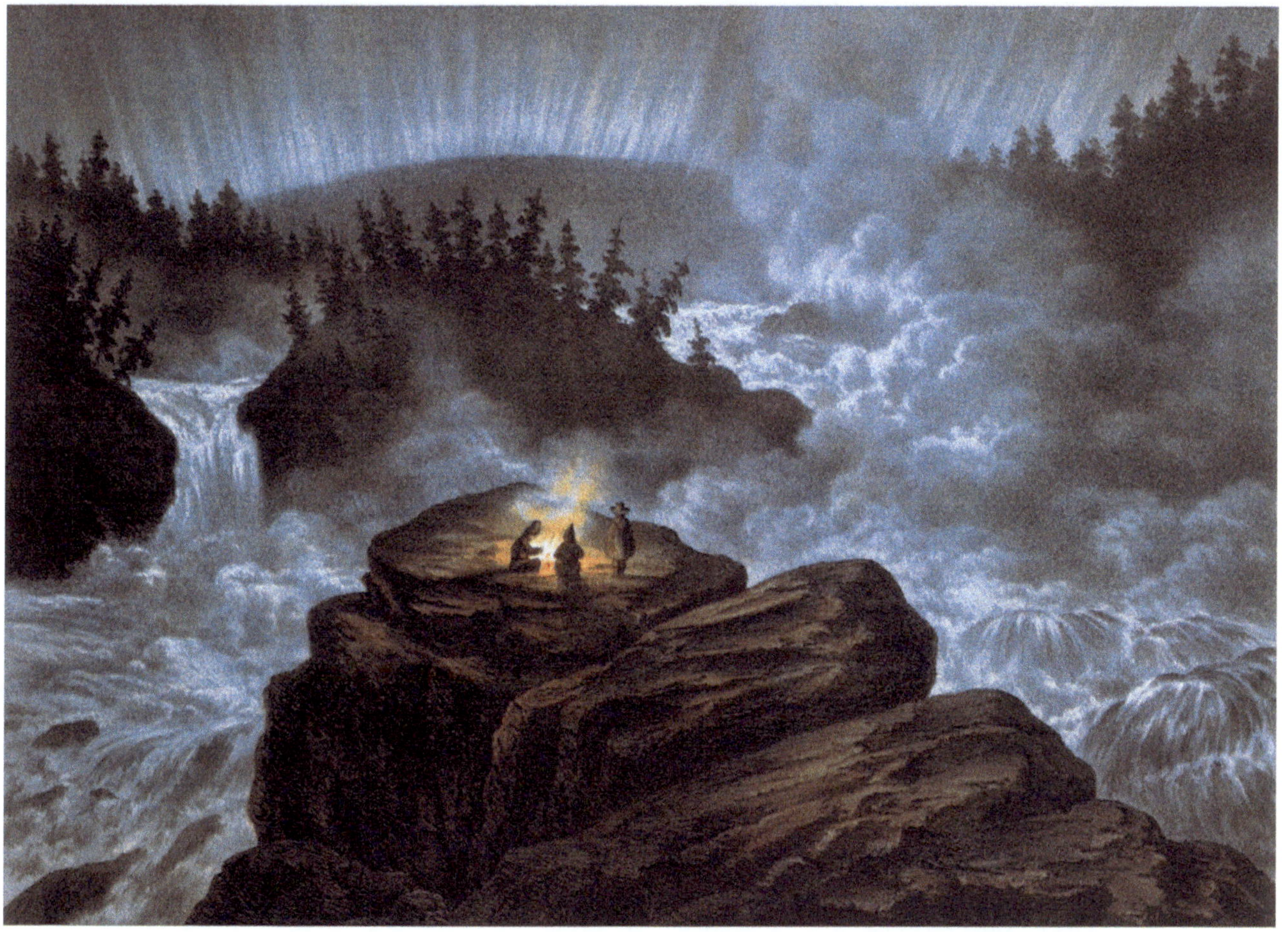

Fig. 4.20 An aurora depicted in 1856 in Sweden

Other atmospheric phenomena were equally of interest to the nineteenth-century mind—in this case, the aurora borealis. In this dramatic image, the three human figures are dwarfed by the power of nature (Fig. 4.20). The waterfall Njommelsaska, known today as Harsprånget, boils around the rocks in a nocturnal scene brightly illuminated by the northern lights. This chromolithograph of 1856 is by the Swedish artist Carl Svante Hallbeck (1826–1897).

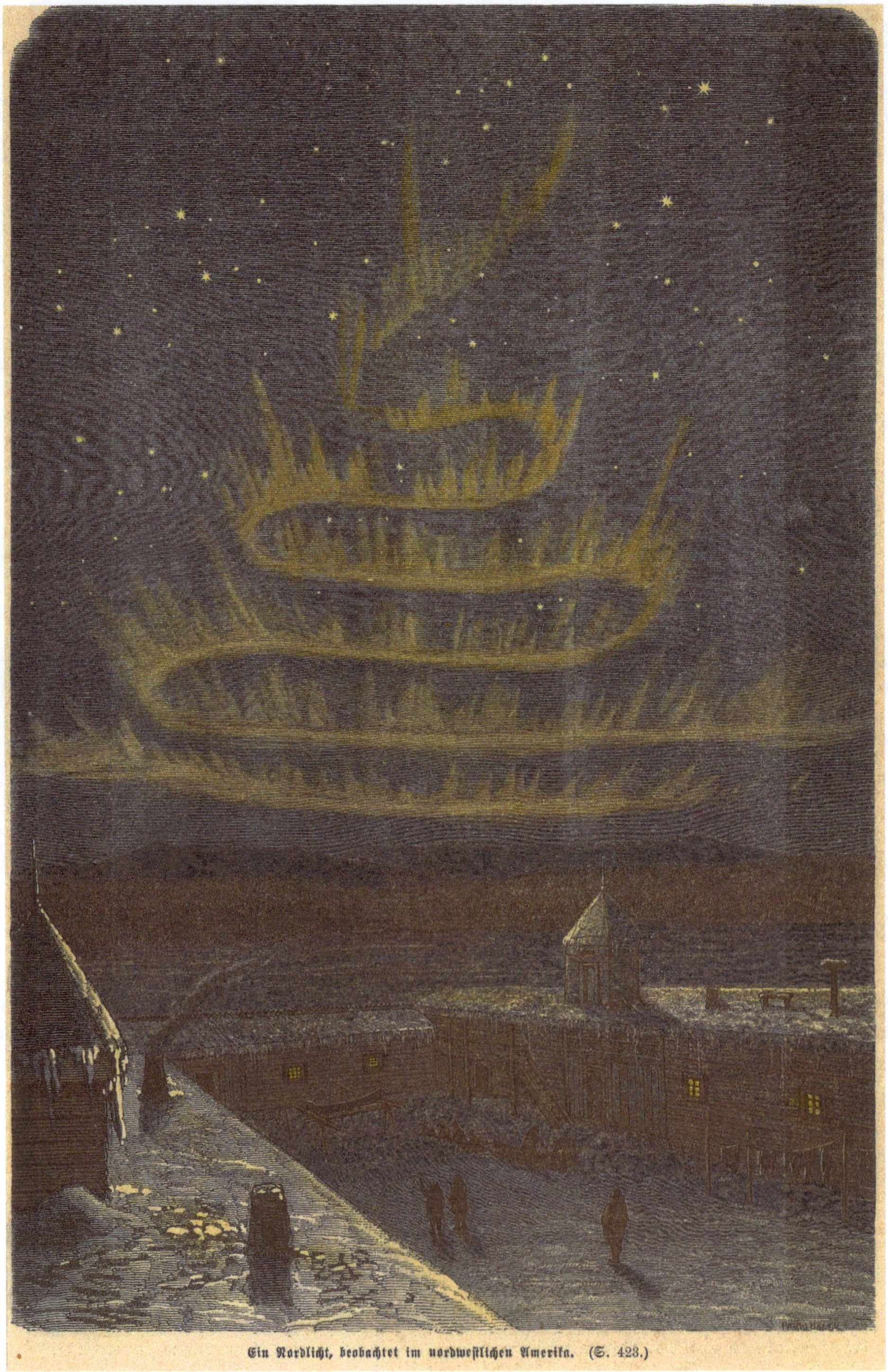

Fig. 4.21 An aurora depicted in 1869 in Alaska

The preceding picture (Fig. 4.20) was from Sweden; this one is from Alaska (Fig. 4.21). It shows the aurora observed at Fort Sitka and was published as a book illustration in Frederick Whymper's *Travel and Adventure in the Territory of Alaska* in 1869. Before the true nature of the aurora was understood, all sorts of ideas were prevalent. The sixteenth-century astronomer Tycho Brahe believed that the lights were caused by sulfurous vapors and were injurious to health.

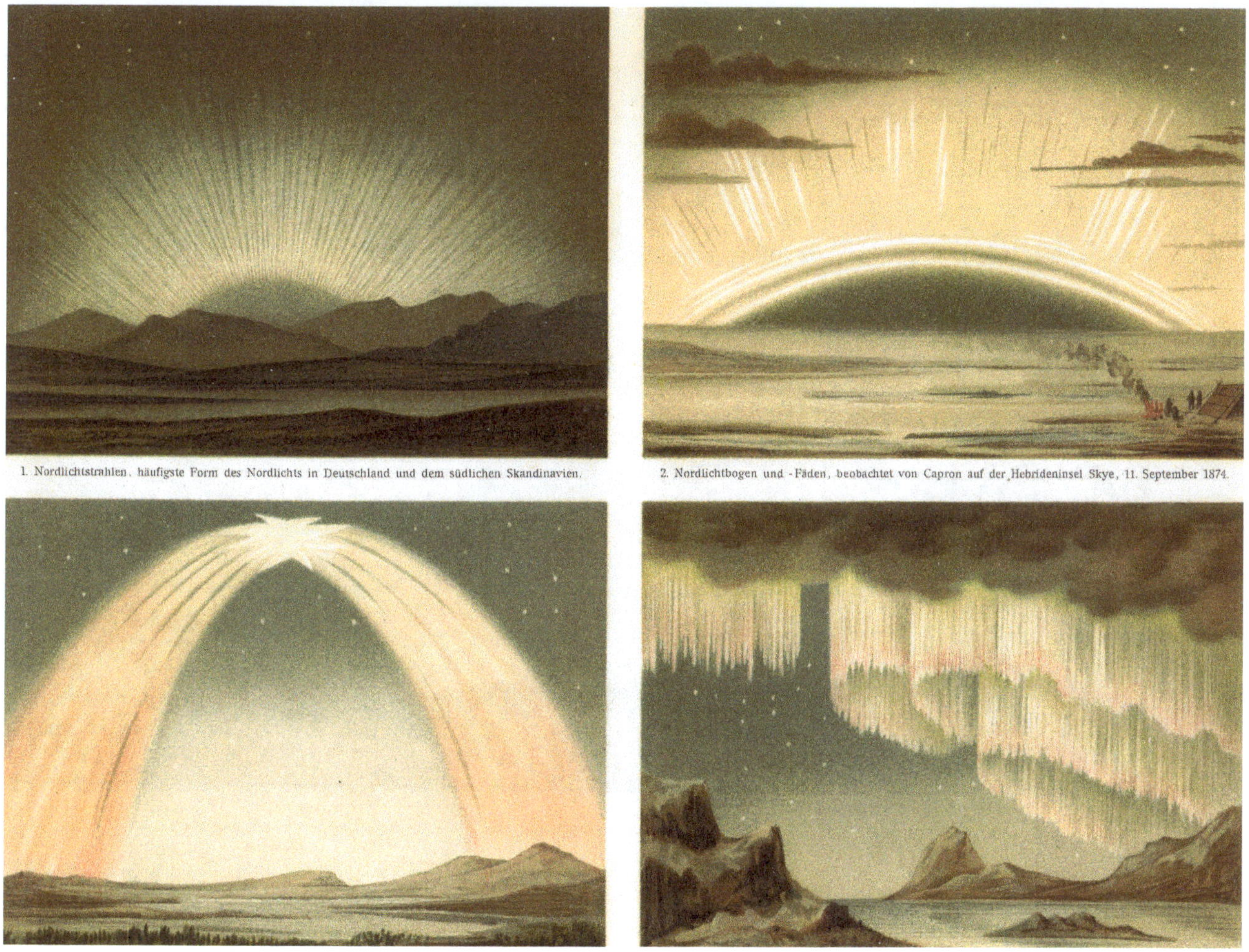

Fig. 4.22 The "polar lights"

This image, from a popular science book published in Wroclaw in 1882, shows some typical manifestations of the aurora borealis, or "polar lights" as they are named here (Fig. 4.22). The picture top right is dated—an observation made on the coast of Skye, Scotland, on 11 September 1874. It was suggested by Benjamin Franklin in the 1779 that the aurora borealis was caused by electrically charged particles, but he imagined that this was due to snow and moisture in the air.

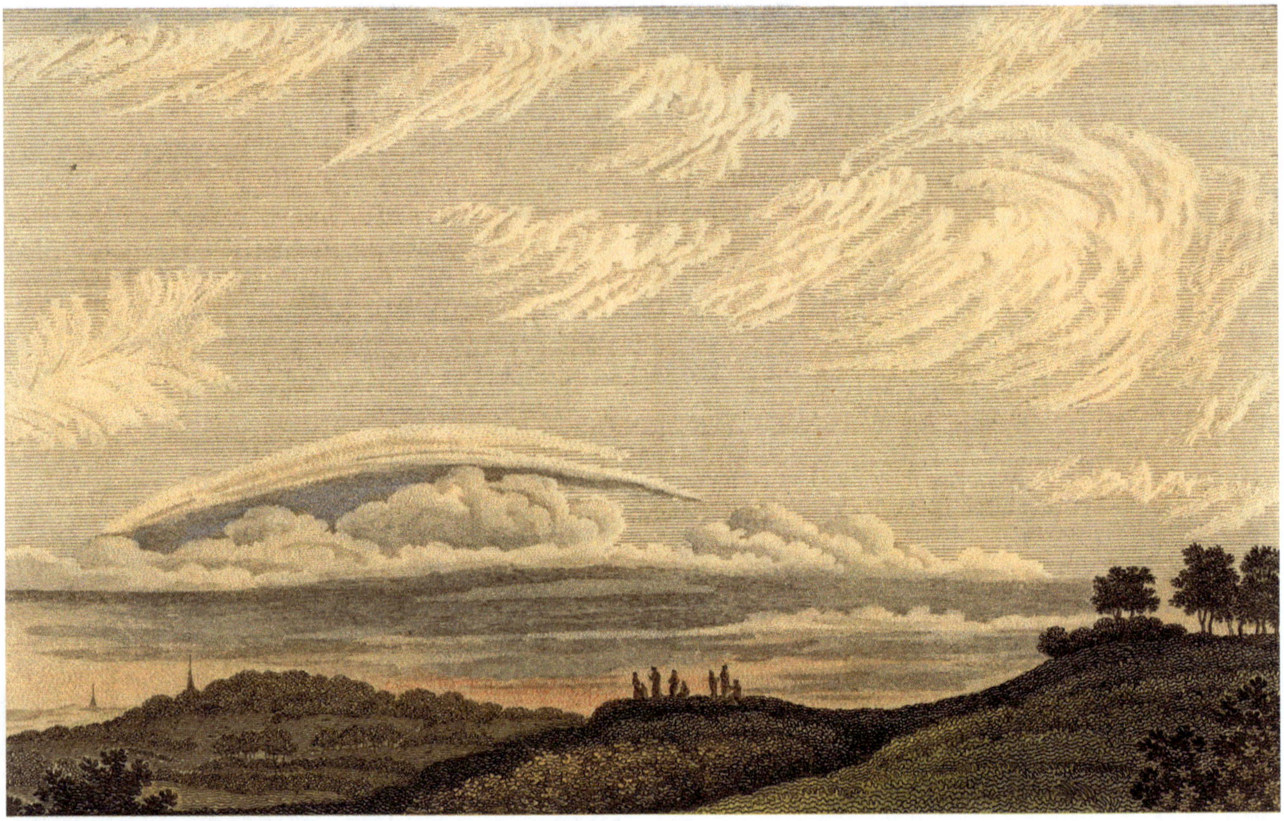

Fig. 4.23 The first nomenclature of clouds

An important landmark in the history of meteorology was arriving at a standard terminology. A system of nomenclature for clouds was first proposed by Jean-Baptiste Lamarck (1744–1829), but his French terms lacked appeal and were not generally adopted. The system of naming by which clouds are now universally known was the work of a London pharmacist and amateur meteorologist, Luke Howard (1772–1864). This illustration (Fig. 4.23)—plate 1 in his landmark book *Essay on the Modification of Clouds* (first published in 1803), and much copied later—has the following description by the author: "In the blue sky, at top, the Cirrus, in some of its fair-weather shapes (but much too near the eye in position). The feather on the left is a rare, but genuine variety of the cloud. Above the horizon are some large Cumulus clouds, receiving on their tops a subsiding Cirro-stratus: the manner in which these modifications unite upon occasion to form, ultimately, Cumulo-stratus." Howard's use of Latinate terms, his emphasis on the mutability of clouds, and a nod towards Linnean principles of classification contributed to the success of this system.

At the beginning of the nineteenth century, the microphysical structure of clouds was not explored, and the physical processes behind different cloud shapes and their development in time were not known. That is why sometimes the artist was not able to properly depict the vertical extent of clouds or the height of cloud occurrence above the Earth's surface.

According to the contemporary classification (International Cloud Atlas), most of the low-level clouds that dominate the middle part of this picture are Cumulonimbus calvus ("Cumulus" means "heap" or "accumulation," "Cumulonimbus" means "precipitating Cumulus," and "calvus" means "bald"). These clouds generate showers. The layered clouds, which Howard called "subsiding Cirrostratus," are probably middle-level, mostly nonprecipitating clouds (Altocumulus). Howard was not right in his explanation that Stratocumulus (low-level, layered clouds) can form by unification of Cumulus and Cirrostratus. Cirrus (which means "lock of hair" or "tuft") clouds are high-level clouds, which are made of ice crystals and bring no moisture to the Earth's surface (Sedlák 2017, personal communication).

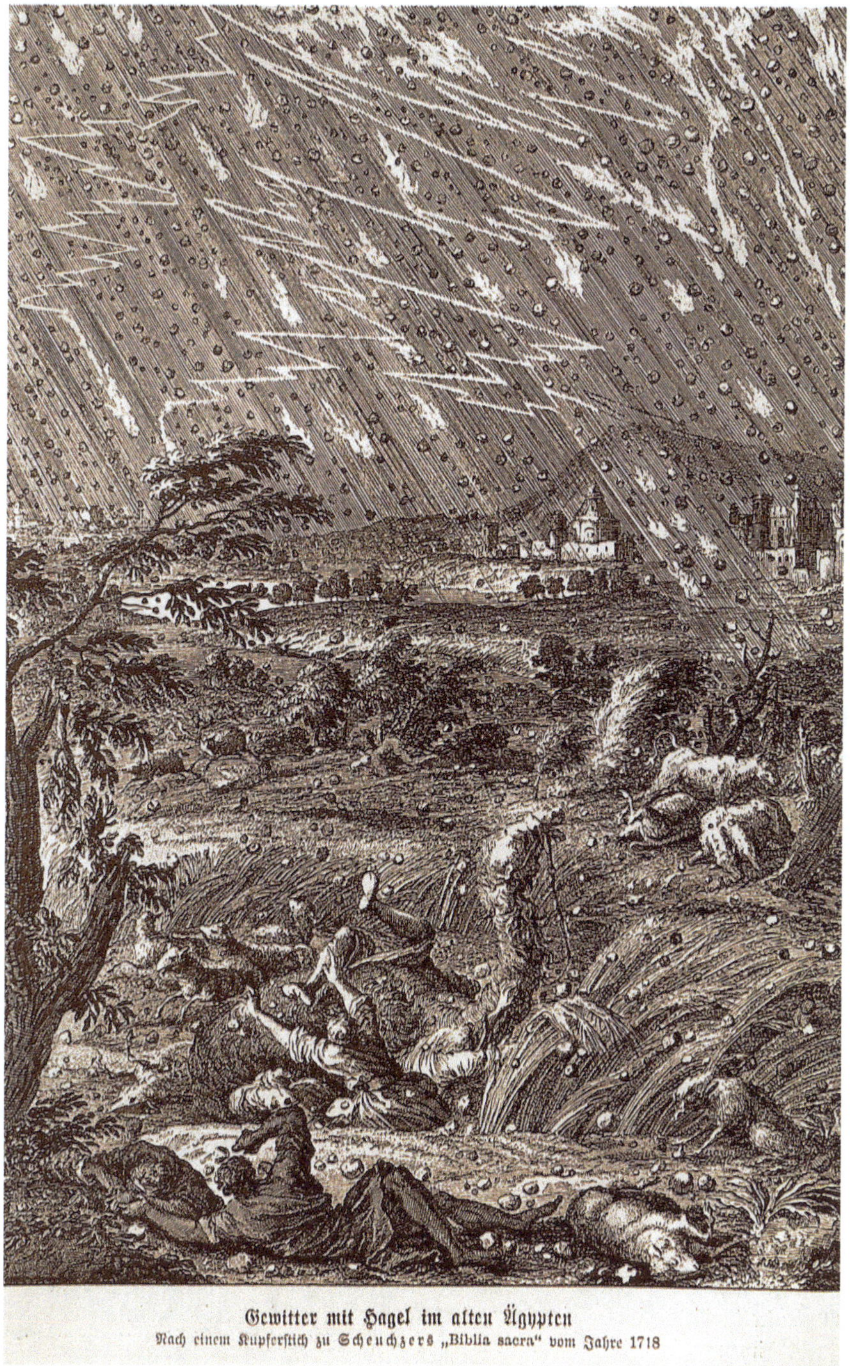

Fig. 4.24 The biblical plague of hail

Hail as a hazard is chiefly to be feared on account of the damage that hailstones do to property and, especially, to crops. It is rare but not unknown for people to suffer injury or death by being struck by hailstones, although individual hailstones up to 20 centimeters across have been recorded. The larger the stone, the greater its terminal velocity; thus, being struck by a monster hailstone is definitely to be avoided. The image here (Fig. 4.24) shows the biblical seventh plague of Egypt, taken from the *Physique Sacrée, ou histoire-naturelle de la Bible* (1732), written by the Swiss naturalist Johann Jakob Scheuchzer (1672–1733). Note the stylized depiction of forked lightning, typical of this sort of production. Recently, a scientist at the Institut für Physik der Atmosphäre (Mainz) has suggested that the biblical plague of hail had a factual basis and was fallout from ash clouds following the eruption of Thera (on Santorini, located in the eastern Mediterranean) in around 1600 BC.

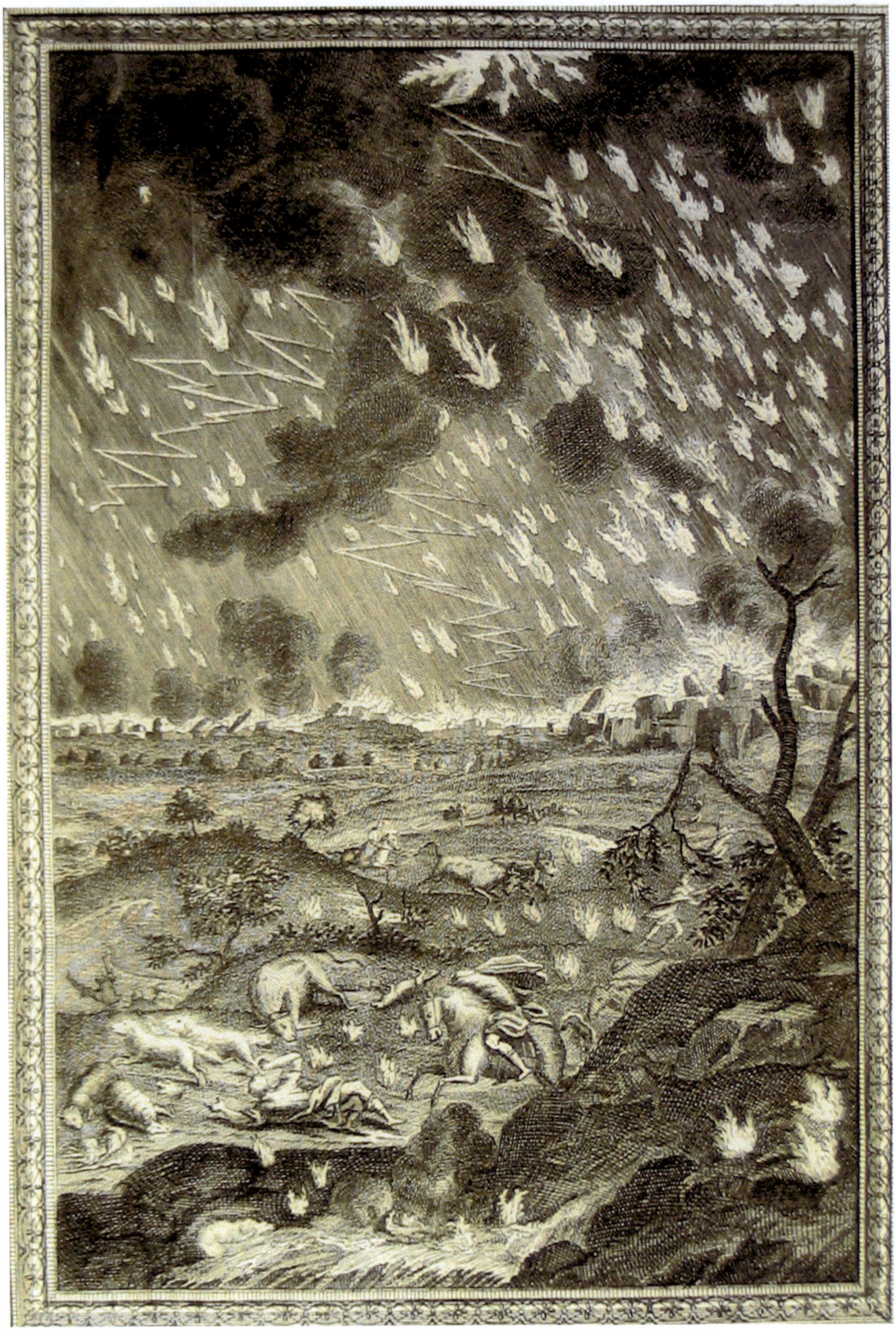

Fig. 4.25 The destruction of Sodom and Gomorrah

Calling down fire from Heaven is a familiar Christian trope: Elijah did it, the disciples James and John wanted to do it, and it was further prophesied in the Book of Revelation. This is another of Scheuchzer's pictures, and the fire from Heaven is falling on the cities of Sodom and Gomorrah (Fig. 4.25). A number of hypotheses have been put forward to suggest scientific explanations for the destruction of these two cities, either volcanic or seismic, but no natural fall of meteorites rains down fire in the way shown here. It may be noted in passing that in the vicinity of Sodom and Gomorrah there were extensive deposits of highly flammable asphalt, and one might have expected both towns to have kept extensive stocks of this material. Therefore, it is not surprising that in all biblical representations of Sodom's fall, all of the town burns rather well.

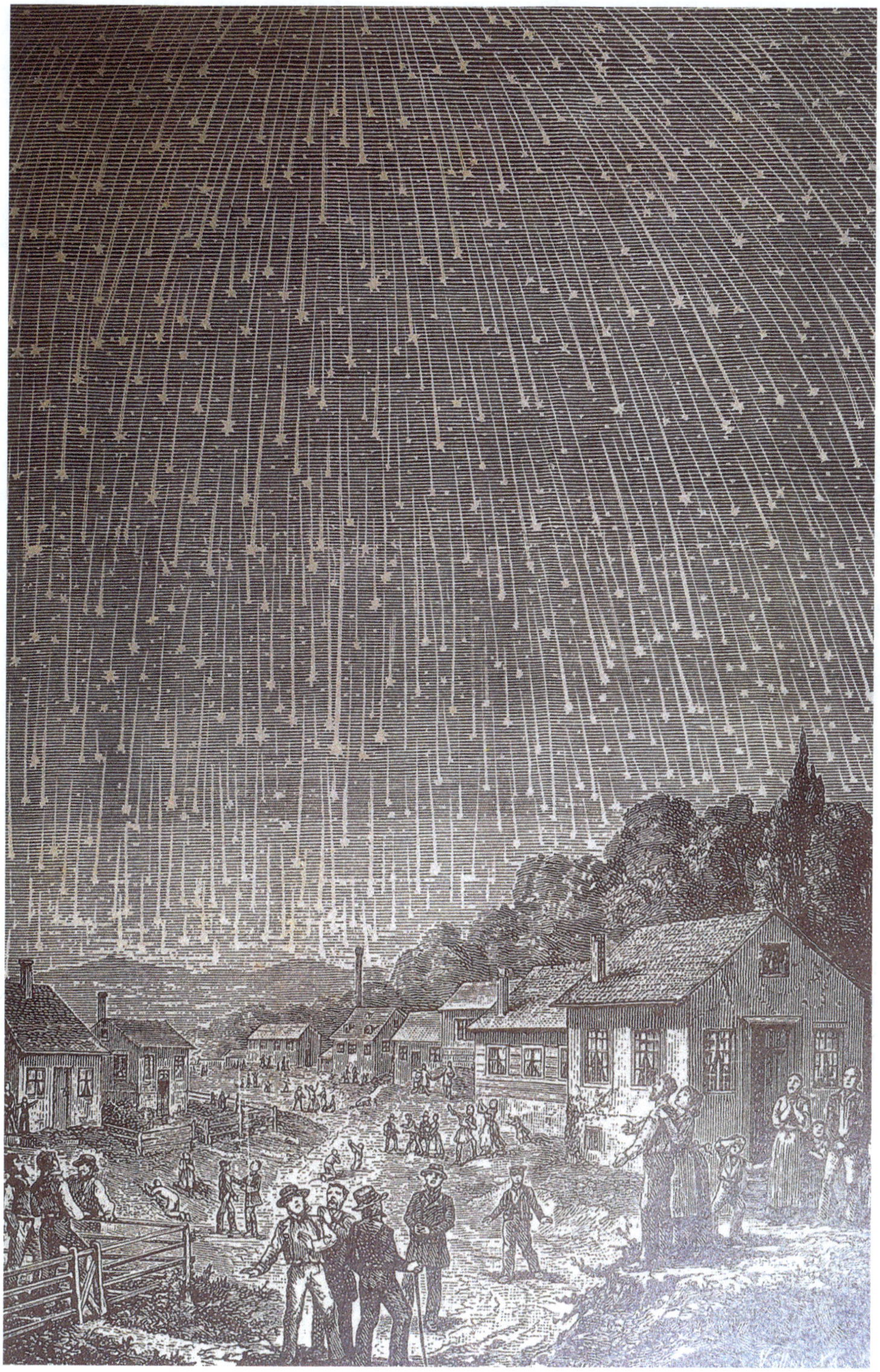

Fig. 4.26 A Leonid meteor shower in 1833

This is a rather more accurate depiction of a meteor shower at its most impressive (Fig. 4.26). Perhaps the most spectacular display of the Leonid meteor shower, scenes like this were observed across eastern and central North America in the early hours of 13 November 1833. The origin of this famous image is elusive; it has been reproduced many times, sometimes cropped, sometimes reversed, sometimes with the wrong dates. It is probably an engraving by Adolf Völlmy, made in 1888 from a painting by Karl Jauslin.

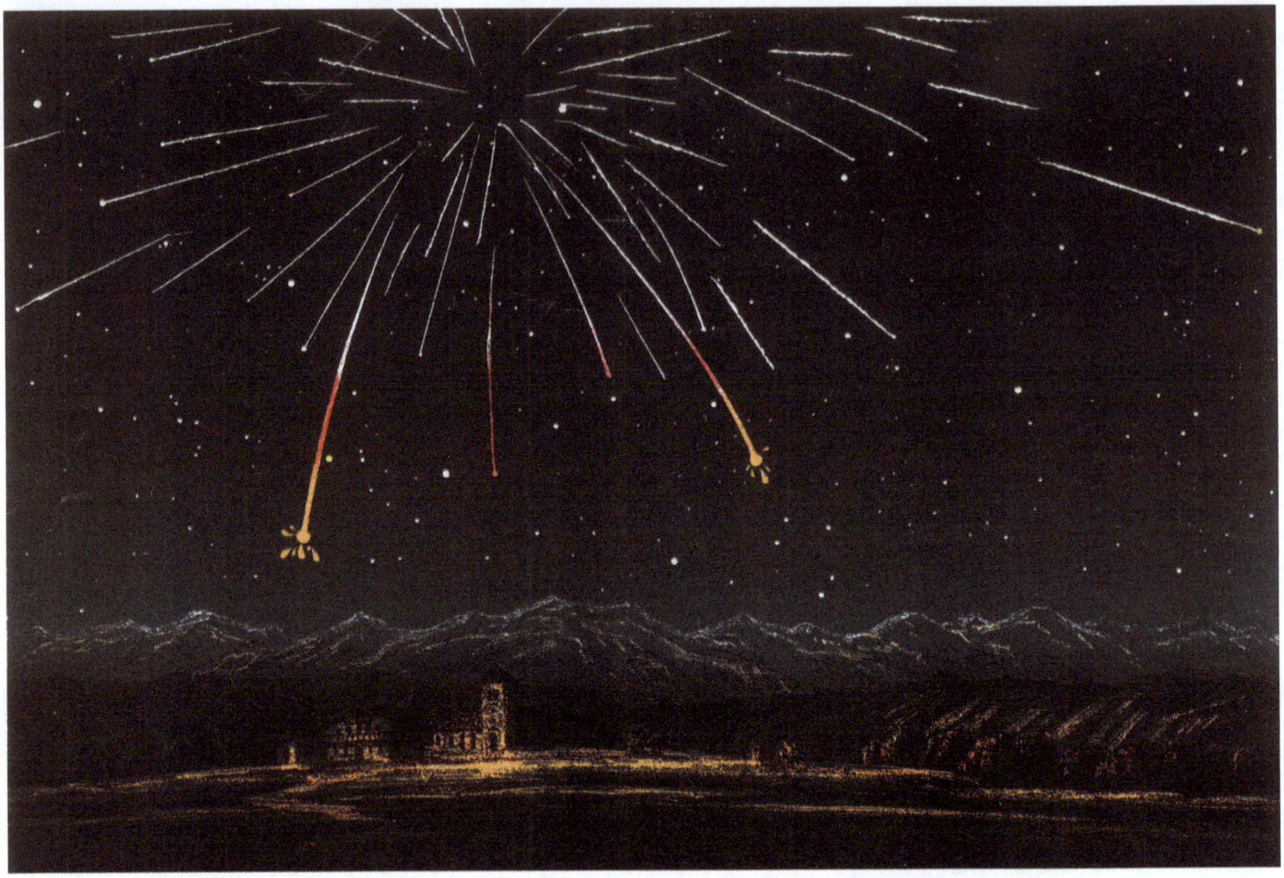

Fig. 4.27 Leonids observed in Europe

In 1864 the American astronomer Hubert Newton, who had calculated the periodicity of the Leonid shower, correctly predicted that its next return would be in 1866. In fact, the Leonids were also observed in 1867 and 1868. This time it was Europe that was favored with the display of stellar fireworks. This image (Fig. 4.27) is a print from the *Atlas der Himmelskunde auf Grundlage der Coelestischen Photographie* by Amand Freiherr von Schweiger-Lerchenfeld, published in 1898 by A. Hartleben's Verlag in Vienna.

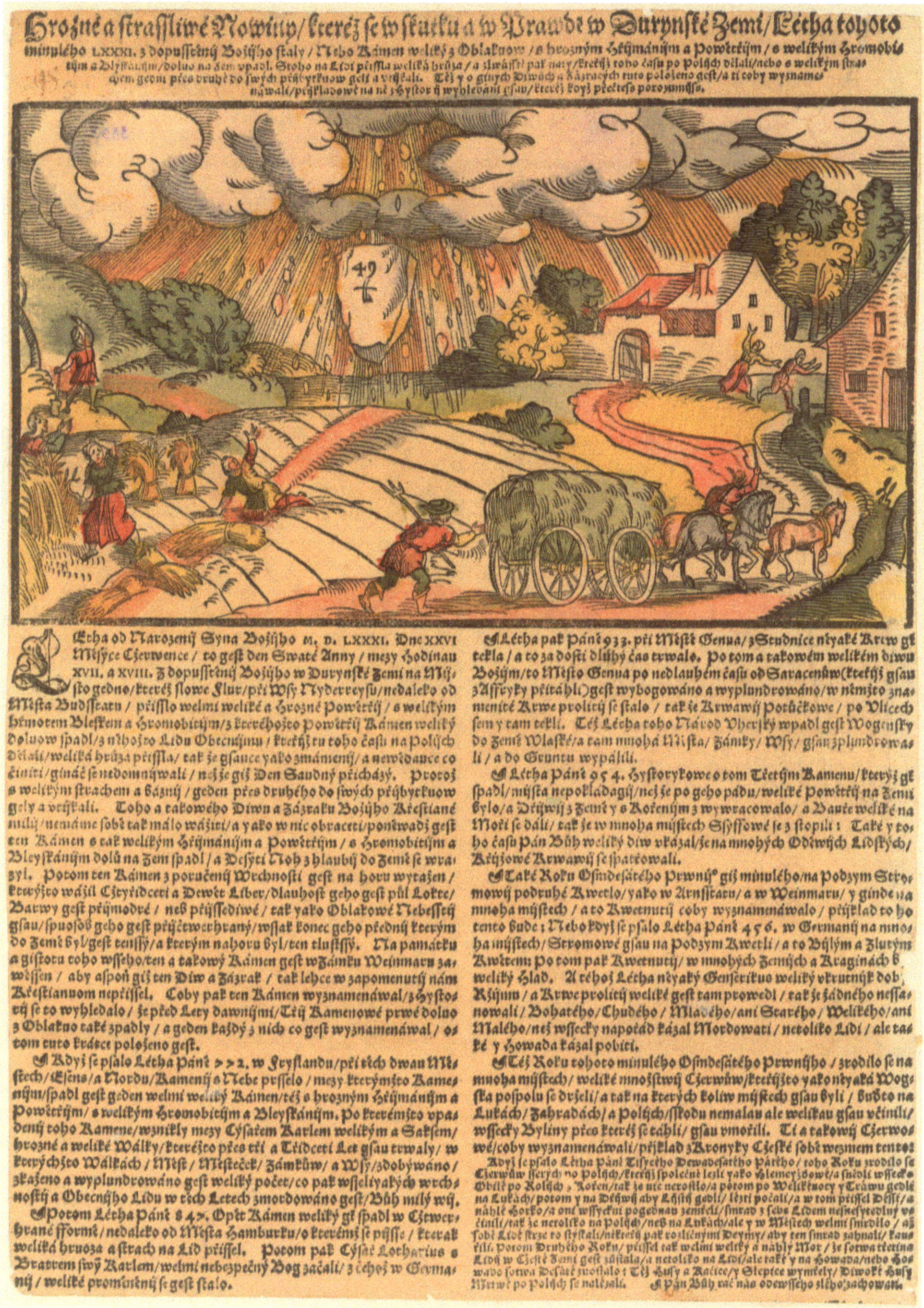

Fig. 4.28 A large meteorite fall in 1581

The largest bodies coming to the Earth from outer space usually have their origin in the collision of asteroids, the remains of which may reach the Earth's surface. Many large impact craters scattered over the world confirm such large impacts in the geological past—notably, the Chicxulub impact at the end of the Cretaceous, believed to have caused the mass extinction that wiped out the dinosaurs (although there are also other hypotheses). This illustration (Fig. 4.28) shows a sixteenth-century leaflet, the Czech text of which is decorated by a hand-colored woodcut pictorial composition showing a bucolic scene of harvesting "in the Thüringen country" in which the harvesters were interrupted by large meteor fall on 26 July 1581. The blue-gray stone weighed about 25 kilograms and its length was described as "half of an elbow," or about 34 centimeters. Evidently, the size of the bolide in the picture is much exaggerated. In the text under the image, a brief list of falls of other stones from the heavens in central Europe is given; the explanation of the unusual phenomenon is purely religious. The occurrence of the sky stone fall predicts the approach of the Day of Judgment, in which Jesus Christ will come to judge all mankind.

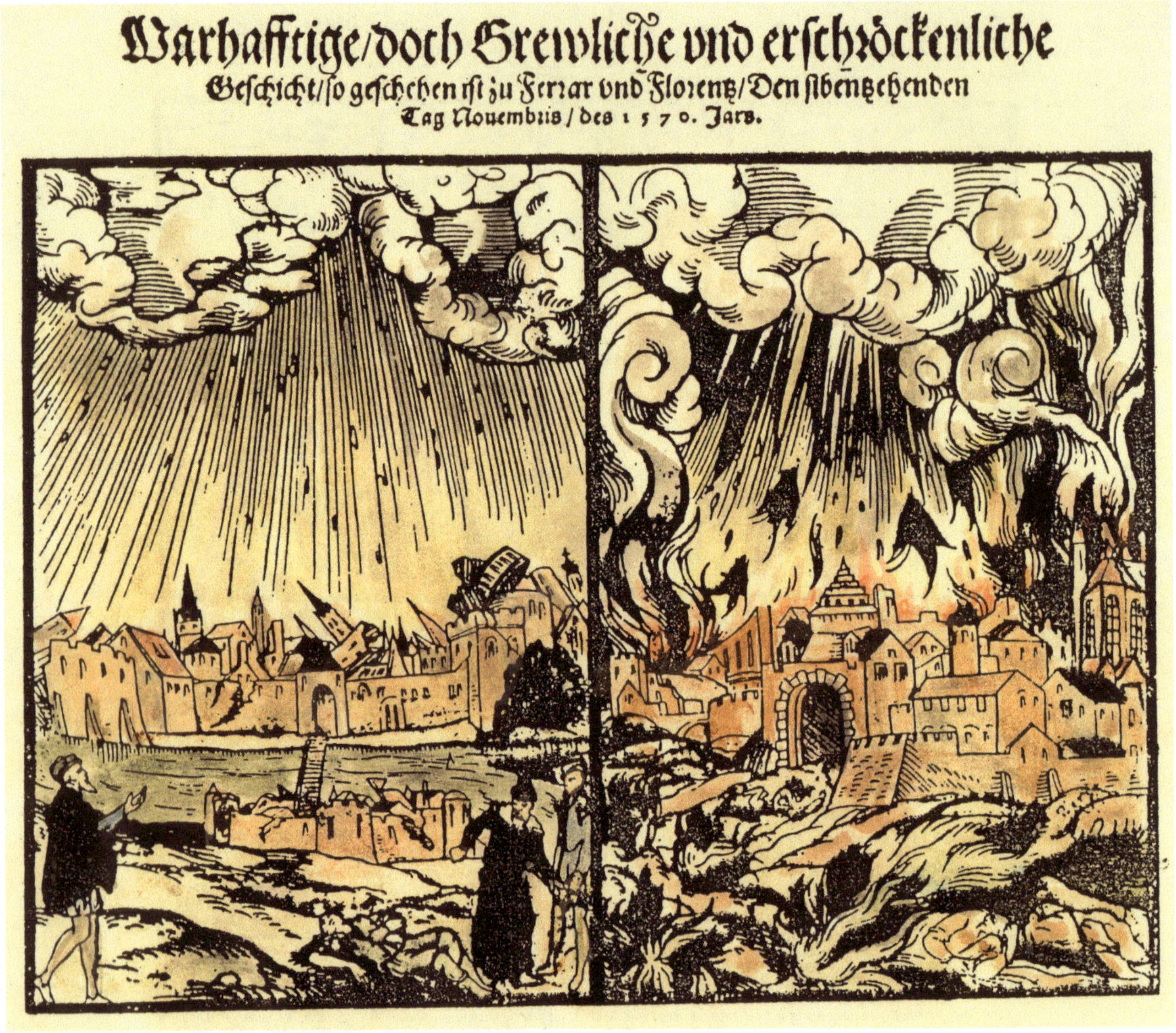

Fig. 4.29 An Italian earthquake in 1570, depicted as an aerial phenomenon

This image (Fig. 4.29) looks visually similar to the previous one, but in fact it shows an earthquake. This is a strong earthquake that occurred on 17 November 1570, hitting the mid-Italian towns of Ferrara and Florence. After four foreshocks the same day, the main shock struck at 7 p.m. and caused heavy damage.

So what is the connection with the element Air? For a long time—in fact, really up to the second half of the eighteenth century—it was very uncertain whether earthquakes were produced in the ground or in the atmosphere. There were strong arguments for the latter: the sound was similar to thunder, and the effects tended to be stronger in the upper parts of buildings. This seems to have been the opinion of the artist who drew this double depiction for a sixteenth-century German leaflet, authored by Daniel Holzman and published by Hans Moser in Augsburg; the damage is clearly shown to be the result of some sort of aerial phenomenon.

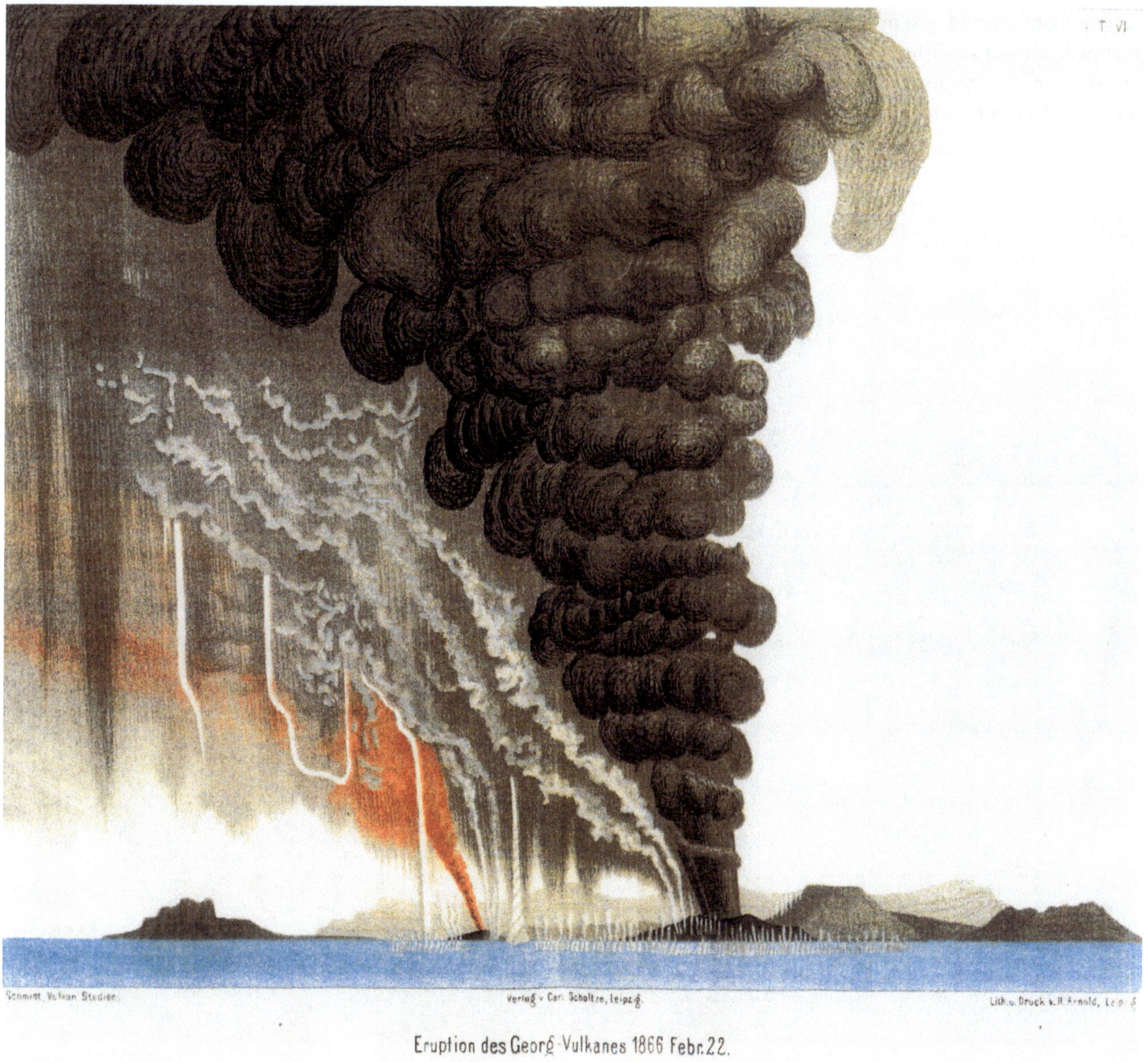

Fig. 4.30 A volcanic cloud in 1866

As we mentioned at the beginning of this chapter, volcanoes have a place here as well. This picture (Fig. 4.30) shows the 1866 eruption of the Georgios volcano rising from the flooded caldera of Thera. The volcano was named after George I, the King of Greece at the time, and is one crater within the Thera complex.

A gigantic volcanic eruption here, which affected the entire region, took place sometime in the middle of the second millennium BC; the present assessment is that about 60 cubic kilometers of subterranean material was launched into the atmosphere during the eruption. The effects on the Minoan civilization on Crete were severe; whether the effects were terminal is still a matter of debate. It seems possible that the climatic effects were experienced as far away as China, leading to famine and the collapse of the Xia dynasty.

A smaller eruption of Thera occurred in AD 197, after which the volcano attained its present shape. Eruptive activity was also recorded and widely discussed in the years 1866–1867. The French physicist Ferdinand André Fouqué (1828–1904) personally visited the island and carried out intensive research, including regular observations and measurement of gas emanations, and he summarized his results in a book (Fouqué 1879). He is regarded as one of the pioneers of the study of volcanic gas eruptions. The other scientist to visit, the Silesian-born physicist J.F.J. Schmidt (later an important seismologist in Greece) studied the Thera eruption in detail and published a book about this event, founded on his own observations (Schmidt 1874), from which the picture shown here is reproduced. The different colors of the volcanic ejecta shown here are intended to depict the contrasting gases and ash, but it has been suggested that the actual colors have been exaggerated to make the distinction clearer.

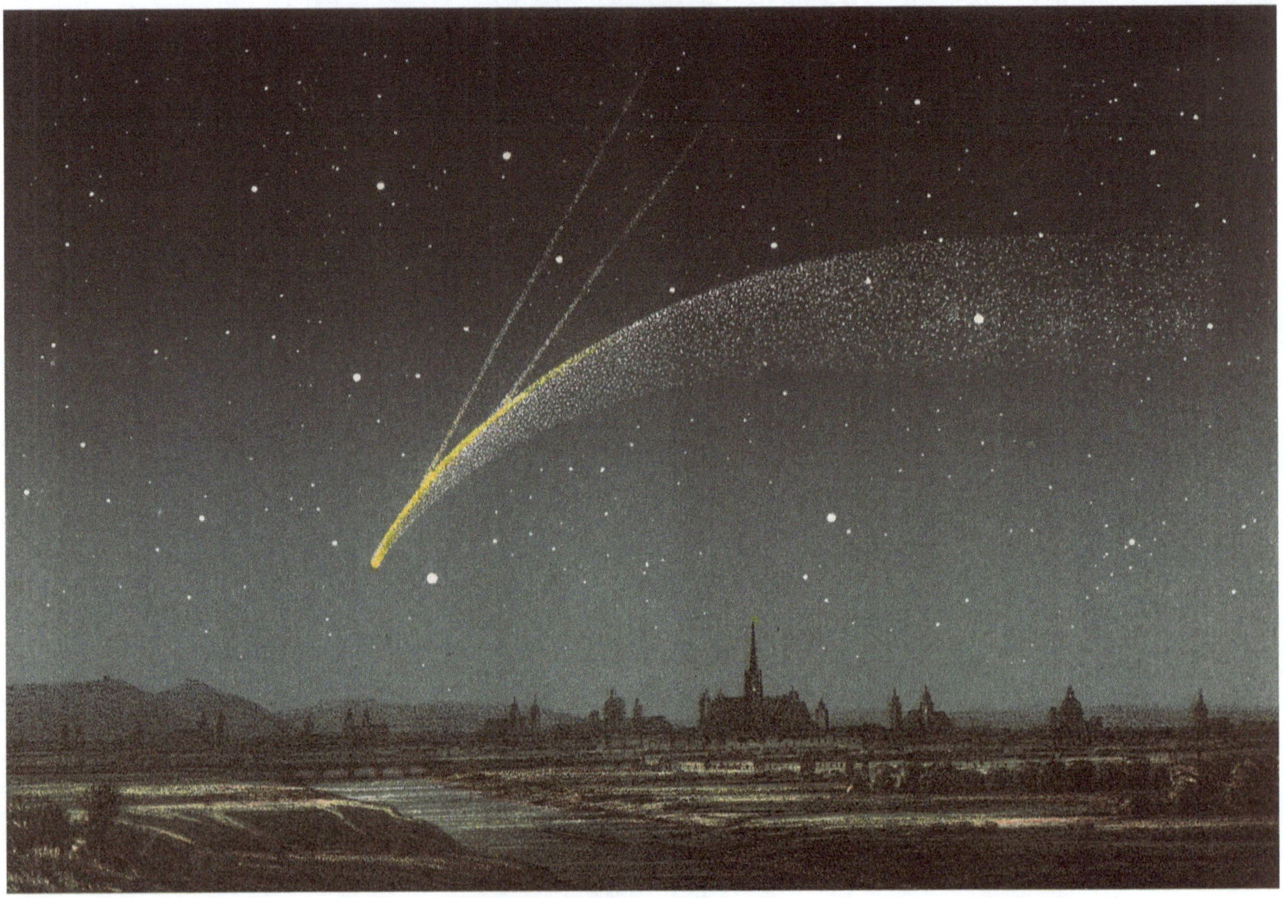

Fig. 4.31 A view of a comet

Is a comet a natural disaster? The famous Tunguska event of 30 June 1908, which flattened trees over an area of more than 2000 square kilometers in Siberia, is thought by many to have been due to a comet entering the Earth's atmosphere and exploding. Had it occurred in a populated area rather than in a remote part of Siberia, the consequences and casualties would indeed have been severe. Fortunately, such events are very rare, and, for the most part, comets merely provide a nocturnal spectacle, as shown here in an image taken again from the *Atlas der Himmelskunde* (Fig. 4.31).

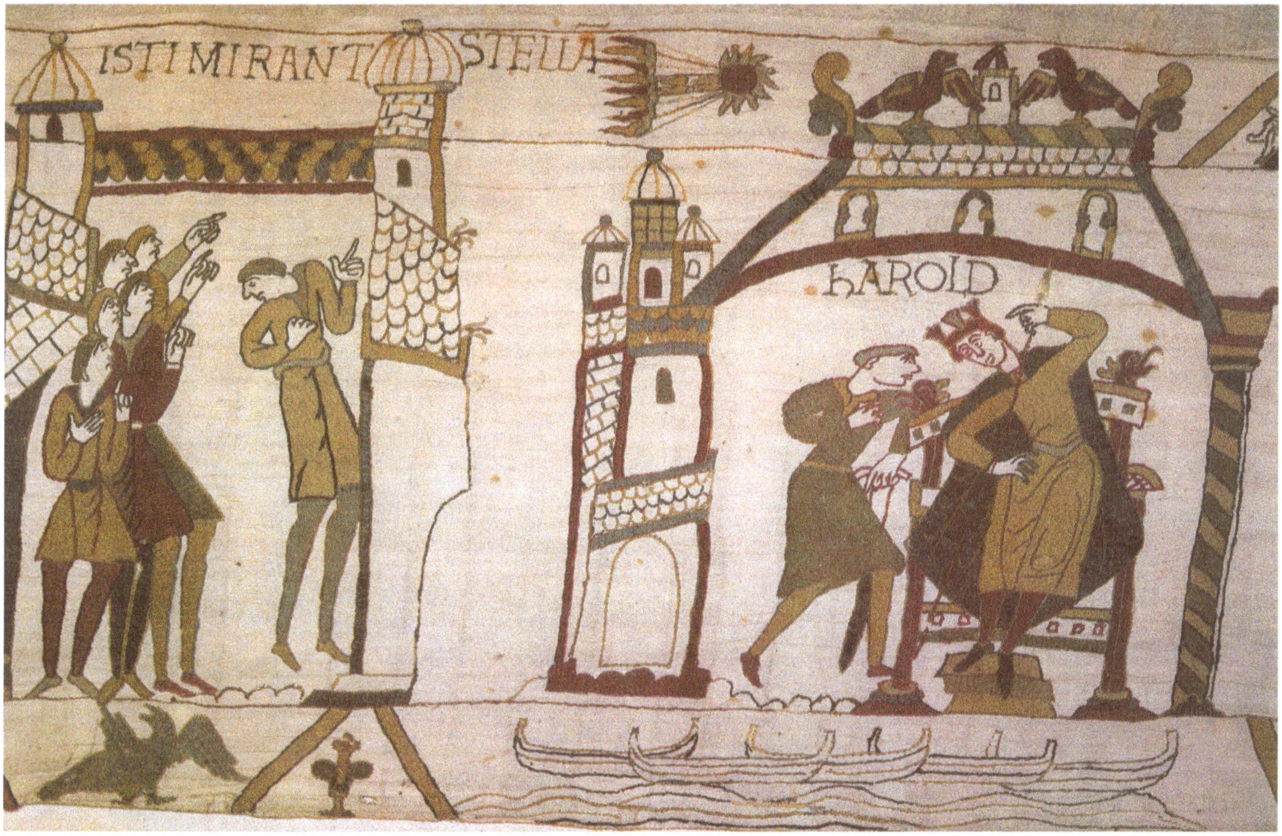

Fig. 4.32 Halley's comet in 1066

If a comet is not itself a disaster, it was at least formerly seen as an omen of disaster, particularly the death of kings. In a world where everything was thought to be connected, the appearance of a comet must mean something, and who could it concern but the most important person in the realm? In this case, the unlucky monarch is King Harold of England. In this scene from the famous Bayeux tapestry, the appearance in 1066 of what later became known as Halley's comet causes alarm, later seen as foretelling the death of Harold at the Battle of Hastings (Fig. 4.32).

A further appearance in 1456 was widely regarded in Europe as foretelling pestilence, famine, and calamity, and in Kashmir as heralding the fall of Sultan Zain al-Abidin, the eighth Sultan (although he managed to hang on for another four years).

It was Edmond Halley (1656–1742), in his book *Synopsis Astronomia Cometicae* (1705), who collated historical sightings of a comet that he believed referred to the same object, which he calculated would next return in 1758. The success of this prediction (which he did not live to see) led to it being named Halley's comet in his honor.

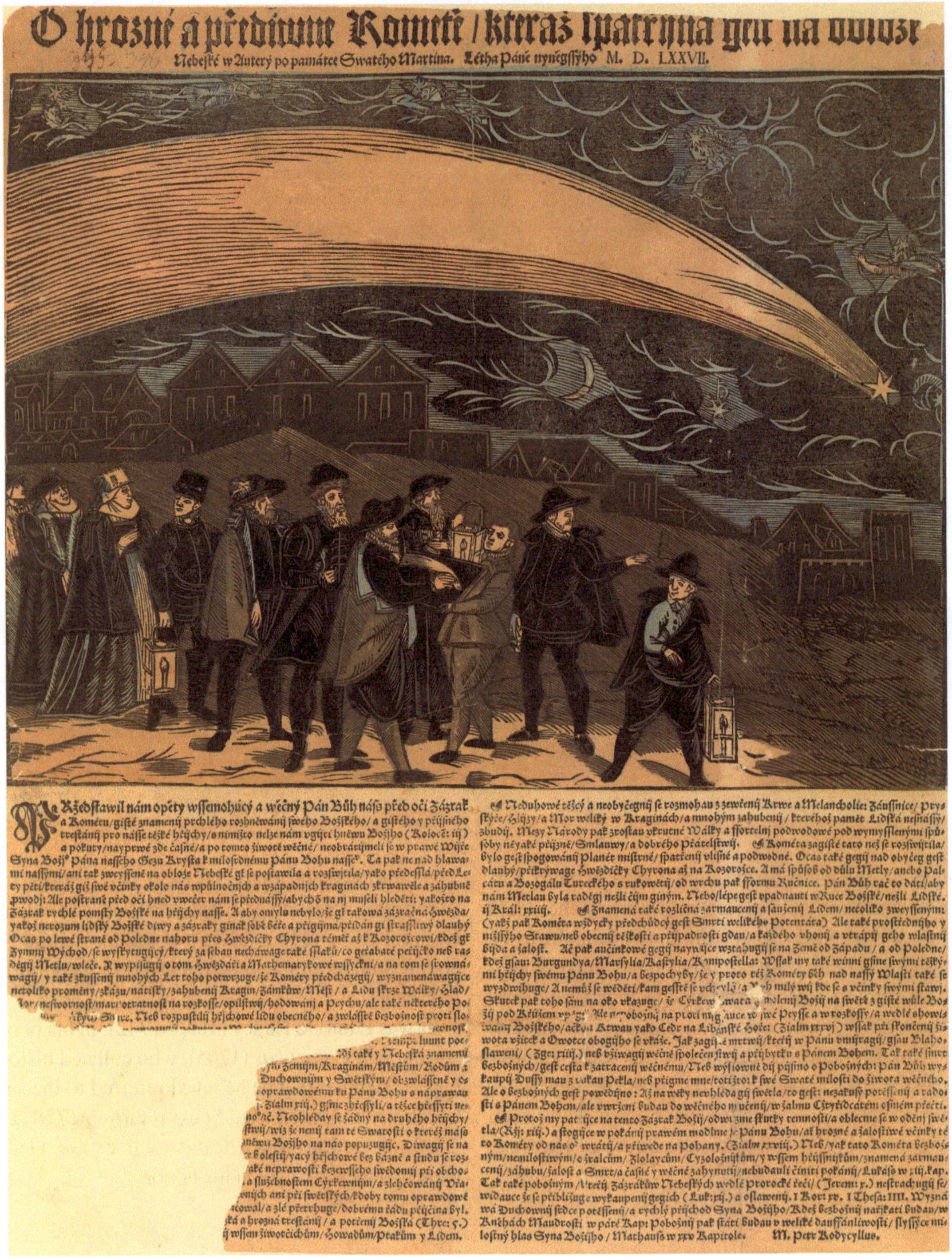

Fig. 4.33 The great comet of 1577

The great comet of 1577 was one of the brightest of all historical comets. It is famous now for the observations made of it by the famous Danish astronomer Tycho Brahe (1546–1601), who was able to deduce that comets travel through space and are not atmospheric phenomena. He also discovered that the tail of a comet always points away from the sun (of course, he had no means of understanding why this is so). The comet was observed over Prague on 12 November, and this picture (Fig. 4.33) is one of two made at the time by two Prague artists, Michael Peterle and Jiří Dačický. In Peterle's Czech version shown here (published as a leaflet by Peter Knížek, a mathematics professor at Prague University), one can see the artist himself, carrying his sketching board and pen, drawing the comet, while his companions hold lanterns for him to have light to draw by. There is only one known copy of this hand-colored woodcut, which is kept in the original sixteenth-century collection of the leaflets known as the *Codex of Dobřenský*; the collection is kept in the library of the Premonstratensian Monastery in Strahov, Prague.

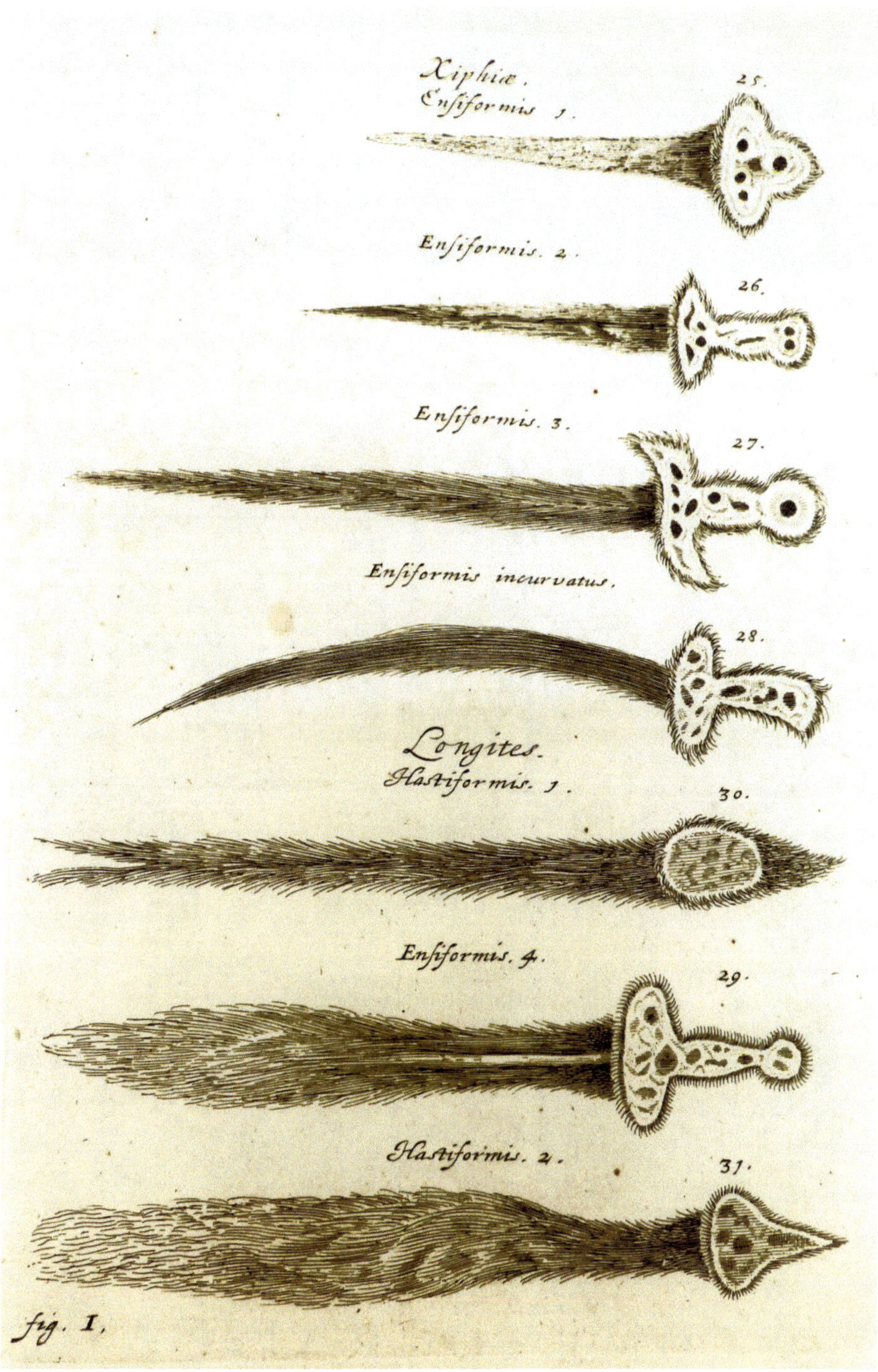

Fig. 4.34 Types of comet

Comets could be omens of natural disasters but might also themselves bring fever, illness, pestilence, death, difficult times, shortages, and times of great famine in their wake, as a fifteenth-century poem suggests. Even as late as the mid-eighteenth century, the English physician Thomas Short was looking at historical records to see if there was any correlation between comet appearances and the plague (Short 1749).

In the seventeenth century, efforts were made to categorize comets into different types; a work by the Polish astronomer Johannes Hevelius, *Cometographia*, was published in 1668 and proved highly influential. The figure here (Fig. 4.34) is from a slightly later work, *The Sphere of Marcus Manilius* by Edward Sherburne, published in London in 1675. Of course, the length of a comet's tail is conditioned by the angle of the solar wind with respect to the direction of observation, but this was not realized in Helvelius's time. The author correctly depicts the most important parts of all comets: the comet core, and the tail, which may have a variable length. The sword-like shape of the tail in the center of the picture corresponds well with real observations.

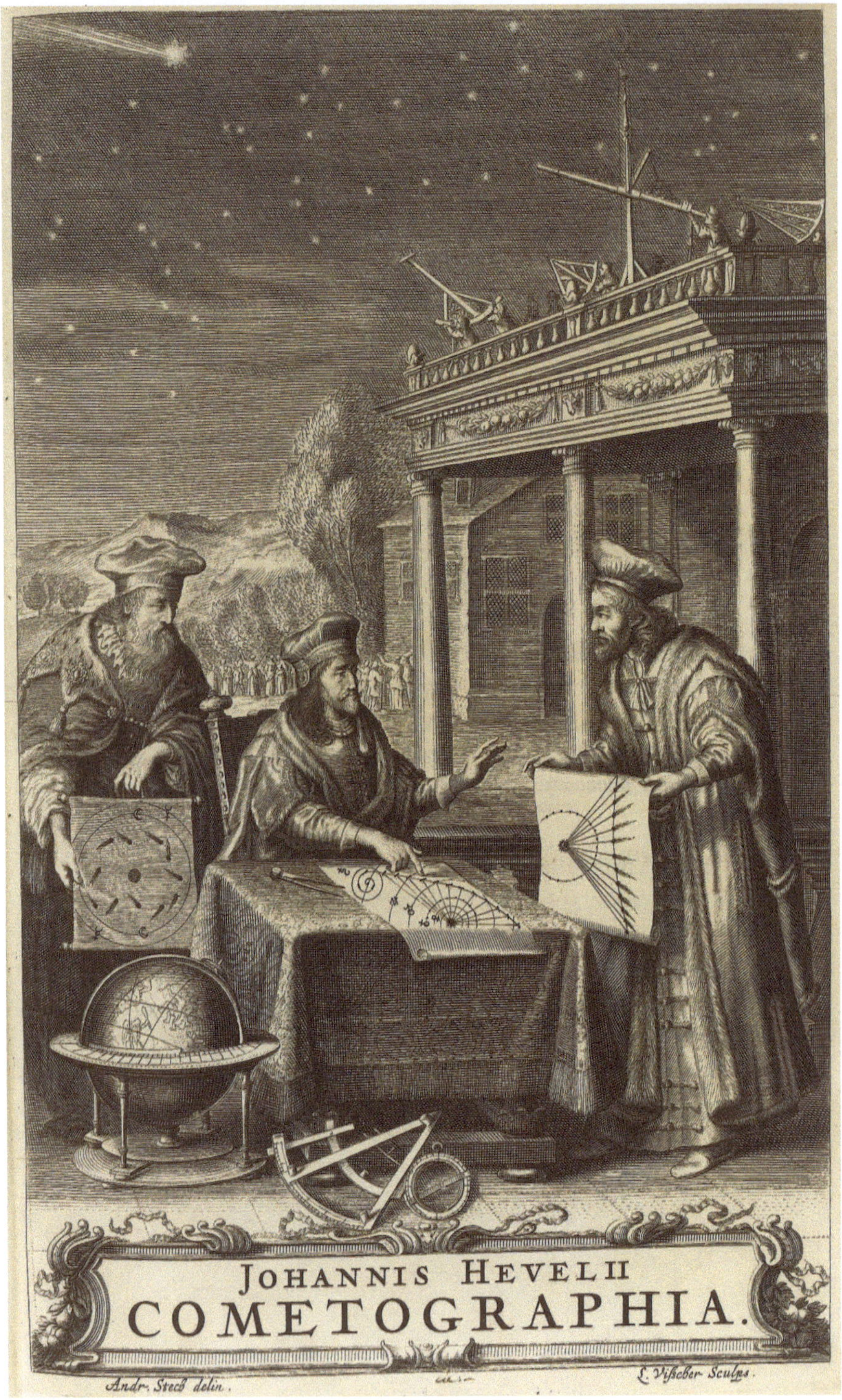

Fig. 4.35 Aristotle, Helvelius, and Kepler discuss comets

This is the title page of Helvelius's work (Fig. 4.35). The figures shown are (left to right) Aristotle, Helvelius himself, and Johannes Kepler. With the aid of charts, they are discussing the trajectories of comets.

Aristotle held that they were a purely atmospheric phenomena—a belief that was decisively refuted by the work of Tycho Brahe in 1577. Caught up in discussion, perhaps they do not notice the actual comet hanging overhead.

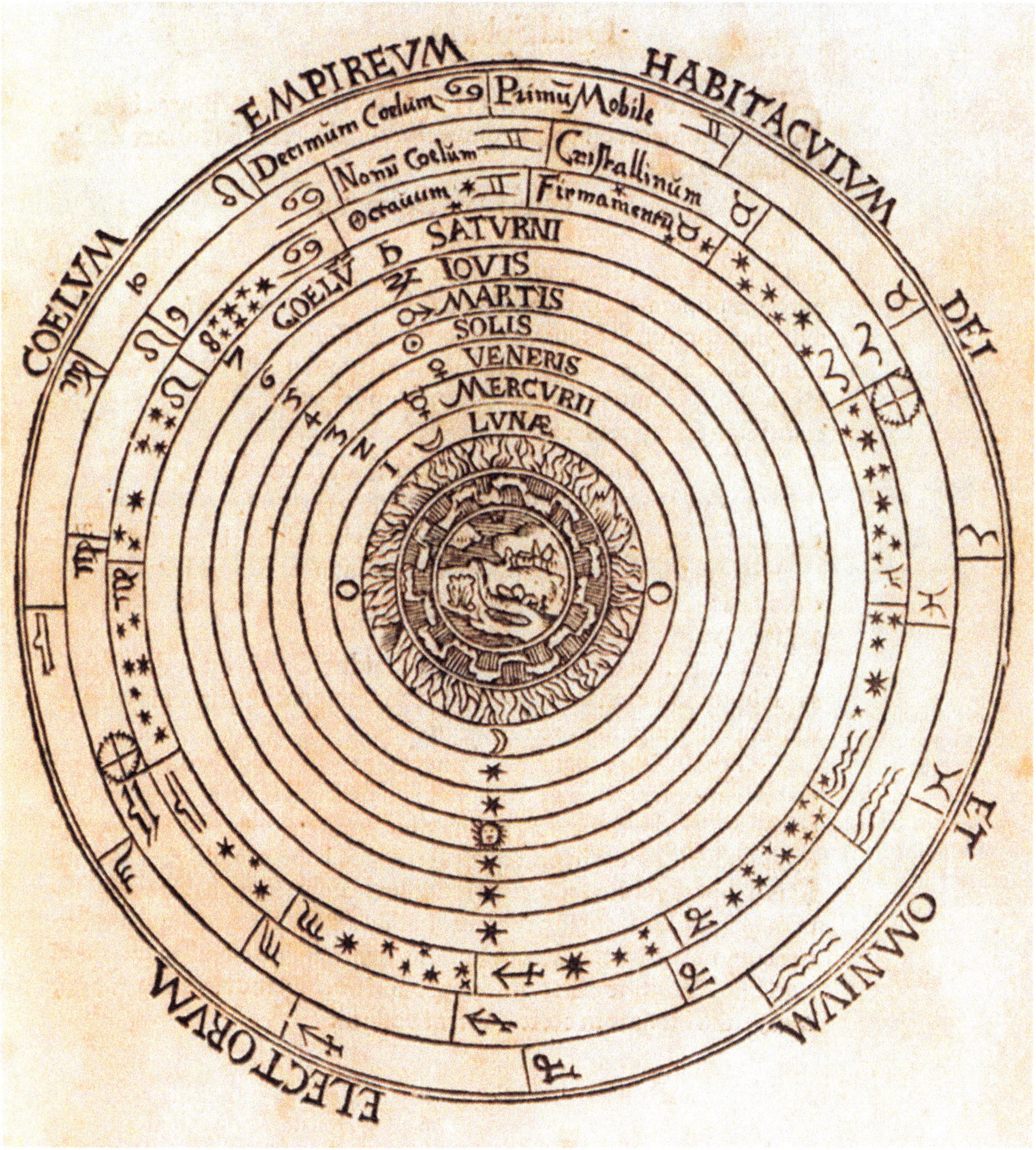

Fig. 4.36 Peter Apian's map of the solar system

The first person to realize that a comet's tail always points away from the sun, and that this controls the length, was the German humanist Peter Apian of Ingoldstadt. He drew this conclusion from observations made of the comet of 1532, a particularly bright comet that was visible from 1 September to 29 December that year. He published his work in the same year as *Ein Bericht der Kurtzer Observation und urtels des Jűngst erschinnen Cometen*. Another work by Apian, his *Cosmographicus Liber* of 1524, shows that his idea of the nature of the solar system was still the old geocentric model, as shown here (Fig. 4.36). This book was very popular and was translated and reprinted multiple times.

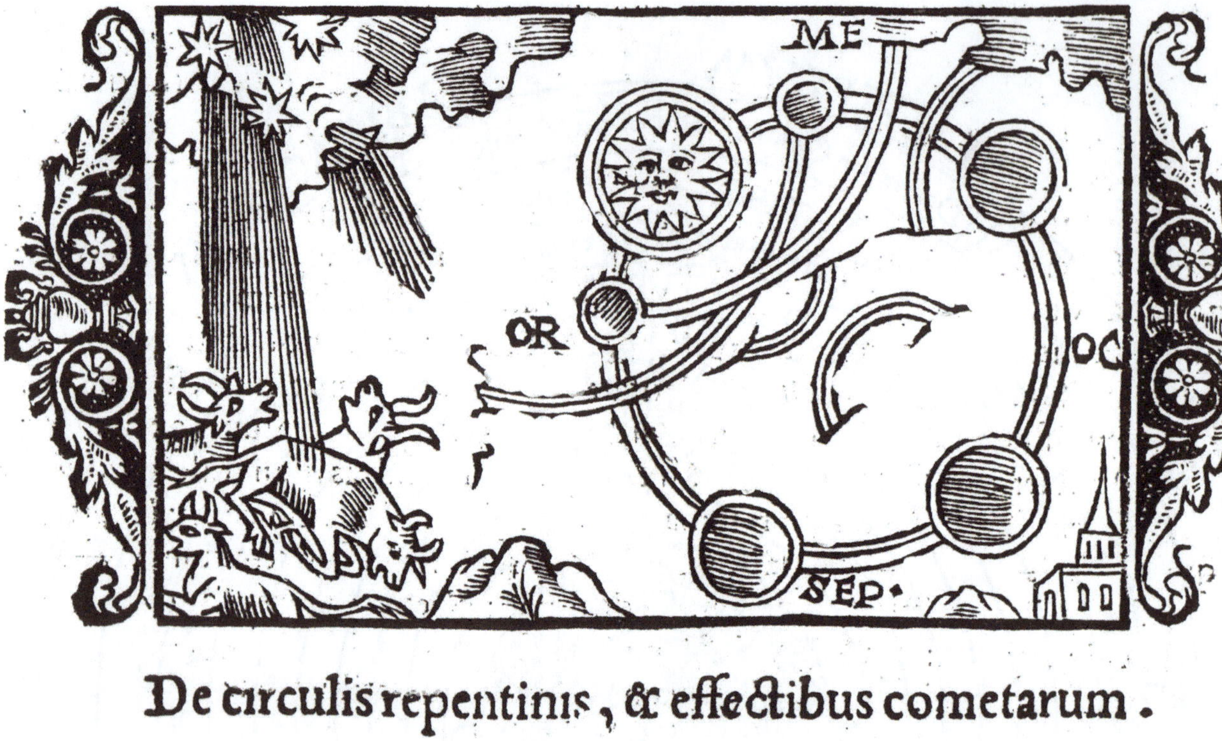

Fig. 4.37 Comets and parahelions in the sixteenth century

This charming little woodcut (Fig. 4.37) is an illustration of the deadly power of comets, as imagined in the sixteenth century. On the left of the image, a comet is directly killing some poor cattle. The curious circles on the right are parahelions, an effect generated by refraction from falling ice crystals high in the sky, perpendicular to the sun's rays. The image is from *Historia de Gentibus Septentrionalibus* (*History of the Northern Peoples*) by the Swedish ecclesiastic Olaus Magnus (Magnus was his name, not an indication of greatness), published in Rome in 1555. He most likely had in mind an observation of parahelions over Stockholm in 1534.

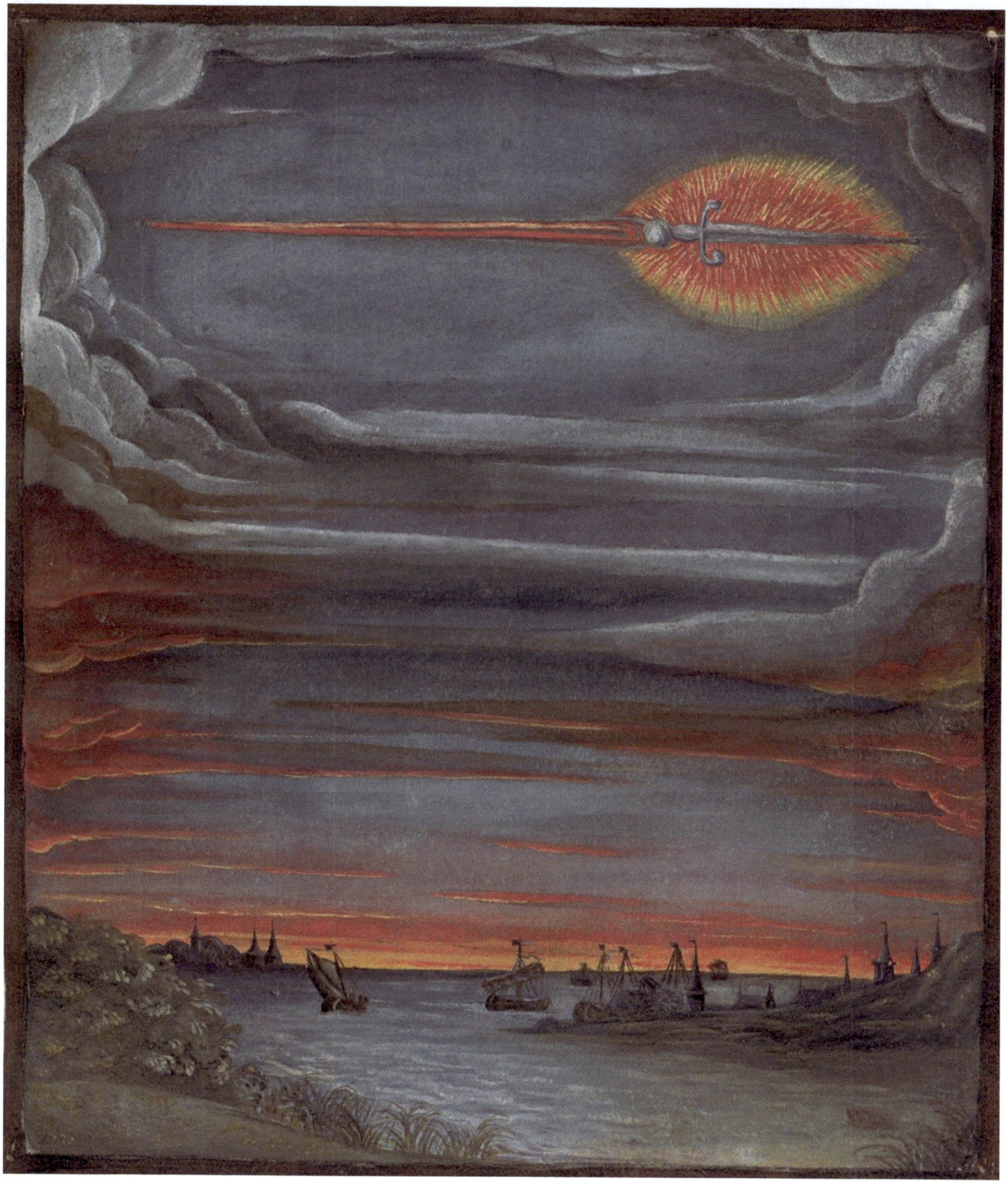

Fig. 4.38 The comet Veru (or Swift–Tuttle)

As a further illustration of comets, this painting of the comet Veru is from an anonymous work, *Kometenbuch*, produced in Flanders in 1587, author unknown (Fig. 4.38). In it, various stylized comets are named, portrayed, and described, together with the evils they bring. The comet Aurora, for instance, triggers wars, fire, and strong winds. Veru, reproduced here, looks very ominous—as well it might—with a head like a sword. Since the author gives it a date of AD 69, this is actually what is now known as the comet Swift–Tuttle, which has been described as "the single most dangerous object known to humanity," given the proximity of its orbit to Earth and the estimate that if it were to hit Earth, its impact would be about 27 times worse than the impact that is supposed to have wiped out the dinosaurs at the end of the Cretaceous—truly a disaster! Fortunately, recent calculations show that this comet poses no threat to Earth for at least the next 2000 years.

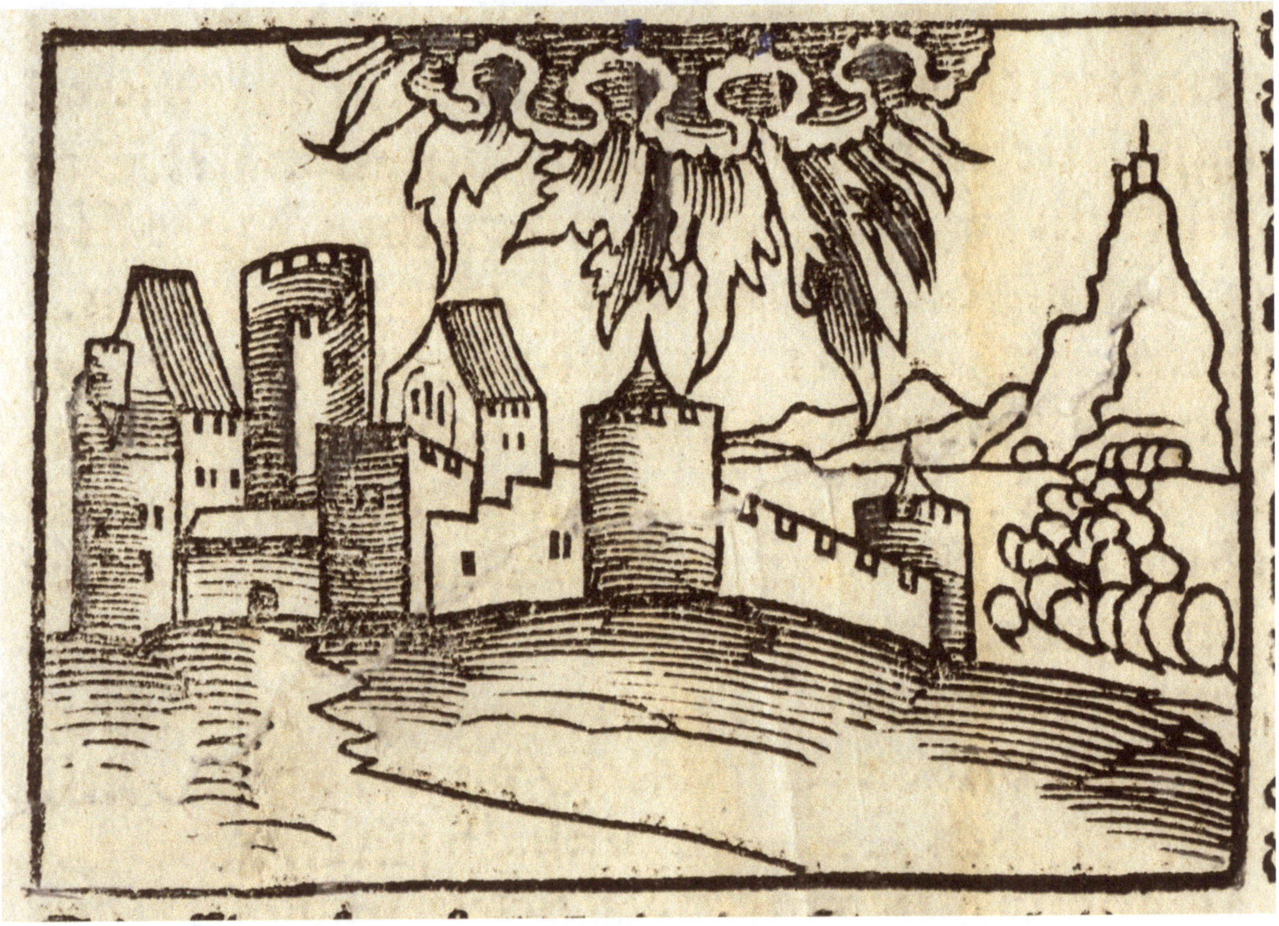

Fig. 4.39 Lightning over Basel in 1455

One might imagine that this woodcut (Fig. 4.39) was another depiction of fire from heaven, this time as tongues of flame. In fact, what looks like fire is very badly drawn lightning. The image is intended to show a storm affecting the city of Basel in 1455. The portrayal of the storm, however, seems to be far from being a true depiction of the situation; it seems to be more like an artistic abbreviation of a storm, a schematic, or even a logo of a storm, as one might say today. Such schematics or allegorical depictions of actual, concrete events were applied frequently at the turn of the fifteenth century, when representation of natural events was in its infancy in central Europe. The picture comes from the remarkable *Nuremberg Chronicle* of 1493, written by Hartmann Schedel, which contains 1809 woodcut illustrations from the Bavarian workshop of Michael Wolgemuth and Wilhelm Playdenwurff, although many are duplications or generic "stamp" pictures showing no real depiction of the actual subjects.

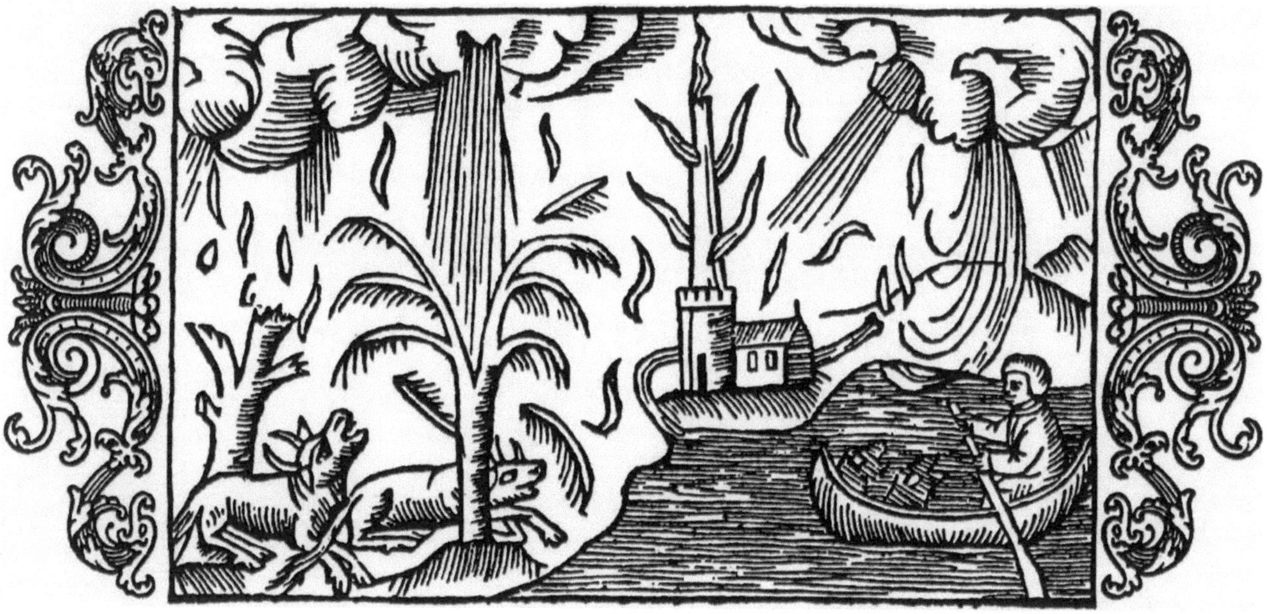

Fig. 4.40 Lightning killing cattle in Sweden

This stylized woodcut (Fig. 4.40) again shows the danger of lightning. Cattle are killed, a tree is struck, a church is set on fire, and a man in a boat is desperately trying to get away to safety. The source, again, is Olaus Magnus. The boat is a typical traditional Swedish design.

4.4 Epidemics and Pandemics

From lightning and other heavenly appearances, we turn to epidemics. To the modern reader, it might seem strange to include diseases, epidemics, and pandemics in with the element Air. However, in medieval times, before anyone understood germs or bacteria, an obvious explanation for disease was impurity in the air. Besides, medicine at that time was permeated by the idea of the four elements, even to the extent that the human body was thought to be conditioned by four humors: black bile (Earth), yellow bile (Fire), phlegm (Water), and blood (Air). Thus, epidemics and pandemics, which are killers on a grand scale, can be fitted in to this chapter as disasters that happened when it seemed that Air went bad.

In general, images illustrating the deadly effects of epidemics and pandemics are not common. The lack of historical pictorial material of this kind is due to the fact that the symptoms of deadly illness are often not clearly seen and cannot so easily be expressed or portrayed by means of engraving techniques. The second reason lies in the fact that such pictures of victims approaching death or in agony have never been favored by the public; on the contrary, they often evoke aversion and repulsion.

It is worth mentioning that epidemics and pandemics are—in contrast to many other dynamic extremes of the Earth's elements, covered here—the only factors carrying exclusively negative effects for mankind. Even earthquakes have a benefit as part of the processes of tectonic uplift that counter erosion.

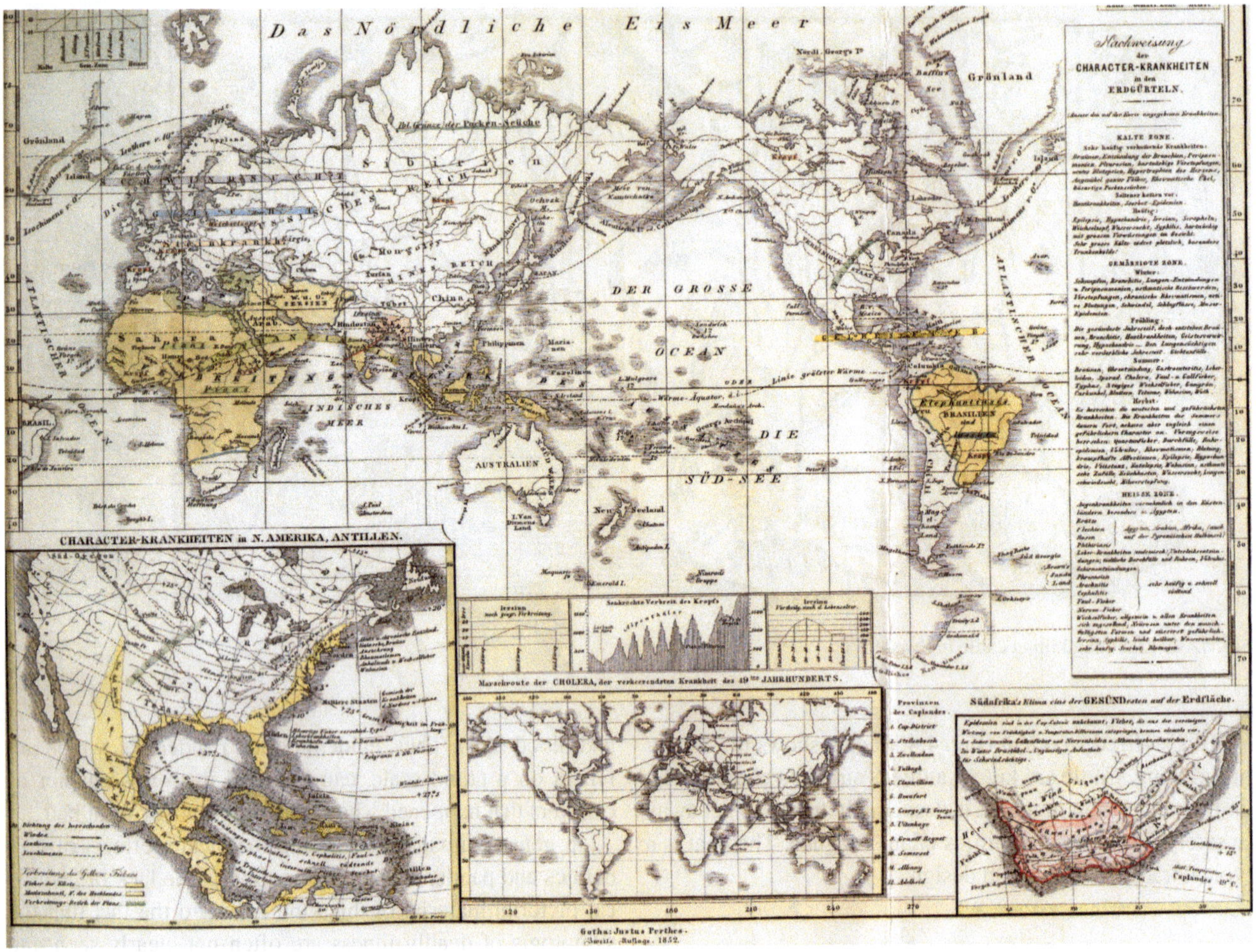

Fig. 4.41 An 1849 world map of diseases

The map here is titled *Planiglob zur Ubersicht der Geographischen Verbreitung der Vornehmsten Krankheiten*, a world map showing an outline of the distribution of principal illnesses (Fig. 4.41). It is another map from Heinrich Berghaus's *Physical Atlas* of 1849 and is another really remarkable production, as is the whole atlas, which contains many landmarks in the study of different phenomena from a global perspective. (This is map no. 2 in the seventh section of the 1852 edition of the *Physikalischer Hand-Atlas.*)

In the world map (in the Mercator projection) there are zones of frequent occurrence of various infectious diseases, marked by the name of the illness and partly emphasized by colors. In the additional table on the right, the diseases are listed, arranged by the degree of their occurrence in three principal types of climatic zones—namely cold, temperate, and hot or tropical zones. The main map is accompanied by six smaller insets, carrying additional information, and three specific maps related to distribution of illnesses in North and Central America and in South Africa, and the migration of cholera outbreaks in the first half of the nineteenth century.

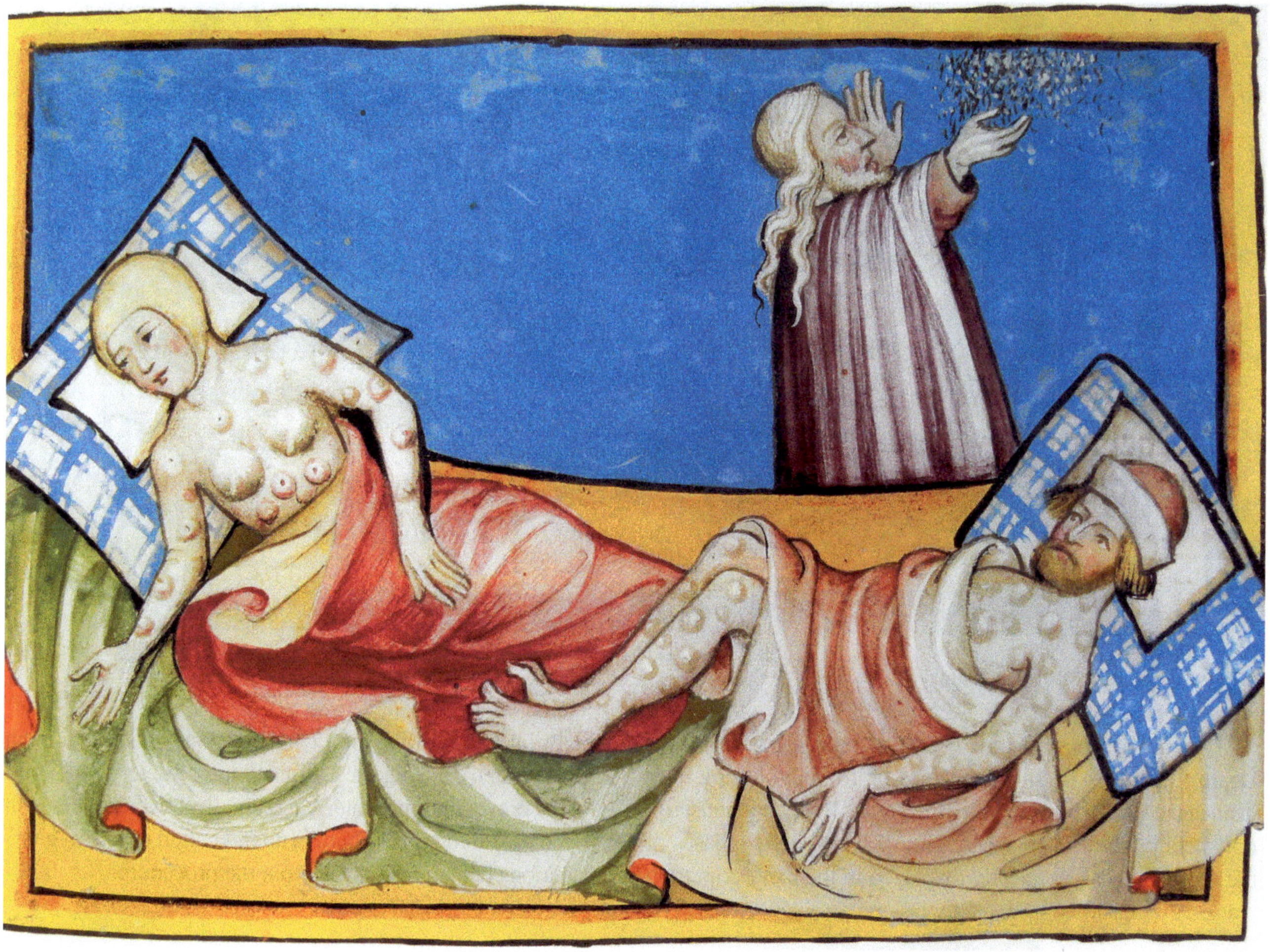

Fig. 4.42 A Swiss miniature of Black Death victims

One of the deadliest of all disasters, the Black Death in 1348–1349 wiped out about half of the population of Europe. The Black Death is generally thought to have been bubonic plague, caused by *Yersinia pestis*, but competing hypotheses have been put forward, including anthrax. This miniature (Fig. 4.42) from the Toggenburg Bible (Switzerland) of 1411 is generally thought to show victims of the Black Death; the symptoms of these two sufferers are plainly obvious. However, it is also suggested that this couple are suffering from smallpox, since the symptoms of bubonic plague do not appear all over the body (as shown here), or perhaps the artist was simply working from inaccurate descriptions.

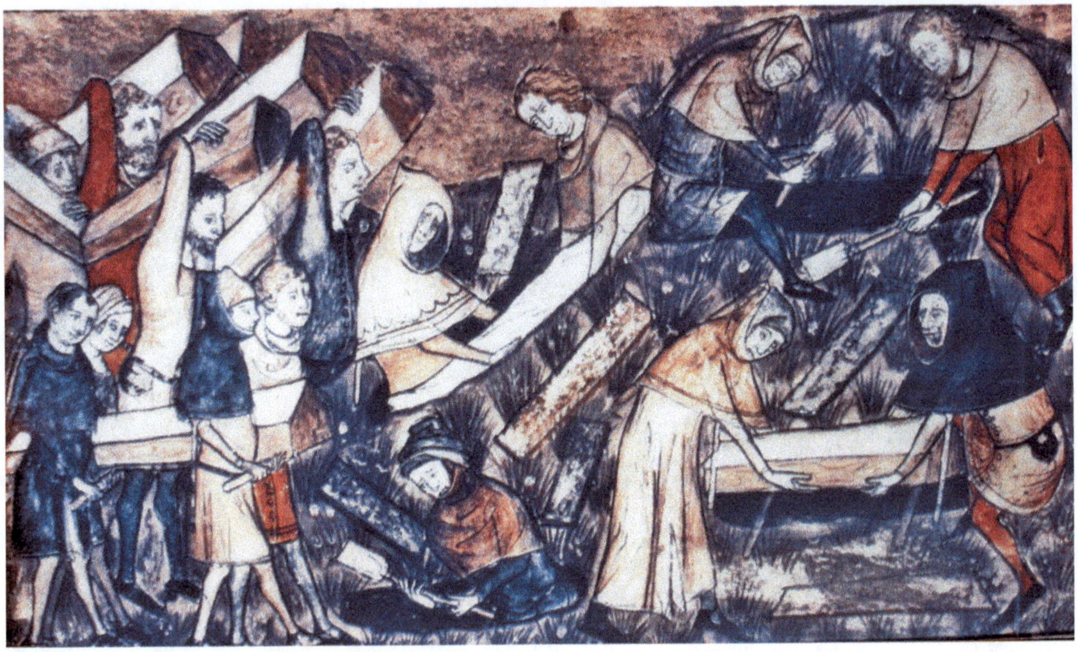

Fig. 4.43 The Black Death in Tournai

This anonymous picture certainly relates to the Black Death (Fig. 4.43). It shows citizens of Tournai burying their dead and comes from a fourteenth-century Belgian manuscript now held by the Diaspora Museum in Tel Aviv. All phases of the burial process are clearly displayed in the picture: putting a body into the coffin (top center), closing the coffin (right upper corner), carrying the coffins to the cemetery (left), digging the grave (bottom center), and, finally, lowering the coffin into the prepared grave. There is no indication of any hygienic precautions being taken throughout the procedure.

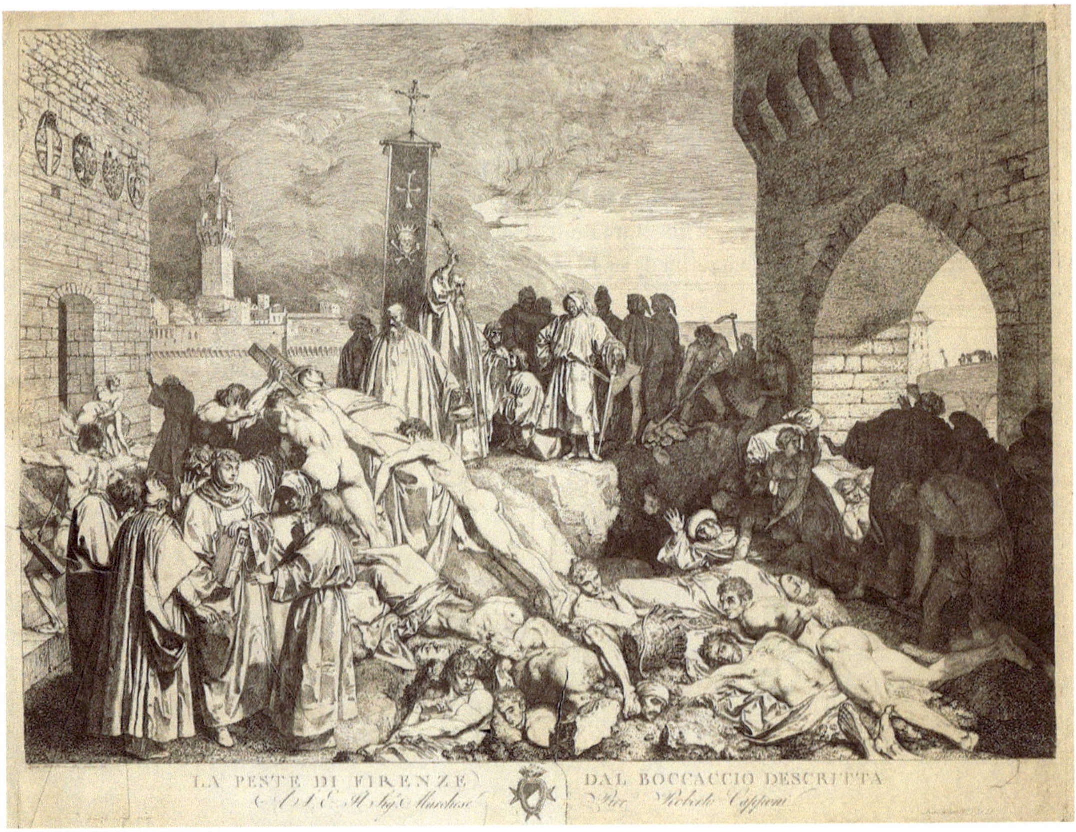

Fig. 4.44 The plague in Florence, as described by Boccaccio

The Black Death also left its mark on literature. One of the most imaginative writers of the fourteenth century, Giovanni Boccaccio, composed his famous collection of stories, the *Decameron*, between 1349 and 1352. The structure of the work involves a tale of ten people fleeing the plague-struck city of Florence and taking refuge in a country retreat. To pass the time, they tell each other stories. This engraving by Luigi Sabatelli (1772–1850) shows the plague in Florence, as described by Boccaccio (Fig. 4.44).

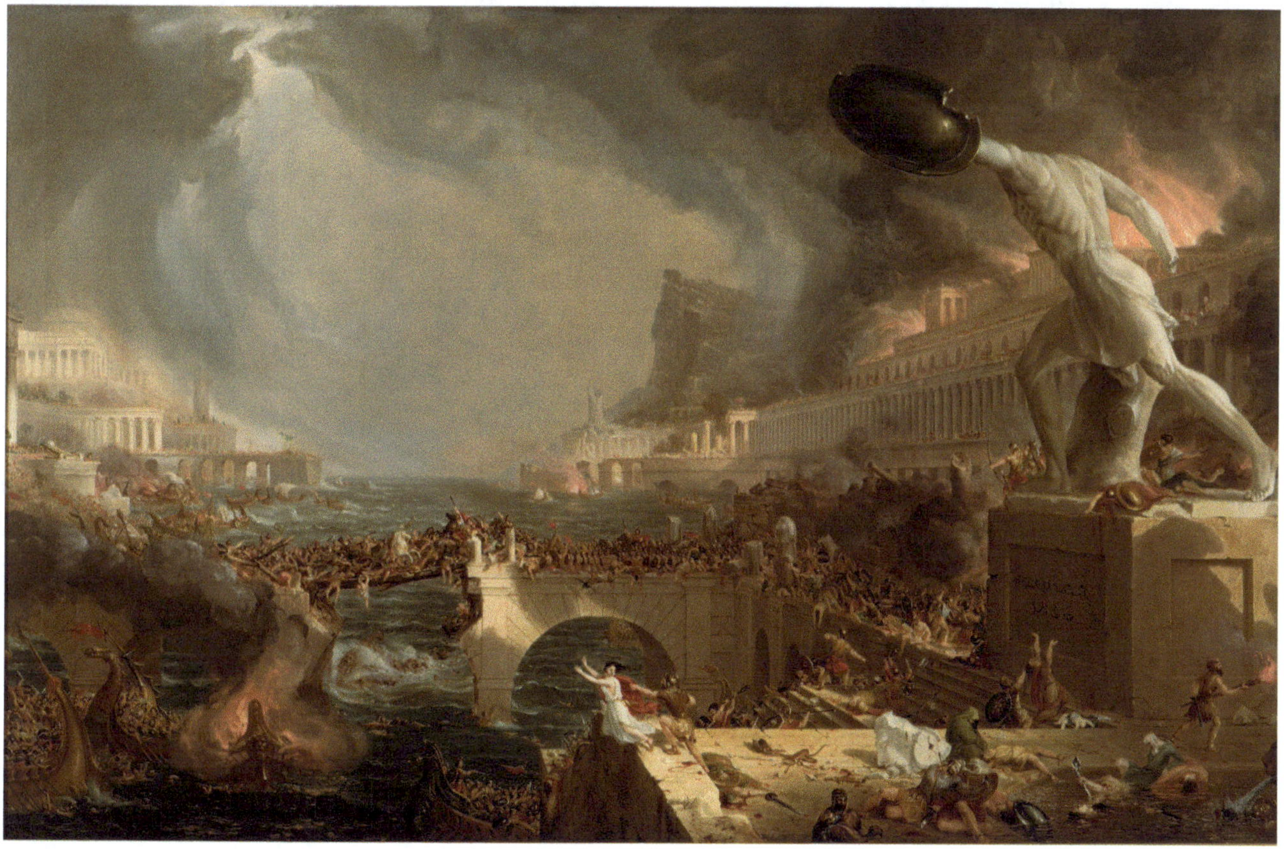

Fig. 4.45 The Plague of Justinian

The first global pandemic was the so-called plague of Justinian (541–542), which particularly affected the Mediterranean region through to Persia and is named after the Eastern Roman Emperor of the time. Modern estimates of the death toll are around 25 million people, or about an eighth of the total world population at the time, with as many again succumbing to recurrences of the plague in the following 200 years. Again, bubonic plague is held to have been responsible, as evidenced by DNA samples taken from ancient grave sites, among other things. It has been suggested that the conditions for the spread of the plague were favored by extreme weather events five years previously, probably due to fallout from a large volcanic eruption somewhere—perhaps Rabaul near Papua New Guinea. This splendidly dramatic and totally unrealistic painting (Fig. 4.45) is by the English-born American artist Thomas Cole; the artist's aim was symbolic rather than literal.

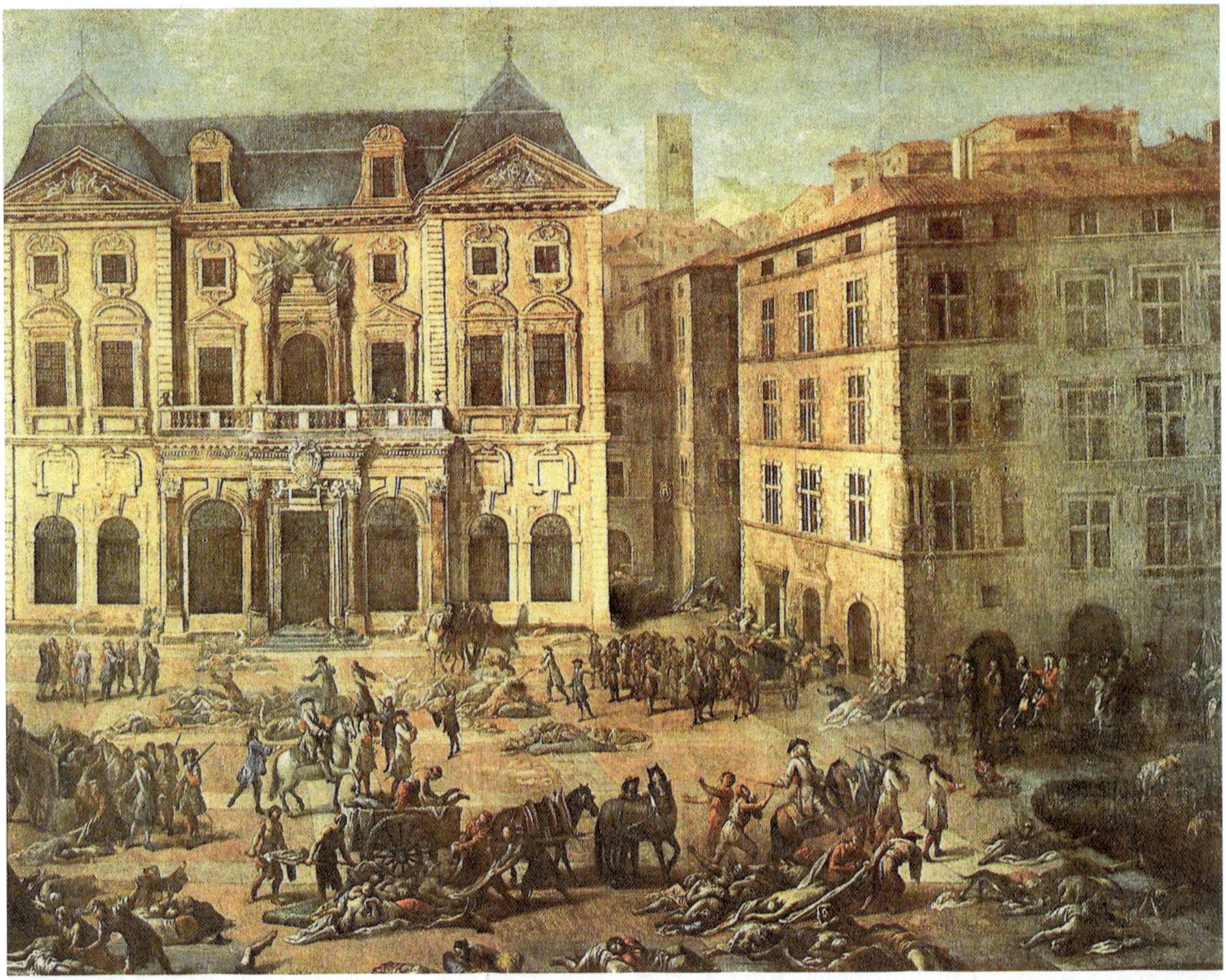

Fig. 4.46 The Plague in Marseilles, 1720

Recurrences of the plague after 1349 were a constant source of concern for Europeans. One of the most notable was the appearance of the plague in London in 1665, which inspired Daniel Defoe's (1722) work *A Journal of the Plague Year*. The third and last pandemic of bubonic plague started in China in 1855 and continued to spread for the next 50 years, killing an estimated 12 million people. However, by this date, modern medicine was beginning to conquer the disease. When the plague struck Hong Kong in 1894, mortality was high among the Chinese population but very low among the British residents, who had access to better health care. The last significant outbreak of bubonic plague in Europe was in 1720, which chiefly affected Marseilles and the surrounding area. The illustration here (Fig. 4.46) is a detail from one of two large paintings made at the time by Michel Serre. It shows the dead and dying being brought out to the square in front of the town hall.

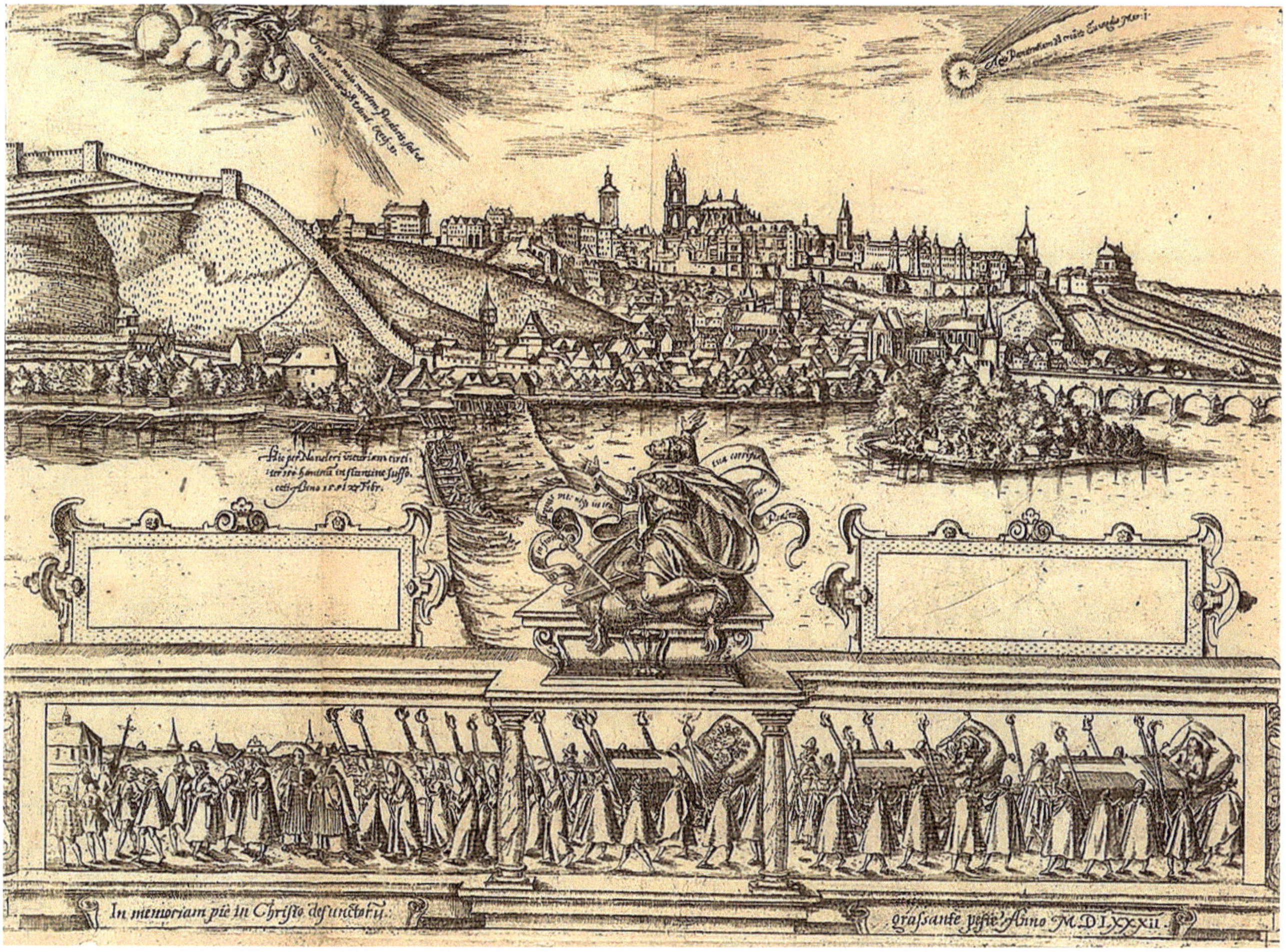

Fig. 4.47 The plague in Prague, 1580

The picture here is taken from a Czech broadsheet and is one of the oldest images of Prague made using the technique of copper engraving (Fig. 4.47). It represents the various disasters affecting the city in the years 1581 and 1582. On 27 March 1581, in high-water conditions, a ferryboat overturned on the Vltava River and, according to the leaflet, "one and a half hundred people died." The capsizing boat can be seen in the right part of the image, with storm winds in the upper left part.

The following year, there was an outbreak of the plague, and thousands of people died. The corpses were taken out of the city, for rapid burial outside the city walls. The bodies of the poor were taken out at night on hay wagons; the rich were taken out on opulently decorated hearses. Such a procession is seen at the bottom of the image, with the victims apparently sitting up in their coffins as if alive. To cap it all, a comet is seen at the top right. According to the leaflet, this was all God's will, confirmed by the appearance of the comet. Only one copy of this leaflet is known, again from the *Codex of Dobřenský* (Opus 317), and is reproduced here with kind permission from the Premonstratensian Monastery in Strahov, Prague.

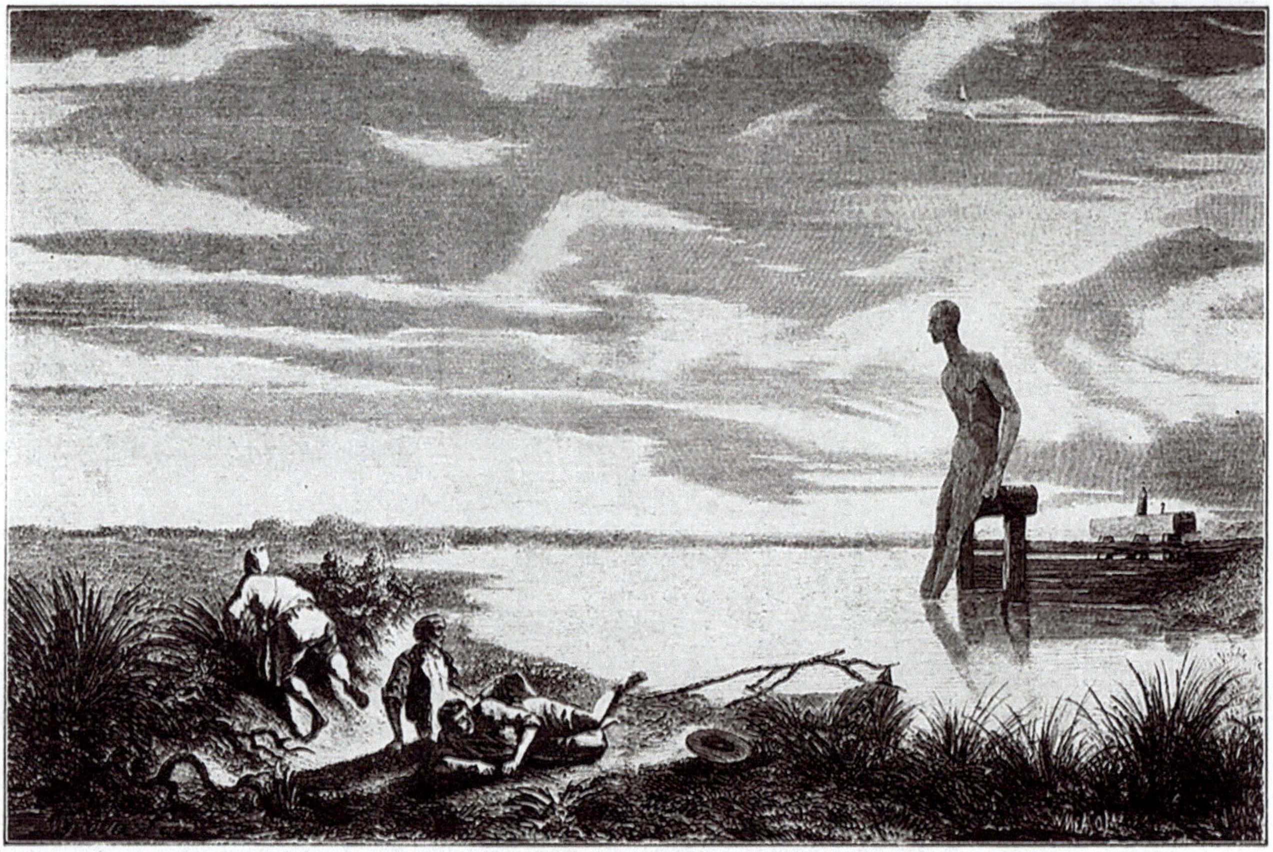

Fig. 4.48 A symbolic image of malaria

That epidemics and pandemics were associated with Air in the premodern mind is amply demonstrated by the very name "malaria" (literally, "bad air"). The disease also went by the names of "ague" and "marsh fever." Malaria is more of an endemic disease than a pandemic, being known in history right back to the days of Hippocrates. Even today malaria typically kills around half a million people per year, mostly in sub-Saharan Africa. The image here (Fig. 4.48) is an allegory of malaria, portrayed as *The Ghost of the Swamp* (the sinister figure on the right side of the picture). The artist was Maurice Dudevant (1823–1889).

5 Fire

We should start by considering the nature of this element carefully, because it is rather different from the other three. In the case of Earth, Water, and Air (or—as we might put it in other terms—solid, liquid, and gas), these are all types of stuff or substance. They all have mass, the physical state of which is determined by conditions of pressure and temperature. It is possible for a particular chemical element or compound to be known in all three states: the substance given by the chemical formula H_2O, which we generally think of as water (in its liquid state) is also commonly encountered as a solid (ice) or a gas (steam). We normally only encounter a substance such as iron in its solid state unless we are involved in smelting, when its liquid state appears. Iron as a gas is unknown on our planet but undoubtedly occurs in the Sun and other stars.

Fire, however, has no substance. It is a chemical process—most commonly, oxidation. Under certain conditions of pressure and temperature, the electrons revolving around the atomic nucleus along its outer paths begin to fall into paths of lower energy. The energy excess resulting from such a transition is then radiated outside the body in a form of radiated light and heat radiation, and we say that the object is on fire; it is burning.

In fact, there are variations on this, since several factors come into play, besides pressure and temperature. The chemical and energetic structure of the given material and its ignition temperature, the degree of accessibility of oxygen, and the intervention of other chemical factors affect how an object burns in particular circumstances. It is convenient to consider these in terms of the speed with which combustion occurs.

The slowest sort of combustion is exemplified by ordinary rusting of iron surfaces at room temperature, which may operate over months or years. This process is spontaneous, and it is so slow that in most cases no heat or light is radiated, and most people would not consider it to be a form of burning at all; nor is it likely to lead to open fires.

Somewhat faster is the process referred to as smoldering, which is undoubtedly a form of burning but is typical of cases where there is insufficient oxygen for an open blaze. It can last for minutes or hours or days—or even longer in the case of some underground coal fires, which have quietly smoldered away for years. A smoldering fire is typified by some heat and a dull red glow. In the days when coal fires were commonly used for domestic heating, one could "bank" a coal fire by covering it with a layer of fine coals (known as "slack"). This deprived the fire of oxygen, causing it to burn only slowly until it was wanted again. One could then disturb the coals with a poker, letting in air, and the fire would flare up again.

What most people think of as fire is a rapid oxidation process involving clear radiation of heat and light in the form of flames. A flame is, in fact, an emission of gas at such high temperatures that it becomes luminous. Within this category, there are still variations in speed. Put a log on a fire and it will burn for an hour or more; set light to a pool of petrol and it will burn off in seconds.

Fastest of all is the oxidation associated with explosives. Military and industrial explosives are created with the aim of achieving the highest dynamic vigor of oxidation: in classical explosives the burning process propagates with a velocity of around eight kilometers per second. One notes here also the role of pressure: gunpowder spread out on a sheet burns when lit, but when confined in a metal shell, it explodes.

One can also consider, as a form of burning, nuclear fusion or fission of radioactive materials. This can be seen as a slow, subcritical, stable process, as in nuclear power plants, or as a rapid and explosive release of heat and light in the detonation of nuclear weapons.

Our historical images of fire and its effects are divided into two parts. First come natural fires, such as those following volcanic eruptions or lightning, and those due to spontaneous ignition. The second group of images shows cases where the fire has been started by human action either deliberately (as in cases of arson or fires resulting from military action) or by accident.

J. Kozák, R. M. W. Musson, *The Illustrated History of the Elements*, https://doi.org/10.1007/978-3-030-21426-5_5

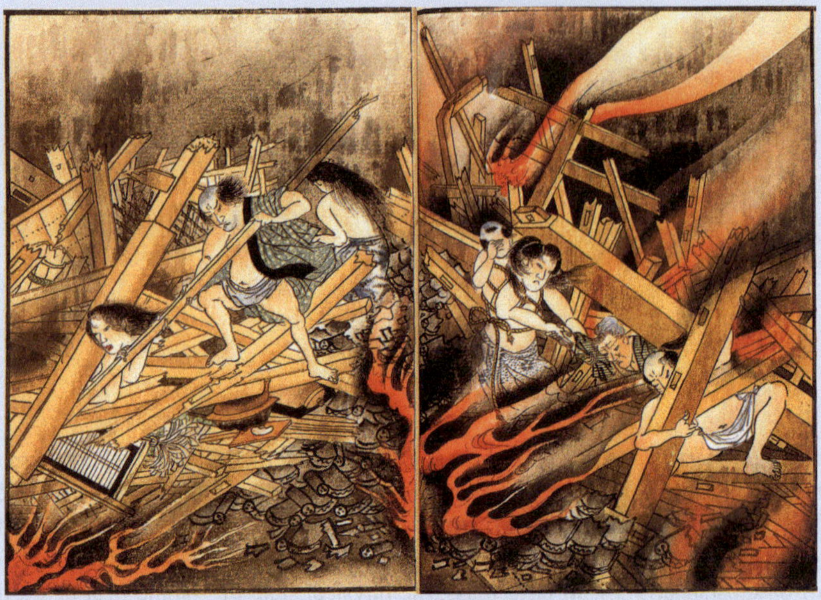

This dramatic composition shows a horrific scene in which a wooden family house has been set on fire as a result of the Zenkoji earthquake of 1847. The image is one of a series produced at the time, showing the painful effects caused by the earthquake: we see the desperate victims of the quake trying to release themselves from the ruins and rescue their nearest and dearest. The painter is named Nagai Zenzaemon

5.1 Natural Sources of Fire

One often thinks of bolts of lightning as being a major cause of fire, especially in cases of drought, where vegetation is tinder dry. Surprisingly, a relatively small number of cases occur where lightning causes fires, despite the enormous heat energy engendered in a lightning bolt. It is estimated that thousands of storms, hurricanes, tornadoes, and other atmospheric extremes are continuously occurring in the Earth's atmosphere, most of which are accompanied by lightning effects. We leave it to the reader to guess how many lightning impacts strike the surface of the Earth in any 24-hour period. The explanation of the relatively low rate at which lightning actually causes a conflagration lies in the required conditions of dry, flammable vegetation. And since lightning is often accompanied by heavy rain, it is clear that the frequency of the phenomenon is disproportionate to the number of cases of lightning-caused fires.

Besides lightning, fires can also be due to volcanic eruptions, during which lava flows (at a temperature of about 1000 °C) may set fire to vegetation or structures along the paths down which they travel. Theoretically, the flying bits of pumice launched into the air during explosive eruptions (volcanic bombs) may also set a fire. In both cases, however, the region endangered by volcanic fire is limited to the immediate environs of the volcano.

Earthquakes often cause fires, and fire is sometimes the chief cause of earthquake fatalities, as in Tokyo in 1923. In historical times, towns frequently consisted of houses constructed of timber; streets were often narrow and roofs were covered by thatch, which blazed if ignited. This could easily happen if seismic displacement broke fireplaces or upset kitchen ovens or heating stoves.

Sometimes even a fallen lit candle could cause a fire, spreading to the whole town, particularly in the case of windy weather. Many towns visited by moderate or medium-sized earthquakes sustained much heavier damage from fire than from the earthquake shaking.

The remaining natural sources of fire do not represent any acute danger: fire—of course—may be initiated by the fall of a glowing meteor or bolide, or by a shower of meteors. The frequency of such flammable impact is low, and in a statistical assessment of risk it approaches zero.

To complete the list of our sources of natural fire occurrence, there is finally the possibility of spontaneous ignition. Flare-ups of this kind happen especially in tropical and subtropical regions, in which long exposure of flammable materials to direct sunshine may result in their ignition temperature being reached. This may happen especially in regions in which naphtha or asphalt outcrops are present on the Earth's surface.

These fires, which have been occurring since long before man ever arrived on the scene, can actually play an important part in ecology. Many plants have evolved in such a way that fire is a precondition of the germination of their seeds. That this has happened illustrates that natural fires are an enduring component of the environment.

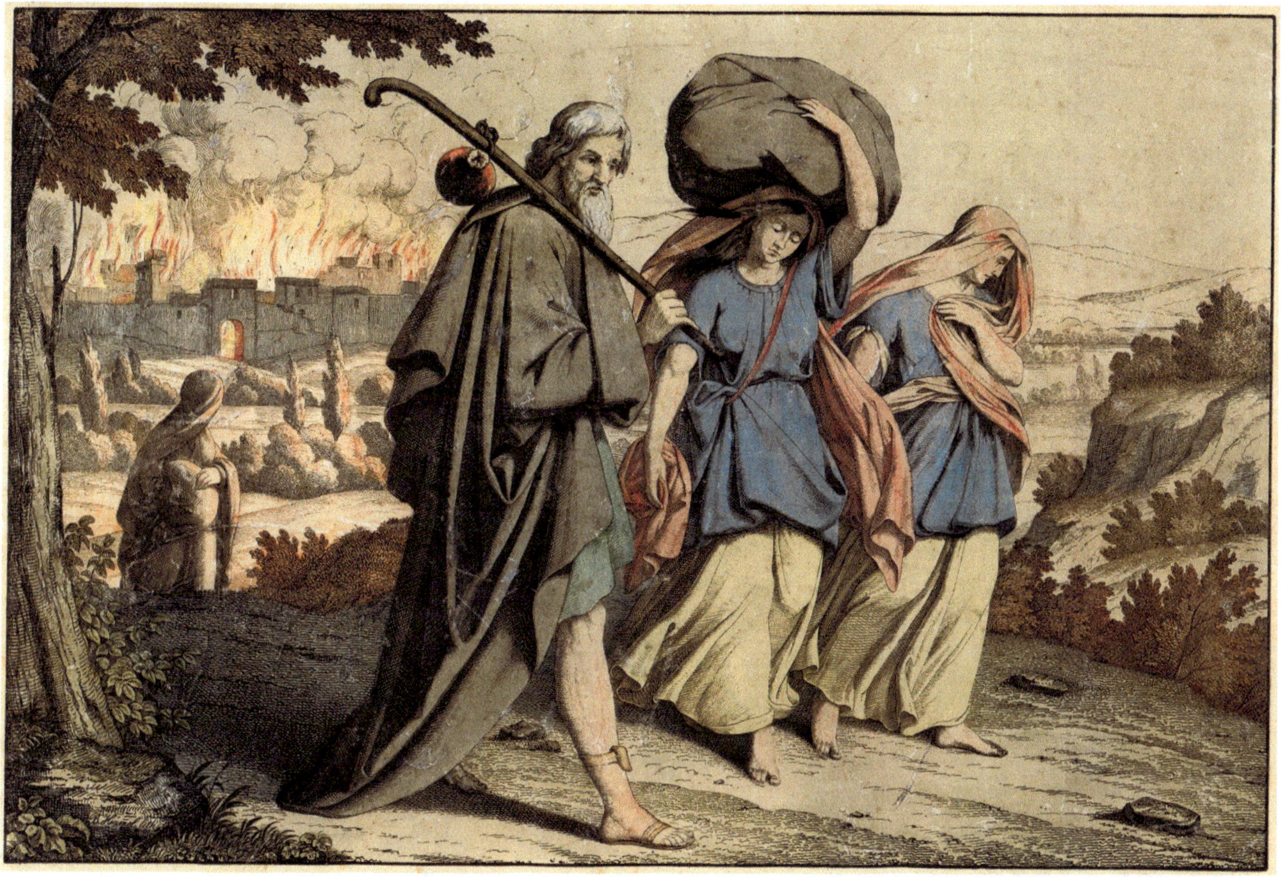

Fig. 5.1 Lot flees from Sodom

The first image illustrating fire might be considered not so much fire from natural causes, but fire from supernatural causes: this is the biblical story of the destruction of Sodom and Gomorrah (Genesis 19) and the escape of Lot and his family (Fig. 5.1). But, as is the case for several other Bible stories, it is possible that behind a tale of supernatural effects, there were natural causes. In this case, it is well known that the location of the two cities, at the south end of the Dead Sea, was a place where naphtha and asphalt could be found at the surface; in fact, both commodities were sold in the market at Sodom for lighting and heating purposes. If these deposits were ignited by lightning, a meteorite, self-ignition, or any other cause (an earthquake has been suggested), the fiery destruction of the two cities could have been the result (see also Fig. 4.25).

The landscape in this folksy nineteenth-century image, published in a series under the title *Cottage Pictures from the Old Testament*, hardly resembles the actual area of the Dead Sea. The burning cities are visible in the distance, but the artist is more interested in the figures of Lot and his daughters. They are resolutely not looking behind them, as instructed by the angel who warned them to flee, unlike Lot's wife at the far left of the image. Unwilling to leave her old life entirely behind, she looked back at her old home and was turned into a pillar of salt.

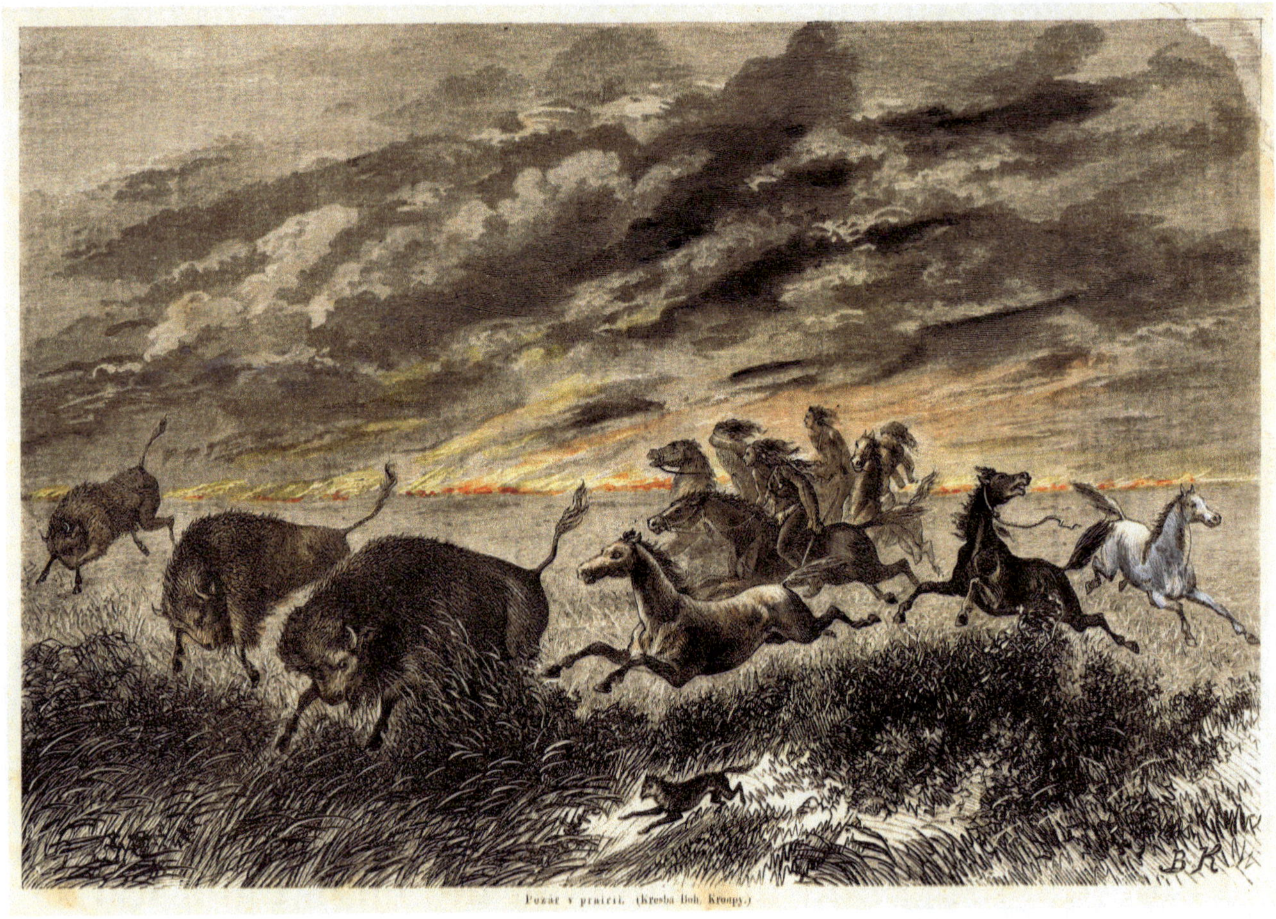

Fig. 5.2 A prairie fire in North America

Images of fire based directly on natural causes such as self-ignition or lightning are not frequent. The picture here (Fig. 5.2) shows an imagined scene of a prairie fire in North America. Terrified animals, horses, and buffaloes attempt to escape from the approaching flames; a wall of fire can be glimpsed on the horizon. The sketch for this xylographic print was drawn by B. Croup, and the image was published in the Czech illustrated weekly *Světozor* sometime in the 1890s. Probably the image was published to accompany a report of a specific prairie fire that was in the news, but it could have been intended for use in a more general article about the hazards of the American West.

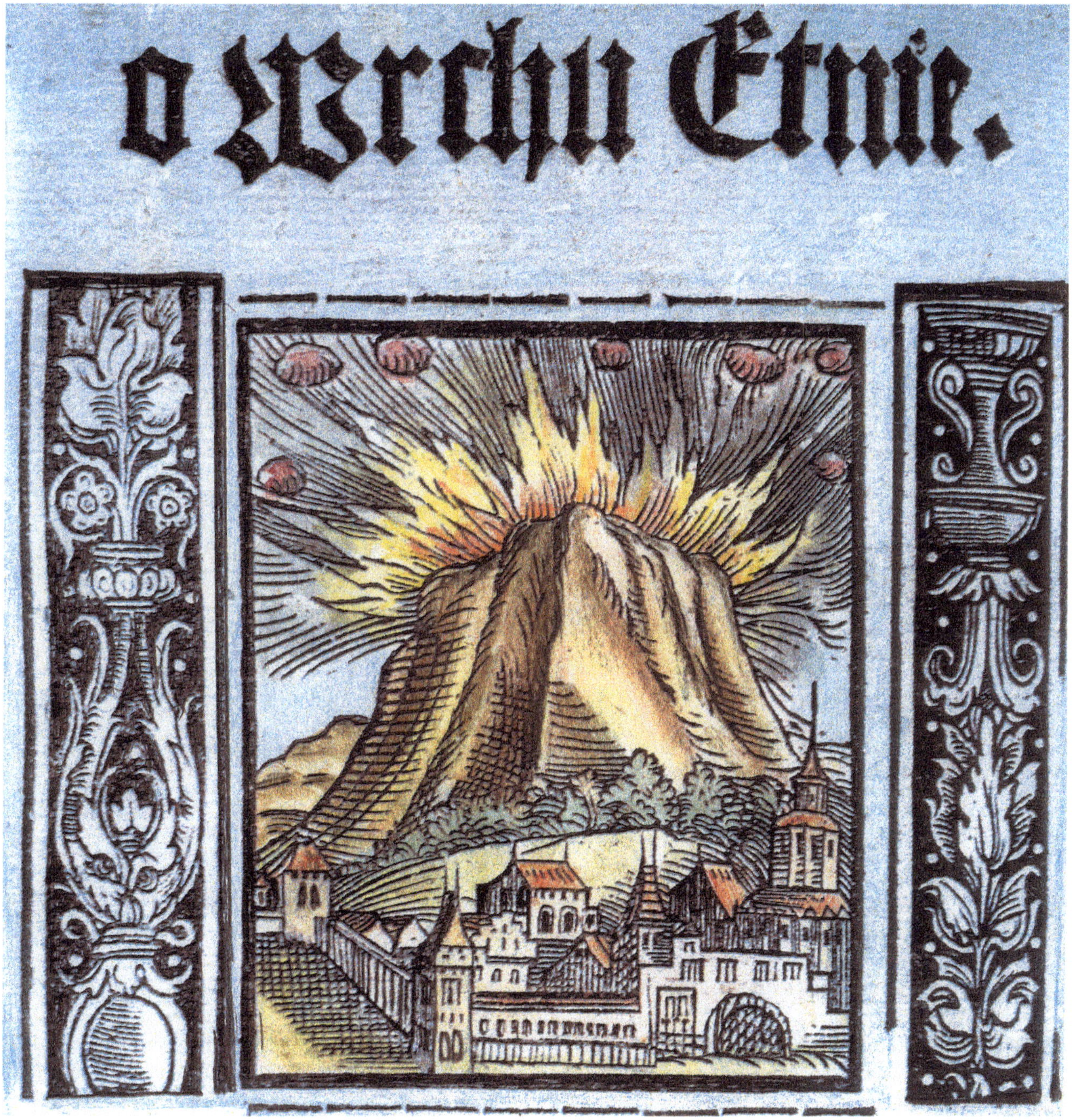

Fig. 5.3 Etna depicted in 1544

This illustration, titled *O wrchu Etnie* (which literally translates as *About Etna Hill*) comes from a book by Zikmund of Puchov, the sixteenth-century savant who adapted and translated Munster's *Cosmographia Universalis* (1544) into Czech. Zikmund's book was edited in Prague and published there in 1554.

The book was illustrated by a series of woodcuts, of which this is one (Fig. 5.3). It shows the town of Catania, Sicily, in the foreground and the "burning" Etna in the background.

The illustration, colored by a later hand, is not at all realistic; all of the distances and dimensions are incorrect (Etna is actually nearly 30 kilometers away from Catania and should appear much smaller). The use of stereotyped symbols in place of faithful images was quite usual in the sixteenth century. In the case of Sicily, few people from central Europe would have had a chance to discover the artist's errors. Also, of course, at that time the mechanics of volcanic eruptions were not understood, and a volcano seemed to be literally on fire, perhaps from the burning of underground coal seams. The flames shooting out from the top of the mountain are grossly unrealistic and far too big, but it does seem that the artist has heard about volcanic bombs, judging by the red objects flying upward.

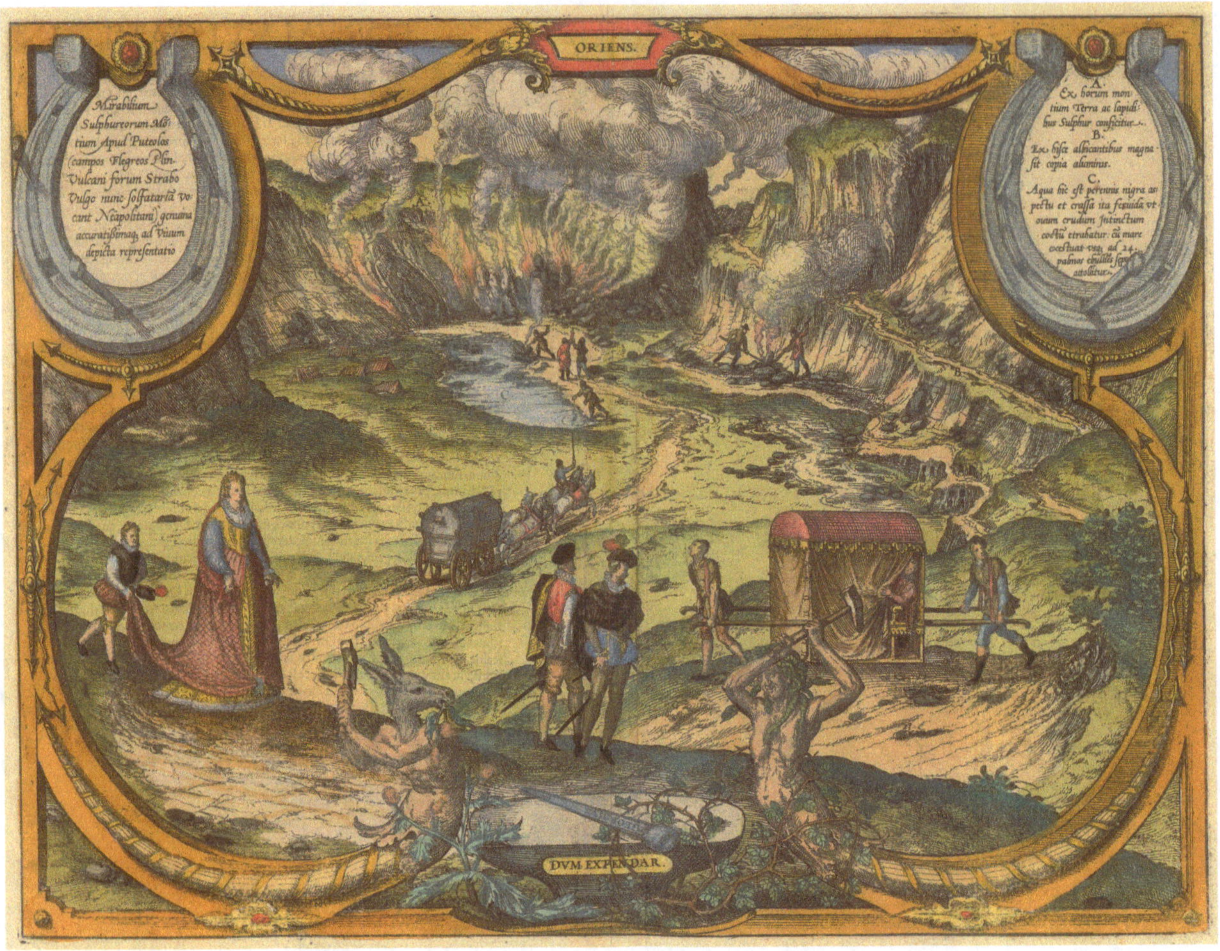

Fig. 5.4 The Phleghrean Fields, shown in 1581

Unlike the better-known Vesuvius, which frequently produced violent eruptions in historical times, the nearby volcanic complex known as the Phleghrean Fields, or the Forum Vulcani, has been relatively quiet. Solfatara is the best known of the 24 craters that pockmark the area around the town of Pozzuoli, just to the west of Naples, and has given its name to a type of volcano characterized chiefly by gas emissions. The most notable historical activity was the birth of Monte Nuevo on 28 September 1538, five kilometers west by northwest of Pozzuoli; the new volcano grew to a height of 150 meters in one week of nonviolent eruption. Otherwise, volcanic activity is limited to the emission of carbon dioxide and hydrogen sulfide, hot sulfurous and mineral springs, and changes in land levels.

Nevertheless, the possibility of a violent eruption in the future is a matter of concern, and the Campi Flegrei (which literally translates as "Burning Fields") are now carefully monitored by the Vesuvius Observatory. The craters seen today are the relics of a catastrophic eruption about 37,000 years ago; the last major activity was about 12,000 years ago.

The picture shown here (Fig. 5.4) is an ornamental composition, published as a hand-colored copper engraving, created from an original sketch by the Flemish painter, draftsman, and engraver Joris Hoefnagel (1544–1600). It appeared in the famous six-volume, folio-sized Renaissance atlas of the cities and famous localities of the sixteenth-century world. The atlas was edited by Georg Braun and Franz Hogenberg; it is known by the title *Civitates Orbis Terrarum*; numerous editions were published between 1572 and 1624. The depiction shown here is titled *Forum Vulcani or Solfatara* and was published as folio 58 in the third volume of the *Civitates*, appearing first in 1581. Hoefnagel made his sketch for later engraving during his visit to Italy in the late 1570s, when he visited central Italy with his friend, the famous Flemish cartographer Abraham Ortelius.

In the image itself, the two figures (bottom center), appearing as part of decorative scrolls, are both smiths (one with an ass's head) hammering away at what may be a sword blade on an anvil. Solfatara was believed to be the home of the god Vulcan (Hephaestus in the Greek pantheon), who was the god associated with smithing. One notes also the two horseshoes worked into the design in the top corners, enclosing some explanatory text.

In the main part of the image, while one can see sulfurous smoke in the distance, the eye is drawn to the various figures loitering in the foreground—clearly some period tourists who find the volcanic phenomena worthy of a visit and who are clearly unperturbed by any thoughts of danger. The lady on the left is clearly wealthy, since she can afford a page to hold her dress off the ground. The one on the right has gone one better, travelling from Pozzuoli in a sedan chair. Some historians have speculated that she may be here to take the sulfurous waters or bathe in them to improve her health or even regain some of her lost youth. The cart in the middle distance is no doubt laden with sulfur, a reminder that tourism was not the only economic activity in the caldera.

Hoefnagel's composition evidently proved popular; in the following centuries, one can find many copies by different hands, often slightly rearranged.

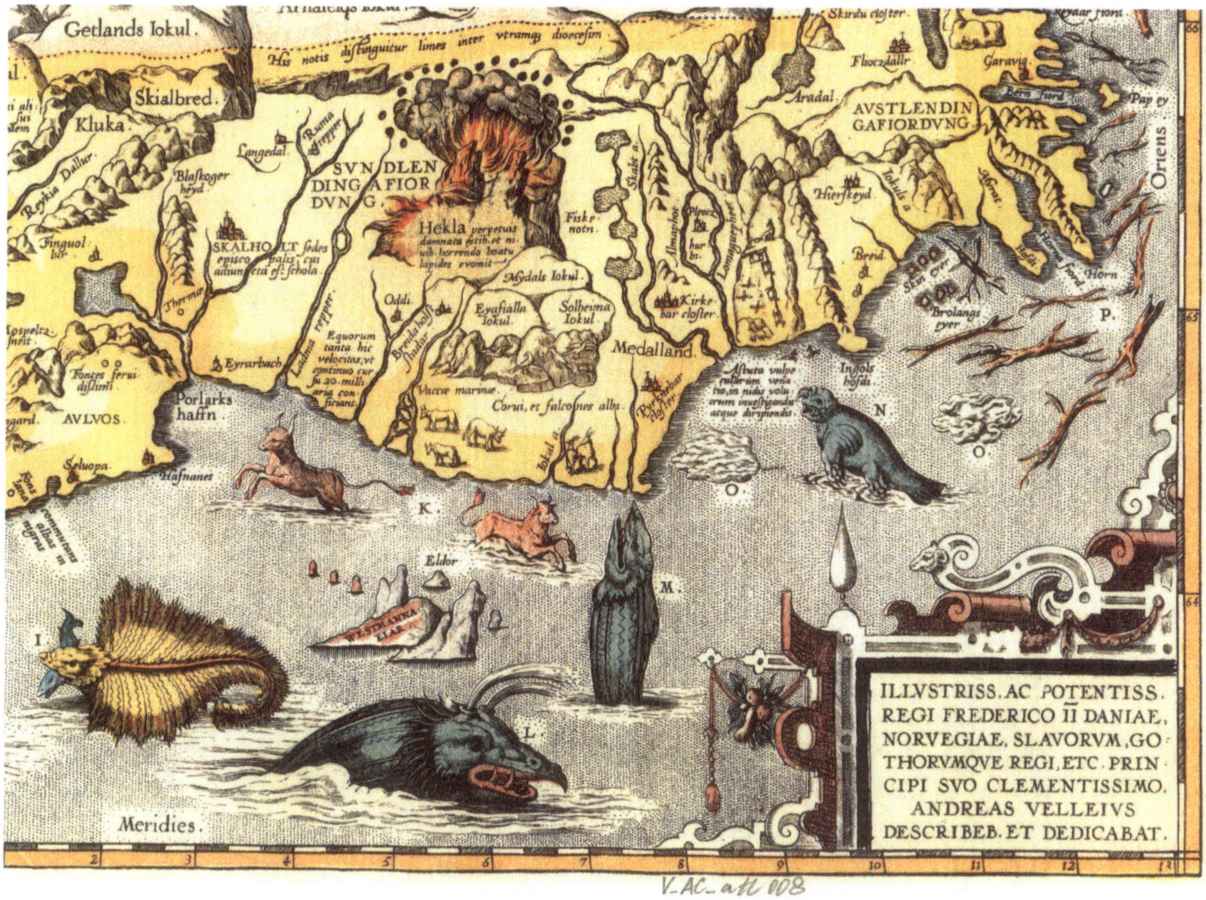

Fig. 5.5 Hekla, mapped in 1586

In terms of geological formations, Iceland is relatively young. Consult a geological map of the country and one finds that, despite the varied colors, every part of it is one sort of basalt or another. The age of geological specimens collected on the island indicate that Iceland was formed slightly less than 20 million years ago, associated with the so-called Icelandic volcanic plume and the rifting that opened up the Atlantic Ocean as North America and Europe drifted apart. There are numerous active volcanoes across the island, accompanied by seismic activity and geothermal phenomena such as hot-water geysers. Geothermal energy is so plentiful that electricity is virtually free, allowing cultivation of such unlikely crops as tomatoes, despite the cold climate.

Hekla is one of the most prominent Icelandic volcanoes, along with Katla, Eyjafjallajökull, Askja, Laki, and Grímsvötn. The image here (Fig. 5.5) is part of a map originally drawn by the Icelandic bishop Gudbrandur Thorlaksson and was edited by the Flemish cartographer Abraham Ortelius, who prepared the original sketch for publication in 1586, in his famous cartographic series, or atlas, titled *Theatrum orbis terrarium* (*World Theater*). The pictorial map shows true data combined with fantasies. Hekla is depicted erupting, accompanied by a note that reads, "Hekla, perpetually condemned to storms and snow, vomits stones under terrible noise." Or, as an English madrigal of the seventeenth century has it: "Thule, the period of cosmography / Doth vaunt of Hekla, whose sulfureous fire / Doth melt the frozen clime and thaw the sky / Trinacrian Etna's flames ascend not higher."

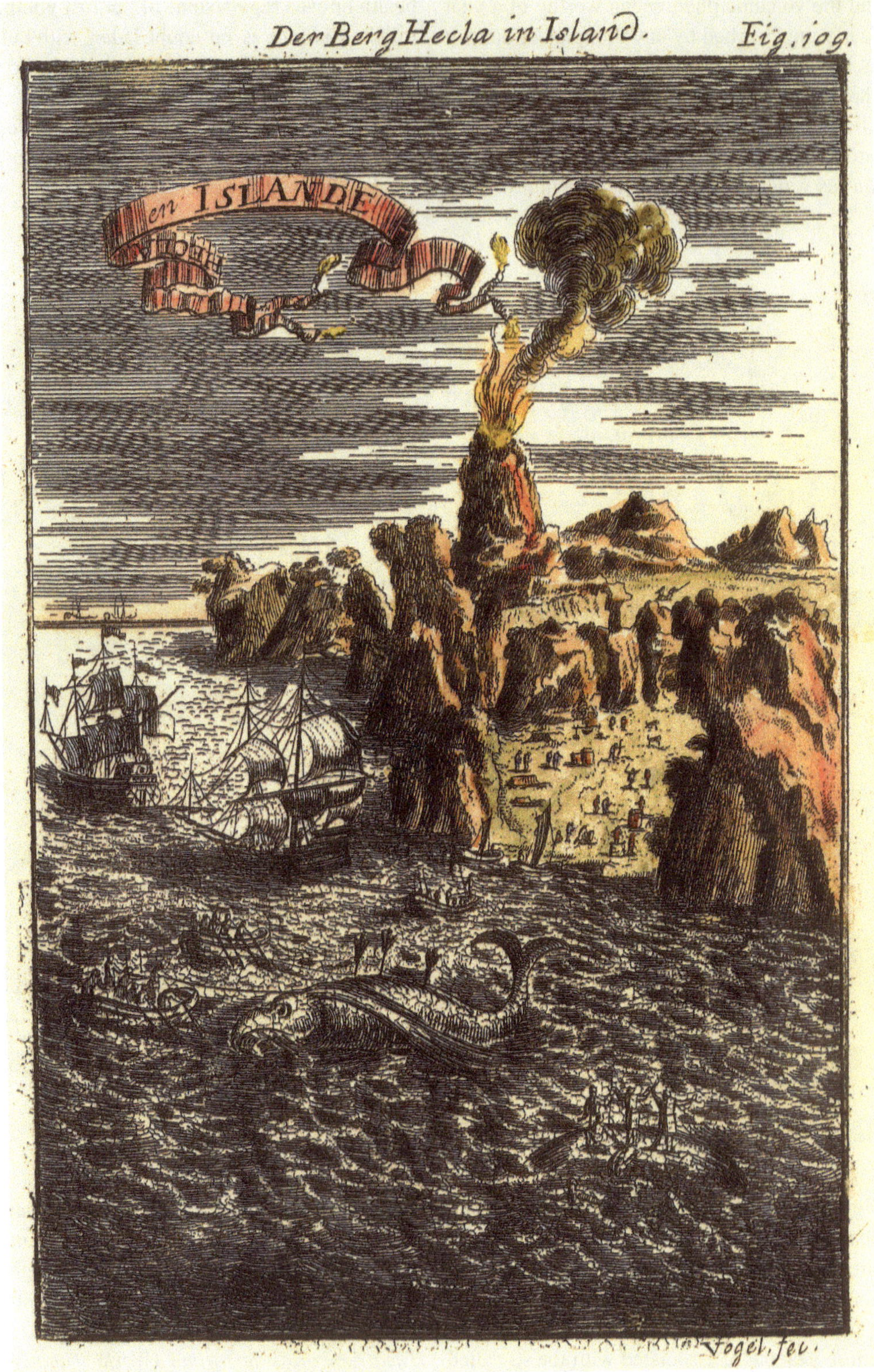

Fig. 5.6 An engraving of Hekla, circa 1700

"The period of cosmography" is where mapping comes to a full stop. At the turn of the eighteenth century, few continental Europeans visited this lonely, cold region with arctic flora, offering nothing but snow, ice, rocks, and volcanic eruptions. Certainly, the artist of this German copper engraving (Fig. 5.6) could not be bothered much to make his image correspond to the facts. Instead of showing Hekla as the mighty inland volcano it is, he depicts it as a candle-like structure on the coast, barely taller than the ships sailing by.

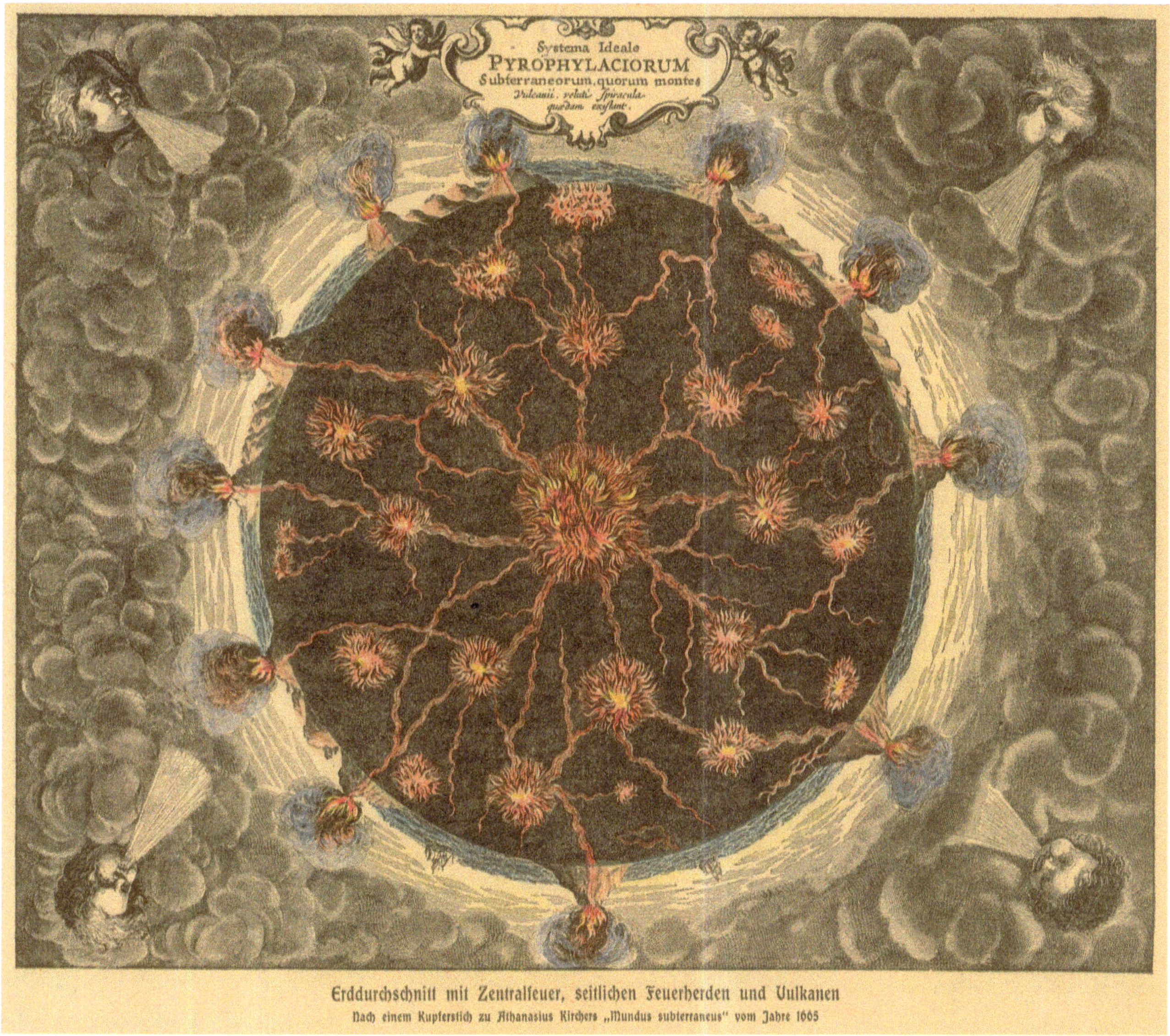

Fig. 5.7 Kircher's idea of the Earth's interior

In the introductory chapter of this book, we mentioned the contribution of the seventeenth-century scholar Athanasius Kircher to the advancement of studies of the Earth and its interior. His speculative ideas on how the planet's "internal fire" could reach the surface are depicted in the rather grand figure reproduced here (Fig. 5.7), in which eleven volcanoes distributed all around the globe are shown as being connected by a series of fiery passages leading down to a central reservoir. Of course, Kircher did not have any data on the Earth's interior at his disposal. But his imaginative scheme certainly provoked other seventeenth- and eighteenth-century thinkers (Immanuel Kant comes to mind) to consider what a model of the Earth's interior should look like.

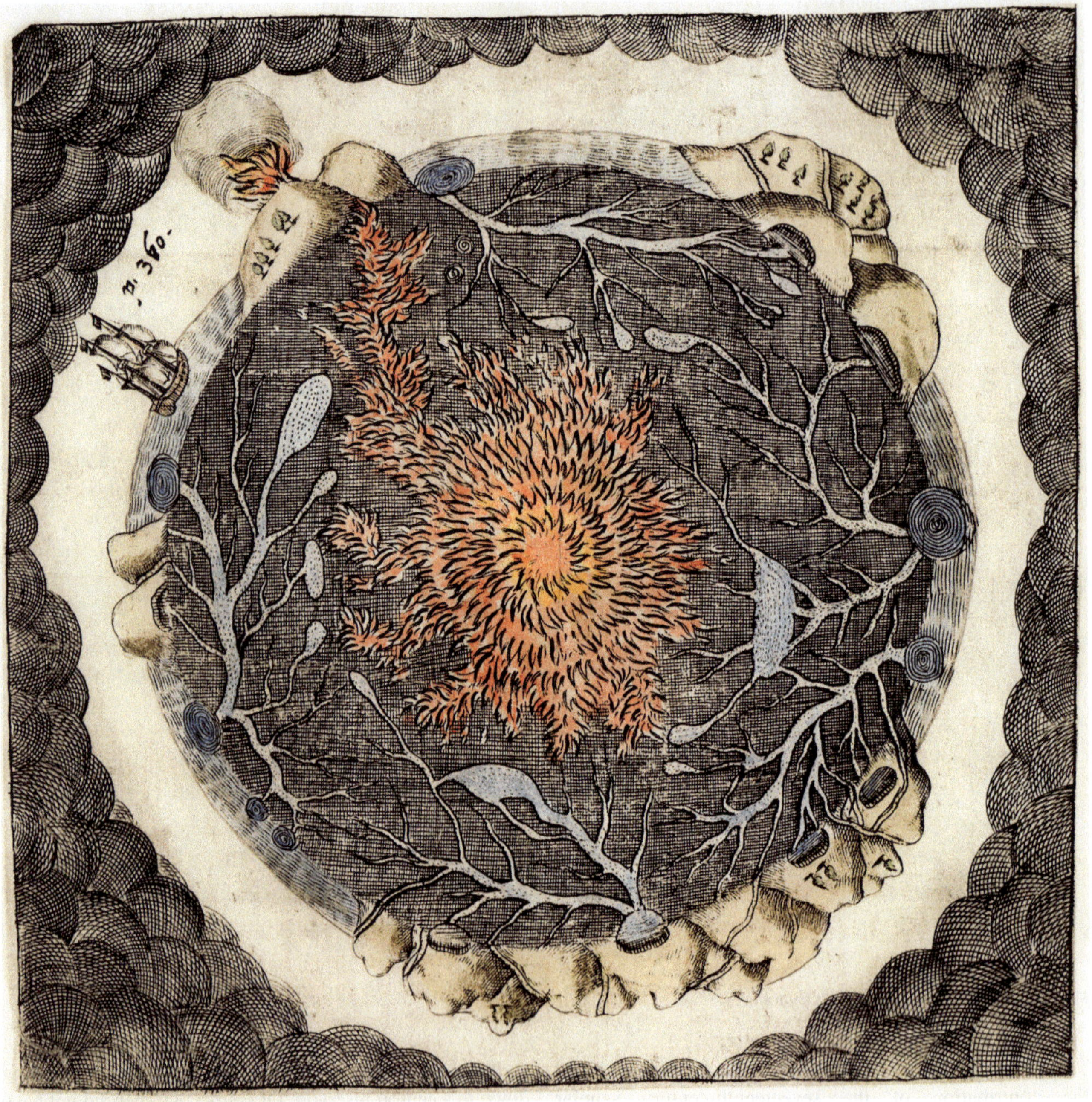

Fig. 5.8 Kircher's influence in the 1690s

To demonstrate how pervasive Kircher's ideas were in his time, the picture here (Fig. 5.8), published in the 1690s, is a poor version of the previous image. It is little more than a simplified torso, incorrectly altered in many details. Only one fiery passage leads to a volcano, and the other underground tunnels seem to hold water instead. Nevertheless, whoever drew it showed the strong influence of Kircher.

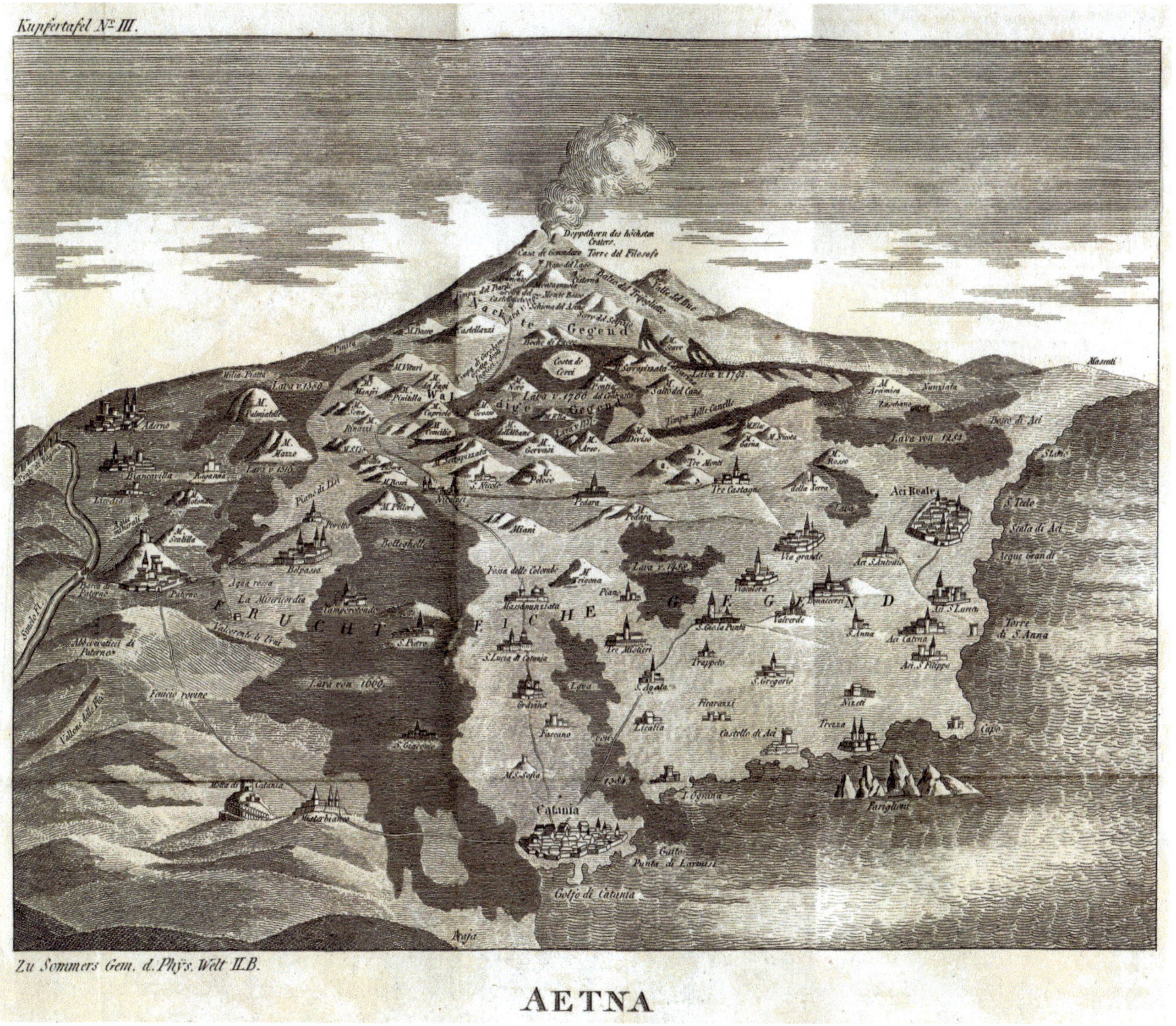

Fig. 5.9 An eruption of Etna in 1669

In contrast to the previous picture of Etna and Catania (Fig. 5.3), the image here (Fig. 5.9) is relatively accurate. It comes from a book titled *Gemälde der physischen Welt* (*Picture of the Physical World*), prepared and published in Prague by J.G. Sommer in 1821. It shows diagrammatically the effects of the major eruption of 1669; the different lava flows and new side craters are shown and even dated. Attention should be drawn to the foot of the largest lava flow, which is shown as just skirting Catania. This is exactly what happened; the town walls deflected the lava away from the main town, and it ran into the sea instead.

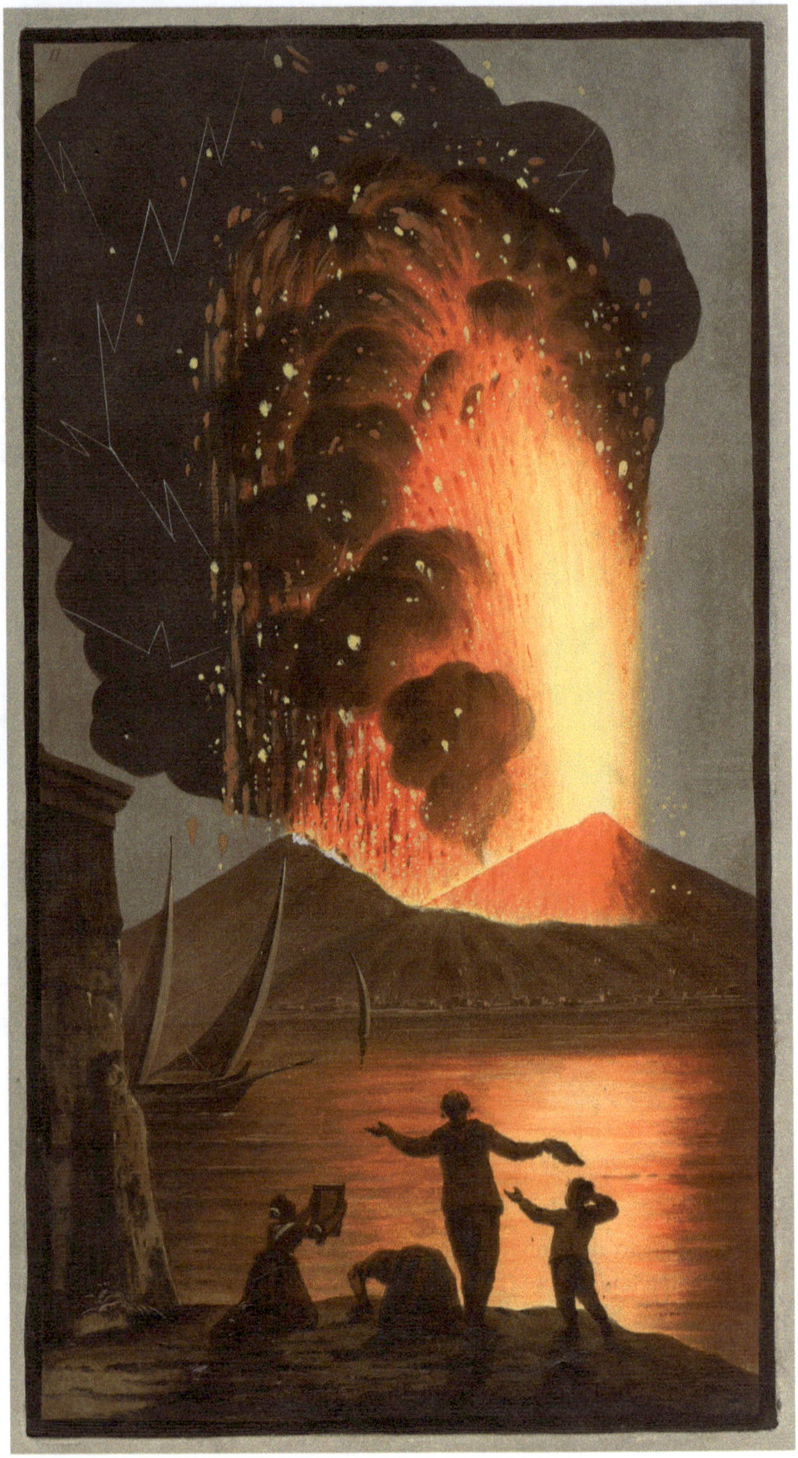

Fig. 5.10 Vesuvius in 1779

Probably the most important European volcano, even more so than Etna, is Vesuvius. A long series of attractive pictures of Vesuvius in eruption were made by the artist Peter (or Pietro) Fabris (1740–1792), of which we reproduce two here. Fabris prepared 54 large-sized contour prints, which were then hand-colored, for a two-volume work by Sir William Hamilton (the British Envoy to the Spanish Court at Naples and an enthusiastic amateur volcanologist). This work, titled *Campi Phlegraei: Observations on the Volcanoes of the Two Sicilies as They Have Been Communicated to the Royal Society of London by Sir William Hamilton*, was published in 1776. A later supplement was devoted to the great eruption of August 1779, and this spectacular image (Fig. 5.10) depicts the night of 8 August 1779. Hamilton sought to show that volcanoes were not just a force for destruction but also contributed to a region's prosperity, thanks to rich volcanic soils.

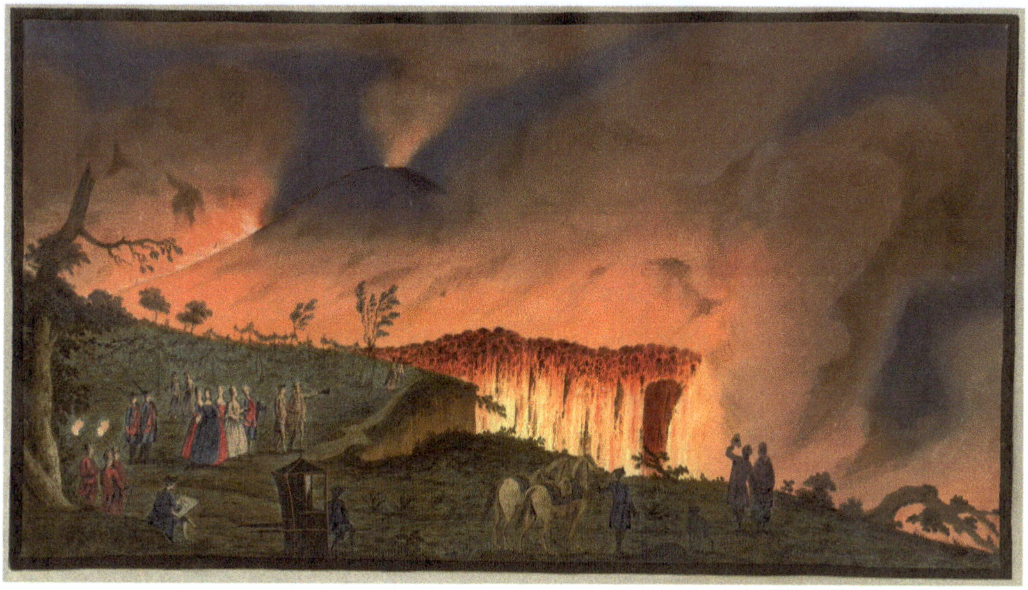

Fig. 5.11 The 1771 eruption of Vesuvius

This second image (Fig. 5.11) from Hamilton's book is a night view dated 11 May 1771. Hamilton is seen here arranging some volcano tourism; he is shown escorting the King and Queen of the Two Sicilies (and some of their retinue) to witness the drama of an eruption at close range. As can be seen, at this point the lava is falling over a cliff before flowing down toward the town of Resina. The plate's description states that the original drawing was taken that night on the spot; in fact, in the bottom left corner, Peter Fabris has included himself, busy sketching the scene. Many of Fabris's pictures include Hamilton as a figure, usually identifiable by his red coat. Hamilton is more celebrated today as a volcanologist than as a diplomat but perhaps is even better known as the husband of Emma Hamilton, the mistress of Admiral Horatio Nelson.

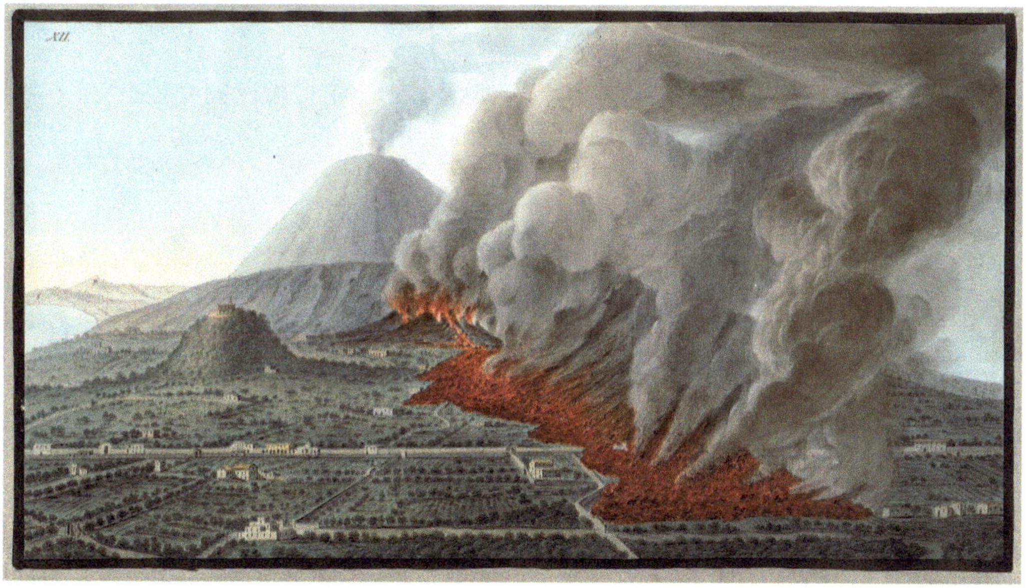

Fig. 5.12 The 1761 eruption of Vesuvius

This is another image from Hamilton's book, this time showing an earlier eruption of Vesuvius, which occurred between 23 December 1760, and 8 January 1761 (Fig. 5.12). Hamilton had this to say about it: "The object of this plate is to show, that those who have asserted that the seat of the fire is always towards the summit, or not lower than the middle of the volcano, have been very ill informed." The lava is clearly issuing from an opening almost at the base of the mountain, running down toward the Bay of Naples and setting fire to all it passes.

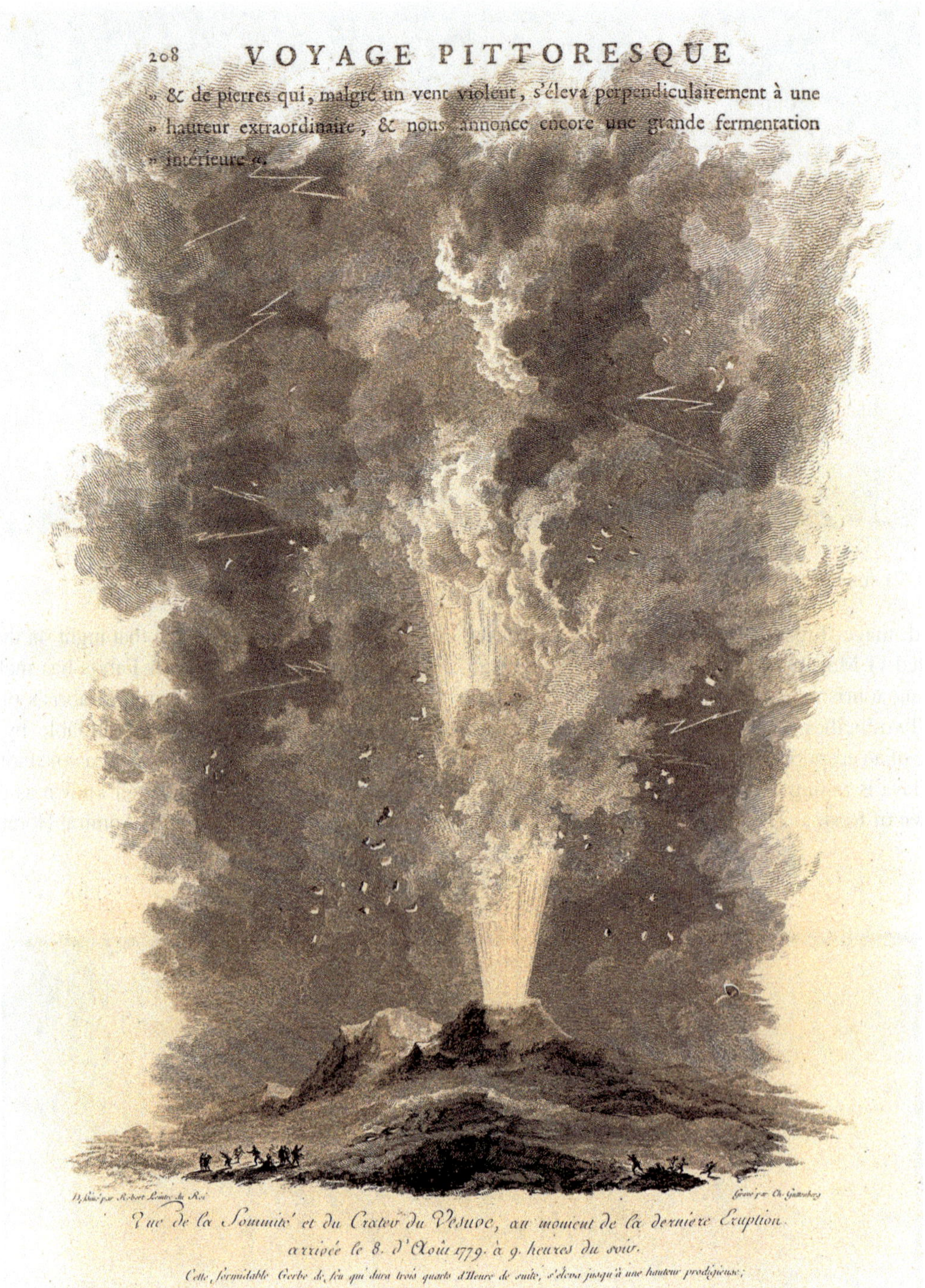

Fig. 5.13 Vesuvius, visited during the 1779 eruption

Jean-Claude Richard, the abbot of Saint-Non (1727–1791), was a French engraver, draftsman, and philosopher, and was a member of an intellectual and artistic group that included such famous names as Fragonard, Franklin, Rousseau, D'Alembert, and Voltaire, among others. In his thirties, he made a trip to Italy, and in 1781 he published his recollections of the trip in a work titled *Voyage Pittoresque ou Description des Royaumes de Naples et de Sicile*. This chapter is devoted to descriptions of Vesuvius and its eruptions, and the picture here (Fig. 5.13) is intended to depict the summit and crater of the volcano at the time of the eruption of 8 August 1779 (at 9 p.m. in the evening). The picture appears on page 208, and the eruptive column of ash actually extends to the top of the page, obscuring the text and even the chapter header!

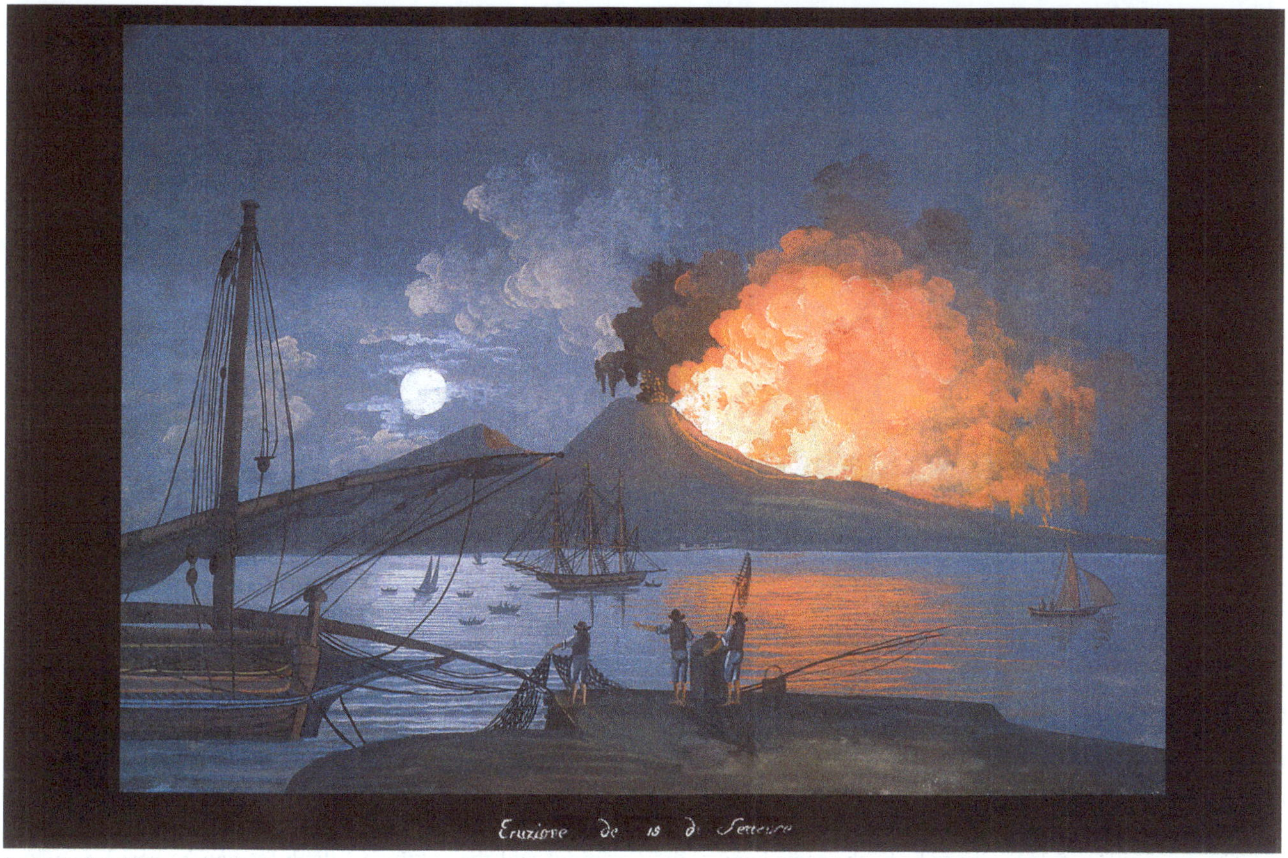

Ignoto. Eruzione del Vesuvio del 18 settembre
Museo dell'Osservatorio Vesuviano

Unknown author. Vesuvius eruption on September 18
Museum of the Vesuvius Observatory

Fig. 5.14 A souvenir view of Vesuvius in 1790

The second half of the seventeenth century proved to be a period of considerable eruptive activity at Vesuvius, and it became part of the "Grand Tour" that wealthy people from northern Europe were expected to make at least once in their lives. The awesome forces of nature made manifest in fountains and streams of lava could be viewed safely from a distance, usually across the bay, and could be counted on to be particularly spectacular at night. This also gave rise to a local industry for hack artists, churning out stylized paintings for sale to tourists. The elements tend to be similar: the volcano viewed from across the Bay of Naples, with boats and local fisherfolk in the foreground and the mountain in the distance, spewing ash and flaming lava, and the whole scene lit by moonlight. Most probably, the eruptive effects are exaggerated to make a more dramatic picture to please the buyer, who would take it home, where his souvenir could be proudly displayed on the walls of his family mansion. The picture here (Fig. 5.14) is a typical example, and the original is in the Museum of the Vesuvian Observatory (reproduced here with kind permission). It is not entirely clear which eruption this is intended to be; the caption only gives the date as 18 September, without a year. Probably, it was in 1790, when repeated lava flows took place between 15 September and 20 October.

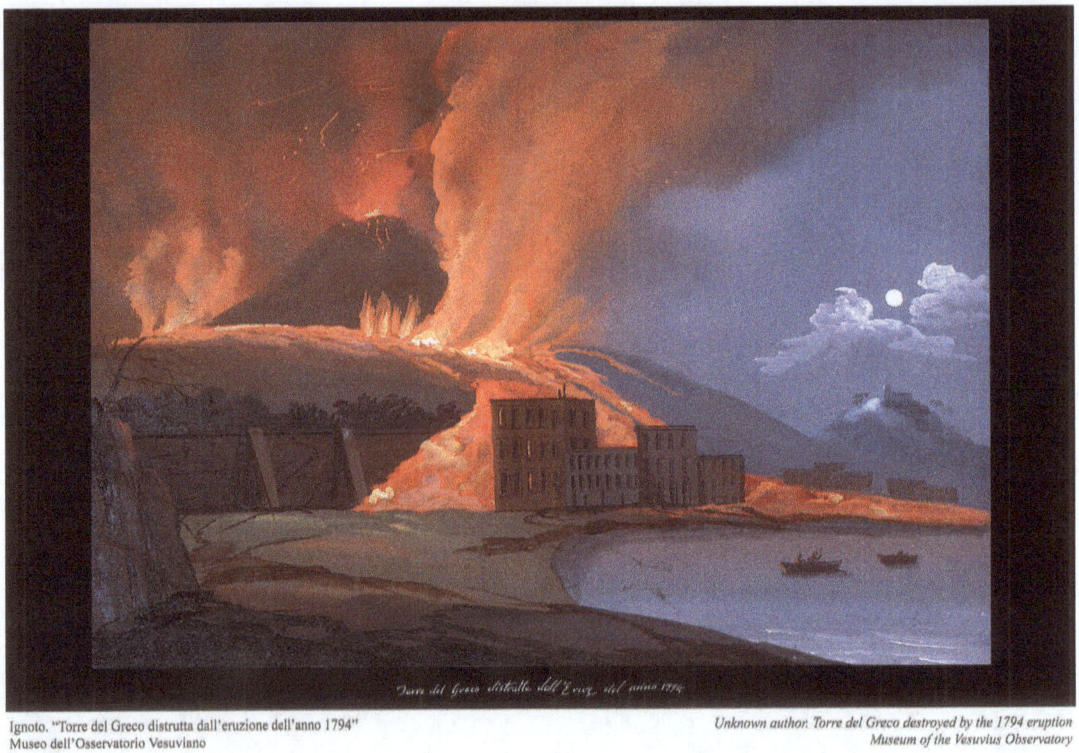

Fig. 5.15 Torre del Greco, destroyed in 1794

This image (Fig. 5.15) looks to be by the same hand as the previous one and is also from the collection of the Museum of the Vesuvian Observatory. This time, there is no doubt that the eruption is that of 1794, an eruption involving both the summit and the southwest flank (as can be seen), in which the village of Torre del Greco was destroyed by the flank lava flow, the center being buried by a layer of lava ten meters deep.

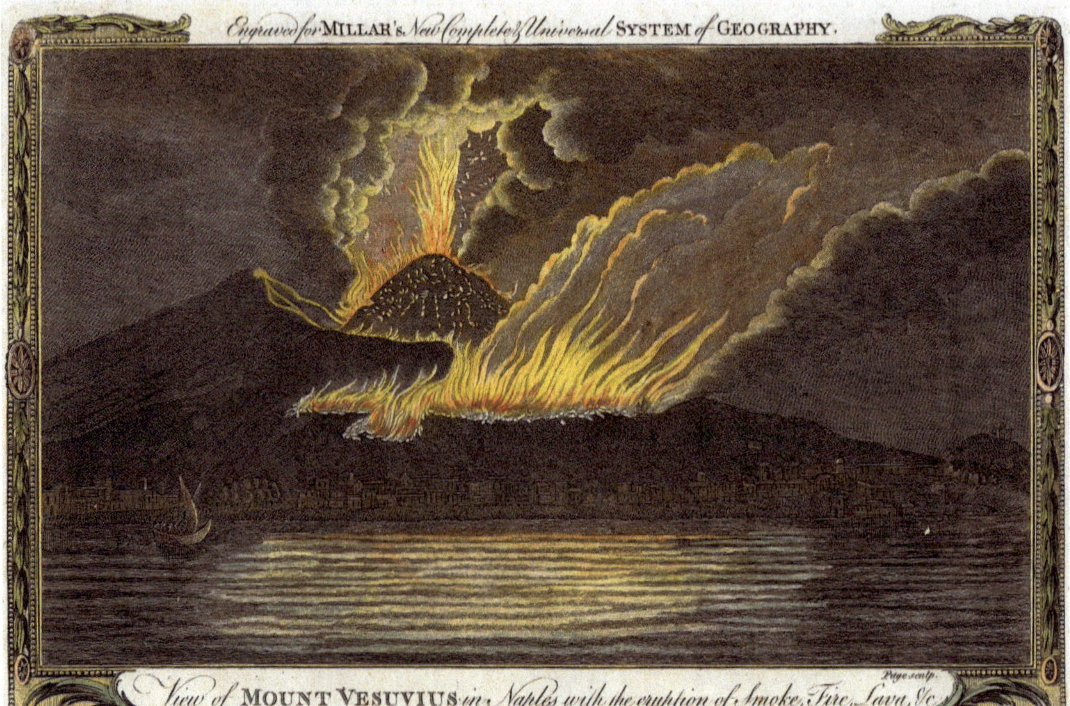

Fig. 5.16 An image of Vesuvius, designed to be backlit

The image shown here (Fig. 5.16), again of Vesuvius, is interesting from a technical point of view, being a copper engraving, colored by hand, and specially printed on glassine paper, so that while in the daytime it could be viewed in natural light, at night it could be viewed illuminated from behind with a candle or lamp. It was prepared originally for a work titled *The New and Universal System of Geography* by George Henry Millar, published in 1782. The title page boasts that "The Whole [is] embellished with upwards of ONE HUNDRED and TWENTY capital Engravings, being the most elegant Set of Copper Plates ever published in a Work of this Kind ..." The original of the view here is another picture from Sir William Hamilton's book and shows the eruption of 20 October 1767. The engraver has greatly simplified the image, leaving out all of the foreground detail.

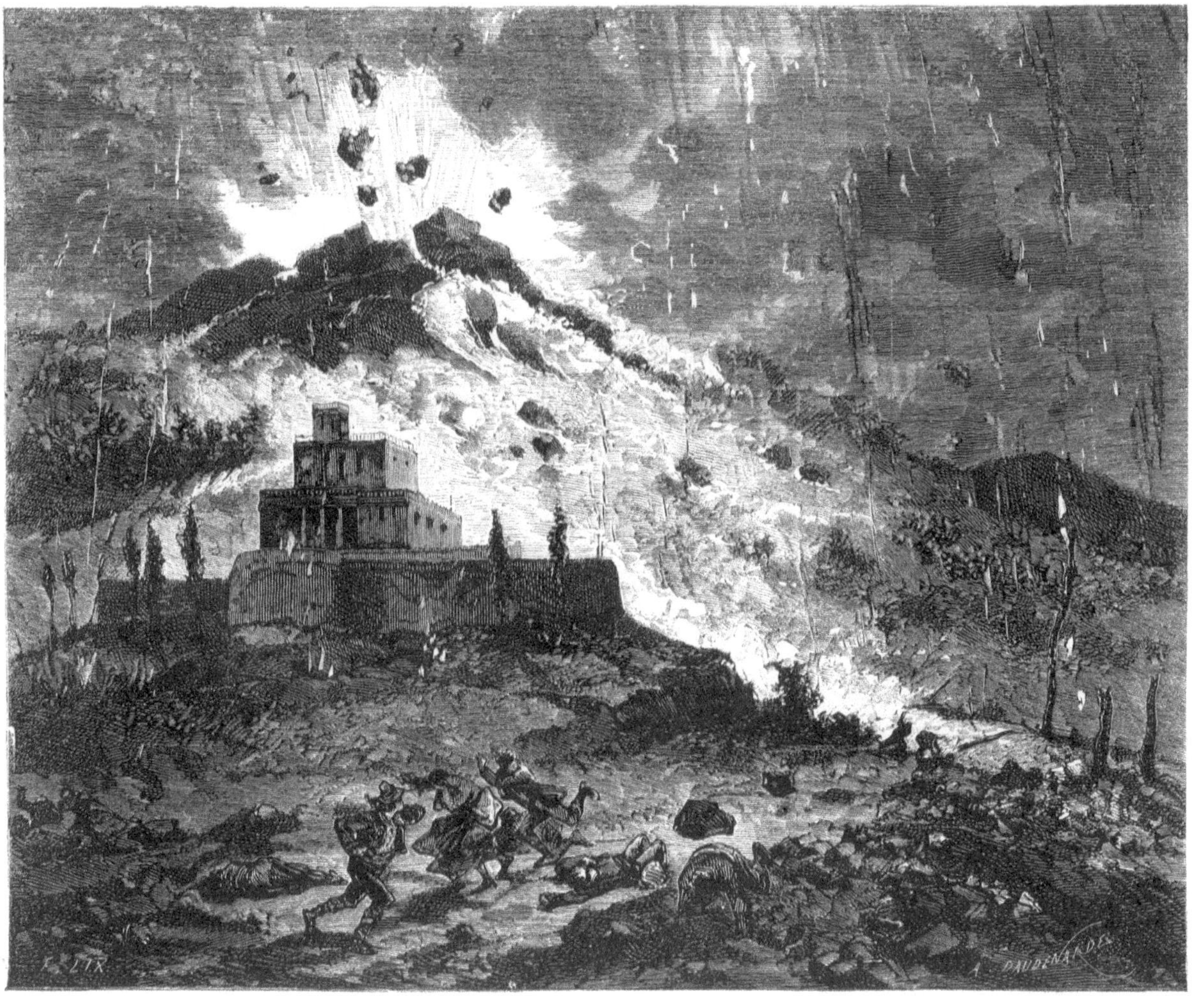

Der Ausbruch des Vesuvs: Die Oeffnung eines neuen Kraters, vom Observatorium aus gesehen.

Fig. 5.17 The Palmieri Osservatorio Vesuviano

The increasing number of visitors to Vesuvius—naturalists as well as tourists—in the earlier part of the nineteenth century prompted the founding of the Osservatorio Vesuviano in 1841. Fourteen years later, in 1855, Professor Giuseppe Palmieri was appointed as the director of the Observatory. For nearly 20 years he carried out fundamental seismological research into the seismic manifestations of the volcano, the aim of which was to find the relationship between the timing of volcanic earthquakes and subsequent eruptions. He monitored seismic activity by means of a seismograph of his own design, the first ever to incorporate a clock mechanism for recording the time of shocks accurately. Today the observatory is named *Palmieri Osservatorio Vesuviano* after him. In the picture here, the observatory is shown in the foreground, while behind it is a grossly exaggerated volcanic eruption, with a veritable sea of lava coursing down the mountainside, while huge chunks of rock are tossed into the sky from the summit (Fig. 5.17). The anonymous xylographic print was published after the mid-nineteenth century.

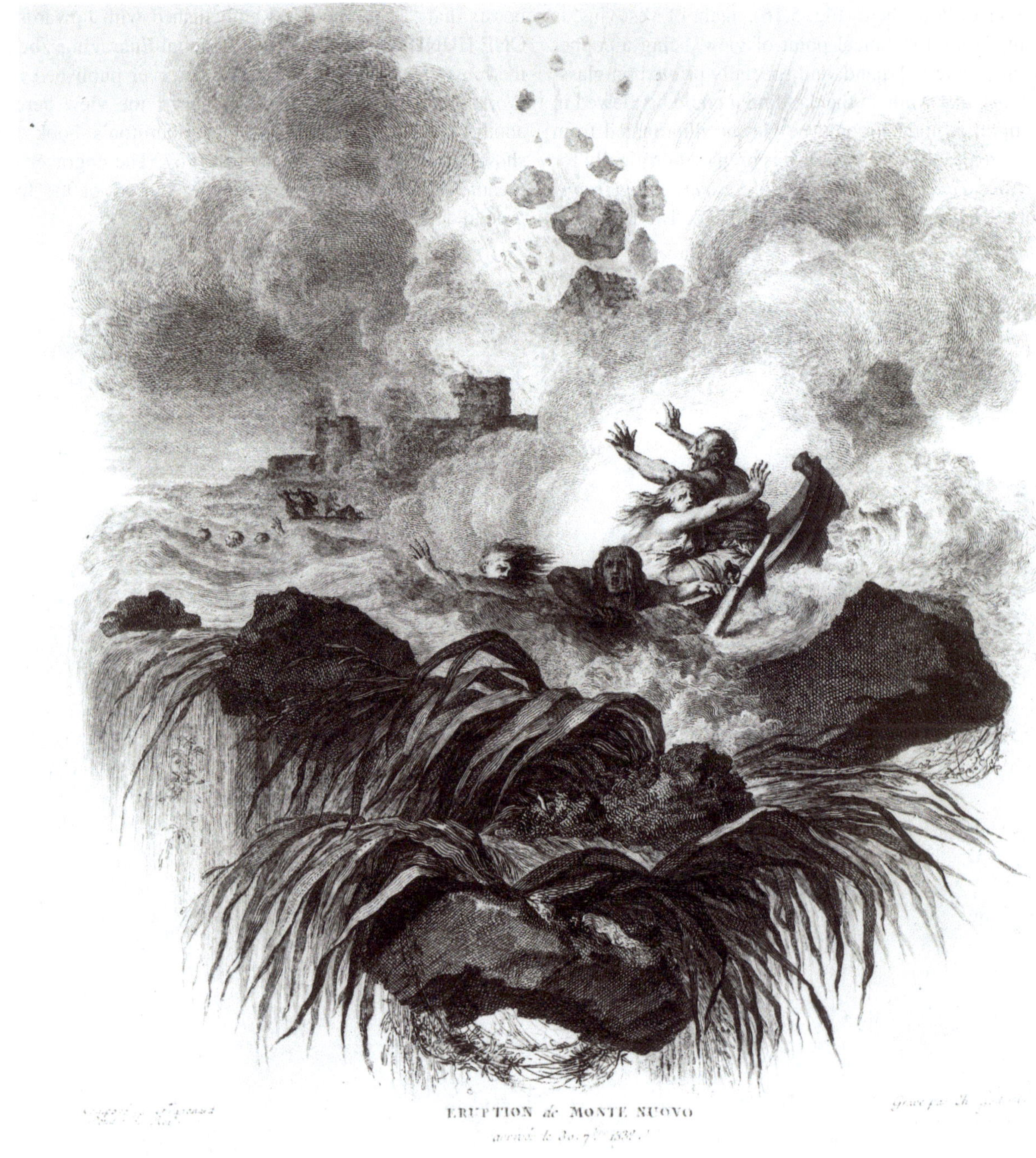

Fig. 5.18 Monte Nuovo, 1538

The birth of a new volcano on what was previously flat ground or shallow sea has always exerted a fascination not only for naturalists but for the common public too. One of the earliest recorded instances was the eruption that formed Monte Nuovo in 1538 (starting on 30 July), which has already been mentioned in the discussion above of the Phlegrhean Fields near Pozzuoli. A famous contemporary depiction of the event has previously been published by Kozák and Čermák (2010); instead, we show here a less well-known, much later image of a completely fantastic nature (Fig. 5.18). It is not entirely clear what is happening, except that the group of terrified peasants seem to be in considerable trouble as the volcano appears to be bursting out under their feet. The artist responsible seems to have been none other than Jean-Honoré Fragonard, better known for rococo paintings of elegant young women; if so, the work was probably conceived in 1760, when Fragonard toured Italy with fellow painter Hubert Robert, and it was subsequently engraved by Charles Guttenberg. The work has also been attributed to Jean-Claude Richard de Saint-Non, who met Fragonard in Rome.

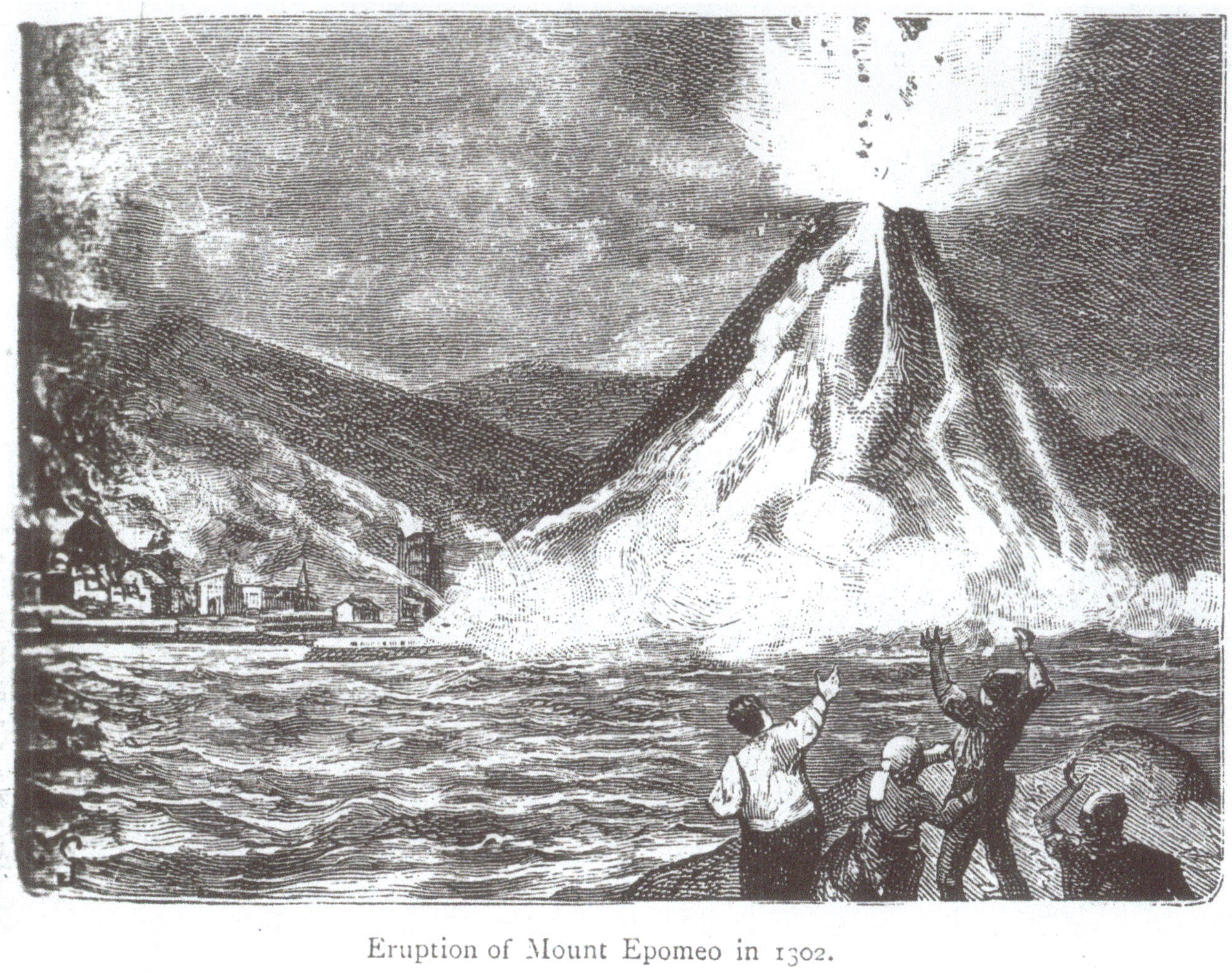

Fig. 5.19 An imagined eruption of Monte Epomeo

The island of Ischia, a pleasant tourist destination at the western limit of the Bay of Naples, is, so far as geological hazards are concerned, better known for its earthquakes than its volcano. The earthquake of 28 July 1883, destroyed the town of Casamicciola, with over 2000 deaths (Figs. 2.28, 2.29, and 2.30), and even a very modest shock in August 2017 killed two and injured dozens. Ischia's volcano, Monte Epomeo, attracts less attention. In fact, the whole island of Ischia is a volcanic complex, and the name Monte Epomeo actually applies to the highest point (789 meters tall). The last eruption was back in 1302 and consisted of a lava flow and spatter cone from the crater of Arso, on the east side of Monte Epomeo. The picture here, from Arnold Boscowitz's 1890 book on earthquakes, purports to show the 1302 eruption but is obviously a work of pure imagination (Fig. 5.19). For a start, it shows a volcanic cone directly on the coastline; Arso is actually about two kilometers inland.

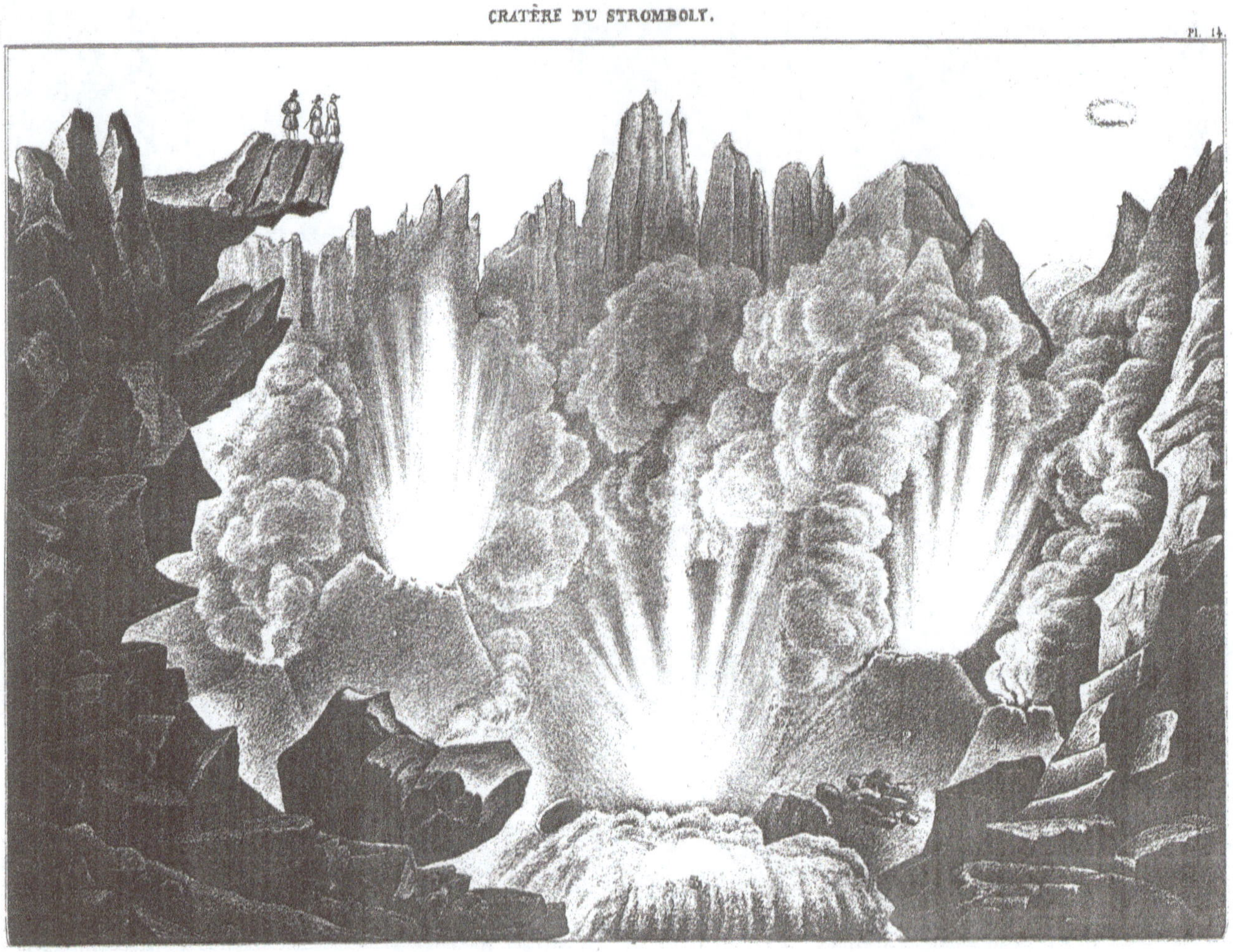

Fig. 5.20 The crater of Stromboli

One of the obscure byways of the history of volcanology is the work of the Count de Bylandt-Palstercamp. His *Résumé Préliminaire de Lóuvrage sur la Théorie des Volcans* in 1833 was followed by 1835 by his *Théorie des Volcans*. In these works he propounded the idea that magma and gas were propelled out of volcanoes by a mysterious magnetic force. Not surprisingly, this idea did not catch on. He attempted to illustrate his theory with cutaway views of volcanic craters—this is one from his second book (Fig. 5.20). The volcano in question is Stromboli, one of the islands of the Lipari Archipelago in the Tyrrhenian Sea, northeast of Sicily. The artist is unknown; it seems that he drew more on his imagination than on reality. Three observers on an exposed rock look down at the explosive eruption from their hazardous perch. Stromboli has given its name to a particular type of eruption involving ejection of incandescent cinders. The volcano itself is in a state of almost continual eruption; it is said that sailors can set their watches by it.

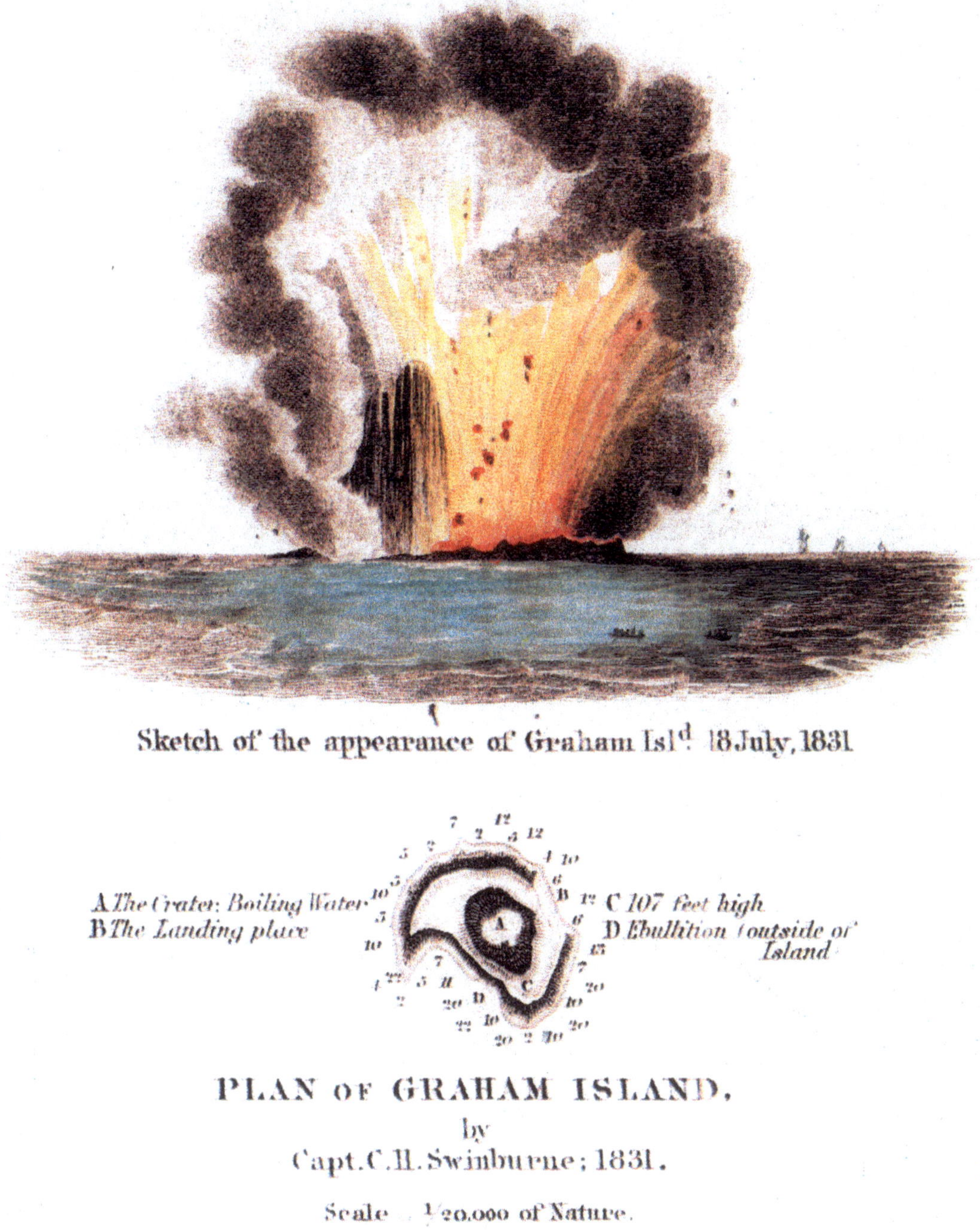

Fig. 5.21 Graham Island rises from the sea

When a new volcano appears at sea, sufficiently far from shore, it can cause some political problems. Such was the case in 1831, when a violent underwater eruption in the Mediterranean waters between Sicily and Tunisia created a new island. The eruptive column was first visible on 10 July, and by 17 July a small island had grown, which was quickly claimed by Great Britain and named Graham Island, after Sir James Graham, the First Lord of the Admiralty. Lying across critical sea routes, it was potentially a base of strategic significance. For the same reason, Naples was quick to claim the barren shore, naming it Ferdinandea after King Ferdinand II of the Two Sicilies. Lastly, the French arrived to lay claim to it also, giving it the name Île Julia, after the month of its appearance. Diplomatic wrangling followed but was terminated by the island itself, which sank back into the sea. By the end of the year, nothing could be seen of it. The image here (Fig. 5.21) is taken from A.K. Johnston's *Physical Atlas* of 1845.

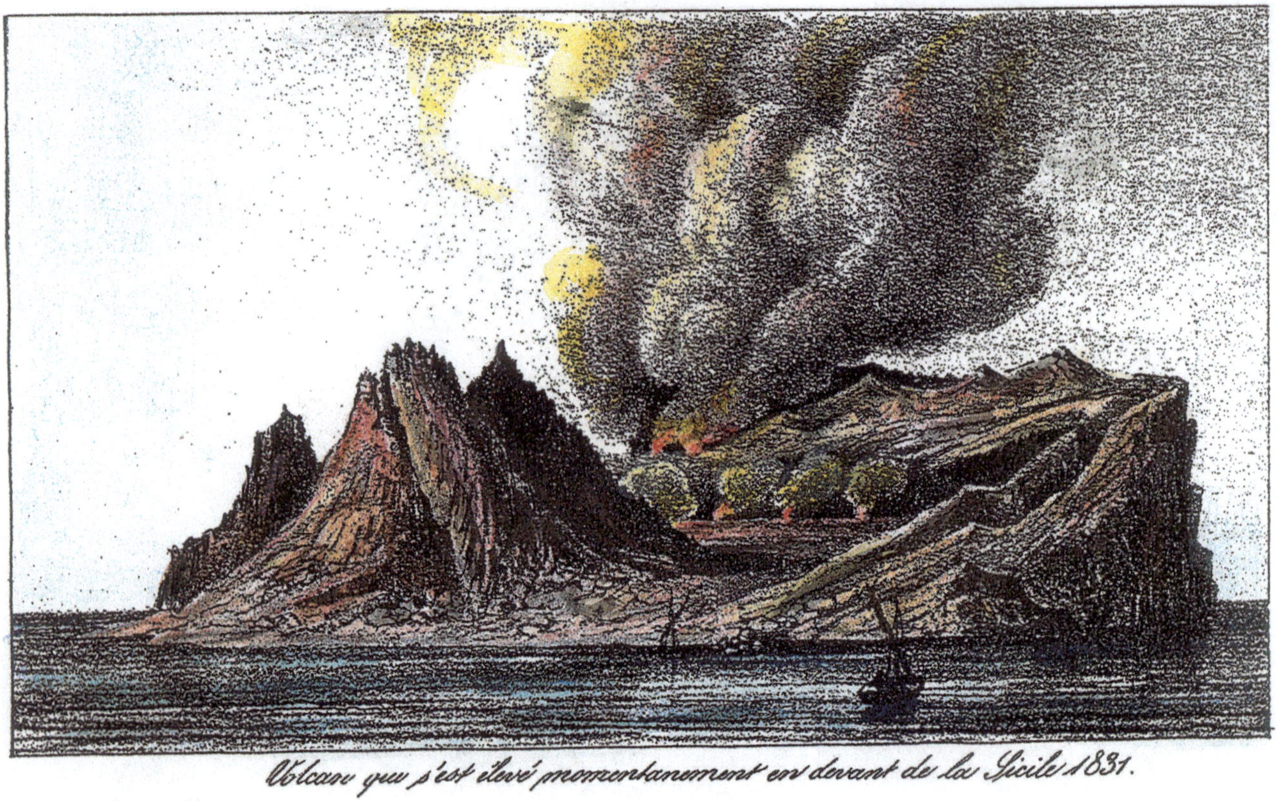

Fig. 5.22 Graham Island shortly before it sank

The colored lithograph shown here (Fig. 5.22) is based on a sketch made by a crew member on board the ship *Il Gustavo*, shortly before the island sank back beneath the waves. It appeared in the published version of the ship's log. At this time, the lava had already cooled down to such an extent that the island could be walked on safely. It became a tourist destination, and there was even talk of setting up a holiday resort. The appearance of the island in 1831 was not its first—it had been seen in Roman times—nor is it likely to be the last; Italian volcanologists speculate that a new eruption may be imminent, and the island may rise again.

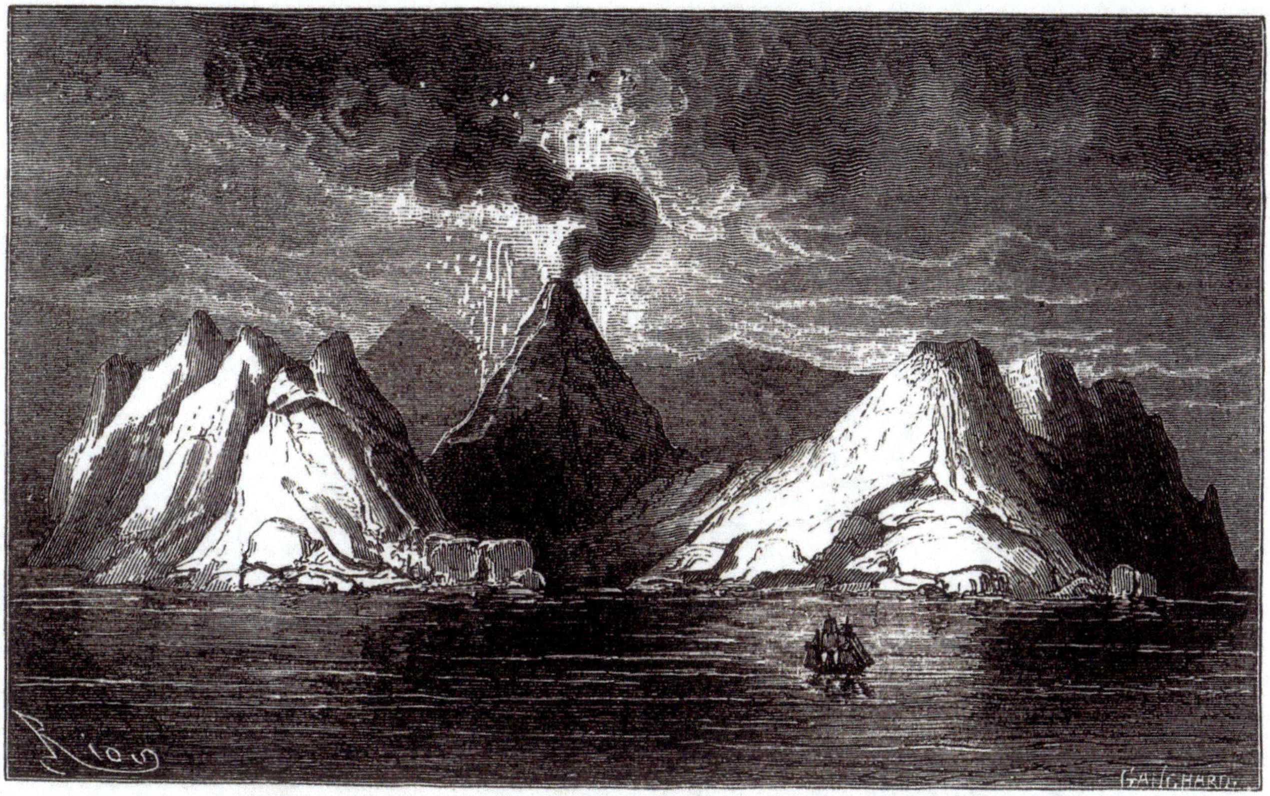

Fig. 5.23 Barren Island

The previous images have all been of European (mostly Italian) volcanoes, which are prominent in the record of illustrations, for cultural and historical reasons. But there are many more in other parts of the world. This is Barren Island, a volcanic island in the Bay of Bengal, located eastward of the Andaman Islands (Fig. 5.23). Barren Island is the upper part of a giant volcano, which rises from the seabed. Its caldera is flooded by the sea; at its center a central cone rises, which has been known to erupt at least five times in the eighteenth and nineteenth centuries, twice in the 1990s, and another six times so far in the present century. It lies along the north-south-trending volcanic arc extending between Sumatra and Burma (Myanmar) and is the only volcano on the edge of the Indian and Burmese tectonic plates that is known to have been historically active. As the name suggests, it is not a hospitable place. The illustration here is taken from *Volcans et Tremblements de Terre* (1868) by Elie Margolle and Frédéric Zurcher. The signatures at the bottom indicate the work of the artist Édouard Riou and the engraver Gauchard Brunier.

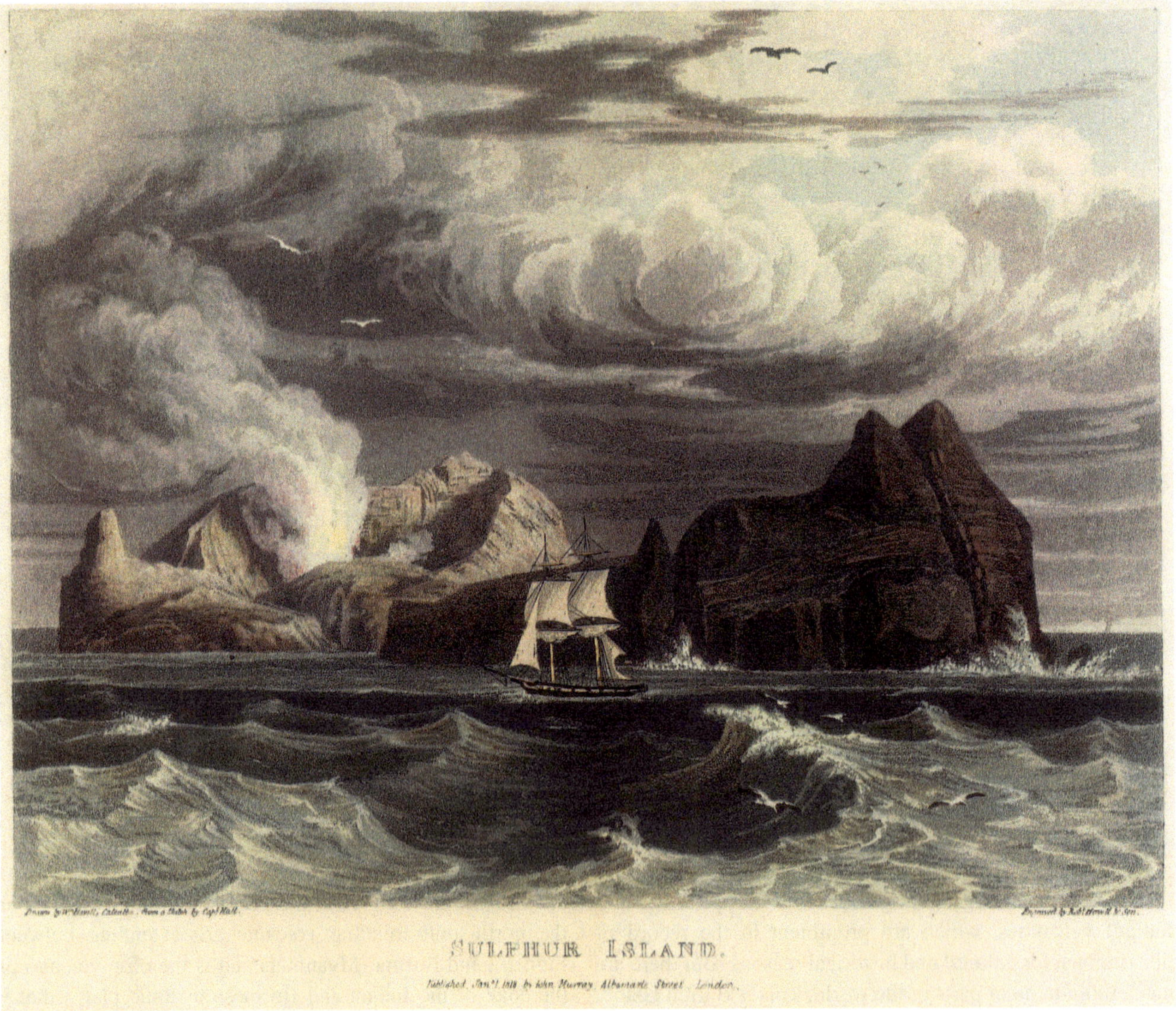

Fig. 5.24 Sulphur Island (Ioto)

Sulphur Island is the name given to one of the Bonin Islands by James Cook in 1779, during his third expedition to the Pacific Ocean in 1776–1780. The name came from the yellow sulfur dust he saw emanating from the cone of the volcano. The island had first been visited by Spanish sailors in the sixteenth century. The island lies some 1200 kilometers south of Tokyo on the Japanese mainland, in the central Volcano Islands portion of the Izu–Marianas arc. It is better known today as Iwo-Jima, the site of a much celebrated and fierce battle between the USA and Japanese defenders in 1945, in which 6800 US soldiers and 19,000 Japanese were killed during the 36 days of the assault. In 2007 the name of the volcano was changed to Ioto. The island has undergone a dramatic uplift in historical times, and the beach where Cook landed is now 40 meters above sea level. This uplift has been accompanied by many very minor eruptions, mostly gas and mud. It has been speculated, though, that a major eruption may occur sometime in the present century. The illustration here (Fig. 5.24) first appeared in Basil Hall's book *Account of a Voyage of Discovery to the West Coast of Corea, and the Great Loo-Choo*, published in 1818.

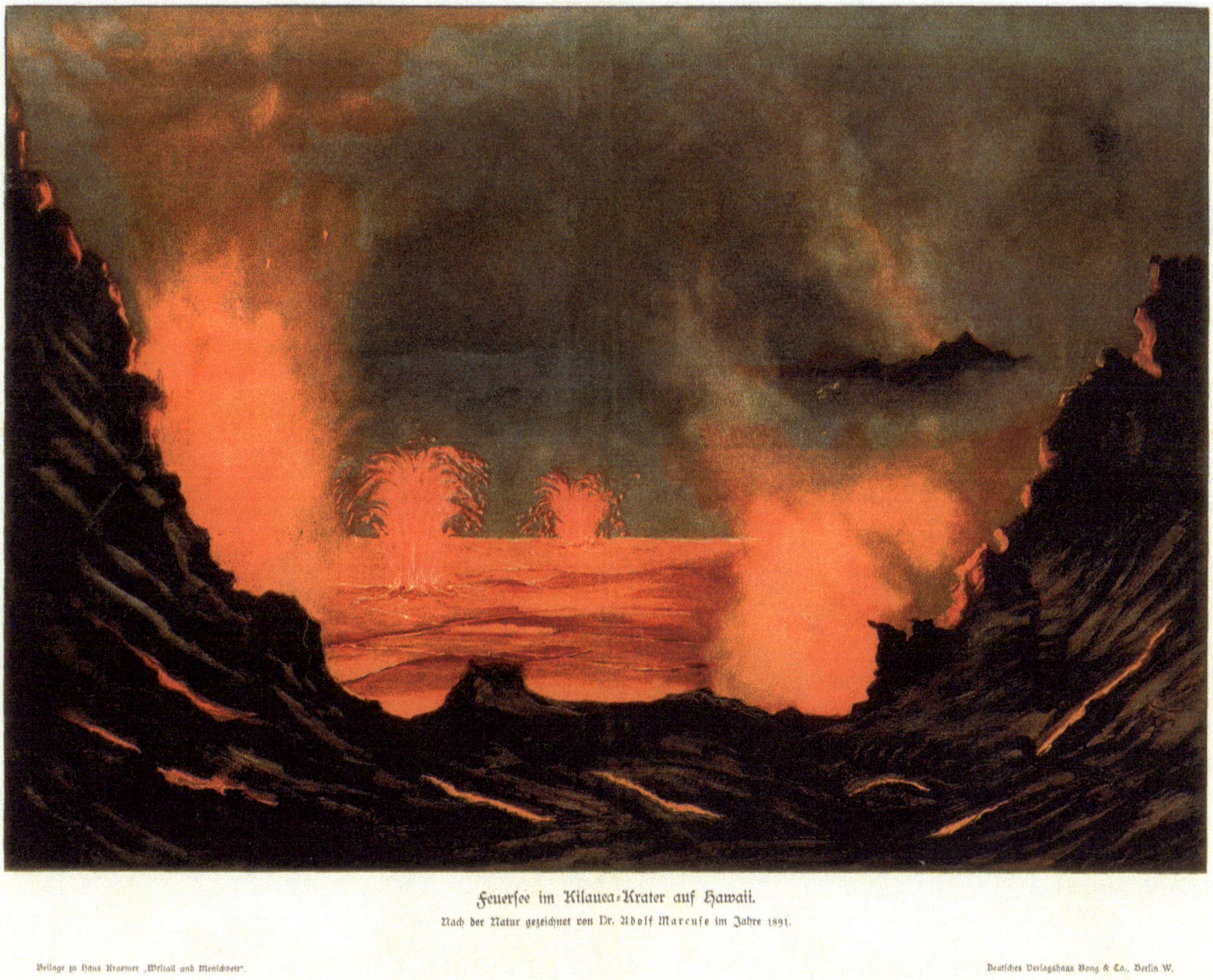

Fig. 5.25 Kilauea

For a guaranteed fiery spectacle, the tourist cannot do better than visit Kilauea, in Hawaii. The Hawaiian island chain consists of a series of mostly submerged volcanic cones, two of which have permanently active craters: Kilauea and Mauna Loa. These are shield volcanoes, quite different from the likes of Vesuvius and Etna. The lava here is alkaline in chemistry, rather than acidic, which means it flows very smoothly. Kilauea has a large, permanent lava lake inside its crater, and when it erupts, the lake rises up and overflows like milk in a pan boiling over. The formation of the Hawaiian Islands is due to a feature known as a hot spot—a point on the Earth's mantle where magma is liable to well up through the Earth's crust to create a volcano. The tectonic plates of the Earth are mobile; the mantle is not. Thus, as the Pacific Plate slowly moves, previous volcanoes that formed over the hot spot fall silent and new ones are formed as a fresh piece of crust moves over the hot spot. Thus, the island of Hawaii itself, where volcanoes are still active, is the youngest of the islands in the chain, which get progressively older as one travels down the chain. The picture here (Fig. 5.25) shows the lava lake of Kilauea; it was executed by the German scientist Adolf Marcuse, who was visiting Hawaii in 1891. From the look of the image, it may have been prepared from a photograph.

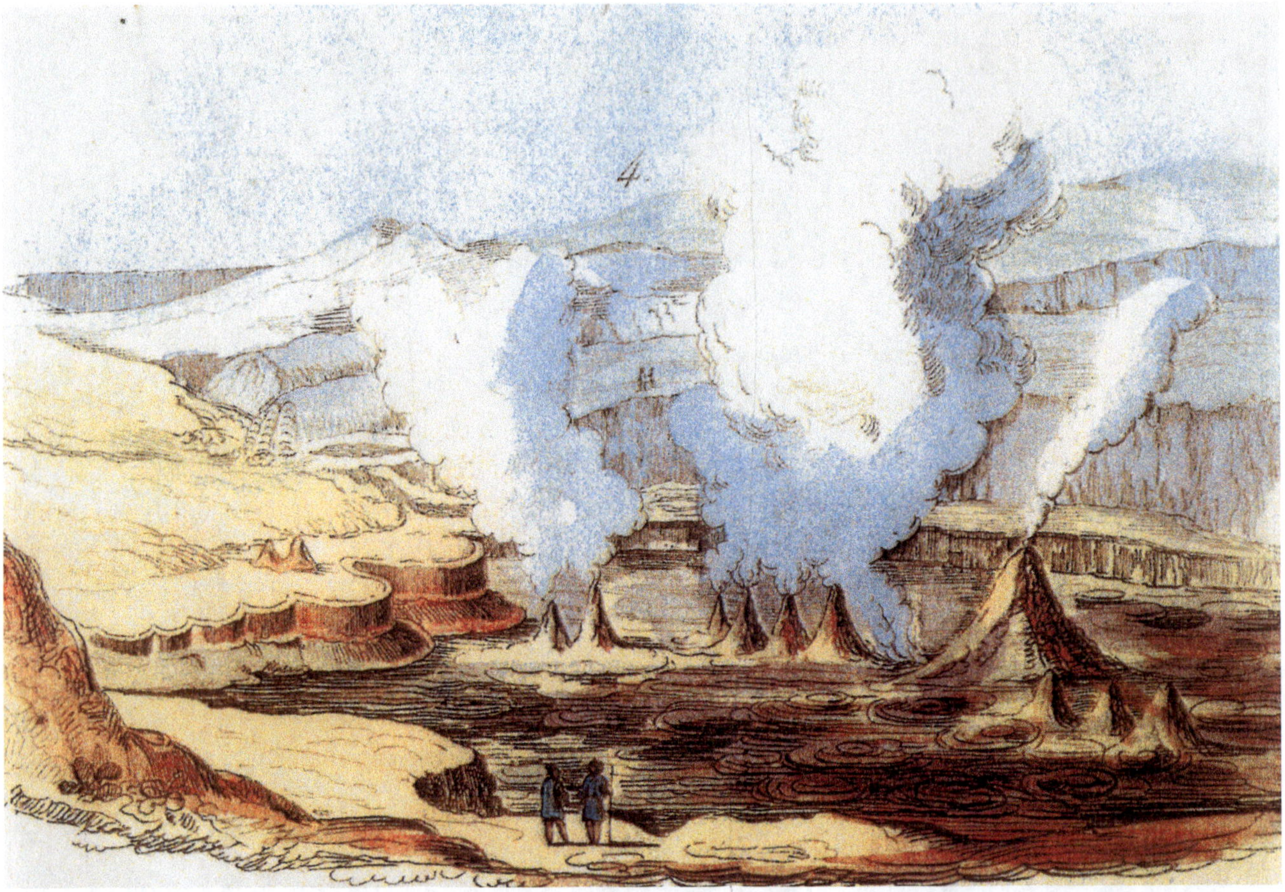

Fig. 5.26 Mauna Loa

This anonymous composition from the mid-nineteenth century shows two visitors to the crater of Mauna Loa on the island of Hawaii (Fig. 5.26). Gas is venting from minor cones, but one assumes that lava is not erupting, or the two figures would be in considerable danger. On the other hand, one can note massive, thick strata of solidified lava covering the crater bottom, a relic of previous eruptions.

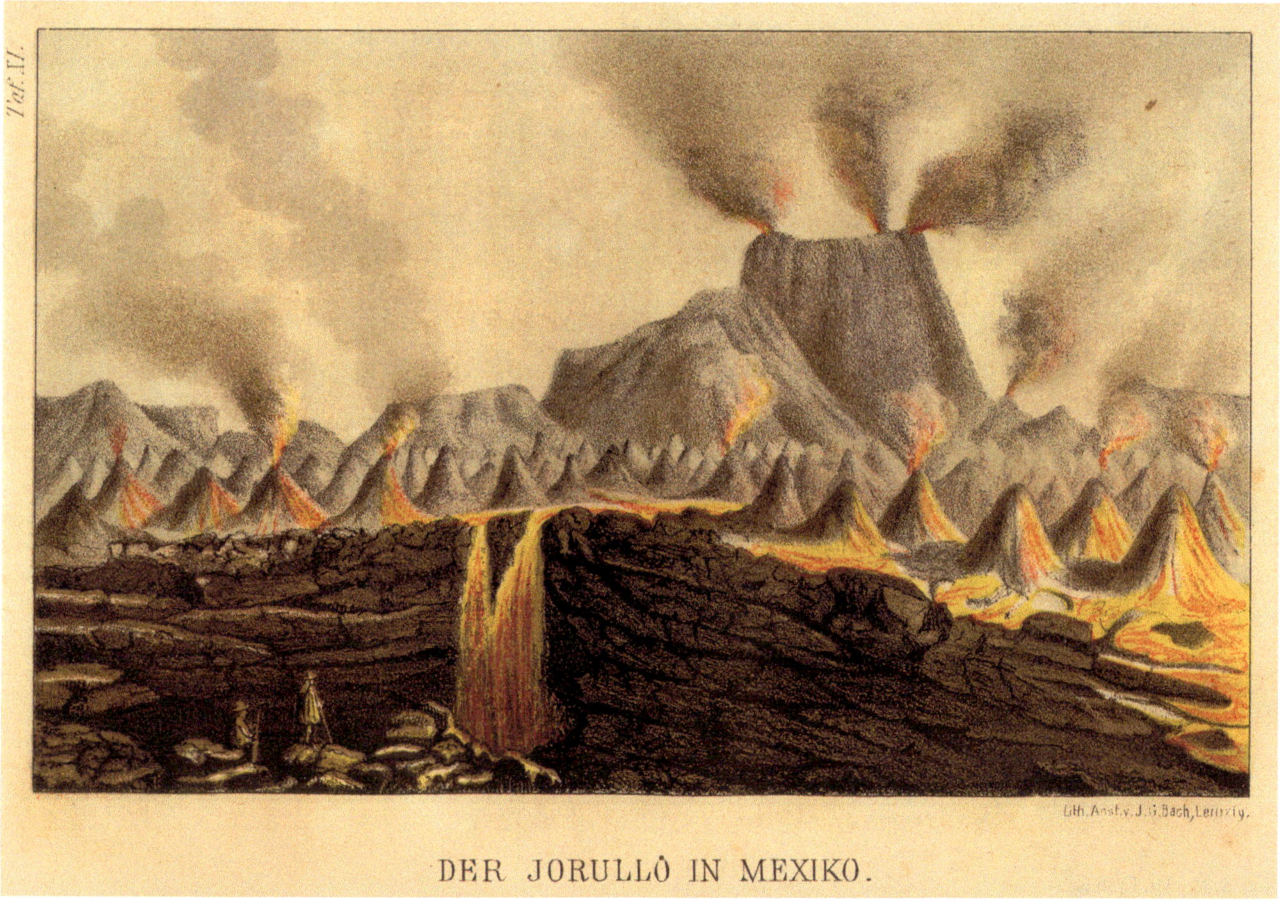

Fig. 5.27 Jorullo

When one thinks of the appearance of new volcanoes, such as Monte Nuovo in 1583 and Ferdinandea in 1831, the better-known examples are modern ones: Paricutin in Mexico in 1943, and Surtsey, off the coast of Iceland, in 1963. But Paricutin is not the only example in Mexico. Jorullo is a volcano in western Mexico, which appeared in a sugarcane field on the night of 28–29 September 1759. The creation of the new volcano was accompanied by a massive eruption of lava. Around the main cone, some smaller craters 400–500 meters high were created, with hundreds of tiny volcanic cones just 3–5 meters high named *hornitos* (little stoves) by the locals, which also briefly erupted lava. This initial eruption of Jorullo lasted for 15 years, between 1759 and 1774. Afterward, the only remaining activity was from fumaroles emitting volcanic gases. The first study of Jorullo's origin and eruptive history was performed by the great Alexander von Humboldt in 1803, when he visited the volcano and made sketches. This color lithographic portrayal of Jorullo erupting (Fig. 5.27) was published in Leipzig at the end of the first half of the nineteenth century by the editorial house of J.G. Bach, based on Humboldt's drawings. The numerous *hornitos* make a striking and unusual image. It appears from this that the slopes of the main volcano were extremely steep, but later measurements showed that the slope's inclination did not exceed 28–34°.

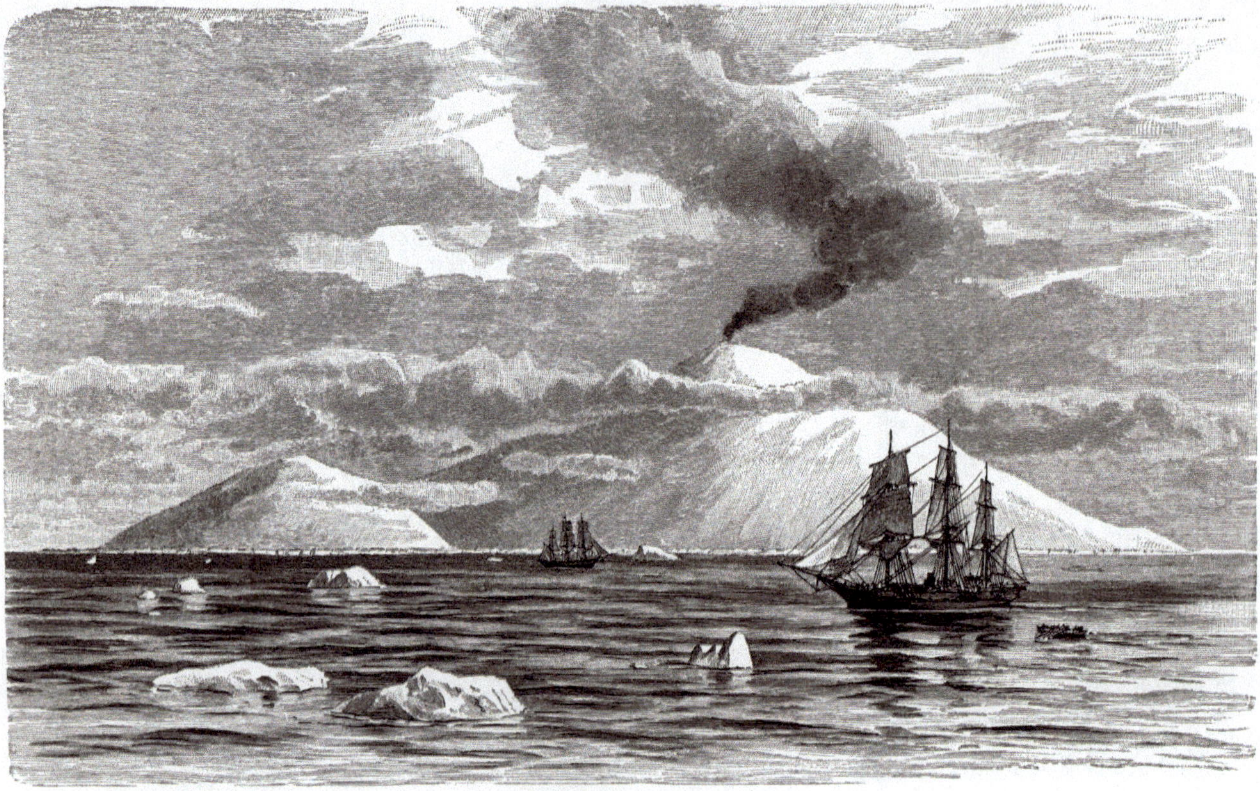

Fig. 5.28 Mt. Erebus

The volcanoes of the seventh continent, Antarctica, were unknown to Europeans until the expeditions of polar explorer Sir James Clark Ross in the 1840s. It was Ross who, in 1841, discovered and named the mountainous island near the large field of ice, which is today called the Ross Ice Shelf. Ross named the new island (consisting of three volcanic cones) Beaufort Island and the highest peak Mt. Erebus (from the Greek for "deep darkness"). It was clear that Erebus was an active volcano, since it continually emitted volcanic ash, accompanied by sound effects akin to explosions. However, it took another 67 years for the first successful ascent of the mountain to be made in 1908. This expedition revealed that on the top of the volcano there is a large caldera, hosting a permanent volcanic lava lake. It appears that the magma chamber deep below the volcano contains a great deal of gas under pressure; bubbles of this gas regularly escape upward until they reach the crater, whereupon they explode, giving rise to the periodic booms heard coming from the volcano. It is believed that this phenomenon is stable even in the long term; the mountain is now continuously monitored by American and New Zealand geophysicists, working in two permanent volcanic observatories.

The woodcut shown here (Fig. 5.28) is a plate from Volume One of Ross's own *A Voyage of Discovery and Research in the Southern and Antarctic Regions, During the Years 1839–43*, published in 1847. It was much reproduced, and the German caption indicates that this was a subsequent German edition.

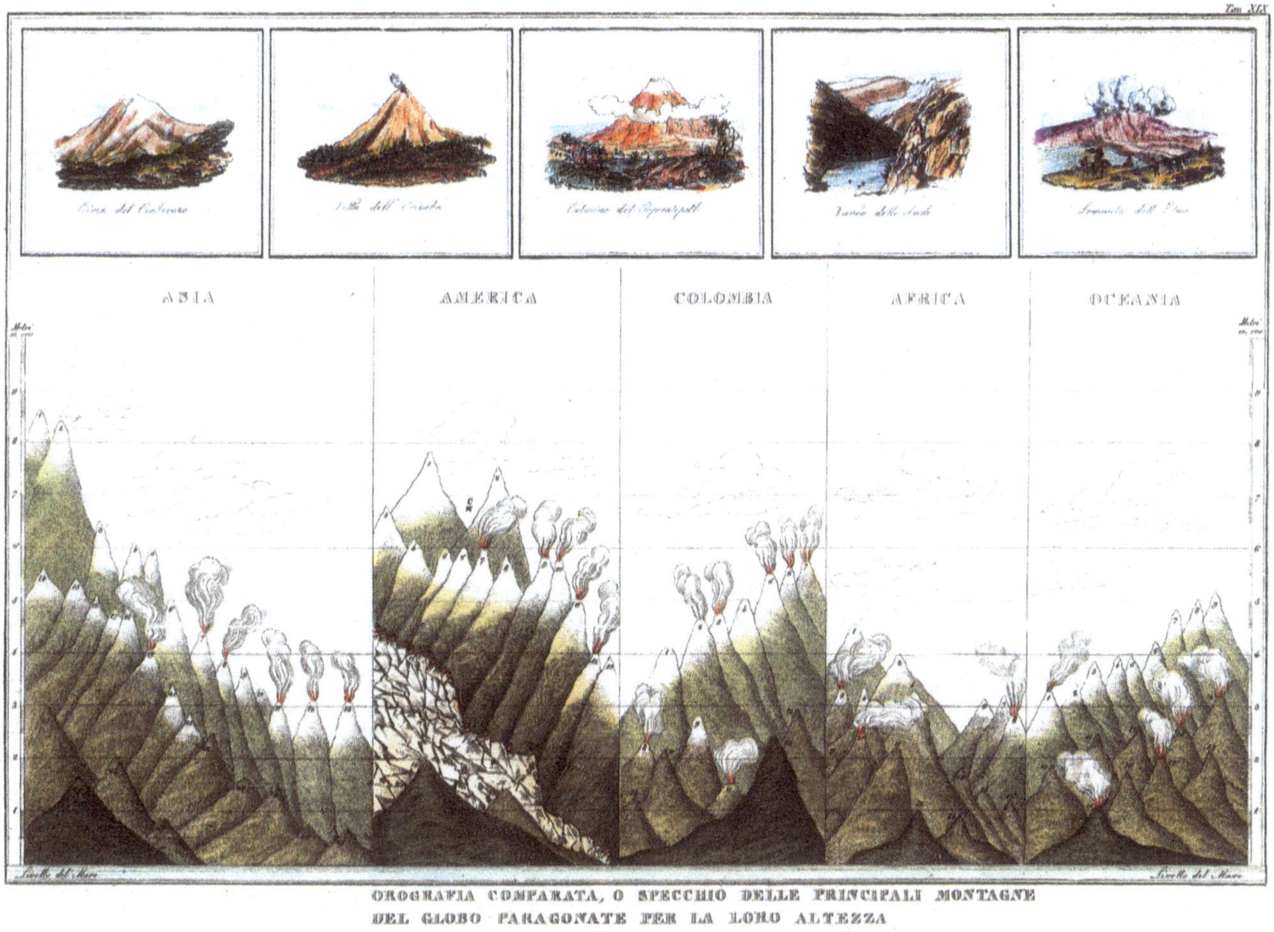

Fig. 5.29 Mountains compared in height

The print shown here (Fig. 5.29) is captioned *Orographia comparata, o specchio delle principali montagne del globo* and appears to have been extracted from an Italian work from the late 1830s or early 1840s. The mountains are grouped by continent (except for Europe) and graphically compared by height. Those that are volcanoes are indicated by puffs of smoke at the summit, and pictures of five particularly notable mountains are arranged at the top. Such comparative images of the tallest mountains, longest rivers, etc., were popular features of atlases in this period, and this shows the growing interest of the European reading public in the statistics of the wider world. Unfortunately, this print, in isolation, does not allow identification of individual mountains; they are identified only by numbers, and the key to the figure would have been in the book. Another version of this print, published in 1842, restores Europe between Africa and Oceania, and increases the number of views from five to nine. The author can be identified as Francesco Constantino Marmocchi (1805–1858), an Italian geographer. The style recalls somewhat the work of Humboldt.

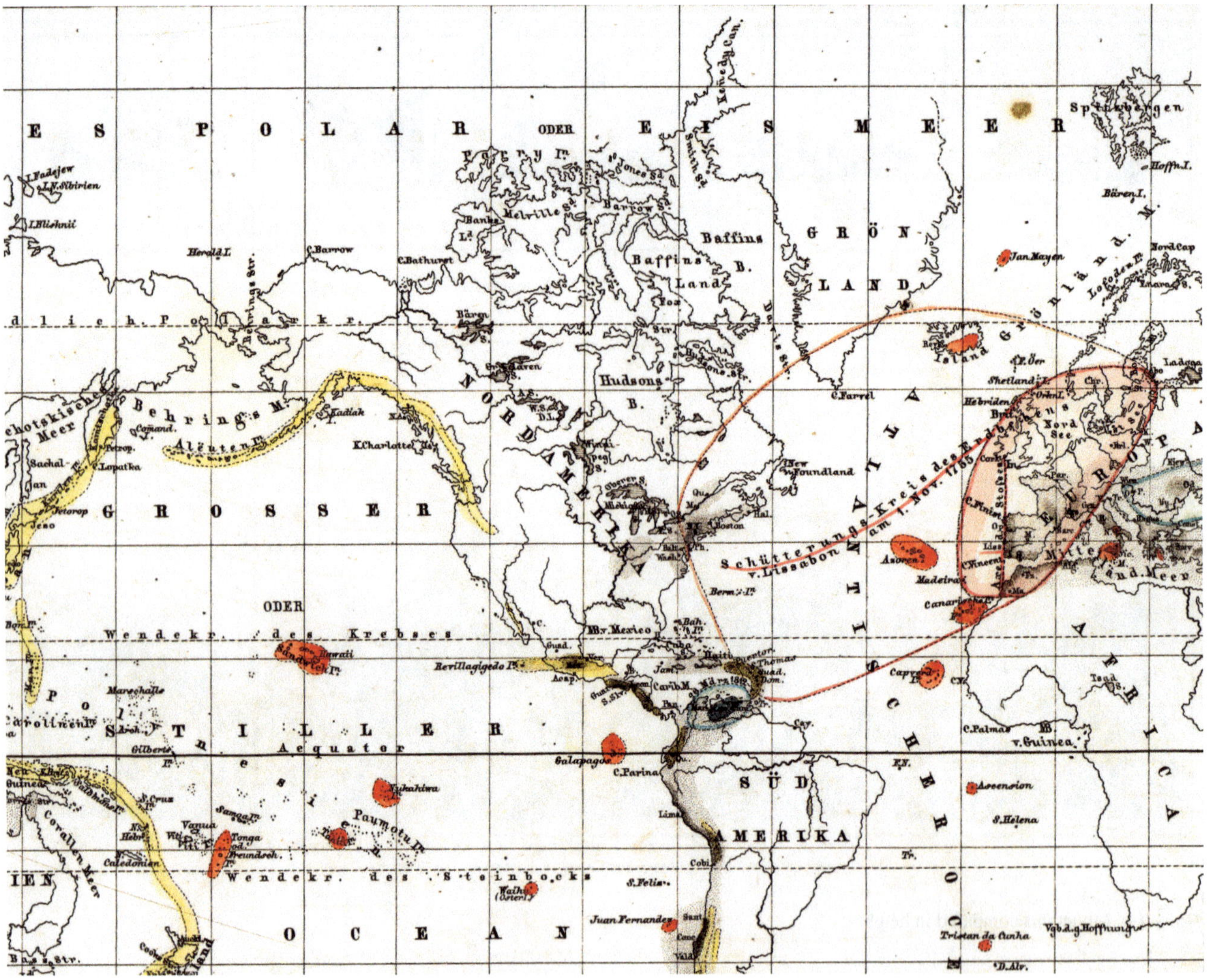

Fig. 5.30 Kiepert's map of volcanoes

The research into volcanism conducted in the first half of the nineteenth century resulted in efforts to construct the first world maps of volcanoes, often using the Mercator projection—a not entirely suitable choice. These were often maps of both volcanoes and earthquakes; those in the atlas of Heinrich Berghaus in 1848 are a well-known example. This is a later example, by the geographer Heinrich Kiepert (1818–1899), published in Weimar in 1860 (Fig. 5.30). The red spots indicate what Kiepert referred to as "central volcanoes," corresponding more or less to hot spots.

The yellow bands are "volcanic arches" prominently showing the "Pacific Ring of Fire"; we now recognize this as the edge of the Pacific Plate. Seismic zones are also shown (not very accurately) by two levels or densities of hachures on the map.

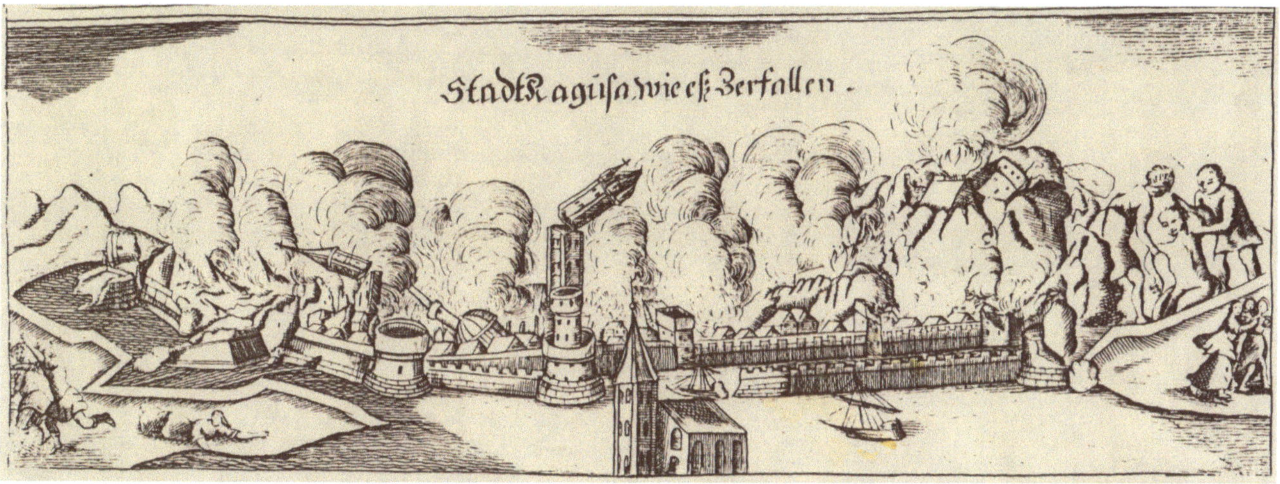

Fig. 5.31 Ragusa in flames in 1667

The east coast of the Adriatic Sea, a region of many attractive towns, is also a plate boundary between the Eurasian and African tectonic plates, and is thus a marked earthquake zone. Ragusa (now Dubrovnik) was a Venetian colony and one of the principal towns on the Adriatic—a center of power, business, and maritime traffic, with a naval base. The town was damaged by strong earthquakes in 1520 and 1637 but was rebuilt. It was once more struck by an even stronger earthquake on 6 April 1667. Many of the townsfolk (some several thousand) were killed, most of the town was ruined, and the port facilities were devastated, with damage to the ships at anchor there. Some sort of accident set fire to what was left of the town, and strong winds fanned the blaze, which continued to burn for some 27 days. The *Theatrum Europaeum*, from which this illustration (Fig. 5.31) was taken, was a journal of events between 1618 and 1718, published in 21 quarto volumes. This is from volume 10, covering 1666–1671 and prepared by Martin Meyer the Younger and Wolfgang Jacob Geiger.

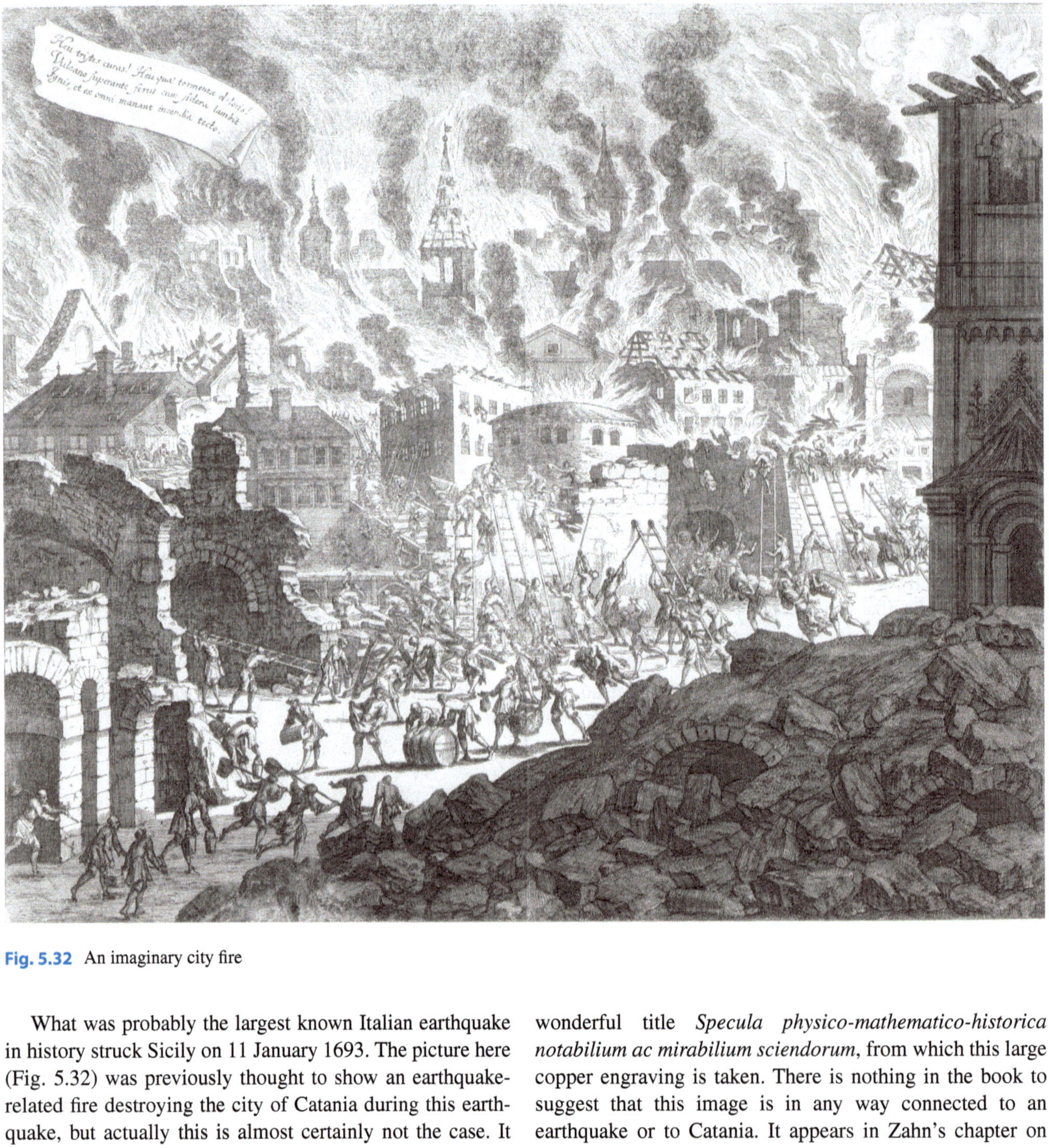

Fig. 5.32 An imaginary city fire

What was probably the largest known Italian earthquake in history struck Sicily on 11 January 1693. The picture here (Fig. 5.32) was previously thought to show an earthquake-related fire destroying the city of Catania during this earthquake, but actually this is almost certainly not the case. It comes from a work by Johann Zahn (1641–1707), who was a member of the Premonstratensian Order of Oberzell near Würzburg. He published a superbly illustrated compendium of mathematics and natural history in 1696 under the rather wonderful title *Specula physico-mathematico-historica notabilium ac mirabilium sciendorum*, from which this large copper engraving is taken. There is nothing in the book to suggest that this image is in any way connected to an earthquake or to Catania. It appears in Zahn's chapter on memorable fires. The legend (top left) gives no details beyond a general description of a severe fire, and the probability is that this is simply a drawing of an imagined inferno consuming some unfortunate city.

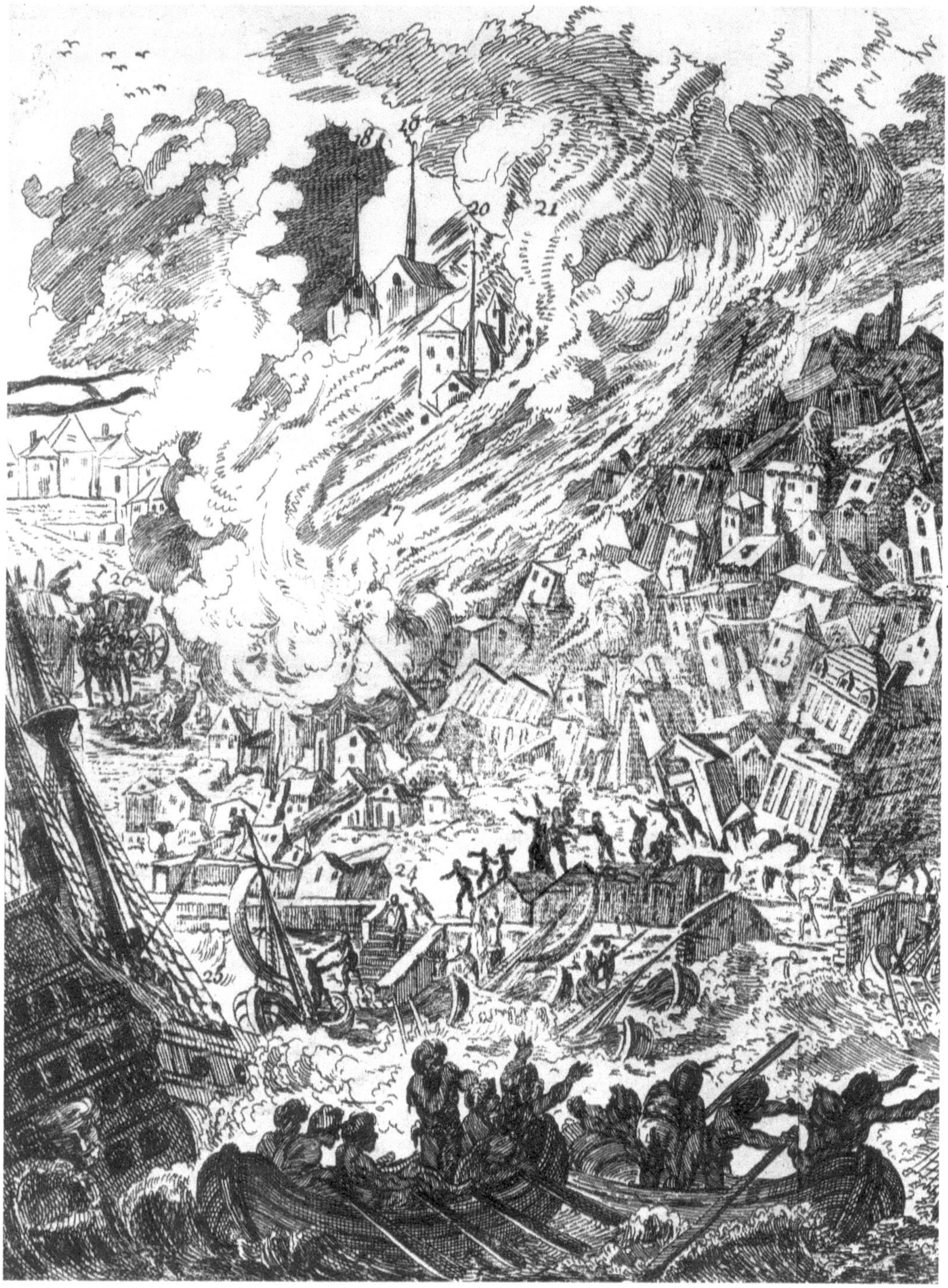

Fig. 5.33 A fire after the 1755 Lisbon earthquake

The most celebrated earthquake fire before 1900 is probably that which aided the destruction of Lisbon in 1755, along with the earthquake shaking itself and the subsequent tsunami. Images of the destruction of buildings directly as a result of the earthquake have already been shown (Figs. 2.11, 2.12, 2.13, 2.14, and 2.15), but fires also broke out in many parts of the city (Fig. 5.33). Since the earthquake struck on 1 November—All Saints Day, one of the most important feast days in the Catholic calendar—a great deal of the city's population was attending morning service in the churches, and the housewives would have been preparing a sumptuous lunch for the whole family at home. So most ovens and stoves would have been lit, and when these were damaged by the earthquake, fires broke out. As a result, many Lisboners who escaped the collapsing buildings and avoided the tsunami perished in the fires, which persisted for several days. In general, fire from earthquakes is one of the major causes of earthquake-related deaths.

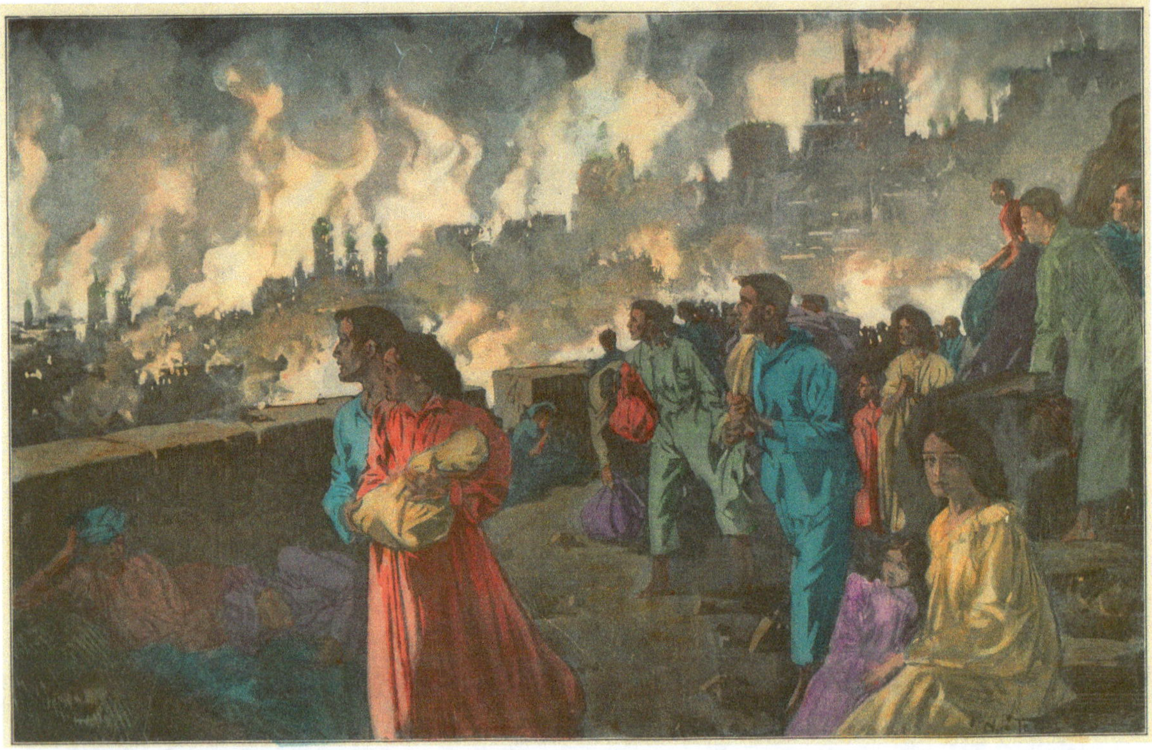

Fig. 5.34 San Francisco in flames in 1906

The famous San Francisco earthquake of 18 April 1906, has already been illustrated and discussed in Chap. 2, in the section on earthquakes. But the resulting fire is of particular interest. In fact, for some time afterward, there were attempts to refer to the disaster as the "great San Francisco fire" on the grounds that the idea that California was prone to earthquakes would deter investment (for the same reason, earthquakes in San Francisco in the nineteenth century were hushed up).

The fire resulting from the earthquake took three days to bring under control. The temperatures were so severe that firemen had difficulty getting close enough to fight the flames. The principal attempts at controlling the fire therefore had to be made by blasting firebreaks. The problem here was a shortage of dynamite, so gunpowder was used. The resulting explosions often reduced buildings to matchwood, which showered into the air and ignited, spreading the fire rather than containing it. One of the buildings chosen for demolition turned out to be a warehouse storing spirit alcohol, and the resulting fireball caused a whole new fire.

One problem for the survivors was that people had insurance against fire but not against earthquakes. So, in the case of a building that was heavily damaged by the earthquake, there was some advantage in setting it alight so that the fire insurance could be claimed. To try and combat this, insurance companies took to offering money for photographs of earthquake-damaged buildings to make sure that arsonists did not try to claim that the destruction was caused by fire. Ultimately, the cost of insurance payouts was borne by Lloyds of London, which greatly strained the firm's finances but established its reputation for dependability.

The picture here (Fig. 5.34) is a lithograph, which was published in Leslie's Weekly and also in the Illustrated London News on 12 May 1906 (page 679). This copy has been subsequently hand-colored (rather clumsily). It was titled *Homeless Millionaires: The Flight from the Flames at San Francisco*, with the caption "Where money was useless: the plight of wealthy San Franciscans during the burning of Nob Hill." The point was made by the paper that these rich people had to flee their homes in nothing but their nightclothes—which is not entirely clear, given the gaudy colors applied by whoever touched up this picture.

5.2 Man-Made Fires

We can turn now to man-made fires: those caused by error, carelessness, or bad luck, and those caused intentionally by arson or military action. The majority of large cities in medieval or Renaissance times suffered major fires repeatedly. Urban construction was largely of inexpensive and easily put-together wooden buildings, often with flammable thatched roofs, and these houses were densely packed within the limited space inside the city fortifications. Vulnerability to fire was therefore high, and once a fire started in one house, it could easily spread through the district, especially given the limited means of fire fighting.

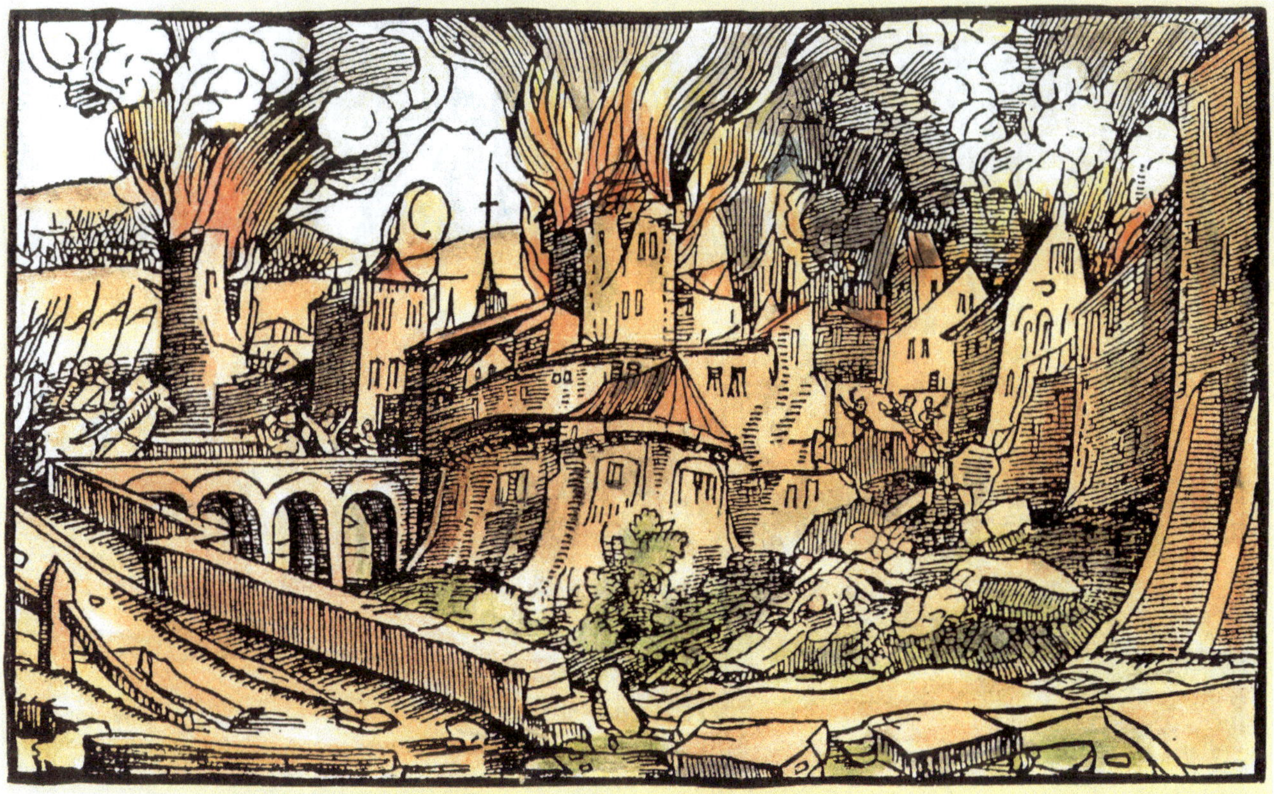

Fig. 5.35 The great fire of Prague in 1541

The picture here (Fig. 5.35) shows the great fire of Prague (then the seat of the ruler of the Kingdom of Bohemia) in 1541. The fire started during repair works on the roof of what is now the Sternberg Palace in the Central Square of the Lesser Town of Prague. A careless workman ignited some of the roofing material, and a strong wind spread the blaze over all of the buildings of the left bank, including the royal castle Hradčany. Few of the houses in the area have survived to the present day; either they were partly or entirely demolished by their owners after the fire, or they simply burned through from the roof to the ground.

The depiction reproduced here is from a contemporary German booklet describing the event. It was published by Heinrich Steiner in Augsburg in 1541. The image itself is rather rare: in the Archives of the Czech Republic, no copy of Steiner's work is registered, and this picture is taken from a copy in the British Museum in London. One can also note, as a matter of historical interest, that this picture is the oldest image of Prague Castle seen from the north or northwest. Moreover, since the Gothic structure of the castle burned completely in 1541, this picture—the first and simultaneously the last—seems to be the only one that shows the old, pre-Renaissance Prague Castle. However, as is so often the case with pictures of this sort, one can ask whether Steiner was actually present in distant Prague during the fire to make a sketch, whether he used the sketch made by some other eyewitness of the fire, or whether he just made the whole thing up.

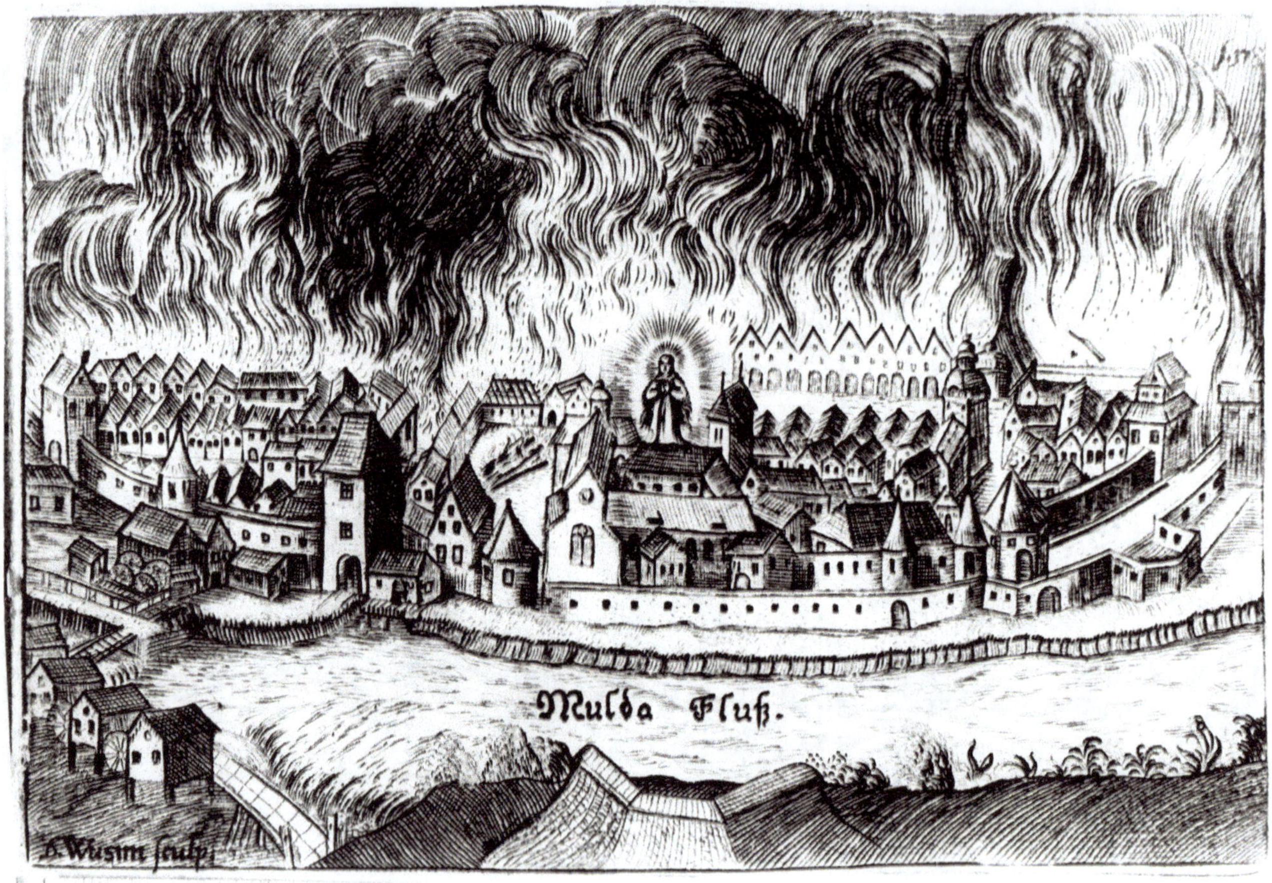

Fig. 5.36 České Budějovice burns in 1641

The fire here was one that ruined the Bohemian town of České Budějovice (Budweiss) in 1641 (Fig. 5.36). Again, the cause was human carelessness. It came in the dying years of the Thirty Years' War, when Central Europeans were worn out by an endless series of tragic incidents reflecting the disasters of the long war. With such war weariness, people became careless and failed to observe fire and other safety measures. The town chronicles relate how the Emperor's garrison troops, who were paid mercenaries, accidentally started a fire, which spread across the town. It destroyed three church complexes and around three quarters of the houses of the town—some 300 dwellings and some schools. As a result of this fire, the town lost its Gothic character, since it was rebuilt in a different style. The black-and-white copper engraving of the 1641 fire reproduced here is the work of Daniel Wusin (1601–1694). The picture shows the first stage of the disaster, when the fire spread first across the roofs in the central square. The shining figure might be St. Nicholas, to whom the cathedral was dedicated; he does not seem to be very successful in quelling the destruction.

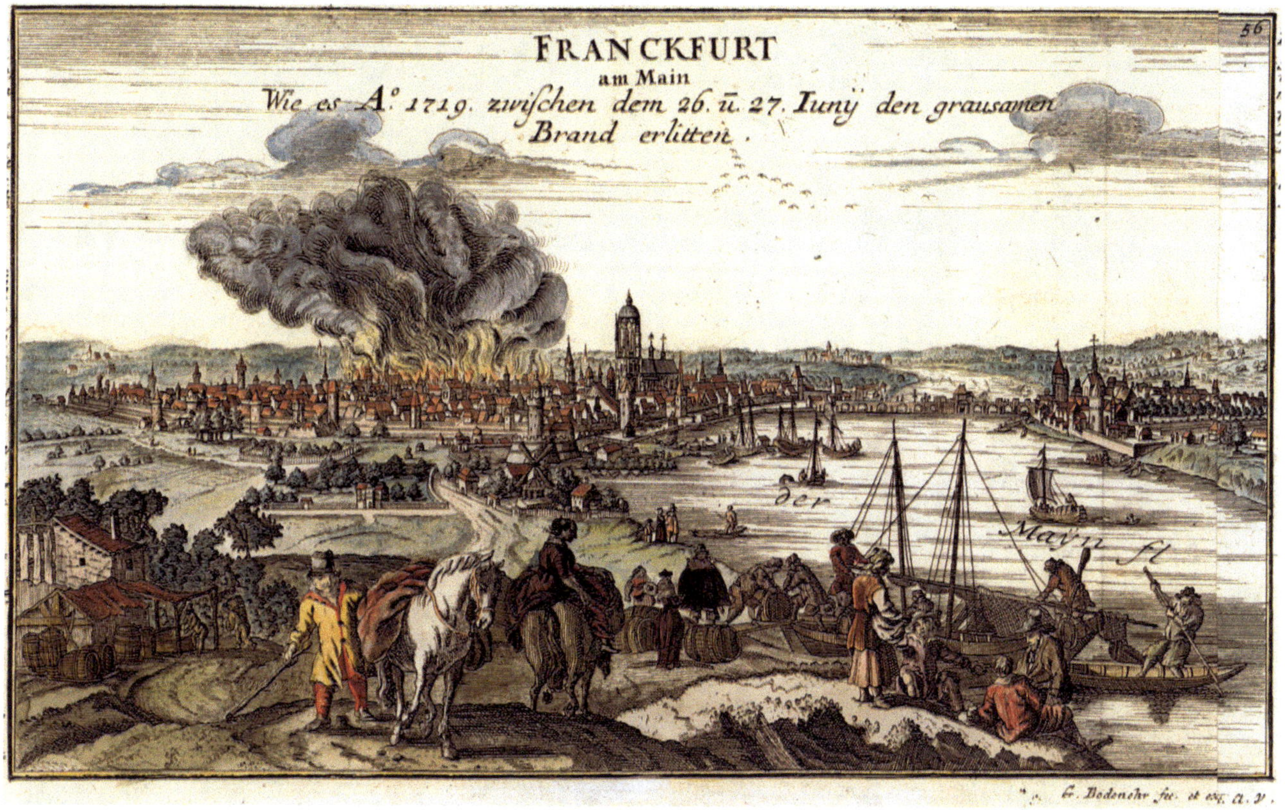

Fig. 5.37 The Frankfurt fire of 1719

One upset candle can be the cause of much destruction, as in the case of the great fire of Frankfurt am Main on 26–27 June 1719 (Fig. 5.37). The story goes that the culprit was a visitor from Dresden by the name of Morgenstern, a wigmaker by trade. He was staying in a small hotel in the town quarter of Jewish craftsmen and traders. Either a candle fell over, or the flame caught some curtains—no one can be sure—but the fire was discovered around midnight. Efforts were made to put it out, but wind had already started to spread the blaze, and there was not enough water on hand. More water was sent for, but by the time it arrived, the strong wind had whipped the fire up into a widespread conflagration. It was not until 13 hours later that the main blaze was brought under control, and it took three more days for all of the smoldering remains to be finally dealt with. Fourteen people lost their lives, and many more were wounded; around 400 houses were ruined, especially in the Jewish quarter, where the fire started (and which had only just been reconstructed after another fire eight years earlier). About 500 families were rendered homeless; it demonstrates how one hitch in the effort to control a fire can result in catastrophe.

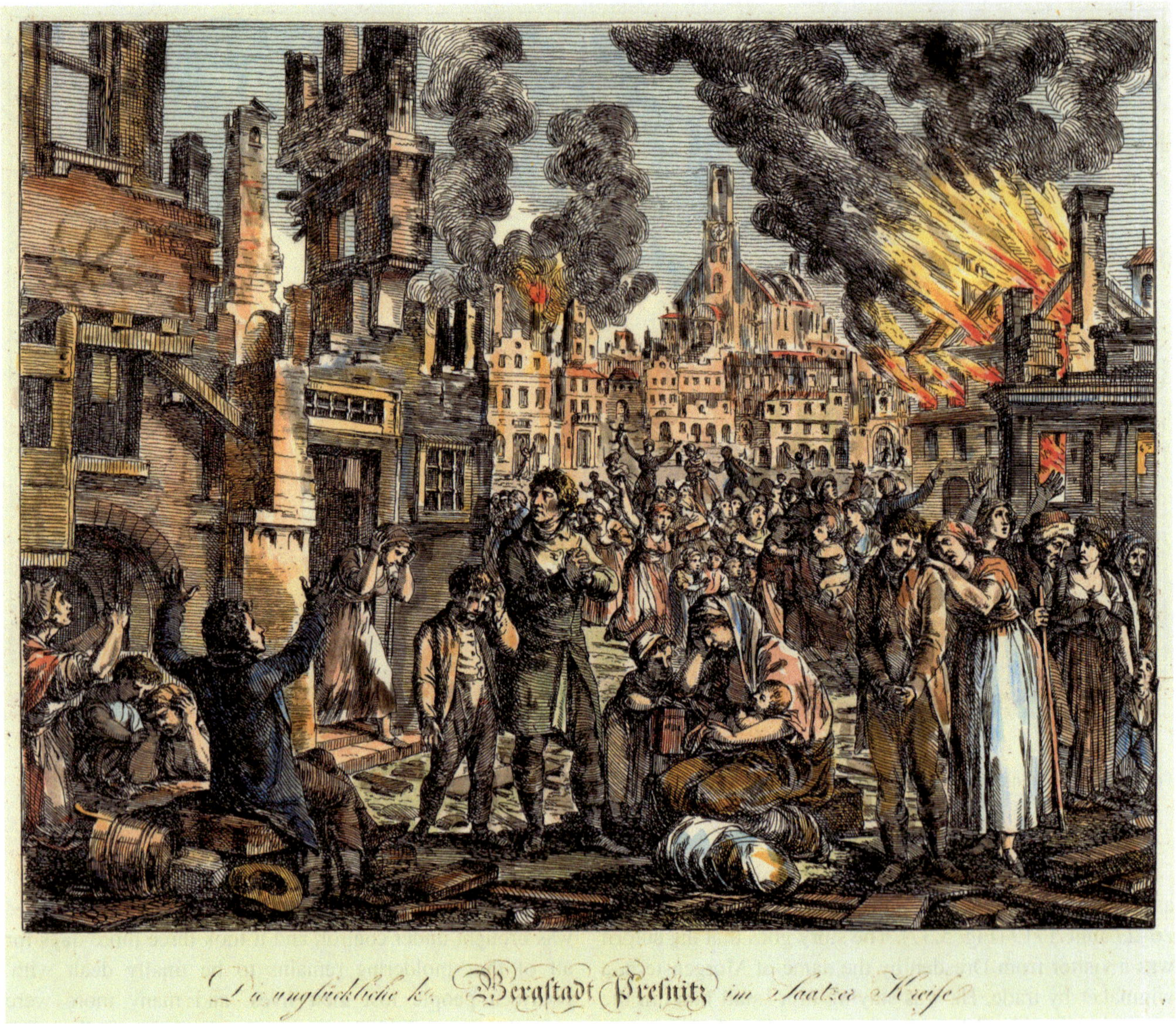

Fig. 5.38 Přísečnice burns in 1811

The illustration here (Fig. 5.38) is another example from central Europe. The town in this case is Přísečnice (Preßnitz in German). The town was founded in 1335 on the Czech/Saxony border; while it was on Czech/Bohemian territory, the population was German. By 1811, the date of the fire shown here, the town was home to around 4000–5000 people and was famous for mining, the manufacture of musical instruments (especially harps), and lace making. The cause of the fire is unknown, but its effects were severe: 11 people died, and of the 400 houses that made up the town, reportedly only 66 survived. The print here is particularly interesting, as it was made specifically to be sold in support of the relief effort for the poor homeless inhabitants. It was made by Joseph Berger, who was the first director of the Czech Academy of Arts in Prague (in the years 1805–1829). This hand-colored copper engraving clearly shows the despair of the locals, milling in the streets amid the ruins of their homes.

There is a sad modern conclusion to the story. Although the town was rebuilt after its destruction in 1811, after World War II the German population of Přísečnice was resettled out of Bohemia. As a result, the town became virtually abandoned and quickly deteriorated. Finally, the construction of the Přísečnice Dam in 1974 flooded the site of the town. Some people say that if one crosses the reservoir by boat on a still day, one can see the tower of the Přísečnice church under the waters of the lake.

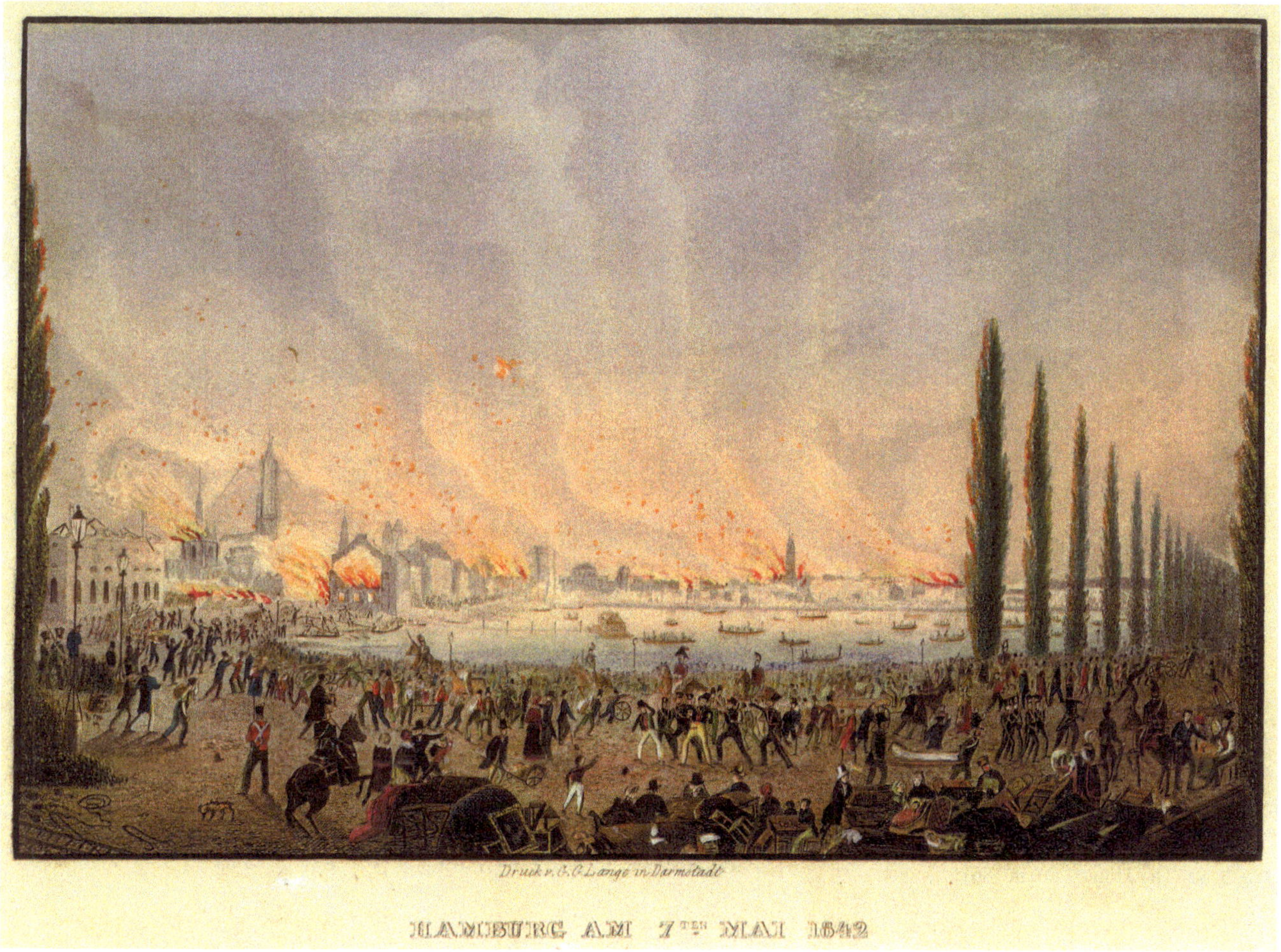

Fig. 5.39 The great fire of Hamburg in 1842

The great fire of Hamburg in 1842 was one of the most famous historical fires in Europe and was the subject of many paintings and illustrations; there even exists a very early photograph of the ruins. The image here (Fig. 5.39) is one of the less well-known pictures of the event and is the work of Gustav Georg Lange (1812–1873). The fire started early on 5 May, in the cigar factory of Eduard Cohen in Deichstraße. The fire was quickly reported to the night watch, but, once again, a strong wind was fanning the flames, and the recent weather had been very dry. The city largely consisted of tall, narrow, wooden or half-timbered houses packed close together, and many also served as stores for merchants and contained stocks of flammable goods. Hamburg was supposedly prepared against fire; there was a system of lookouts and volunteer fire brigades, and few European cities have such an abundance of water (the estuary of the River Elbe) to hand. But the river was low because of the dry weather, and the pumping system was inefficient. The next days saw a frantic struggle to contain the fire, which was made harder by collapsed buildings blocking the streets and canals. Efforts were made to contain the fire by blasting firebreaks, and firefighters arrived from outside the city to help the efforts. The picture here shows the scene two days after the outbreak of the fire; many people are in a state of panic, and some are looting. In fact, the local militia had to stop their efforts to fight the fire in order to crack down on looters. Early the next morning, the fire was finally extinguished, but 51 people had died, including 22 firefighters; about a third of the city had been destroyed. The financial losses severely strained insurance companies, and it was this that led to the first reinsurance company, Kölnische Rückversicherungs-Gesellschaft, or Kölnische Rück (Cologne Re), chartered in 1846.

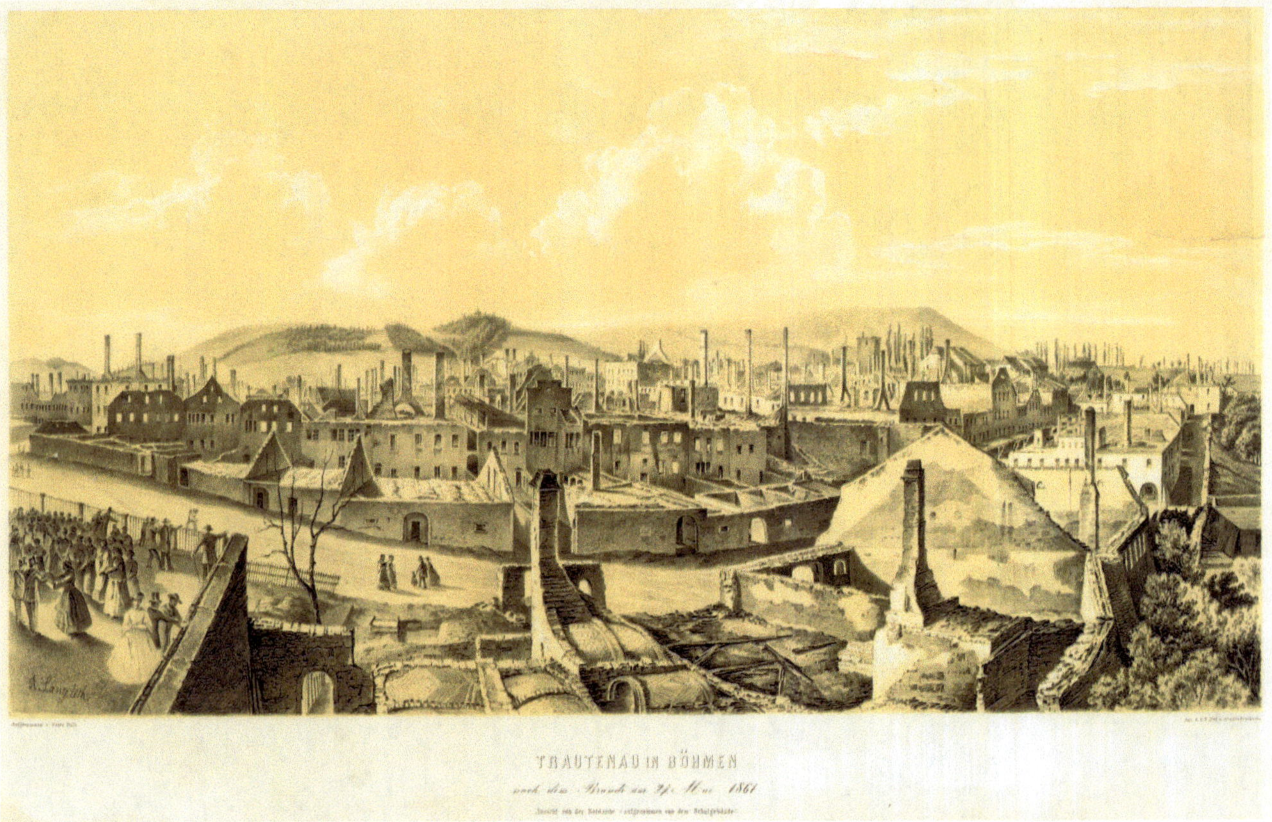

Fig. 5.40 The Trutnov fire in 1861

The case here is similar to the 1811 fire in Přísečnice; the town here is Trutnov (Trautenau in German), a town in the Sudeten region of East Bohemia, populated by both Germans and Czechs. Fire broke out on 27 May 1861, and largely devastated this small old town of about 400 dwellings (Fig. 5.40).

This time, the dominant construction of the town greatly affected the nature of the damage. The houses consisted mostly of masonry ground stories, with one or two wooden stories above. When the fire broke out, it was mostly the wooden parts of the houses that were damaged, while the ground floors survived. In this toned lithograph (signed by R. Lang) seen here, the townsfolk in the bottom left corner survey the ruins of their town, uncertain of their future. It can be seen that the chimneys of the houses are largely intact, being made of stone. In fact, the masonry parts of the houses were so little damaged that reconstruction consisted largely of carpenters' work, which could be completed relatively quickly. Nearby towns and settlements helped out with charity collections, and the town was back on its feet quite soon.

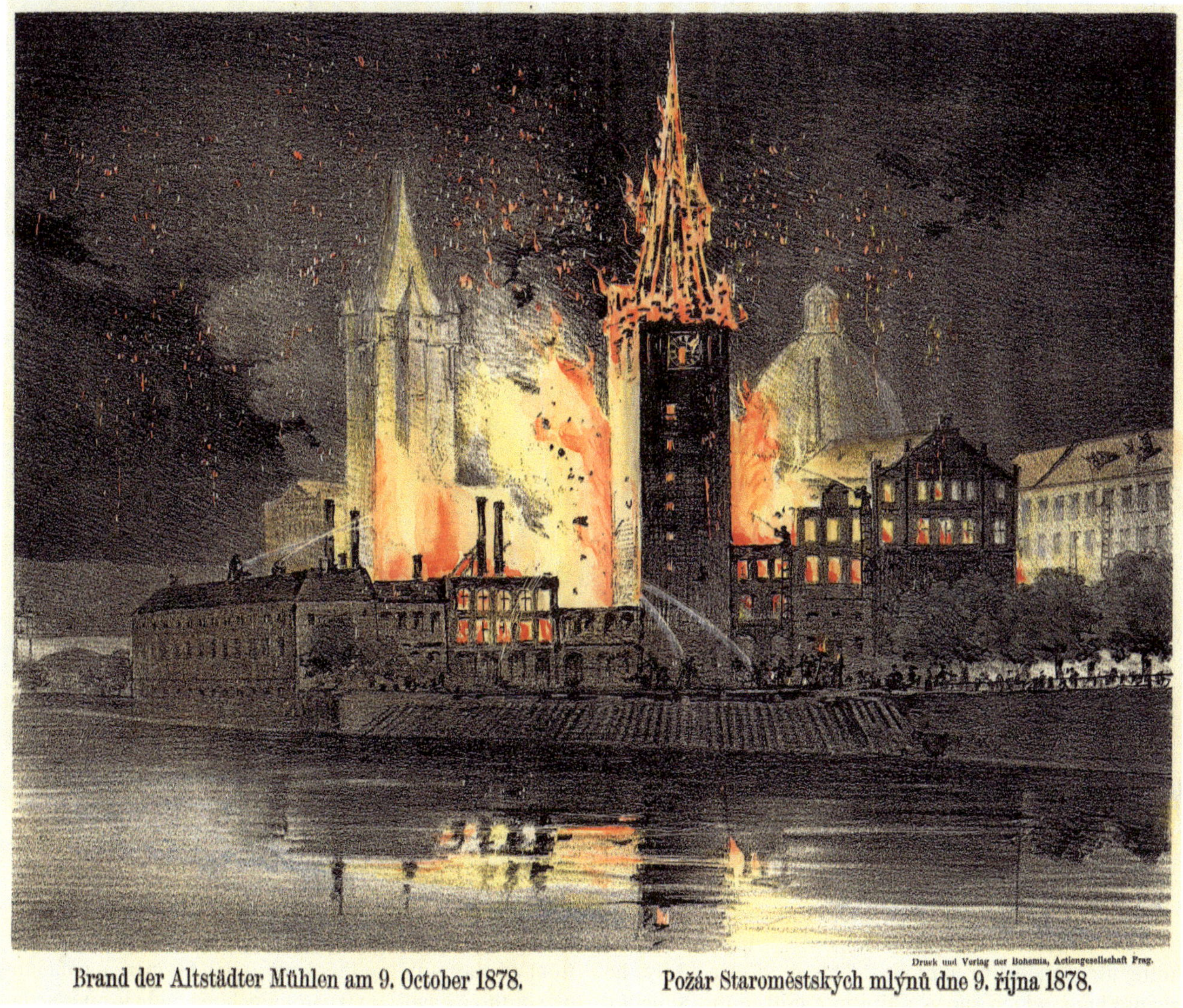

Fig. 5.41 A burning water tower in Prague in 1878

Some types of structure present a particular fire hazard; one might not have expected that water towers would be among them. Water towers have been constructed in many large towns of Europe since the late Middle Ages. They were often attached to river mills, since the mills supplied enough energy for transporting water up to the highest level of the water tower, where it would be kept in a cistern. From there, water was fed down by gravity into the palaces and residences of the city potentates. In northern Europe, however, there were problems with the operation of this system in winter; the pipes would simply freeze up in any strong frost. The only way to keep them warm was an open fire. Since the early towers were built of wood, the consequences can be imagined.

Later, stone towers were constructed, which were much safer. The wooden floors could still catch fire, but the walls themselves were constructed in such a way that replacement of the burned floors was fast and easy, which is why most of the water towers in central and northern Europe constructed in the Gothic period have survived until our time, regardless of suffering a series of fires. On the other hand, houses and workshops adjacent to the water tower were definitely at risk, as seen in this hand-colored lithographic print (Fig. 5.41) portraying a fire at the Prague mills on the night of 9 October 1878.

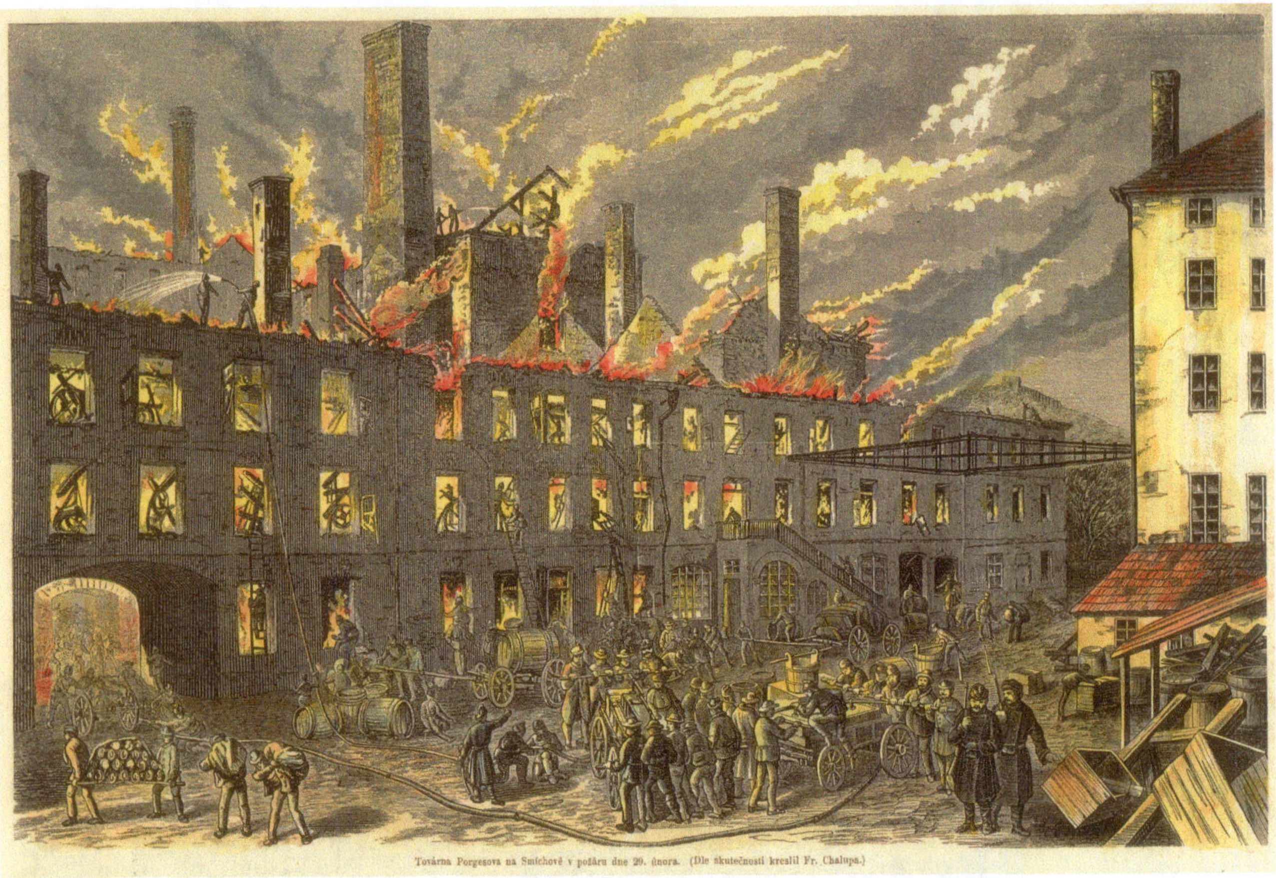

Fig. 5.42 A mill fire in Prague in 1872

As the eighteenth century proceeded, the organization of firefighting gradually improved. This is illustrated by the picture here, which shows a fire in a mill in the Smichov district of Prague on 29 February 1872. It was sketched on the spot by the famous Prague illustrator František Chalupa (1828–1897) and published on 15 March 1872, in the Czech illustrated weekly *Světozor* (Fig. 5.42). The fire started at 4:30 a.m. in a textile mill owned by the Porges family of Partheim; the cause was unknown. In the picture the roof has already collapsed; it came down with a deafening noise, burying much valuable equipment. One notes the large number of firemen operating inside the burning building. In front of the building there are other firemen surrounding two or three hand-pump carts. Each machine is operated by two groups of firemen, supplying water to their colleagues fighting the flames inside the burning building. The nosy public is excluded, kept back by an army guard, seen in the right bottom corner. The entire picture speaks of a well-organized, systematic, professional performance of the local fire service. Actually, the first service of 30 professional firemen in Prague was established in 1853, and in 1866 their rights and duties were strictly defined.

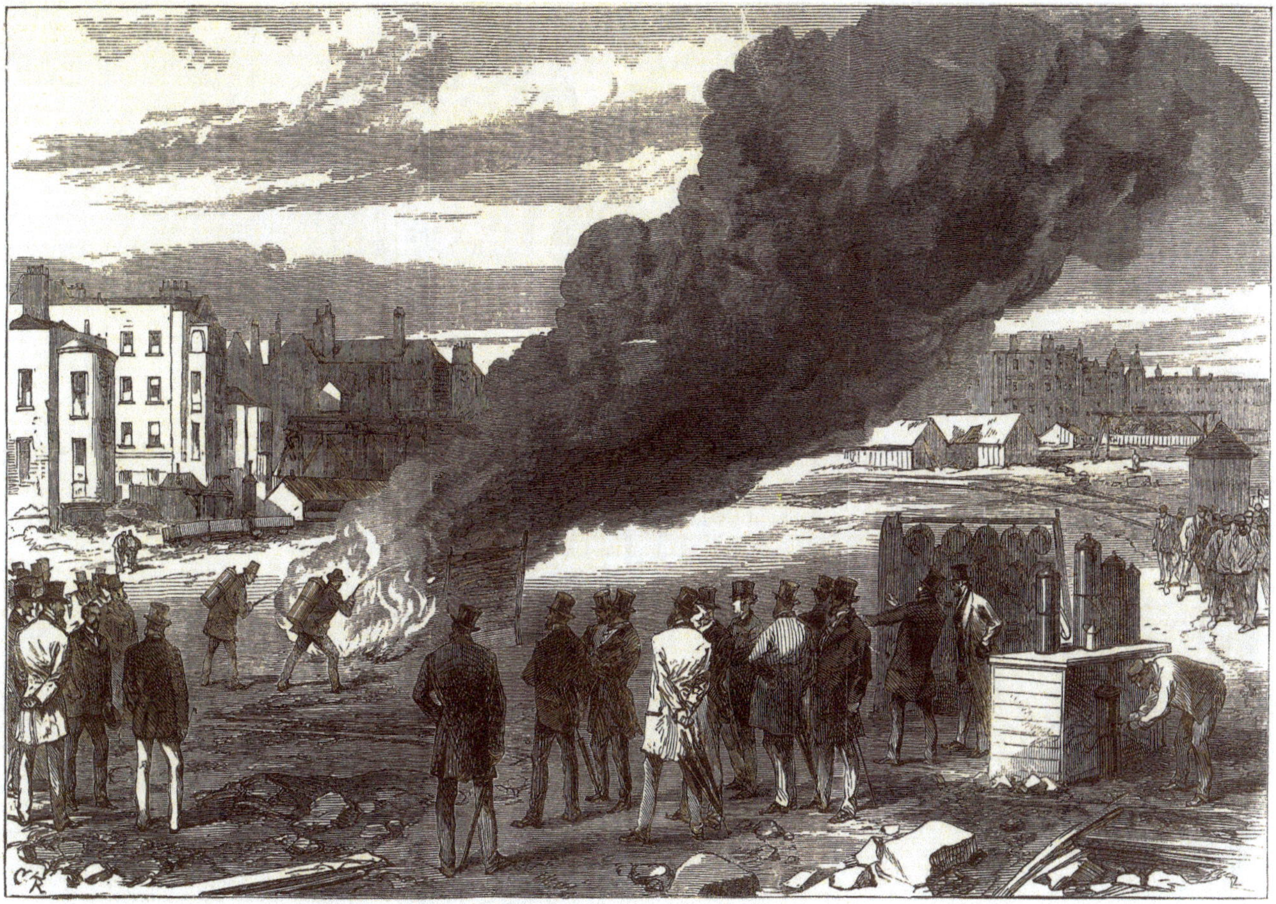

EXPERIMENTS WITH L'EXTINCTEUR ON THE THAMES EMBANKMENT BEFORE THE PRINCE OF WALES.—SEE PAGE 217.

Fig. 5.43 Testing a CO_2 fire extinguisher in 1868

The nineteenth century also saw technical innovations in firefighting. Anyone who undergoes a fire safety training course today will be familiar with different types of fire extinguisher, water being only one. The picture here shows an early application of carbon dioxide in fighting fires. It depicts a demonstration of Dick's Portable Self-Acting Fire-Engine (nicknamed L'Extincteur). It consisted of a metal cylinder to be carried on the back, filled with water supersaturated with carbon dioxide. When the hose was activated, the liquid shot out in a stream, and the gas discharged itself, being no longer under pressure. The CO_2 then stifled the flames. The demonstration here took place on the Thames Embankment, London, in the presence of the Prince of Wales and the Duke of Sutherland, both of whom had an amateur interest in the subject of firefighting. Three wooden structures were put up on this bit of wasteland and filled with various flammable materials saturated with tar and naphtha, likely to burn with an intensity that would defeat the conventional attempts at putting out a fire in this period. The fires were set and blazed fiercely, fanned by a wind, with thick clouds of black smoke. But all three fires were quickly extinguished, much to the astonishment of the spectators. This illustration was made by C. Robinson for the *Illustrated London News* of 29 February 1868 (Fig. 5.43).

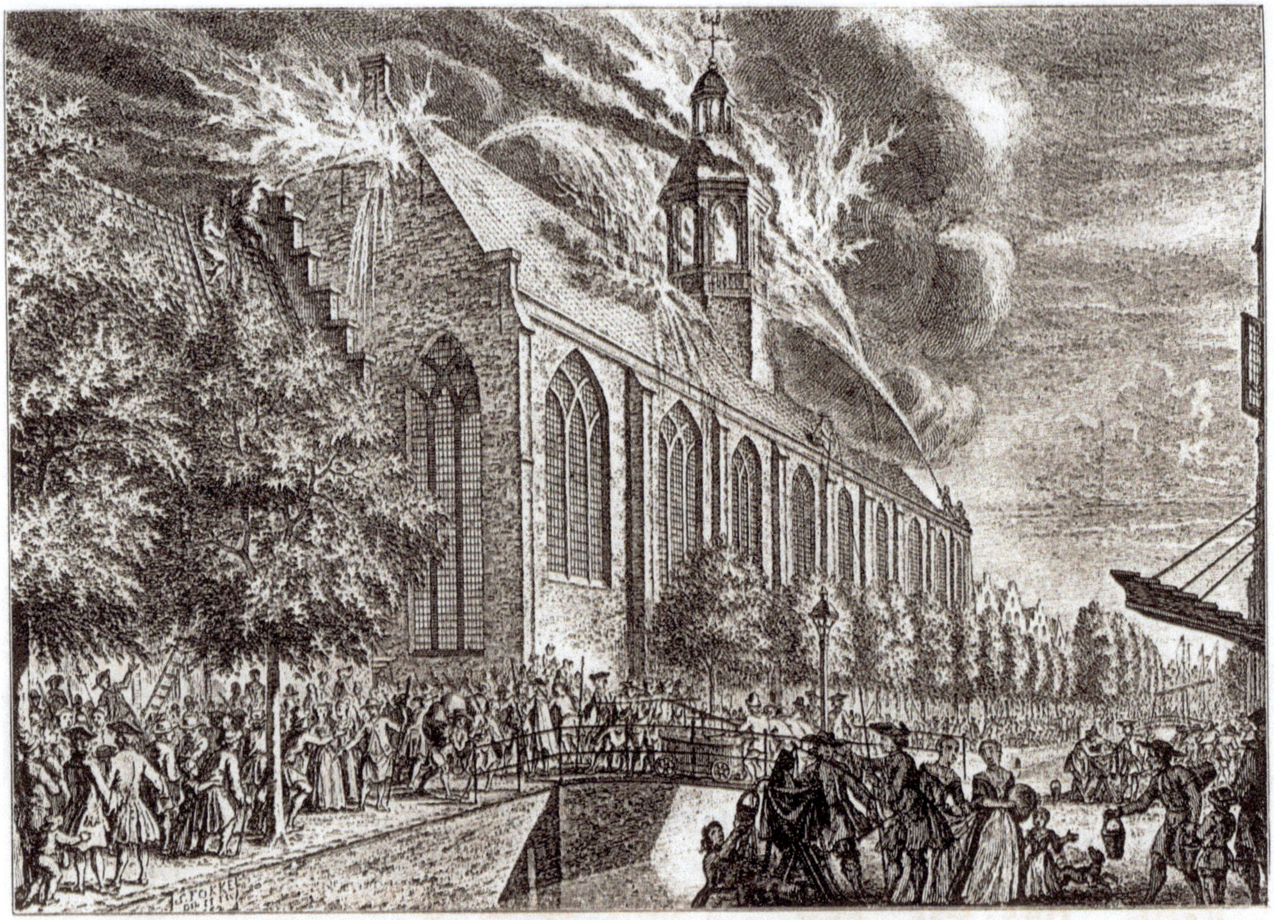

De Kapel, te Alkmaer in brand 1760. Nach einem alten Kupfer.

Fig. 5.44 The Alkmaar fire in 1760

Chapels and churches built of stone are still vulnerable to fire if they have wooden roofs. This print, quite popular in the Netherlands, shows a fire in a chapel roof in the Dutch town of Alkmaar on 21 August 1760 (Fig. 5.44). The print is the work of Simon Fokke, after Albert Janszoon van Panders, and published by Frans Houttuyn. The original is in the Amsterdam Rijksmuseum. The fire caused severe damage to the church interior, despite the efforts of fire-fighters, seen here with what appear to be rather powerful fire hoses.

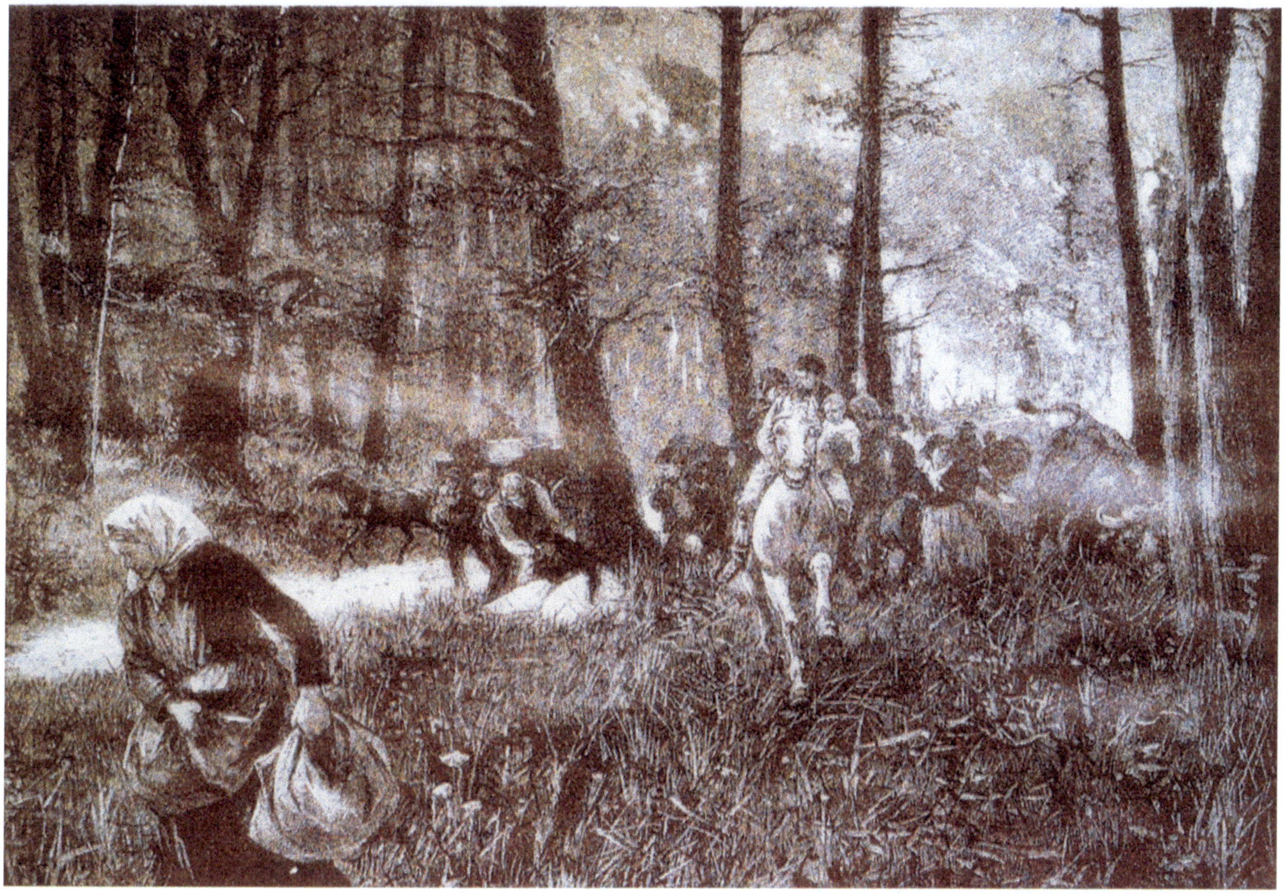

Fig. 5.45 Villagers flee a fire

The reader will have noticed that the preceding images all deal with fires in towns and cities. What, then, of the countryside, in small villages and rural settlements? Obviously, one is less likely to find pictorial material for such fires; a fire in a small village is not as newsworthy as one in a major city. Nevertheless, it is evident that concerning the building-up of a well-organized fire protection service, with brigades of trained firemen (both professionals and volunteers), urban communities in Europe and North America did much better than settlements in the countryside. This is probably not very surprising, given the resources available to a city and the potential for a larger disaster. Effective urban fire protection, purposefully developed in the course of the nineteenth century, contributed dramatically to the mitigation of fire damage in towns, in contrast to what anti-fire measures were available in villages. In central Europe the first established volunteer fire brigades did not appear in rural communities until the end of the nineteenth century. Additionally, the technical equipment of these village firemen was simple and less effective than what was available to their urban counterparts. The traditional village firefighting generally consisted of getting one's neighbors to form a bucket chain, passing water from a stream or pond to be thrown on the burning house. If that failed and the conflagration was severe, the best choice might be to gather what portable property one could, gather domestic animals, and flee. This is what can be seen here; the fire is so intense that it seems to have started to spread to the adjacent woodland as well. This image (Fig. 5.45) is imaginative rather than depicting an actual incident; it appeared in in the Czech illustrated weekly *Zlatá Praha* in 1896 to accompany an article on fires.

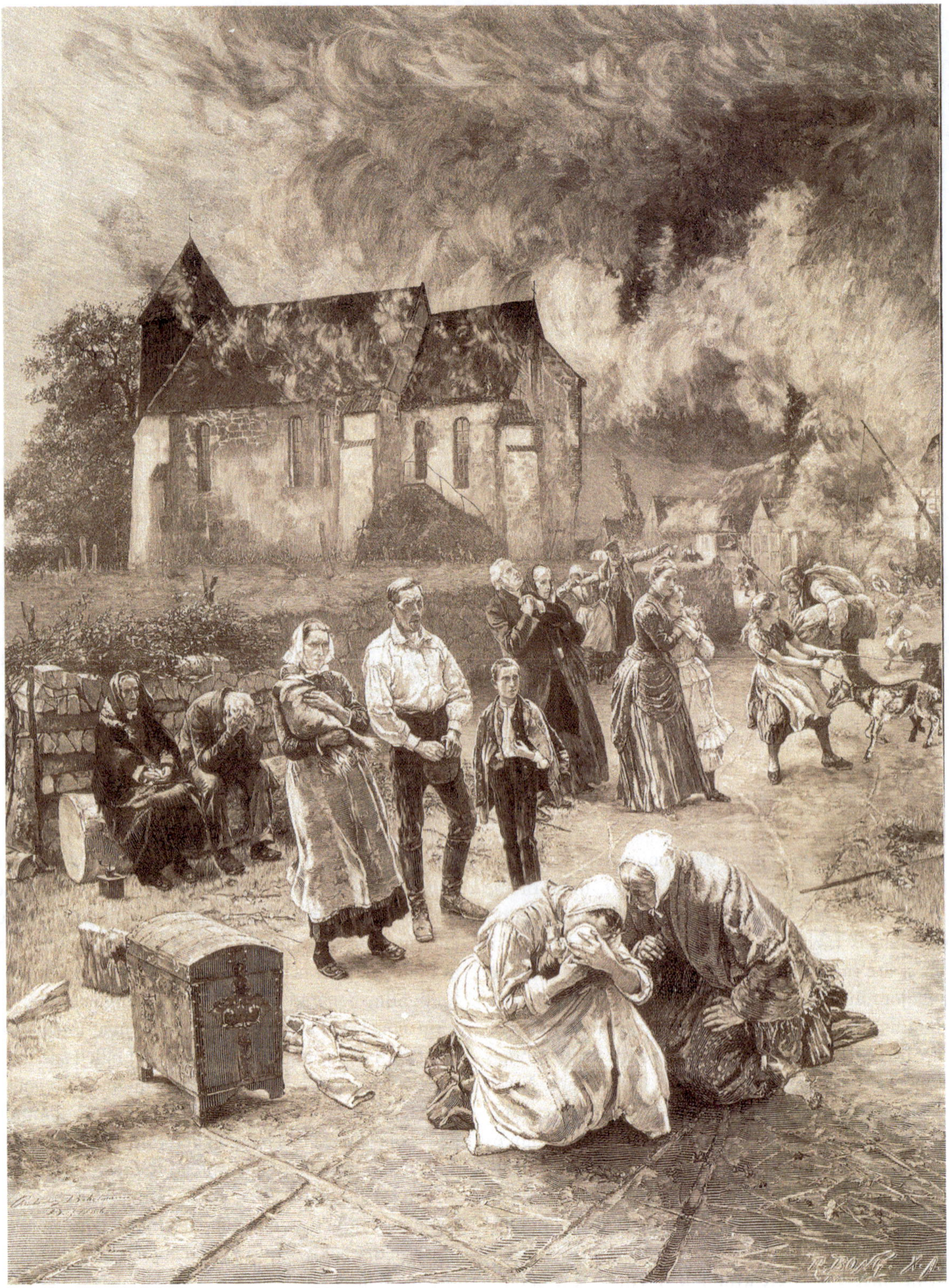

Fig. 5.46 A village chapel on fire

The image here (Fig. 5.46) is from the same source as the preceding one; distraught villagers gather in a safe open space as fire consumes their village. Note that in the case of the chapel, the fire is confined to the wooden roof.

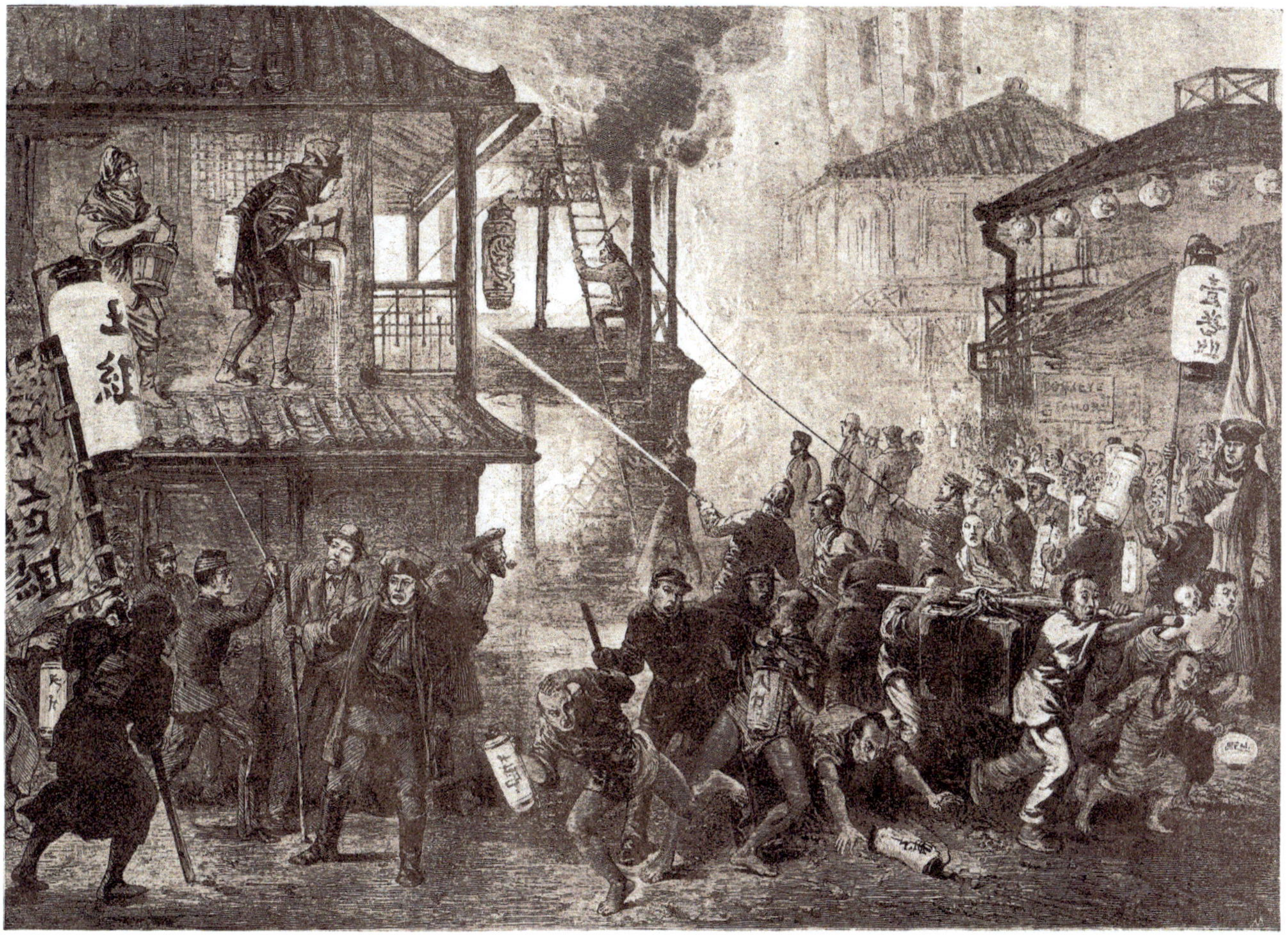

Fig. 5.47 A satirical engraving of a supposed fire in Japan

This picture was published in the *Illustrated London News* on 16 November 1878, with the title *A Fire in Japan* (Fig. 5.47). It is almost certain that the picture does not refer to any particular incident but was one of numerous pictures created to satisfy the curiosity of Europeans about the mysterious land of Japan, only recently opened to the West after the Meiji Restoration in 1868. Fires in Japanese towns must have been common, as they were overcrowded settlements in which light, wooden houses (with numerous paper components) were divided by narrow, dark streets. With lighting often supplied by paper lanterns with a burning candle at the center, it is hardly surprising if fires were a frequent peril. The picture clearly expresses a European contempt for Asiatics; in the foreground, the local Japanese are fleeing in fear, while, closer to the fire, some figures in European clothing watch the blaze calmly with impeccable sangfroid. Behind them is a tailor's shop with a sign reading "Cockeye Tailor," suggesting poor quality of the goods therein or possibly implying that the tailor is a stupid man who thinks that "cockeye" means "sharp-sighted" when in fact it means "at a crazy angle." And one can contrast the effective firemen with hoses in the center of the picture, who wear modern European helmets, with the two locals (top left), who edge nervously toward the blaze with their old buckets, from which water is carelessly spilling. Today this picture would be considered too offensive to print in a newspaper, but it reflects the prejudices of the time. One can compare it to the image of volcano tourism in Chap. 1, in which stalwart British visitors to Vesuvius are unperturbed by the eruption, while the locals run in terror.

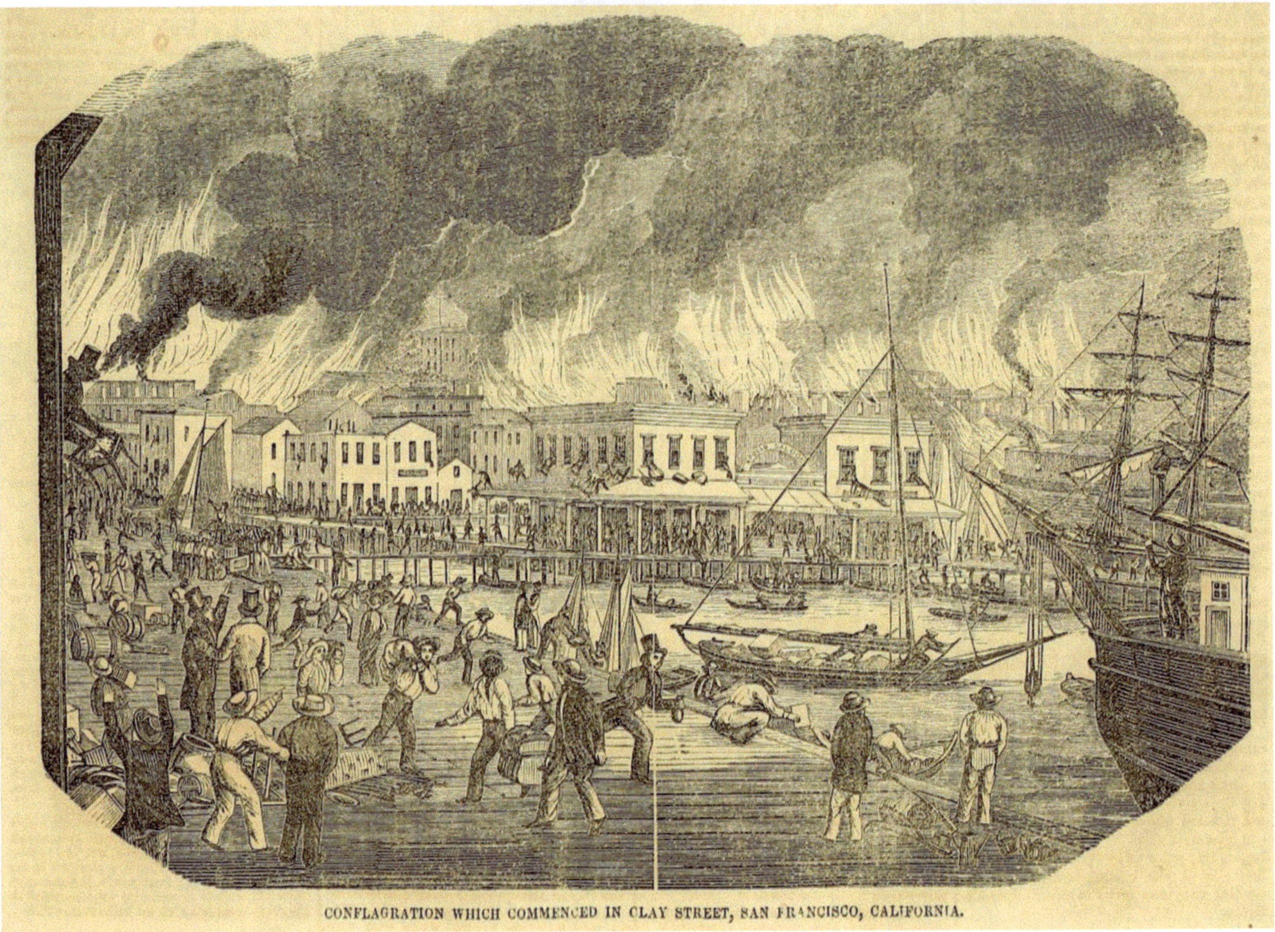

Fig. 5.48 The San Francisco fire of 1851

From Japan we can cross the Pacific Ocean to California and San Francisco. The town was said to have grown up "overnight" after gold deposits were found nearby along the Sacramento River. The new town spread out fast and chaotically, without any urban design or planning; it is no wonder that in the first years of its existence, it suffered several large fires. At the time of the first fire, on 24 December 1849, much of the town consisted of tents and other light shelters that burned easily. A few years later, the town already consisted of thousands of wooden cabins and European-style houses. In the short span of time since the founding of the settlement, civic buildings had been erected and boardwalks laid. In May 1851 the sixth and greatest great fire occurred and destroyed some 2000 newly-erected properties (about three quarters of the whole city). The fire broke out at about 11 p.m. on 3 May in a paint store above a hotel on the south side of Portsmouth Square. Once again, strong winds aided the spread of the conflagration. This hand-colored woodcut (Fig. 5.48) shows a view of the fire on 4 May from the waterfront. It was published in Gleasons' Pictorial on 30 August 1851 (p133).

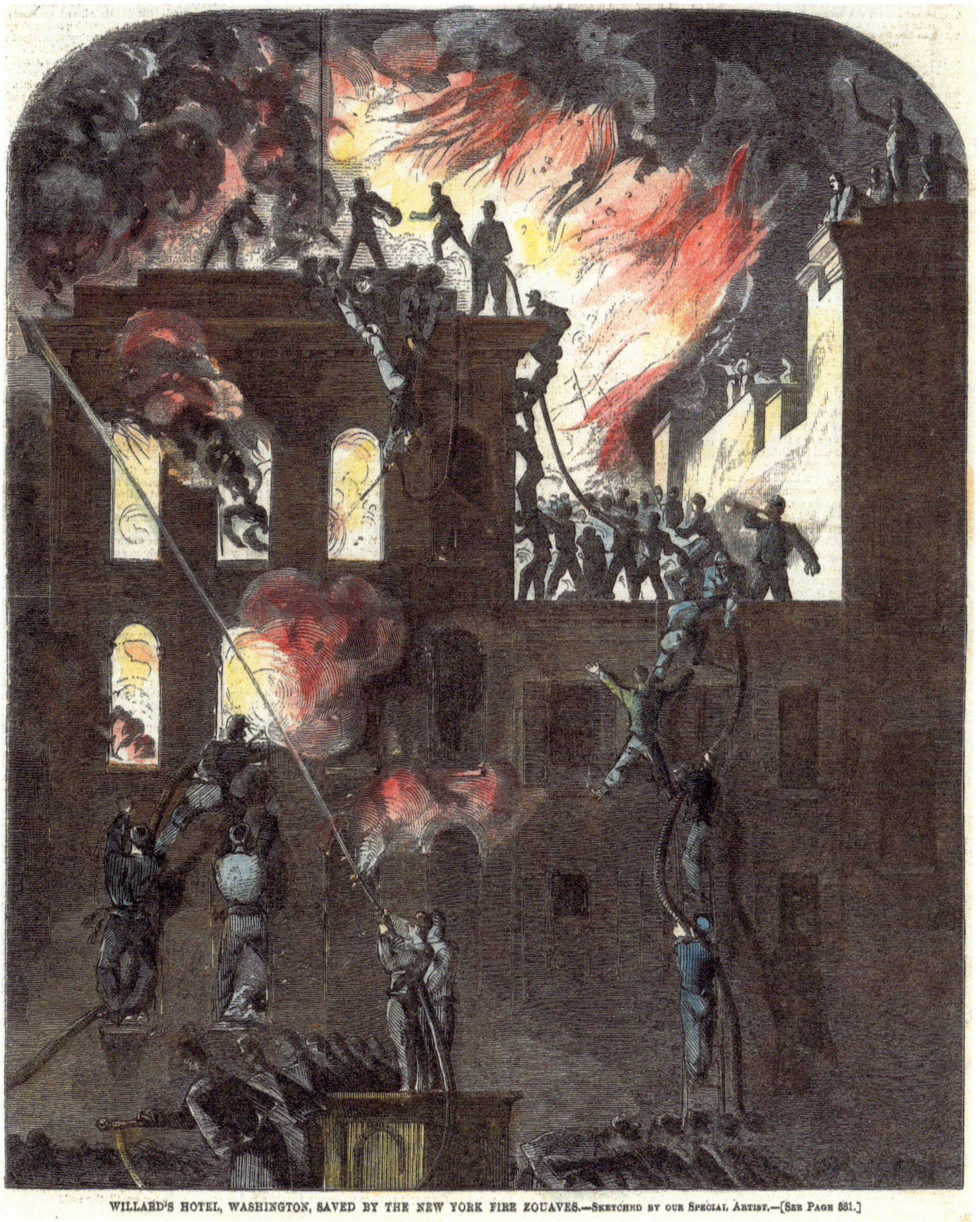

Fig. 5.49 The "Fire Zouaves" tackle a blaze in Washington, DC

The image here is the front page of *Harper's Weekly Journal* of 25 May 1861; the black-and-white print has subsequently been colored by hand (Fig. 5.49). It shows a fire at the famous Willard's Hotel in Washington, DC; the fire actually started in a tailor's shop nearby and spread to the hotel.

Combatting the fire and trying to save the hotel are men of the 11th New York Fire Zouaves. The term "Zouave" originally applied to a class of light infantry regiments of the French Army serving in French North Africa, as well as some units of other countries, modelled on them. In the USA, a drill company called the "Zouave Cadets" was formed by

Colonel Elmer Ellsworth, and "Zouave" units were raised by both sides in the American Civil War. The "Fire Zouaves" were specifically trained as firefighters. In May 1861, just after the start of the war, the Zouaves were quartered in the US Capitol building (then still under construction) and got a bad reputation for rowdy behavior.

However, they redeemed themselves by their heroic efforts in fighting the Willard Hotel fire, as shown here. As with the illustrations from Europe, here in America in this period, one sees hand-powered pumps being used to power the hoses and attack the flames on the upper floors.

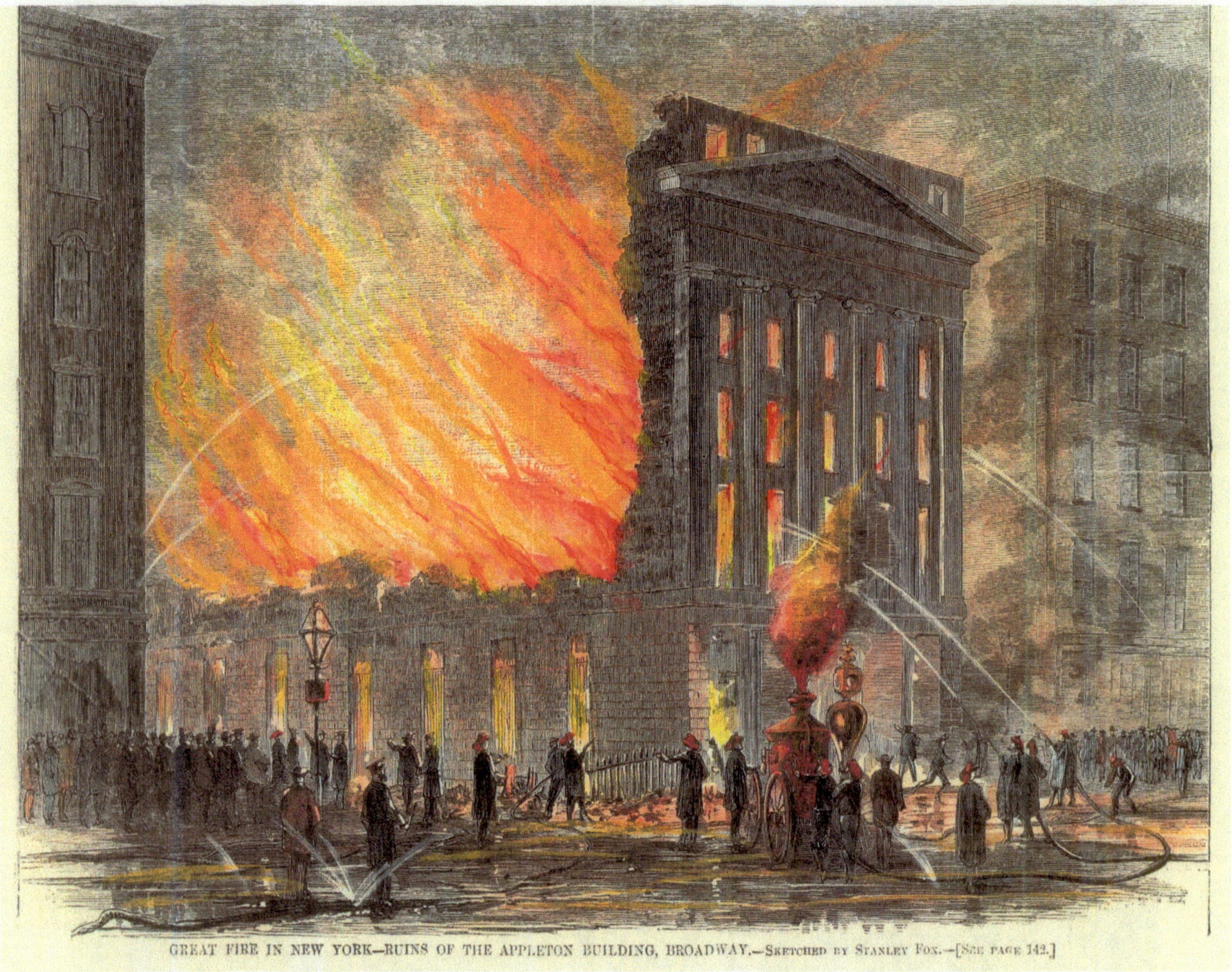

Fig. 5.50 The New York City fire in 1867

However, in 1867, steam pumps had already begun to be used to replace human muscle power for pumping water. This picture (Fig. 5.50) shows the burning of the Appleton Building, Broadway, New York City, on 12 February 1867. This large four-story building was occupied by E.S. Jaffray & Company and was used as a dry goods store. Despite the efforts of the firemen, the building was a total loss. The steam engine pump can be seen in the foreground, just right of center. The sketch was made by Stanley Fox; like the previous picture, it was published as a black-and-white print in *Harper's Weekly* and subsequently hand-colored.

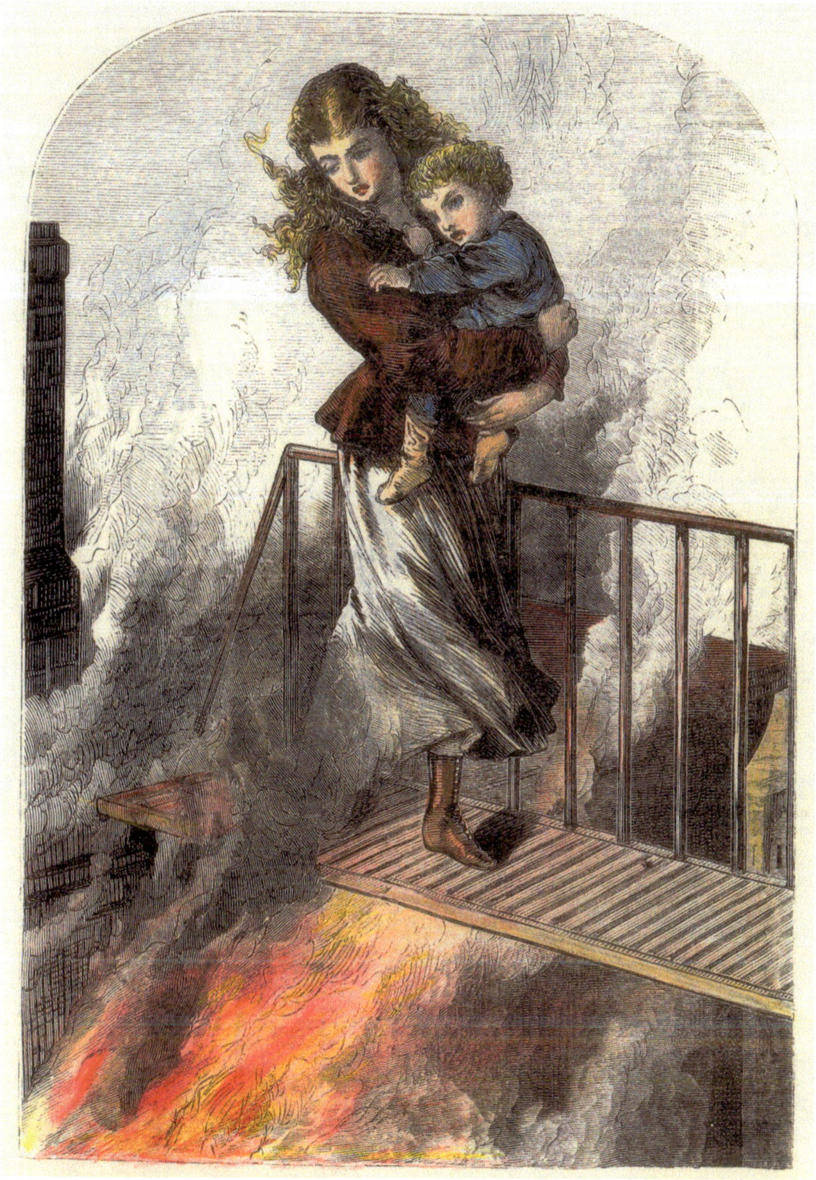

Fig. 5.51 Fleeing via a fire escape

In contrast to European towns, which often have a complex street plan due to their long historical development, newly founded American towns were usually planned and constructed according to a rectangular grid street pattern. From the late nineteenth century it also became a habit to construct tall buildings in American town centers. Skyscrapers of 30, 50, or more stories began to replace classical European-style town buildings of four, five, or six floors. This new approach to urban development laid new requirements with regard to fire protection for the people living or working in such buildings. The problem of water reserves for extinguishing fires was solved by putting giant water cisterns at the top; these had a double function: supplying water for the whole building when the mains supplies were interrupted, and serving as a water reserve for handling fires.

The second problem was one of safe evacuation of people from a burning skyscraper. Using lifts for solving this problem was out of the question: crowded lifts could very easily turn into death traps. Stairwells could also be highly unsafe. The solution was to create exterior fire escapes running down the side of the building. This was not free of problems: the means of access to the ladders connecting the individual balconies were usually locked (to prevent burglars from getting in, but, lamentably, in the case of workplaces, also sometimes for stopping employees from absconding before the working day was over). The illustration here (Fig. 5.51) shows a young woman attempting to escape a burning building, clutching her toddler. It is clearly an illustration from some serialized fiction; the style suggests it is again from *Harper's Weekly*, but the exact publication details have not so far been traced.

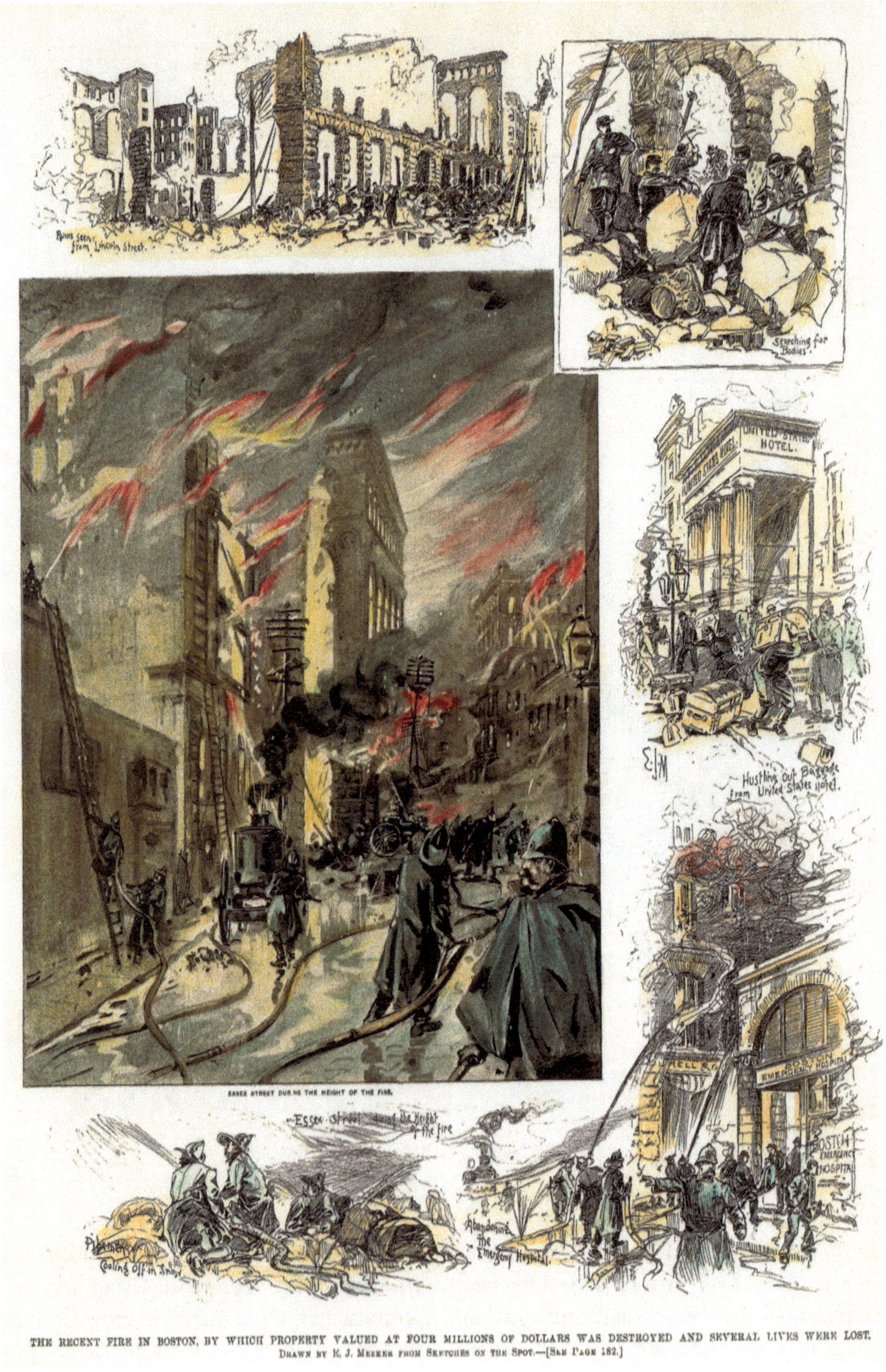

Fig. 5.52 The 1893 fire in Boston

1893 proved to be a bad year for Boston with regard to fires, with three separate blazes on 10 January, 10 March, and 19 March. The worst of these was that on 10 March, which affected principally the area between Kingston Street, Essex Street, and Lincoln Street. The fire started on the third floor of the Partridge Building, situated on the corner of Essex Street and Lincoln Street, at 4:25 a.m., and although the blaze was under control by 7 a.m., the rubble continued to smolder long afterward. The composite image here (Fig. 5.52) shows the scene on Essex Street at the height of the blaze as the main image; the various vignettes around it show (clockwise from top) the ruins viewed from Lincoln Street, searching of the ruins for bodies, removal of luggage from the United States Hotel (further down Lincoln Street), evacuation of the Emergency Hospital (adjoining the hotel), and firemen cooling off in the snow. An editorial in the *Boston Daily Globe* the following day concluded with the call, "There must be no more great fires in Boston"—a wish that was frustrated a mere eight days later.

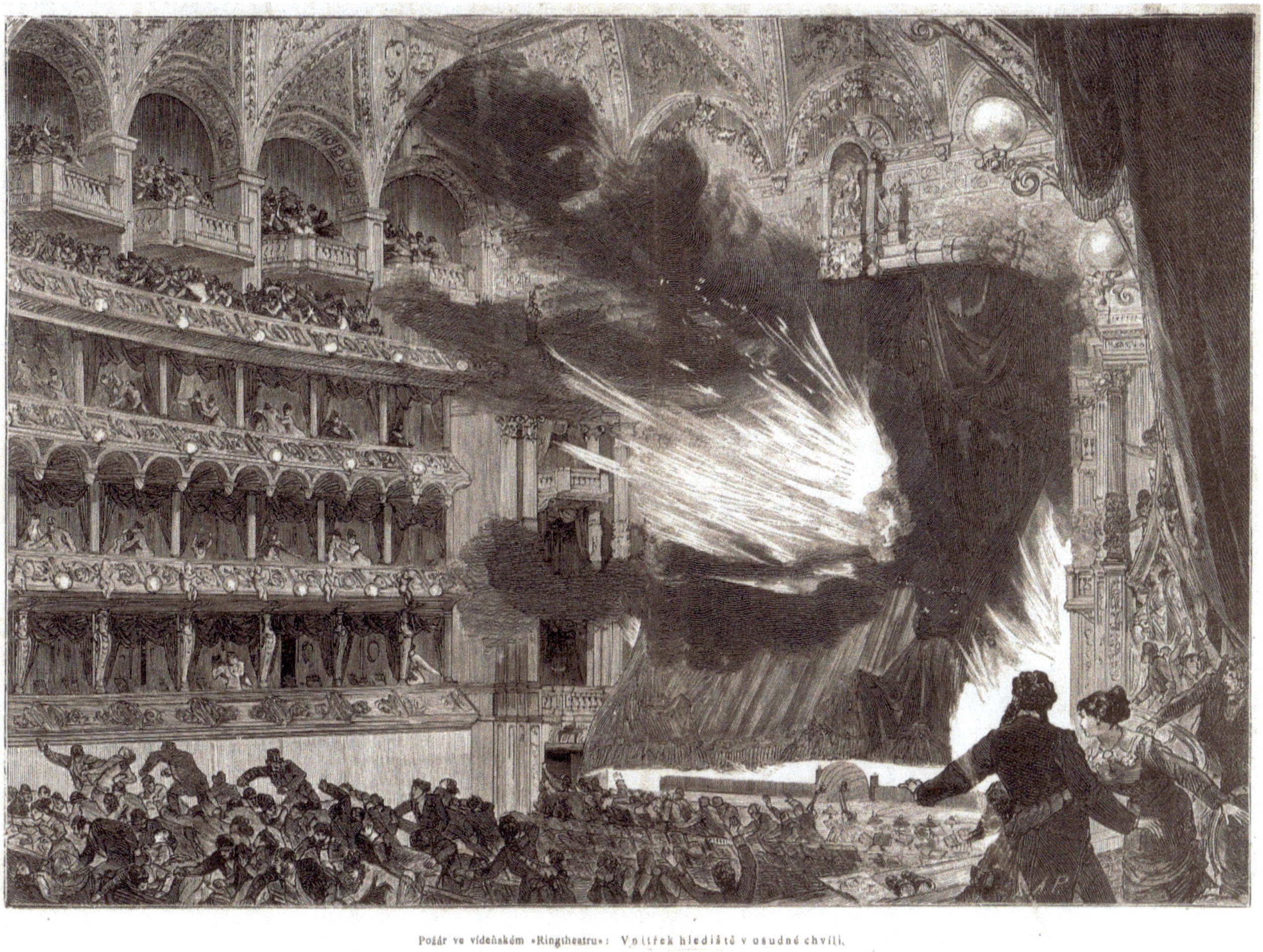

Fig. 5.53 The Ringtheater fire breaks out in Vienna in 1881

This survey of pre-1900 fires would not be complete without mention of a special kind of hazard: fires in theaters. One can think of the common saying, when discussing the limits of free speech, that no one has the right to shout "Fire!" in a crowded theater. This harks back to a particular hazard, more severe in past times than today. In a theater, many people are gathered in a restricted space; the exits are usually few, and dim light may complicate and slow down the evacuation of the auditorium in the event of a fire. Secondly, an awareness of this, conscious or unconscious, may provoke a tendency for some members of the audience to panic should they feel endangered. Many early theaters consisted of mostly wooden, easily flammable constructions. One thinks, for instance, of Shakespeare's Globe Theatre, built of timber with thatched roofing. On 29 June 1613, a cannon fired as a special effect set fire to the roof, and the whole building burned down. Given that there was only one exit and the fire would have spread rapidly (the whole building burned down), there must have been a terrible panic to escape.

Even in the nineteenth century, before electricity, lighting was limited to candles or gas lamps. Theaters could be death traps for the audience, and any fire occurring during a popular show was usually marked by casualties. The illustration here (Fig. 5.53) shows a particularly notorious case, the *Ringtheaterbrand* (Ringtheater fire) of 8 December 1881, in Vienna, before a performance of Offenbach's *The Tales of Hoffman*. The gas lights failed to ignite properly at first, and gas leaked out. The second attempt to ignite the lights caused the leaked gas to explode, as seen in the picture. The evacuation of the auditorium was botched; the emergency lighting failed, the emergency doors only opened inward (preventing easy exit), and officials called out that everyone had been saved when there were still many people trapped. The official death toll was 384, although it is believed that many more died. The engraving is from a contemporary Viennese newspaper.

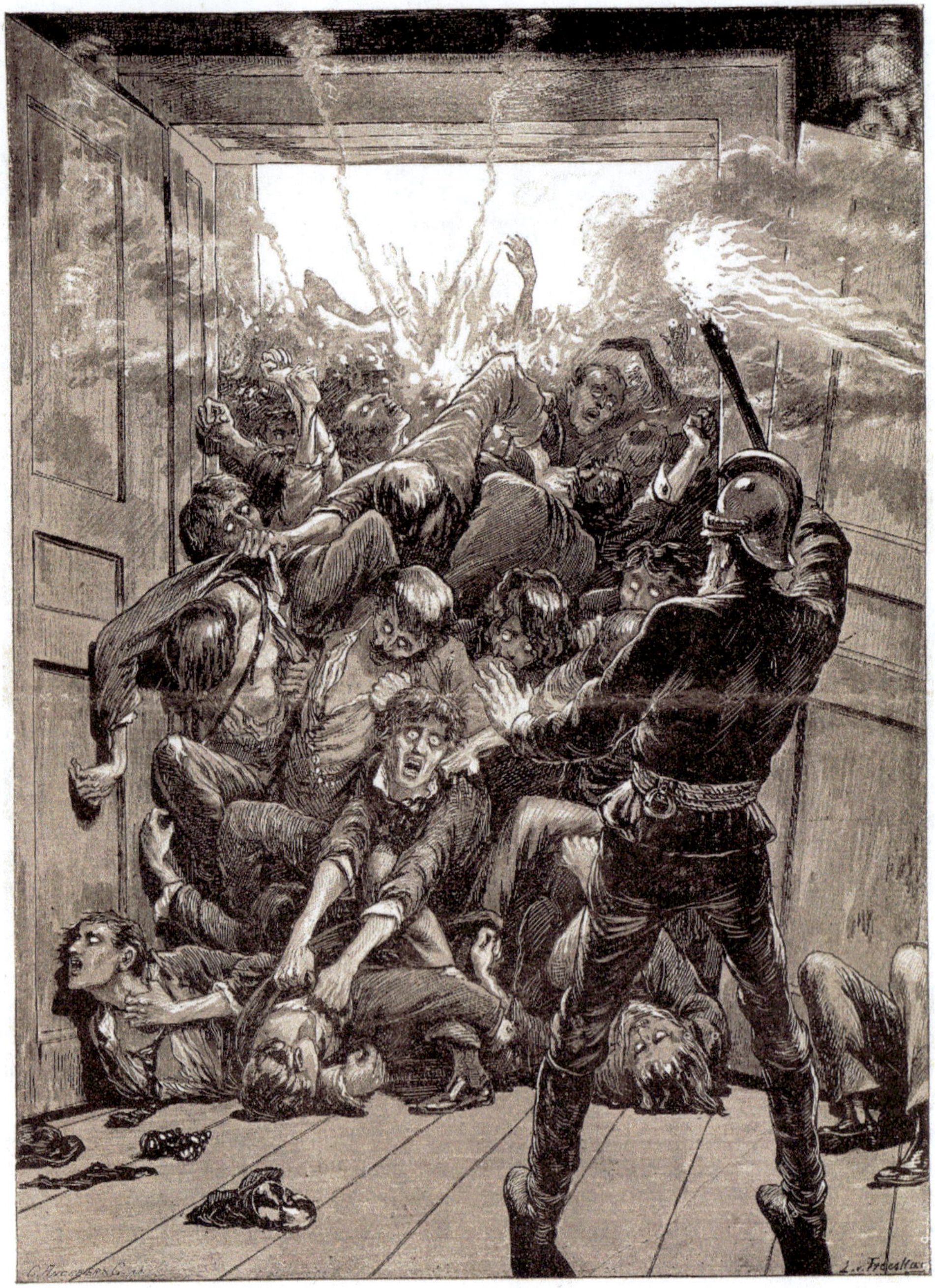

Fig. 5.54 Patrons attempt to flee the Ringtheater fire

This is another image (Fig. 5.54) of the tragedy at the Ringtheater in Vienna, this time showing the scene at one of the fatally badly designed exit doors. The mass of panicked humanity struggling to escape is wedged in the doorway, unable to go forward or back. In many theater fires the largest group of casualties were often the people trampled to death, rather than burned. The single fireman trying to calm down the fugitives in their dreadful situation does not have much chance of succeeding in his efforts. The following year, those thought to have been responsible for the disaster were put on trial, including the mayor of Vienna and the theater director, Franz Jauner (who was jailed). After this, new theater regulations were brought in, and every stage had to be provided with a metal safety curtain that could protect the audience from any fire starting backstage.

There is a little bit of musical history attached to this fire. The composer Anton Bruckner had a ticket to the performance that night, but, feeling indisposed, he did not attend. However, as he was lodging just across the street from the theater, he witnessed the whole catastrophe and was mortally afraid of fire for the rest of his life.

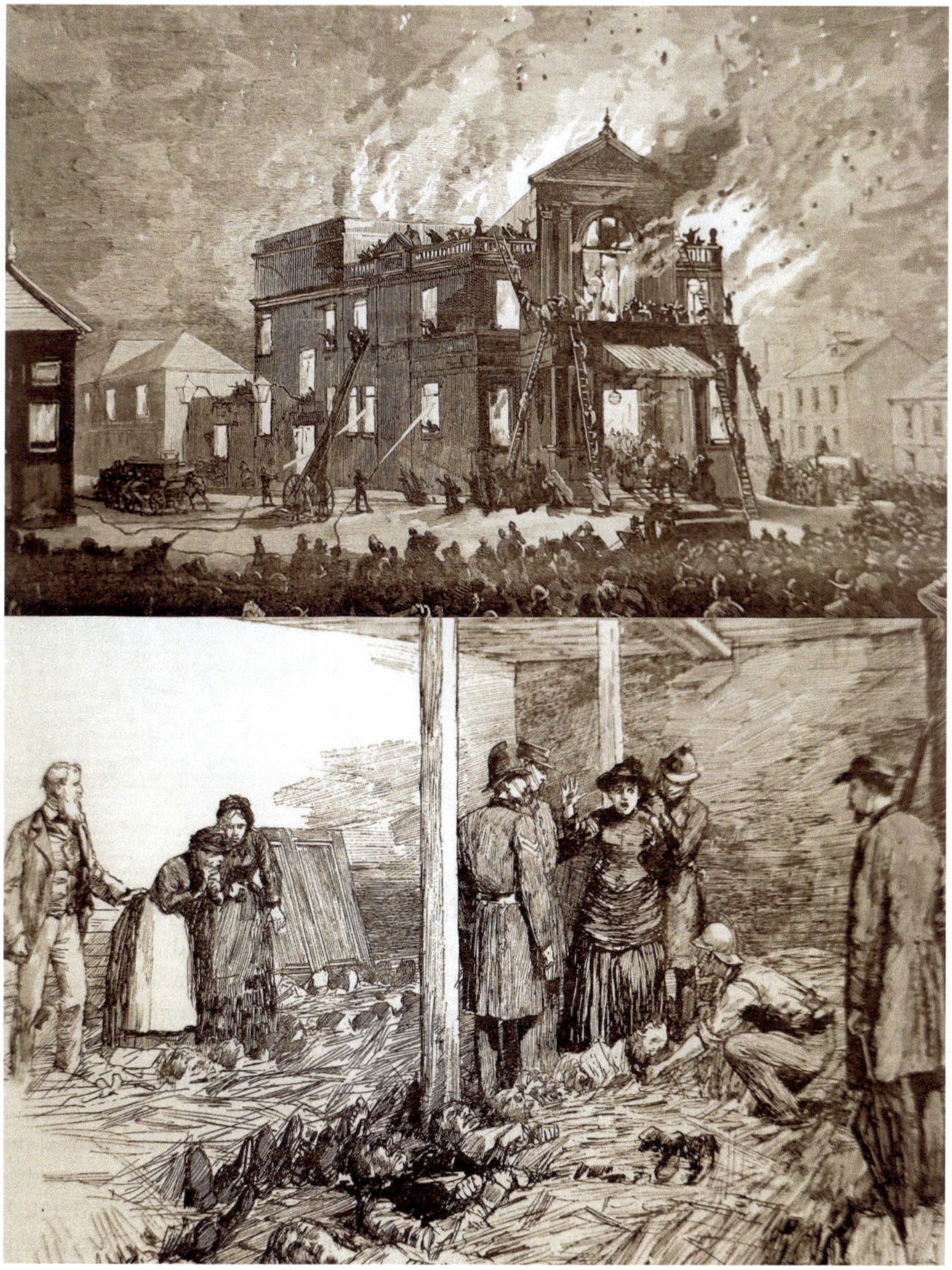

Fig. 5.55 The Royal Theatre fire in Exeter in 1887

After the Ringtheater fire of 1881, one of the most notorious theater fires was that which destroyed the Royal Theatre, Exeter, on the night of 5 September 1887. On this occasion, the fire started backstage during the performance, and it was only when the backdrop collapsed and smoke emerged that the audience realized what was wrong. Those in the stalls, the dress circle, and the boxes exited the theater quickly and without trouble, but it was a different matter for the gallery, which was packed with working-class people enjoying a rare night's entertainment. They clambered down the stairs leading to the emergency exit, only to find the door bolted. The weight of those behind crushed and suffocated those in the lead, and there was no escape then from the smoke and flames. Altogether, about 150 lost their lives. Despite the efforts of the local fire brigade, the theater was quickly destroyed.

The composite image here (Fig. 5.55) is based on the front page illustrations of the *Illustrated London News* of 10 September 1887. The upper part shows the building in flames; it was not sketched on the spot but probably reconstructed from a photo of the ruined building. The lower part is an uncharacteristically gruesome image of bodies laid out for identification. Many were so badly charred as to be unrecognizable. As one journalist wrote at the time, "... shilling customers must not expect any care for their lives in these places of public entertainment. The occupants of the boxes, who paid higher prices, had their safety well provided for; and so it is at most of our theatres."

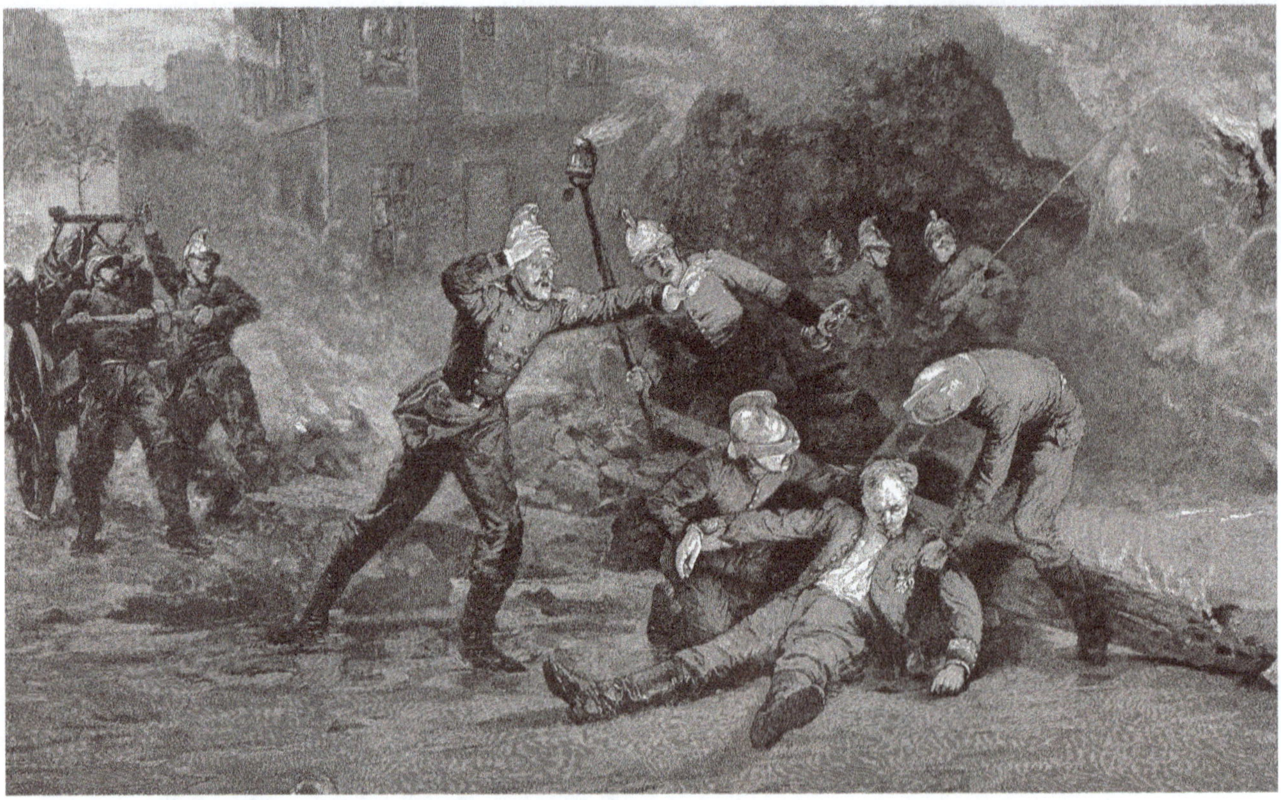

Fig. 5.56 A dying fireman

The casualty figures for major nineteenth-century fires often included the loss of life among those who were attempting to extinguish them. Being a fireman was a dangerous job. This picture (Fig. 5.56) shows a poor fireman lying dead on the ground, surrounded by his comrades. The title of the image is *Death in the Name of Honour*; it appeared in the Czech illustrated weekly *Zlatá Praha* and is based on a painting by F. Rennard, not on any particular incident. The other firemen seem stricken with grief at the depth of the tragedy, in which a young fellow fireman offers and finally loses his life in his effort to help other people in their time of peril. Today we cannot but appreciate greatly the heroism and goodwill of those many young men (mostly volunteers) who were ready to lay down their lives in the service of other people. One should also recollect how hazardous was the task of the nineteenth-century fireman, given the poor equipment of the time. No means for mutual communication; no respiration or air-filtering device for breathing; no eye protection; no asbestos, nonflammable, or heat-isolating suit; no gloves even! Yes, they had a helmet, but it was more suitable for decoration than for protection. And light in dark rooms? It was a simple burning link like the one carried by the left of the two kneeling figures in the picture.

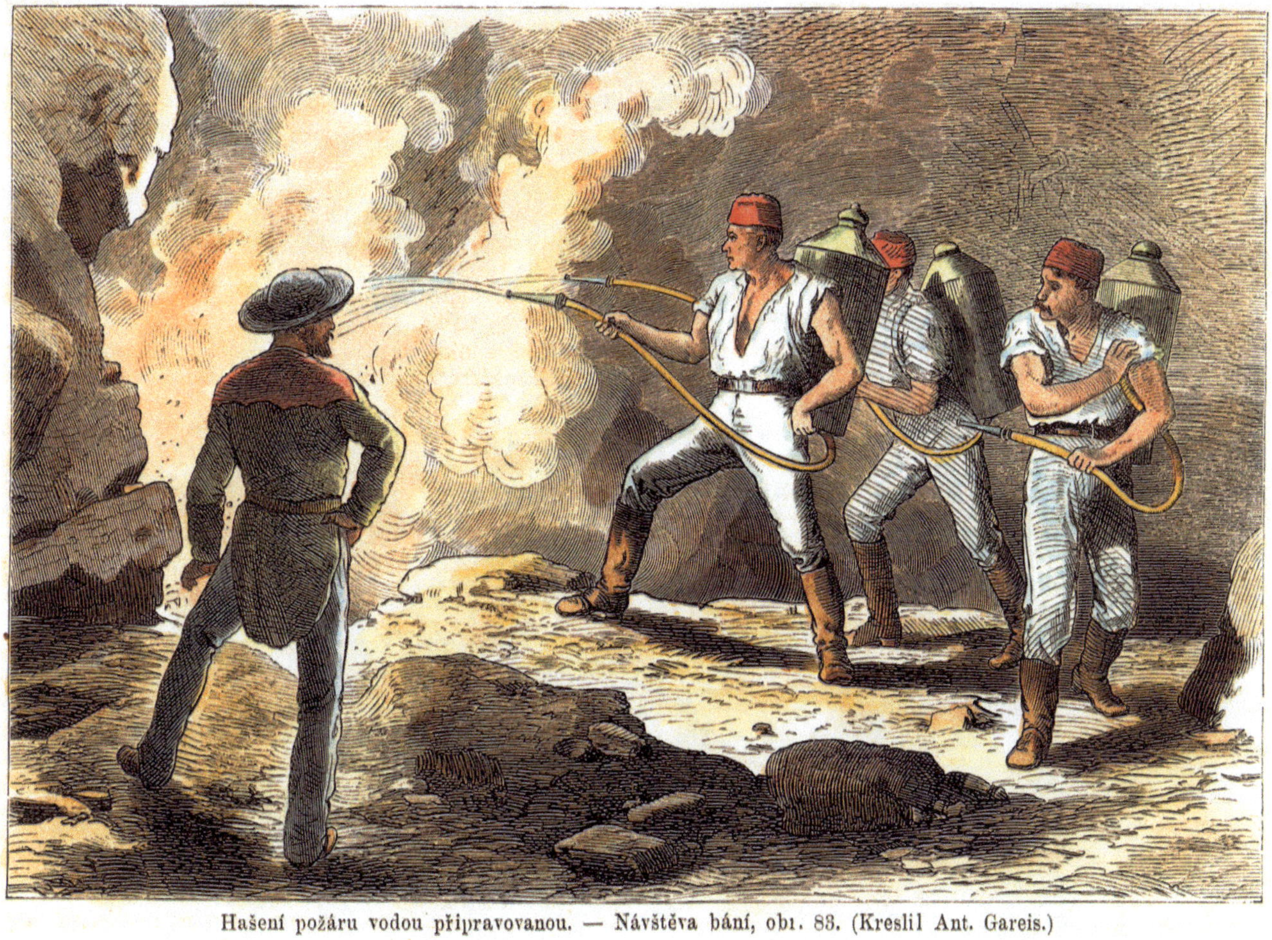

Fig. 5.57 Firefighting in a mine

Fires in mines present a particular peril, different from those on the surface in both cause and effect. Fire has always been a miner's nemesis; in a coal mine, methane is frequently released and builds up in pockets in the mine workings. Known in English as firedamp, methane is not just flammable; it is actually explosive. The danger has always been that the candle a miner uses will ignite the gas; the resulting explosion not only is directly harmful but also may cave in some of the mine passages, trapping the workers. One means of combating this hazard was to carry a canary (or a mouse) in a cage down the mine. The bird is more sensitive to the presence of gas than a human; if the bird faints, the miner knows there is danger. After 1815, miners had the protection of safety lamps, originally invented by Sir Humphry Davy. In these, the candle was enclosed by wire mesh, which prevented the flame from igniting gas; also, the flame turning blue was a sign of gas being present. However, the wire mesh was easily damaged, and a small hole in it rendered it ineffective; mine explosions continued, often with huge loss of life.

The picture here (Fig. 5.57), a hand-colored woodcut based on a sketch by Antonin Gareis (1837–1922), shows the difficulty of fighting a fire in the confined spaces of a mine working. With no sources of water down the mine, water for firefighting had to be brought down in prepared tanks worn on the back, as seen here. The miners are hardly well equipped for firefighting and, given the warm temperatures down the mine, they are only lightly dressed.

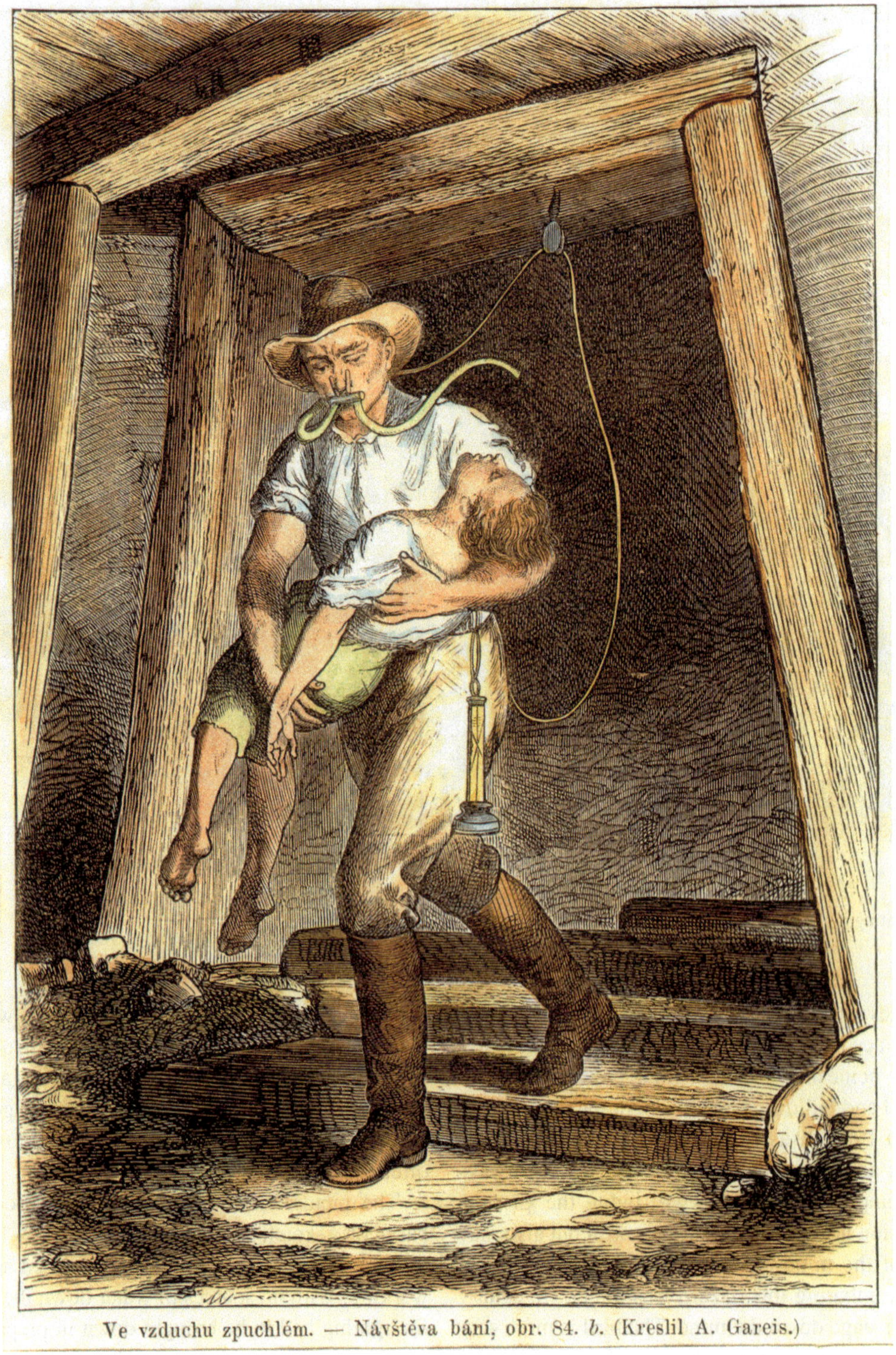

Fig. 5.58 Early breathing apparatus for use in mines

Gas was actually a hazard down mines whether it ignited or not. Carbon dioxide and carbon monoxide were a menace, as was methane. This sketch (Fig. 5.58)—like the previous one, the work of Antonin Gareis—shows a primitive form of breathing apparatus. The first self-contained apparatus for suppling oxygen for breathing was invented in 1853 by Theodore Schwan (1810–1882) and was used especially in cases of mine rescues and firefighting. The scene here is one of rescue, as a miner carries a boy to safety out of the fume-filled corridors. It also serves as a reminder that child labor in mines still existed in Europe well into the nineteenth century and even into the early twentieth century.

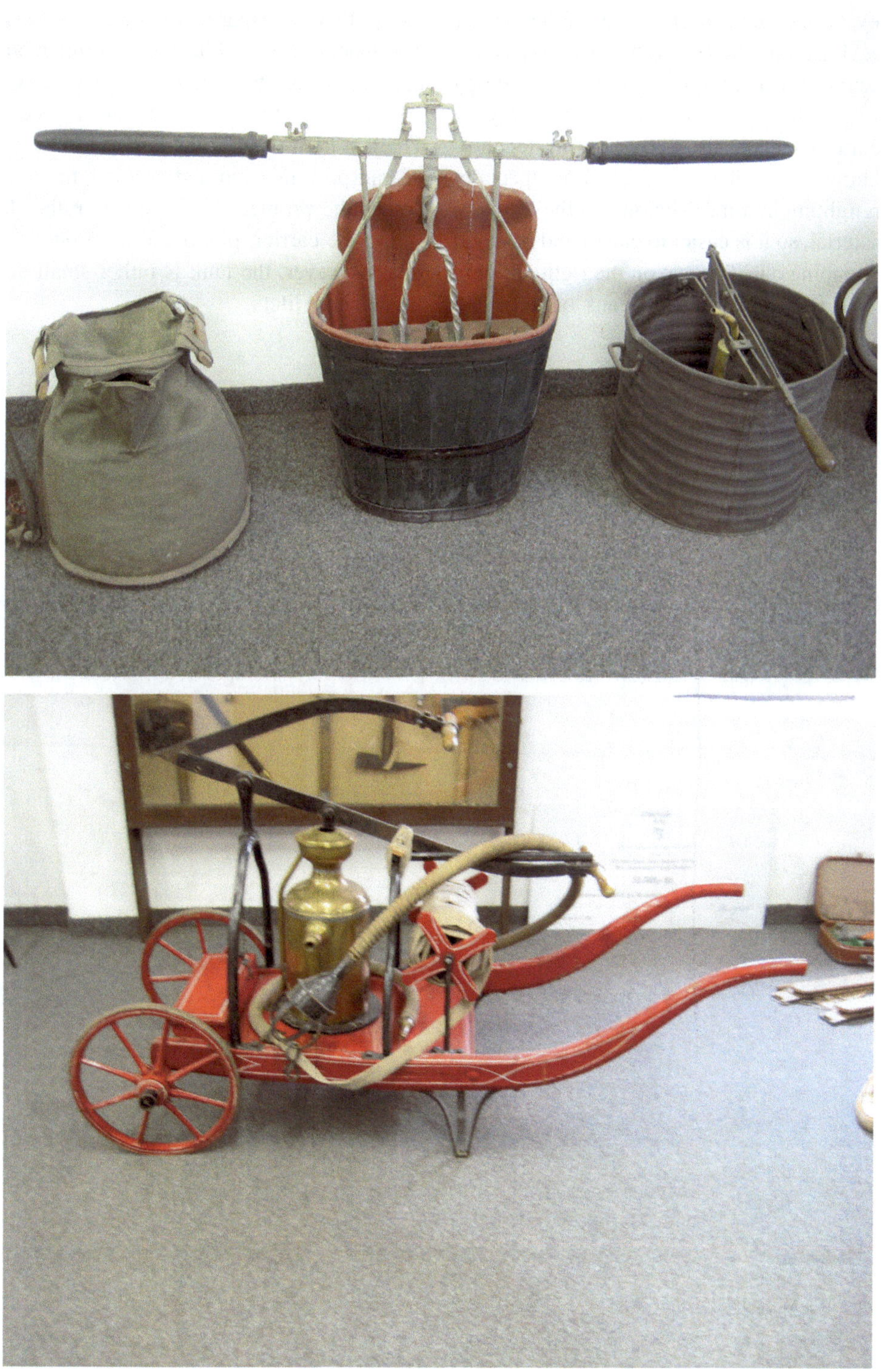

Fig. 5.59 Early examples of firefighting equipment

In the last series of images, it has been possible to make out something of the development of firefighting techniques over the period before 1900. In practice, there are three ways to quell a fire. The first is to remove the fuel, as, for instance, in the blasting of firebreaks in Hamburg in 1842. The second is to remove the oxygen, as in the use of carbon dioxide in 1868 (for small fires, sand can also be used for stifling the flames). The last, and most common, is to reduce the temperature by spraying cold water on the fire, as seen in all of the other pictures of firemen presented here. The first tool for firefighting has always been the humble bucket—something that is found in any village home and can be used for carrying water from a stream to where the fire is. Modern buckets are usually plastic or aluminum; in earlier times they would have been made of leather or wooden slats. The illustration here (Fig. 5.59) shows exhibits from the Museum of Vintage Fire-Fighting Technology in Chrastava, northern Bohemia (see http://www.jizerky.cz/dr-en/1178-museum-of-vintage-fire-8209-fighting-technology-in-chrastava.html). The three buckets are specially designed with firefighting in mind. The one on the left is of leather (a light material, so it is easier to carry) and has a lid to prevent water slopping out. The one on the right not only transports water but also ejects it by means of a special pump that is capable of spraying to a height of over 3.5 meters. The middle one is similar but with two pump handles so that two men can pump at once; it holds about 30 liters of water, so it would need to be refilled less often. The last item dates from 1790 and integrates the functions of transporting water and hosing a fire with it. Also, it overcomes the problem that water is rather heavy to carry by fitting the carrier, pump and hose onto a sort of wheelbarrow. However, the tank is rather small and probably holds less than 20 liters.

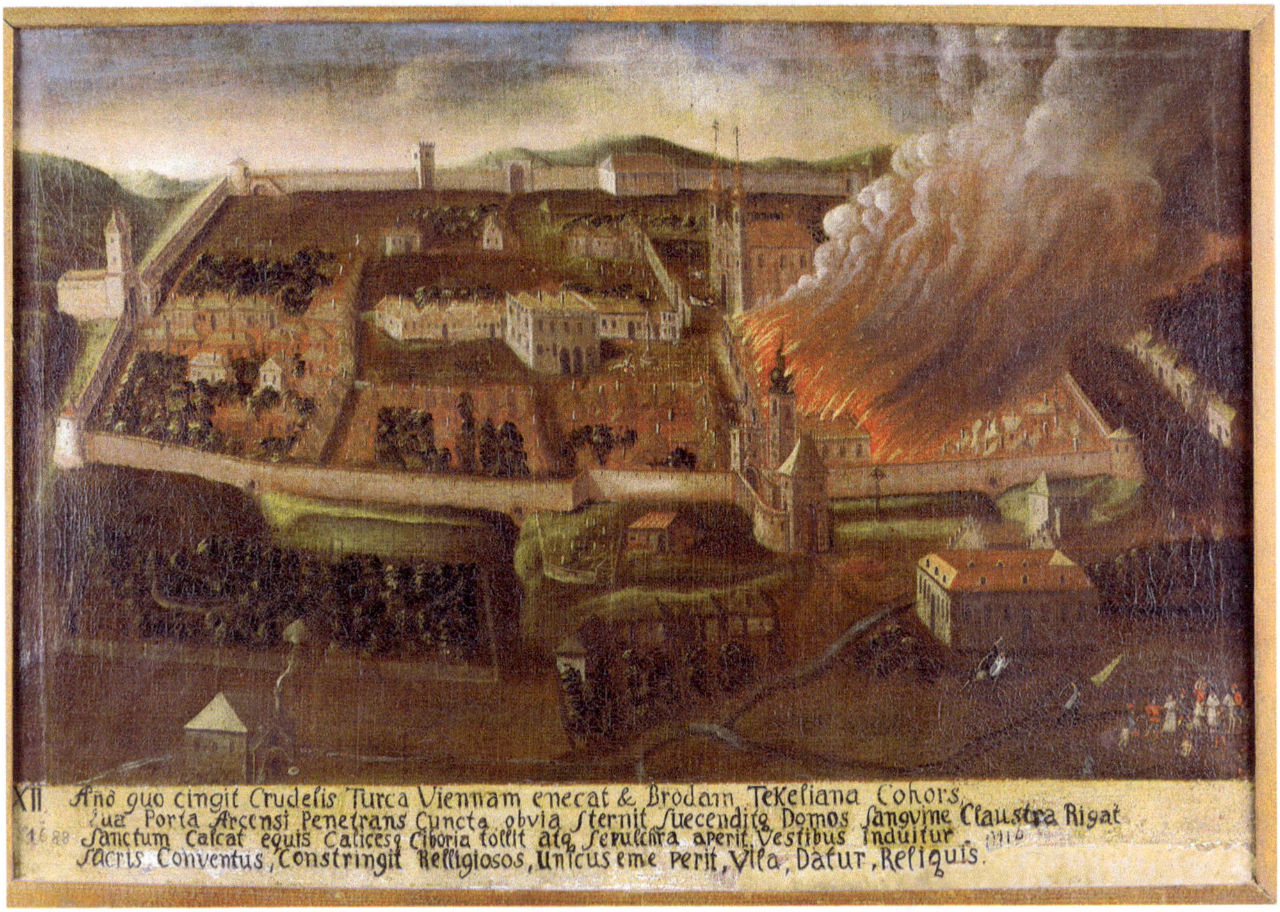

Fig. 5.60 The sacking of a Moravian town in 1683

Lastly, we can consider fire as a weapon of war. Throughout history, a common practice of invading armies has been to set captured villages, towns, and cities blazing. The passage of an army through enemy territory might be marked by noting a trail of columns of black smoke across the countryside. In the twentieth century, aerial attack made fire an even more fearsome weapon, as witnessed in the destruction of Hamburg, Dresden, and many other cities by bombing. A sad case is that of the little town of Kleve, a place of no military significance, which was set alight in World War II simply because the attractive historical wooden houses would burn easily. The number of images that could be presented here is legion; as they are deliberately man-made disasters, we shall content ourselves with three.

The first is a little-known oil painting on canvas (Fig. 5.60), kept in the Comenius Museum in Přerov, in the Olomouc Region of the Czech Republic. It depicts an obscure incident in the 1683 Ottoman campaign to capture Vienna, the high-water mark of the Ottoman march into Europe. Their forces were defeated at the very gates of Vienna in a battle in which the Polish–Lithuanian military divisions proved critical to the Turkish defeat. In the shifting loyalties of the period,

Prince Emeric Thököly de Késmárk, with French encouragement, rebelled against the Hapsburg Empire. He then threw in his lot with Sultan Mehmed IV, who recognized him as the King of Upper Hungary (partly equivalent to modern Slovakia) and aided the Ottomans in the Battle of Vienna (providing the Turks with a useful scapegoat to blame for their defeat). On 14 July 1683 (with the climax at Vienna still some months away), Thököly's forces, some 5000 strong, sacked the south Moravian town of Uherské Hradiště located 120 kilometers northeast of Vienna. The painting shows part of the town in flames; 368 of the inhabitants, including women and children, were massacred.

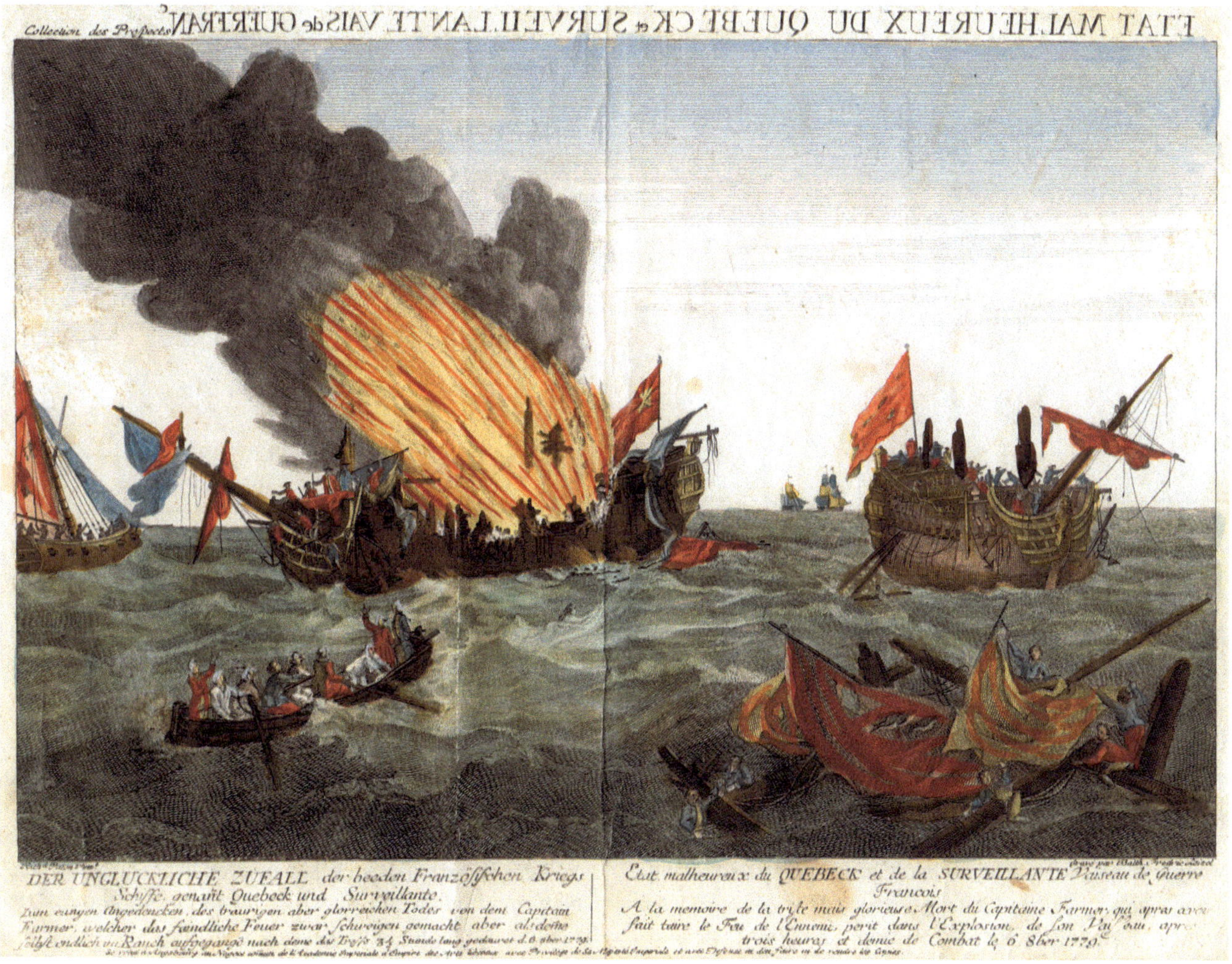

Fig. 5.61 HMS Quebec attacks the frigate Surveillante off Ushant in 1779

Fire was an important weapon of naval warfare in the days of wooden ships. One common and potentially devastating tactic was the use of fireships to attack a fleet at anchor. It needed only a few men to crew an old boat filled with flammable material, guide it into the enemy position at night, and set fire to it. It could be very difficult for the fleet so attacked to maneuver away in the confined space of the harbor. The picture here (Fig. 5.61) depicts an action that occurred during the American Revolutionary War. Both France and Spain joined the war on the American side. British naval strategy in home waters was to keep their major forces in port, while a screen of frigates kept an eye on any naval traffic leaving French ports. On 6 October 1779, one of these frigates, HMS Quebec, spotted a French frigate, Surveillante, off the island of Ushant and joined battle, despite having lost her heavier guns in an earlier accident. In a furious engagement, both ships took heavy damage and became locked together. Fire was ignited by the cannon fire, affecting both ships, but while Surveillante was able to break free and extinguish the flames, on Quebec the fire reached the powder magazine, and the ship blew up. The engagement became a popular subject for illustration, and many pictures of it survive. This one is a zograscope copper engraving, colored by hand, and was made in Augsburg, Bavaria, in the 1780s. The original print comes from a private collection.

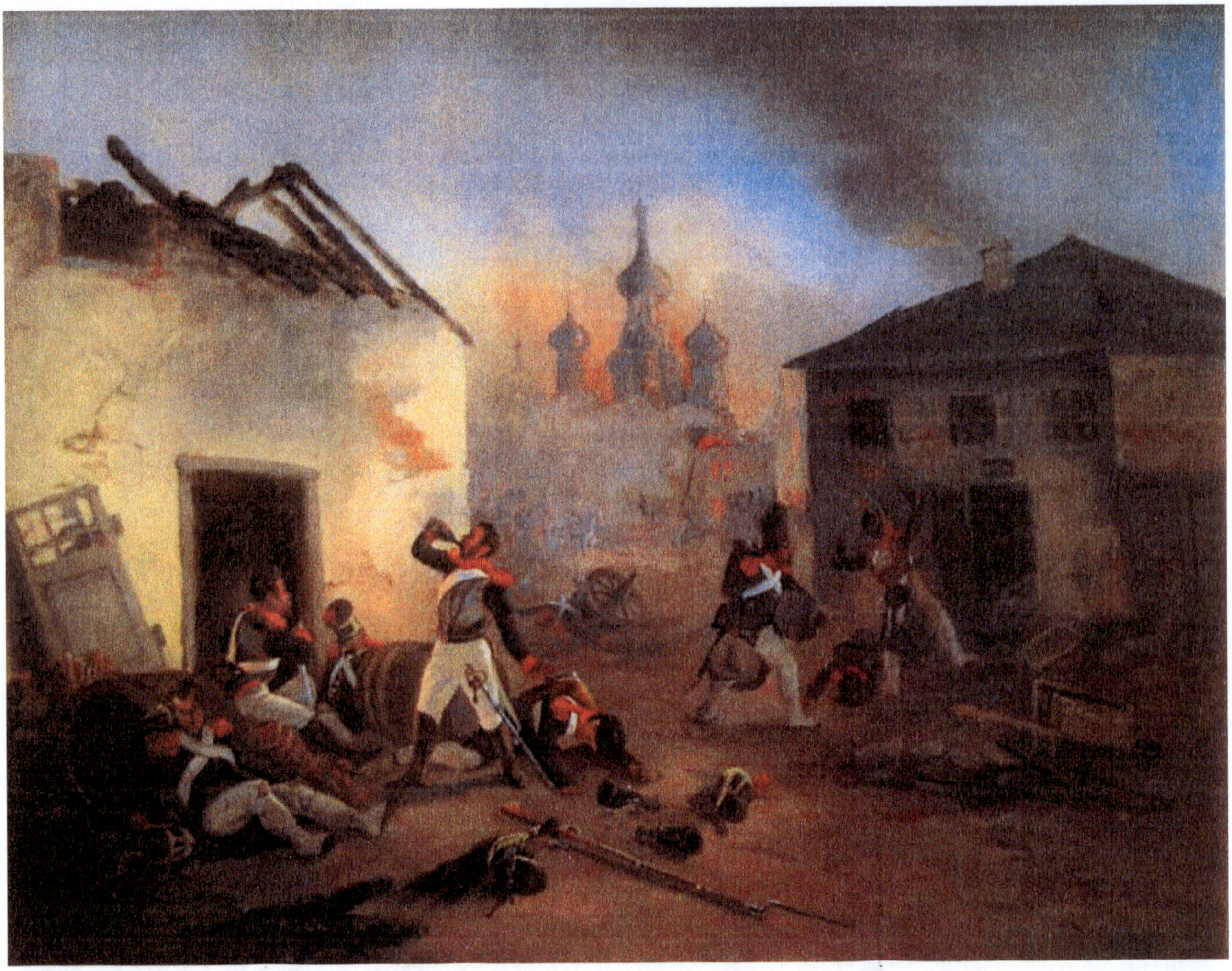

Fig. 5.62 Moscow burns in 1812

The story of Napoleon's attack on Russia in 1812 is too well known to require much elaboration here. He eventually brought the Russian army to battle at Borodino, achieving a French victory but not a decisive one. The Russian commander, Kutuzov, withdrew, allowing the French to enter Moscow. Napoleon had assumed that taking the Russian capital would force an end to the war. In this, he was mistaken. Most of the inhabitants fled before the French troops arrived, and Napoleon found himself master of an eerie, empty city. Almost immediately, on 14 September, small fires began to flare up in different parts of the city. The causes remain uncertain; some may have been accidentally caused by the French troops, but probably most were Russian acts of sabotage. These small fires then merged to create a huge one, which engulfed the mostly wooden buildings of the city (the Kremlin and stone-built palaces generally survived). The retreating Russians had carefully removed or destroyed any firefighting equipment. Napoleon himself retired to a castle outside the city, whereupon the French soldiers lost all discipline and started looting and pillaging, as seen in this anonymous painting (Fig. 5.62) of soldiers carousing in the street as the city burns behind them.

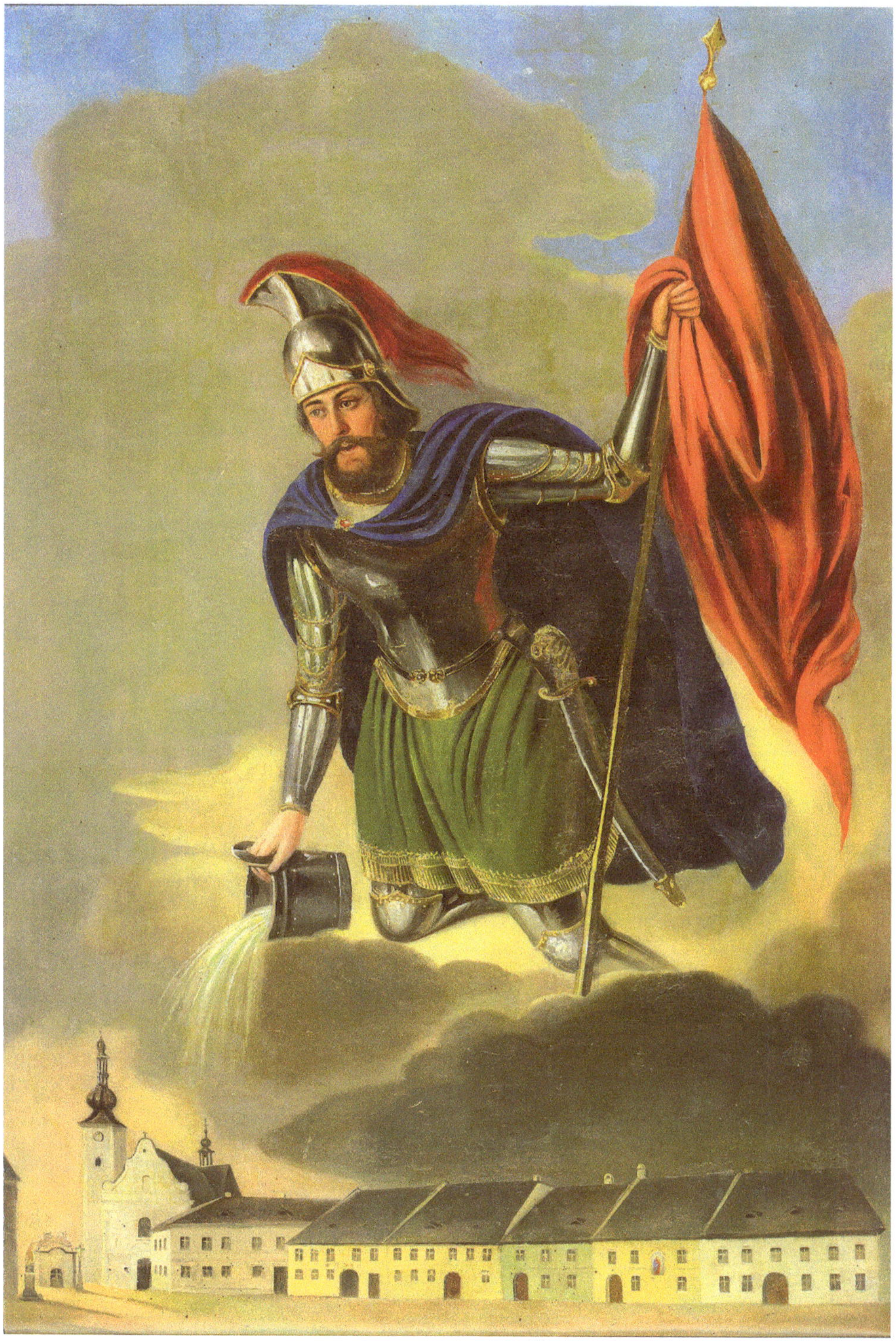

Fig. 5.63 St. Florian, the patron saint of firemen

If all else fails, pray. The appropriate saint is St. Florian, the patron saint of firemen, believed to come to the aid of those threatened by fire. He was originally a Roman soldier stationed near the River Enns in Upper Austria at the end of the third century AD. Shortly after converting to Christianity, he was imprisoned by the local Roman tribunal and sentenced to death; he was drowned in the waters of the Enns in AD 304 and was subsequently declared a martyr. In central Europe he is usually depicted pouring water (from the River Enns?) onto the roofs of burning churches and houses. The painting here (Fig. 5.63), showing the saint wearing a fireman's helmet and looking rather Wagnerian, is from the Comenius Museum in Přerov.

Conclusions

If someone is cast out in the wilds with no shelter, we say that they are "at the mercy of the elements"; or, if they find shelter, they are "protected from the elements." It is clear that the mercy on offer is not very great. After our varied excursion through some 238 pre-photographic images of various kinds of historical natural disasters, it seems from such an extensive inspection of the nature of our planet's manifestations that one is justified in seeing the interplay of the "angry elements" with the human race in mostly negative terms.

The reader may feel entitled to completely reject the ancient paradigm of four material elements allegedly ruling the universe. On one hand, we have to appreciate the complex approach to natural disasters of the ancient Greek philosophers, especially Empedocles (490–430 BC) and Aristotle (384–320 BC), who first formulated the four-element theory. They properly selected and named fundamental categories that were limited at the time by the only means that these savants had at their disposal for direct observation and examination; they were dependent on human senses: sight, touch, taste, smell, and hearing. Consequently, it appeared to these ancient philosophers that the majority of human troubles that people had to undergo in the course of their lives in the natural world must be controlled by these four elements, especially in those situations in which the dynamic and often merciless faces of these four categories were made manifest.

On the other hand, no one doubts that mighty natural forces from both the Earth's interior and the extraterrestrial space around our planet have repeatedly acted on the Earth's surface in such a way as to endanger life. The nature of the resulting dangers and the possible damage caused by individual types of natural disaster have often differed from one other, as the disasters themselves differed. And we still think, to some degree, of these disastrous elements in classical terms. When we say that a man is at the mercy of the elements, we are not thinking of hydrogen, helium, lithium, and the rest of the periodic table.

Some extreme cases can be cited. One might cite the long-lasting, high floods of the principal rivers in China. During the floods on Huang He River, millions of people died in the years 1887 and 1937, and death tolls in the hundreds of thousands also marked the floods in 1911 and 1935. Also striking was the vain attempt of people to escape from the high-velocity avalanche that followed an eruption from the snowy crater of the distant volcano Nevado del Ruiz in Colombia on 13 November 1985. Hot lava from the crater melted the snow covering the top of the volcano, and soon pyroclastic material, stones, mud, and melted snow created a giant combined avalanche (a lahar), which hurtled down the valley and erased all of the settlements in its path, wiping out communications and life in the lower part of the valley. People were caught as in a trap, without any chance to escape. In the town of Armero alone, some 28,700 people died.

Extensive floods and massive lahars do not represent the only natural disasters killing people on a large scale; tsunamis reaching flat coastal regions may—under specific conditions—also kill thousands of people. The death toll from the Sumatran earthquake and tsunami in 2004 was some 230,000 people; in the case of the Tohoku earthquake in Japan in 2011, there were over 15,000 casualties (almost entirely from the tsunami, and very few from the earthquake). Widespread drought and famine may strike an entire region, killing over a period of years. Finally, the most serious scourges of history have been the impacts of epidemics and pandemics. Some of these struck tens of millions of victims, as in the case of the Spanish influenza of 1918–1920 (which had around 25–50 million victims all over the world) or the plague pandemics of the mid-fourteenth century, which killed between a third and a half of the European population. Even today, pandemics kill despite modern advances in medicine: the World Health Organization reports an average yearly toll of 100,000 casualties from cholera, and malaria may kill four times that number.

However, to be objective, we have to admit that the most tragic situations occur infrequently; against most major disasters, one can compare numerous, much weaker events that cause small material losses and few casualties, if any. It is probably a reasonable approximation to say that for every event that kills 1000 people, there are ten that kill 100 people and 100 that have only ten casualties.

When a medium-sized earthquake occurs in sparsely inhabited zones (such as the Aleutian Islands or Kamchatka) the material damage and death toll will be much lower than

J. Kozák, R. M. W. Musson, *The Illustrated History of the Elements*, https://doi.org/10.1007/978-3-030-21426-5

with the same situation in more densely inhabited zones. It is notoriously the case that a map of the world's largest earthquakes has few common points with a map of the world's deadliest earthquakes. It follows that from the point of view of casualties, the most fatal situations usually appear when disasters occur in those densely inhabited regions in which the protection of the population is limited by both technical restrictions and the natural situation. The map of deadly earthquakes also shows a concentration of casualties in places with poor building construction, and this tends to be regions where the materials needed to construct safer houses—chiefly, wood—are lacking.

We can conclude that the four-element model presented by Empedocles and Aristotle in the middle of the first millennium before Christ generally lacked utility. Certainly, it neither explained the reasons for natural disasters nor allowed prediction of them; fear of the power of the elements remained. Even in ancient times, people did make some effort to protect themselves against the natural world. Having in mind the data from our pictorial series incriminating the ancient "Earth elements" as major societal enemies, we must admit that their dynamic manifestations did real harm, especially in ancient times. The ignorance of ancient people about natural processes often resulted in human resignation and a passive awaiting of the punishment from heaven for all of their sins.

Today the situation has altered; modern society has learned much about the great games of Nature and is ready to mitigate or neutralize a number of them; in some cases, we are able to predict them or to recover from or compensate for the troubles these elements cause. With modern technologies, one can see an intensive effort to mitigate the effects of strong natural disasters. The newly organized international projects for protecting coastal areas against tsunamis in the Mediterranean Sea region, in the Caribbean, and in the Pacific are examples. With regard to malaria, modern programs have reduced the mortality rate by around 30% in recent years. Bubonic plague, once so feared, yields easily to modern medicine and is no longer the killer it once was.

If the elements are no longer to be feared quite so much, it becomes easier to paint another, much more friendly portrayal of the elements of Nature, regardless of the fact that many questions still remain unanswered and many problems remain unsolved at present. If everything is made of the four elements, it follows that all of the good things in the world—as well as the bad things, such as disasters—are the product of these elements. As we described in the introduction, the elements help us in many ways. We are no longer driven to view the elements purely in such negative terms, as the progenitors of such disasters as are depicted in this volume.

Johann Zahn (1641–1707) is best known today for his work on optics, but he also published (in 1696 in Nuremberg) a work dealing with natural history, titled *Specula Physico-Mathematico-Historica Notabilium ac Mirabilium Sciendorum*, which includes this copper engraving, showing monks from the vicinity of Campi Phlegrei (the town of Pozzuoli is in the background), cooking their meals using the hot gases rising from volcanic steam holes. Figure 5.32 is taken from the same source

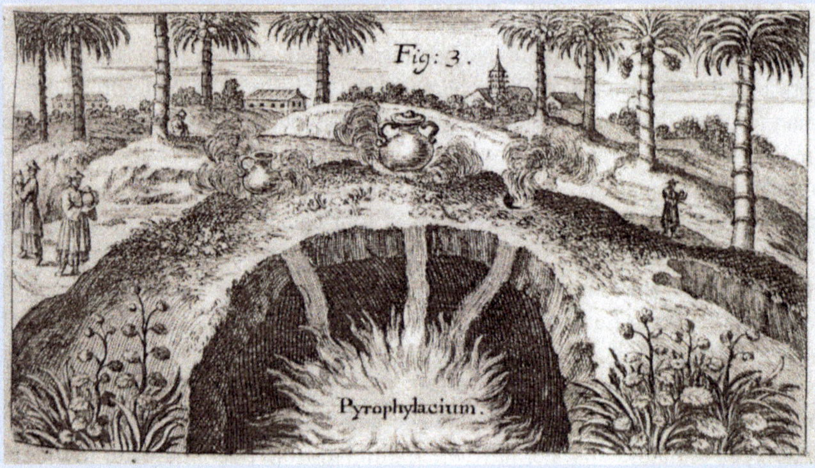

This image shows the use of volcanic hot water for bathing in mineral water at Lake Rotomahana in the North Island of New Zealand. A detailed volcanic and geological survey of the territory of New Zealand was made by the German–Austrian geologist Ferdinand von Hochstetter in 1859, at the request of the New Zealand Government. The picture here is a copy of one of Hochstetter's sketches that appeared in the Czech illustrated weekly publication *Zlatá Praha*

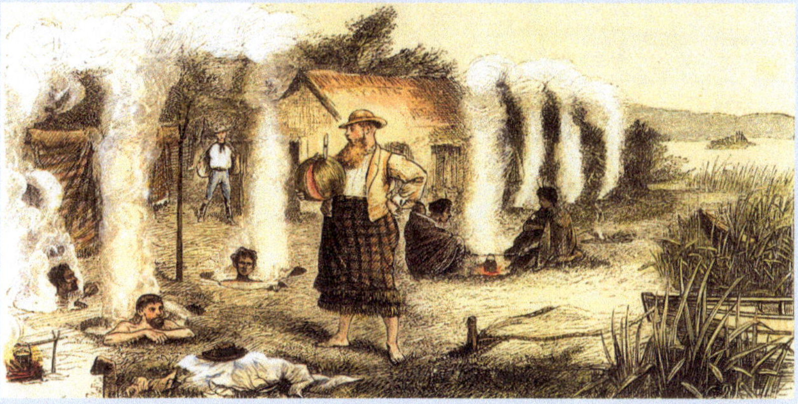

Here is another example of cooking on a volcano; this time a group of tourists is preparing their breakfast on the summit of Vesuvius, using the hot gas coming from a volcanic fumarole. The picture documents the relationship between volcanoes and the European human community at the end of the eighteenth century, with the growth of volcano tourism. With the development of inexpensive international travel, volcano tourism is now even more popular than at any previous time. How many millions of people repeatedly visit volcanoes, hot springs, geysers, and other breathtaking volcanic features? And should you wish to try volcano-cooked food, a restaurant in the Timanfaya National Park, Lanzarote, offers you the opportunity, should you visit

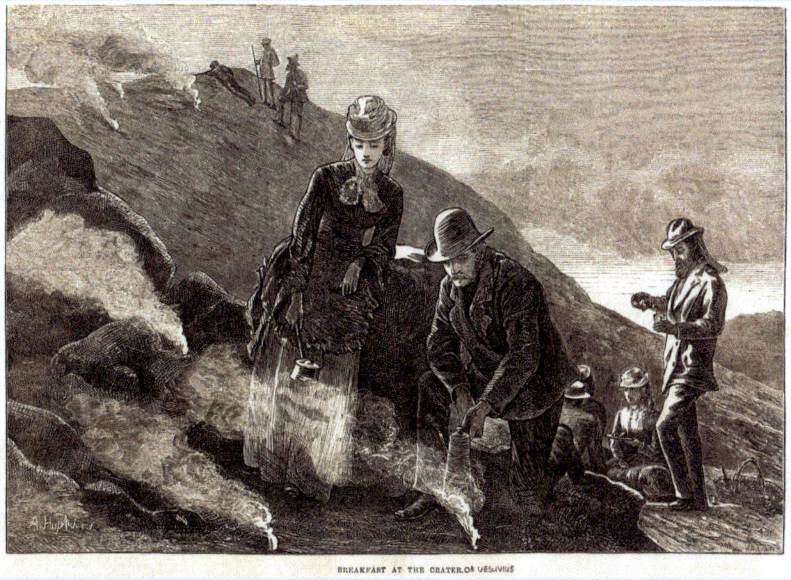

This color lithograph (signed "CHAM") is titled *Young Ladies Learn to Enjoy Being Lifted by a Sea Wave*. The Belgian publisher Jules Géruzet was active in Brussels and Ostend in the mid-nineteenth century, producing souvenir pictures such as this one. It shows that the Europeans of the time had already learned that large masses of water did not necessarily mean flood disaster; they could also serve as a source of amusement and a new kind of enjoyment

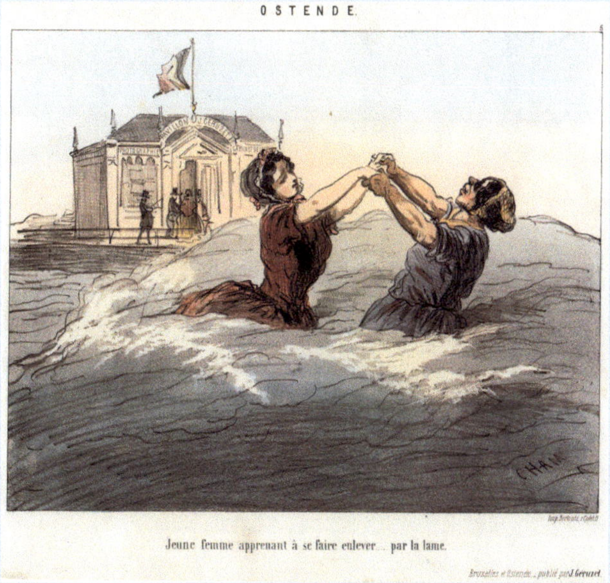

Postscript

Putting aside disasters, we can finish with a flight of fancy. Suppose we were to summon Empedocles to the twenty-first century, and suppose we gave him an overview of the advancement of our scientific knowledge in the earth and planetary sciences, reached in the last few centuries, and then invited him to define anew the family of the four natural elements. What might he come up with in the light of the new scientific discoveries?

Keeping in mind that our thoughts about the endless and mysterious universe are made by that small, negligible, and very rare dot of life that is modern human life, such considerations will necessarily be distorted by subjectivity. But we can speculate.

To start, it could be advantageous to consider and discuss the subject of the Earth's elements under two headings. On one hand, we have the familiar objective categories representing an indubitable common denominator of all of the worlds of the universe—in other words, what most people think of today as elements: the 100-odd chemical elements of the periodic table, which are the building blocks of the molecules that constitute all matter.

On the other hand, we can still posit more-or-less subjectively conceived additional categories linked with the universe's rare anomalous manifestations, such as life and its subsequent development. So our modern Empedocles might reason as follows.

To start, there are evidently two common elements or categories of the universe: mass and cyclicity. Let us discuss each in turn.

First, we should not make the mistake of declaring matter in any form or state (or mass, in physical jargon) to be the fundamental category of all that exists in the Universe. It might be thought that "mass" should be identified with the ancient "Earth"; however, such an equalization would not be quite correct, since the wider meaning of the term "mass" (as it is understood here) is much more complex and versatile.

The present meaning of mass—besides its standard, commonly understood characteristic, that it is the standard material base of any object in the universe—seems to include additional miraculous properties and capabilities. The first of them, the mass duality with energy, must be addressed. It should be noted that the transformation between energy and mass may run along different paths: chemical syntheses (such as oxidation), mechanical transformation (through gravitation and inertia), radiation (heat, light, radioactivity), thermonuclear fusion, and so on.

Gravity, as one special property of mass, should be highlighted, as it is this (together with centrifugal and centripetal forces) that enables the creation all stable and quasi-stable structures in the universe, from those at atomic scales to entire galaxies.

Another remarkable property of mass, as we know it from our terrestrial experience, lies in its diversity. Over a hundred chemical elements and thousands of chemical compounds and mechanical mixtures offer variants of mass with an extremely wide spectrum of physical and chemical properties, occurring in three modes (solid, liquid, and gas); this fantastic diversity is, of course, projected into the diversity of our world and life.

It should be noted that the element mass is still not yet a fully known and explored subject. For instance, so-called dark matter—invisible to human eyes (even those armed with modern techniques)—must exist if other important astrophysical phenomena are to be explained.

Let us move on. Do there exist—besides mass—some other subjects or categories that might rule the universe in such a full spectrum of sizes? In other words, is there any other common universal element that exists? After due consideration, we may say "yes": it is the existence of mass in cycles (or in generations), which is more easily understandable in the world of living units. In the genetic makeup of both plants and animals, the automatic encoded principle of eternal self-reproduction is really remarkable; the simplest seaweed has an urge to impart life to its future descendants that is equal to our own.

A question that remains unsolved is whether the principle of cyclicity also works in an unanimated world. Maybe it does. Every so often, astronomers report the explosion of a star (and the termination of its solar system); the expelled mass and released energy very probably will be rearranged in due course into a new structure. One can also point out the cyclic behavior of tectonically active planets (including our

J. Kozák, R. M. W. Musson, *The Illustrated History of the Elements*, https://doi.org/10.1007/978-3-030-21426-5

own), where the crust is endlessly created at midocean ridges and endlessly destroyed in subduction zones, with the material being recycled into new crust.

Let us consider someone whose life is still at the scale of the biblical "three score years and ten"; his or her natural impression of time scales certainly does not extend for thousands of years. When we consider the possible cyclicity of inanimate subjects, our comprehension of time periods of the order of billions of years (the presumed lifetime of the structure of the universe) will certainly be limited. Similarly, a bacterium in a human colon does not know the world of its host, and, if it were conscious and intelligent, it could hardly imagine the life length of its host. Is this not a similar picture to that of a modern person speculating on distant bodies in the universe?

But while our modern Empedocles might propose mass and cyclicity as two elements in his new scheme, they are clearly not enough. These two common categories penetrate the whole universe and its processes, but it is appropriate also to focus attention on the environs of our small home: our solar system and the Earth. Here, we may find two more natural elements or categories that rule not only our solar system but also the fate of all mankind. Let us name them: the Sun and Life.

The winner of the 2001 Nobel Prize in Physiology or Medicine, Sir Paul Nurse, remarked that "We on this planet are slaves to the Sun." Perhaps, though, there are two remarks that should be added to this quotation. First, without our sunny slave owner (or without any other star of similar astrophysical properties), we people would not exist. Secondly, our slave owner provides for us all of the conditions necessary for living (namely, the initiation of Life and its rapid development): a reliable working planetary system, and optimum delivery of heat, water, oxygen, and other commodities, all of which are necessary for the occurrence of fauna and flora, and for creation of the chain of nutrition. Also, the daily cycles and annual cycles due to the Earth's rotation around its axis, and the rotation of the Earth (in its oblique position) around the Sun, create the important biological sun clock that links the Earth and Sun and the human race.

While the Sun is the special natural element that is important—first of all—for the bodies of our Solar System, the last element, Life, is the specific element specially linked with Earth itself. The reasons, causes, and biological procedures accompanying the occurrence of living units on our planet have always been loaded with great uncertainty and doubts. Many have remained unsolved, regardless of whether they should be settled by invoking some higher will or some unknown consciousness (as many religiously oriented people have believed), or whether life was created spontaneously by chance, as it is often explained at present. Only one point is, however, quite clear and without any doubt: life really has occurred or, one can say, Life is here. At present, numerous researchers are trying to find out how the same chemical elements that we find as the components of rocks, soils, rivers, deserts, and every other inanimate structure somehow alone composed themselves into the pyramid of a living structure, at the top of which there is humanity.

Since we do not have enough real and true data at our disposal to solve this question, we have to leave it open; nor is the question resolved by the idea of panspermia: that life arrived on our planet from elsewhere, borne on meteorites or comets. This just puts the question back a stage, since life must have originated somewhere for the first time.

Anyway, Life on Earth must be taken as a precious gift and the most valuable element presented to us all.

One more question related to the occurrence of life is of major interest: does life exist on other planets? Is it common in the universe or very rare? Are we alone? No signs of life have been recorded elsewhere in our solar system, and the nature of the other planets and their moons suggest that conditions there are inimical to all but the most primitive organisms. Messages about life elsewhere in the universe are not at our disposal yet, and not for want of looking; one can refer, for instance, to the Search for Extraterrestrial Intelligence (SETI) program. It can be argued that even if the probability of life forming on a distant planet is extremely small, the sheer size of the universe means that this tiny probability must have been satisfied many times. It has often been pointed out that a study of our Earth itself shows impressively that numerous important parameters of an astronomical and planetary character, and especially the physical situation on the Earth's surface, seem carefully composed in such a way as to enable the occurrence of life.

This is the so-called "Goldilocks paradox": Earth seems to be "just right" for us. Any slight change in many parameters would have prevented mankind from ever evolving. For some, this seems evidence of a guiding intelligence, but it is obvious that only on a world where everything was "just right" would there appear observers to comment on it.

If these things came together by chance, then how small was that chance? On Earth it seems that life has colonized every conceivable niche, no matter how hostile; is there some similar bias toward life happening, if at all possible, on other worlds? Are there billions of life-forms, residing elsewhere in the universe, that will never be able to communicate with one another, because of the immense distances involved?

In the previous pages, the authors—making use of a gradual transformation of the ancient model of four elements—demonstrated the advancement of astronomy, astrophysics, geology, and the newly born geophysics in the course of 2000 years or more.

The development of astronomy and astrophysics has particularly benefited from recent expeditions into near space, which discovered many important data concerning the behavior and interaction of distant bodies of the universe. (One can mention the Hubble telescope and the International Space Station.) These achievements allow us to draw new, more accurate insights into the nature of the universe.

However, many other important questions remain unsolved, especially those linked with the occurrence of Life on Earth and (possibly) elsewhere. Why was it that in the giant, frosty realms of space, one small paradise planet appeared, on which the all-important physical conditions were combined in such a way as to prepare a pleasant home for Life? Besides "why," "how," and "by whom," many other questions related to the unique subject of nature—namely, "life"—have not yet been resolved. At present, humankind is not yet ready to find the proper answers for these enigmatic questions, and possibly we never will be.

So, with these considerations, Empedocles might turn to us and announce that four is still a good number for elements. But Earth, Water, Air, and Fire are no longer the best selection. Instead, we can think now of Mass, Cyclicity, Sun, and Life.

References[1]

Albini P, Rovida A, Scotti O, Lyon-Caen H (2017) Large eighteenth–nineteenth century earthquakes in Western Gulf of Corinth with reappraised size and location. Bull Seismol Soc Am 107(4):1663–1687

Ambraseys NN, Karcz I (1992) The earthquake of 1546 in the Holy Land. Terra Nova 4(2):254–263

Blöschl G, Gaá lL, Hall J, Kiss A, Komma J, Nester T, Parajka J, Perdigão RA, Plavcová L, Rogger M, Salinas JL, Viglione A (2015) Increasing river floods: fiction or reality? WIREs Water 2:329–344

Buckland W, Conybeare WD (1840) Ten plates comprising a plan, sections, and views, representing the changes produced on the coast of East Devon, between Axmouth and Lyme Regis by the subsidence of the land and elevation of the bottom of the sea, on the 26th December, 1839, and 3rd of February, 1840. John Murray, London

Camassi R, Caracciolo CH, Castelli V, Slejko D (2011) The 1511 eastern alps earthquakes: a critical update and comparison of existing macroseismic datasets. J Seismol 15:181–213

Eisinger U, Gutdeutsch R, Hammerl C (1992) Historical earthquake research—an example of interdisciplinary cooperation between geophysicists and historians. In: Gutdeutsch R, Grünthal G, Musson R (eds) Historical earthquakes in central Europe, vol 1. Abhandlungen der Geologischen Bundesanstalt, Vienna

Ferrari G, McConnell A (2005) Robert Mallet and the "great Neapolitan earthquake of 1857". Notes Rec R Soc 59:25–64

Fouqué FA (1879) Santorin et ses éruptions. Librairie de l'Academie de Medecine, Paris

Graham M (1824) Journal of a residence in Chile during the year 1822; and a voyage from Chile to Brazil in 1823. Longman/Hurst, London

Graziani L, Maramai A, Tinti S (2006) A revision of the 1783–1784 Calabrian (Southern Italy) tsunamis. Nat Hazards Earth Syst Sci 6:1053–1060

Hammerl C (2017) Historical earthquake research in Austria. Geosci Lett 4:7

Hough SE, Page M (2011) Toward a consistent model for strain accrual and release for the New Madrid seismic zone, Central United States. J Geophys Res 116:B03311. https://doi.org/10.1029/2010JB007783

Kozák J, Čermák V (2010) Illustrated history of natural disasters. Springer, London

Kraemer H (1910) Weltall und Menschheit. Bong, Berlin

Larroque C, Scotti O, Ioualalen M (2012) Reappraisal of the 1887 Ligurian earthquake (Western Mediterranean) from macroseismicity, active tectonics and tsunami modelling. Geophys J Int 190(1):87–104

Margottini C, Di Buduo G (2017) The geological and landslides museum of Civita di Bagnoregio (Central Italy). Landslides 14:435–445

Mayer-Rosa D, Wechsler E (1991) The Basel 1356 earthquake, effects on contemporary buildings in the town of Basel. Paper presented at the Symposium on Historical Earthquakes in Europe, Prague

Musson RMW (2012) The million death earthquake. Palgrave Macmillan, New York

Musson RMW (2013) A history of British seismology. Bull Earthq Eng 11(3):715–861

Oldham T (1882) The Cachar earthquake of 10th January, 1869, by the late Thomas Oldham edited by R.D. Oldham. Mem Geol Surv India 19:1–98

Raška P, Zábranský V, Brázdil R, Lamková J (2016) The late Little Ice Age landslide calamity in north Bohemia: triggers, impacts and post-landslide development reconstructed from documentary data (case study of the Kozí vrch Hill landslide). Geomorphology 255:95–107

Scaramellini G, Kahl G, Falappi GP (1988) La frana di Piuro del 1618: storia e immagini di una rovina. Associazione Italo-Svizzero Per Gli Scavi Di Piuro, Borgonuovo di Piuro

Schmidt JFJ (1874) Studien über Vulkane und Erdbeben. Georgl, Leipzig

Sciblia F (2015) Terremoto e architettura storica. Palermo e il sisma del 1726. IRIS, Palermo

Short T (1749) A general chronological history of the air, weather, seasons, meteors, etc. Longman & Miller, London

Szeliga W, Hough S, Martin S, Bilham R (2010) Intensity, magnitude, location, and attenuation in India for felt earthquakes since 1762. Bull Seismol Soc Am 100(2):570–584

Tatevossian R, Albini P (2010) Information background of 11th–15th century earthquakes located by the current catalogues in Vrancea (Romania). Nat Hazards 53:575–604

Vogt J (2008) Earthquake effects on nature and macroseismic intensity scales. In: Fréchet J, Meghraoui M, Stucchi M (eds) Historical seismology: interdisciplinary studies of past and recent earthquakes. Springer, Dordrecht, pp 303–312

Winthrop J (1755) A lecture on earthquakes; read in the chapel of Harvard-College in Cambridge, N.E. November 26th 1755. In: On occasion of the great earthquake which shook New-England the week before. Edes & Gill, Boston

Wulf A (2015) The invention of nature: the adventures of Alexander von Humboldt, the lost hero of science. John Murray, London

[1]For ease of reading in what is not intended as an academic publication, reference citations have been kept to a minimum and, where possible, bibliographic details of sources used, especially for pictures, have been given in full in the captions.

J. Kozák, R. M. W. Musson, *The Illustrated History of the Elements*, https://doi.org/10.1007/978-3-030-21426-5

MIX
Papier aus verantwortungsvollen Quellen
Paper from responsible sources
FSC® C105338

If you have any concerns about our products,
you can contact us on
ProductSafety@springernature.com

In case Publisher is established outside the EU,
the EU authorized representative is:
Springer Nature Customer Service Center GmbH
Europaplatz 3, 69115 Heidelberg, Germany

Printed by Libri Plureos GmbH
in Hamburg, Germany